Printed Antennas for Wireless Communications

Telephone (+44) 1243 779777

Email (for orders and customer service enquiries): cs-books@wiley.co.uk
Visit our Home Page on www.wiley.com

Other Wiley Editorial Offices

John Wiley & Sons Inc., 111 River Street, Hoboken, NJ 07030, USA

Jossey-Bass, 989 Market Street, San Francisco, CA 94103-1741, USA

Wiley-VCH Verlag GmbH, Boschstr. 12, D-69469 Weinheim, Germany

John Wiley & Sons Australia Ltd, 42 McDougall Street, Milton, Queensland 4064, Australia

John Wiley & Sons (Asia) Pte Ltd, 2 Clementi Loop #02-01, Jin Xing Distripark, Singapore 129809

John Wiley & Sons Canada Ltd, 6045 Freemont Blvd, Mississauga, ONT, Canada L5R 4J3

Wiley also publishes its books in a variety of electronic formats. Some content that appears in print may not be available in electronic books.

Anniversary Logo Design: Richard J. Pacifico

British Library Cataloguing in Publication Data

A catalogue record for this book is available from the British Library

ISBN 978-0-470-51069-8 (HB)

Typeset in 10/12pt Times by Integra Software Services Pvt. Ltd, Pondicherry, India
Printed and bound in Great Britain by Antony Rowe Ltd, Chippenham, Wiltshire
This book is printed on acid-free paper responsibly manufactured from sustainable forestry in which at least two trees are planted for each one used for paper production.

In memory of Professor J. James and his pioneering work in printed antennas

Contents

About the Editor

Rod Waterhouse received his BEng, MS, and PhD in Electrical Engineering from the University of Queensland, Australia, in 1987, 1989, and 1994, respectively. In 1994 he joined RMIT University as a lecturer and become a Senior Lecturer in 1997 and an Associate Professor in 2002. In 2001 he took a leave of absence from RMIT and joined the venture-backed Dorsal Networks, which was later acquired by Corvis. In 2003 he left Corvis and resigned from his position at RMIT and worked for Photonic Systems Inc. as a principal engineer. In 2004 he cofounded Pharad, a broadband wireless communications company, where he is now a Vice President. His 2003 book on printed antenna design follows more than 200 publications and ten patent applications in the field of antennas, electromagnetics, microwave, and photonics engineering. Dr Waterhouse is an associate Editor for *IEEE Transactions on Antennas and Propagation*. He chaired the IEEE Victorian MTTS/APS Chapter from 1998 to 2001. Dr Waterhouse is an adjunct Senior Fellow within the Department of Electrical and Electronic Engineering at the University of Melbourne. In 2000 he received an IEEE Third Millennium Medal for Outstanding Achievements and Contributions.

List of Contributors

EDITOR

Rod Waterhouse
Pharad
Maryland
USA

AUTHORS

Amin M. Abbosh
School of Information Technology and Electrical Engineering
The University of Queensland
Brisbane, Queensland
Australia

J. T. Aberle
Department of Electrical Engineering
Arizona State University
Tempe, Arizona
USA

Jennifer T. Bernhard
Electromagnetics Laboratory
University of Illinois at Urbana-Champaign
Urbana, Illinois
USA

Marek E. Bialkowski
School of Information Technology and Electrical Engineering
The University of Queensland
Brisbane, Queensland
Australia

Chi-Chih Chen
ElectroScience Laboratory
Electrical and Computer Engineering Department
The Ohio State University
Columbus, Ohio
USA

Jose A. Encinar
Polytechnic University of Madrid
ETSI Telecomunicación
Madrid
Spain

Karu Esselle
Electronics Department
Macquarie University
Sydney, NSW
Australia

Yuehe Ge
Electronics Department
Macquarie University
Sydney, NSW
Australia

Peter S. Hall
Department of Electronic, Electrical and Computer Engineering
University of Birmingham
UK

Hing K. Kan
School of Engineering and Physics
The University of the South Pacific
Suva
Fiji

E. Lee
Department of Electronic, Electrical and Computer Engineering
University of Birmingham
UK

Dalma Novak
Pharad, LLC
Columbia, Maryland
USA

Dean Pavlickovski
RMIT University
Melbourne, Victoria
Australia

Wayne S. T. Rowe
School of Electrical and Computer Engineering
RMIT University
Melbourne, Victoria
Australia

C. T. P. Song
Mobile Devices
Motorola
Birmingham
UK

G. A. Taylor
Department of Electrical Engineering
Arizona State University
Tempe, Arizona
USA

Tatsuoh Itoh
Department of Electrical Engineering
University of California Los Angeles
Los Angeles, California
USA

Andrew Y. J. Lee
Nanyang Technological University
Singapore

Kevin M. K. H. Leong
Department of Electronic Engineering
University of California Los Angeles
Los Angeles, California
USA

S.-H. Oh
Department of Electrical Engineering
Arizona State University
Tempe, Arizona
USA

Sanjay Raman
Bradley Department of Electrical and Computer Engineering
Virginia Polytechnic Institute and State University
Blacksburg, Virginia
USA

Daniel F. Sievenpiper
HRL Laboratories LLC
Malibu, California
USA

Richard M. Svitek
Renaissance Wireless
Pittsburgh, Pennsylvania
USA

F. C.-E. Tsai
Wistron NeWeb Corporation
Taiwan
ROC

Rodney Vaughan
School of Computing Science
Simon Fraser University
Burnaby, British Columbia
Canada

Rod Waterhouse
Pharad
Maryland
USA

John Volakis
ElectroScience Laboratory
Electrical and Computer Engineering
Department
The Ohio State University
Columbus, Ohio
USA

J. A. Zornoza-Ramirez
Antenna Centre of Competence
EADS Astrium Ltd
Stevenage
UK

Preface

In 2004 I was approached by Professor James to write a book on printed antennas for wireless systems. After much debate (internally and with Professor James!), we decided on a book that could provide the reader with a valuable reference for printed antennas for many wireless applications and also a glimpse into the future of wireless systems and the role to be played by printed radiators. To achieve this goal, we needed to rely on a wealth of information, probably too much for a single person to accumulate, so we set about assembling a team of experts to give valuable insight and design methodologies for many forms of printed antennas as well as how this technology could be incorporated into advanced wireless concepts. I believe, because we have assembled some very good engineers (from academic and industry backgrounds) who are contributing in areas in which they have published significant papers on the particular topics, there is no book of a similar nature to the one presented here that covers the material. Having said that, the book is similar to some well-known and highly praised books that typically fall into the category of handbooks. Importantly the reader will be given opinions from various experts and so it will not be a collage of one person's efforts and their subsequent, and perhaps somewhat biased, opinion. I believe too that the addition of the final section of this book gives the reader a glimpse at the state of the art in wireless technologies and the direction of future communication systems.

The objective of this book is to provide the reader with an understanding of how to design printed antennas and associated technologies relevant for state-of-the-art wireless systems. Experts from around the world have contributed to this book to give the reader valuable insight into important technological breakthroughs in these areas. One common observation throughout this book is the utilization of rigorous analysis tools to help understand the phenomena associated with the radiating structures and their surrounding environments.

After the introduction chapter, which presents the latest systems and new platforms being investigated for wireless technology, the book is divided into three parts: (I) fundamental wideband printed antennas for wireless systems; (II) small printed antennas for wireless applications; and (III) advanced concepts and applications in wireless systems. The first part really focuses on printed antenna solutions that are directly applicable to present-day and future wideband systems (such as ultra wideband (UWB) applications) as well as antenna platforms where multiple wireless communication interfaces are required. The five chapters in this part provide a thorough summary of the five fundamental wideband printed radiators and how these antennas can be designed for a variety of applications. In each chapter practical examples are given.

In Part II we focus on a wireless area where printed antennas have helped accelerate the advances in wireless systems: small antennas for portable terminals. In these chapters we present all the relevant approaches to making small printed antennas, multilayered and uniplanar, and how these radiators can be integrated in WLAN equipment and other platforms. In these chapters the impact of the environment in which the antenna is mounted is also presented. In Part III we present advanced concepts and new applications that require printed antenna technology. In these chapters the usefulness of the previously described printed antenna solutions is highlighted. A thorough examination of printed antennas in reflectarrays is given. We also present ways that artificial magnetic conductors can be used with printed antennas and how printed antennas can be integrated with active devices for concepts such as wireless systems on a chip as well as configured in advanced wireless systems such as software defined radio and multiple-input multiple-output (MIMO). In all the chapters experimental data will be provided when appropriate and theoretical models will be applied to present design performance trends as well as to give the reader an in-depth coverage of the relevant area.

It should be noted that each chapter presents new previously unpublished material. Although some of the concepts presented in the book are available in the public environment, several of the presented designs for a variety of wireless systems are not. To my knowledge some of the presented design procedures are also unavailable in the public literature. There are also several chapters in this book related to topics that have not been addressed before in printed antennas, especially for wireless applications. These include using metamaterials (high-impedance ground-planes) and the integration with software-defined radios and MIMO systems. Also the chapters on the various wideband technologies have not been presented in one collective document before, which will be extremely useful for the reader.

I sincerely acknowledge the contributions from all the authors of the chapters in this book; they should be proud of their efforts. As I said before, I believe such a collection of concepts and insightful views could not be accumulated by one person; at least not in one lifetime. As a technologist I have already used some of the design philosophies and approaches highlighted in this book to yield practical antenna solutions. I hope the reader will get as much out of this book as I already have!

Rod Waterhouse

1

Wireless Systems and Printed Antennas

Rod Waterhouse and Dalma Novak
Pharad, Maryland, USA

1.1 INTRODUCTION

Wireless systems are playing an ever increasingly important role in society. Whether they are used to assist in the distribution/collection of large amounts of information or to make home entertainment systems more convenient, wireless systems are becoming more and more integrated into daily activities. The utilization of wireless technology is not currently confined to either the commercial or the military sectors. In fact, it seems that both sectors are striving for wireless solutions to enhance their resources, or simply to make their product more readily accessible and useful to the consumer.

One of the important enabling technologies for wireless communications is the transducer, which converts guided energy to radiated energy (and vise versa): the antenna. Although advances in antenna engineering cannot be credited for the globalization of wireless technologies, it still plays an important role, whether it provides an aesthetically pleasing solution or helps improve the overall radio frequency (RF) link budget, or allows multiple users to utilize the single interface and thereby increase the capacity. It is undeniable that good radiator engineering allows for better wireless systems.

Figure 1.1 shows a 'snapshot' of present and future commercial wireless systems, highlighting cellular and satellite communications, as well as the extremely popular wireless local area network (WLAN) and short-range connectivity systems such as Bluetooth. Also shown is the quandary facing next-generation systems and how they were interface with a variety of technologies including computers and small handsets.

Figure 1.2 shows the frequency spectrum indicating current and next-generation wireless systems between 300 MHz and 100 GHz. As can be seen, there are a plethora of wireless systems utilizing various advantages of the different spectra. For example, systems at the

Printed Antennas for Wireless Communications Edited by R. Waterhouse

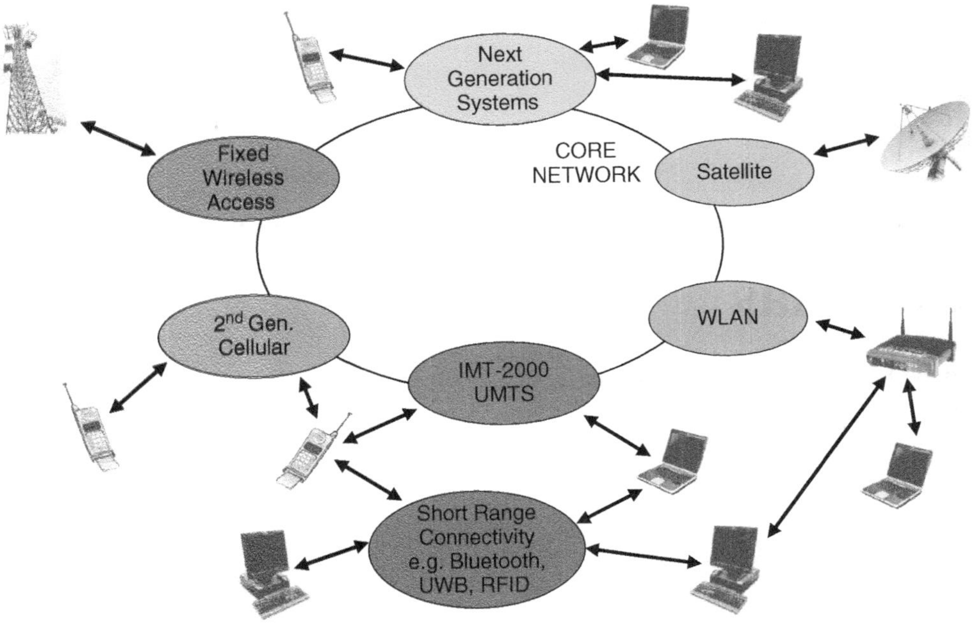

Figure 1.1 Schematic showing a variety of present-day and future wireless systems.

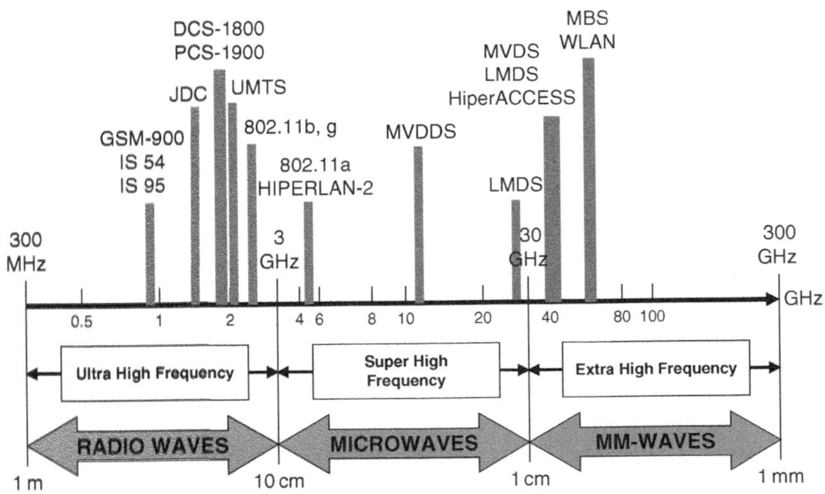

Figure 1.2 The frequency spectrum showing many wireless systems.

lower end of the frequency spectrum have better propagation and mobility characteristics, whereas the higher frequency applications can transmit significantly more information. As can be seen from Figures 1.1 and 1.2, there are many issues with future wireless systems, not just related to antenna engineering. Not only do these wireless networks have to be seamlessly interconnected to wired networks, but perhaps with different radio networks and interfaces. Ideally the next-generation radio network interface will have a degree of flexible control that can adapt to different modulation formats and air interface access formats.

This chapter serves as an introduction to the remainder of the book. In particular the authors will focus on aspects and issues that the antenna engineer needs to be aware of to ensure the ultimate solution is obtained for the problem at hand. In Section 1.2 some present-day examples of wireless systems will be presented. They will focus on some less conventional wireless applications to give the reader an appreciation that the commercial and military sectors are not just interested in cellular systems. From this investigation, it will be seen that there are many and sometimes conflicting requirements for the radiator, and as the antenna is typically the final component to be designed in the RF link, the antenna engineer is usually burdened with an extremely difficult task that may involve attempting to break the laws of physics. In Section 1.3 some new platforms are presented that require efficient wireless solutions: unmanned vehicles and body wearable solutions. Both platforms present very challenging issues to the antenna designer. Based on the reviewed systems and platforms, in Section 1.4 the authors will focus on some general requirements for antennas. Section 1.5 will give a brief review of printed antennas and summarize the characteristics of the different classes. Finally in Section 1.6 the intention of the rest of the book will be summarized, showing how the technologies presented can help the antenna engineer deliver the best radiating solution for a particular wireless system.

1.2 EXAMPLES OF WIRELESS SYSTEMS

Conventional wireless communications systems (cellular/mobile) continue to be the cornerstone of society, with more bandwidth hungry applications being offered to the public (for example the streaming video), and presently these applications are readily being accepted by the public as 'must have' technologies. Satellite systems, for communications, broadcasting, and positioning (global positioning satellite (GPS) system), seem to be experiencing a bit of a renaissance, with recent applications adding to the overall importance of wireless technologies. The WLAN is another example of wireless technology that is having a dramatic impact on society, enabling low-cost wireless solutions that are easy to install in coffee shops, homes, and communal areas for consumers. The unlicensed IEEE 802.11.x systems have opened the door to low-cost wireless systems that are inevitably challenging 'conventional' systems. Excellent reviews of these wireless systems can be found in References [1] to [4]. There are other wireless communication systems that are receiving a lot of attention and here a summary of a few of these is provided. After reading these systems/applications, it will hopefully be apparent to the reader that there are many diverse applications requiring all forms of radio technology, some applicable to both military and commercial interests.

1.2.1 RFID/Asset Tracking

The location and tracking of inventory and assets is a critical activity for many businesses. Passive and active RFID (radio frequency identification) is a well-established technology that provides a means of contactless identification and is now used extensively in a range of industries. Organizations are quickly recognizing the benefits of RFID as a solution for inventory control (identification and tracking) that can bring a rapid return on investment (ROI). In addition, RFID tags are already being used for applications as diverse as: theft prevention; the identification of animals and humans; automating production systems; electronic toll collection and traffic automation; intermodal container identification and tracking; control of access to vehicles, buildings, and parking areas; car immobilizing (security); and automatic automobile fuel dispensing.

Rapidly expanding markets for RFID technology such as supply chain management have already emerged, with the US market for such technology estimated at \$1.1 billion in 2001. This is expected to increase dramatically over the next few years, with the market expected to more than double to \$2.3 billion by 2005. Major commercial companies such as WalMart and Tesco have recently announced their intention to deploy RFID tag systems extensively throughout their organizations. In addition, recent US Department of Defense (DoD) policy has indicated implementation requirements for RFID technology which commenced in early 2005 [5, 6].

RFID is an electronic labelling system that uses radio frequencies to identify goods quickly and precisely. Information from the tag attached to an item or set of goods is obtained from a reader or interrogator via a radio link. The system provides much greater flexibility over bar-code inventory control systems since there are no line-of-sight communication requirements and physical contact between the tag and reader is not required. The tags themselves differ substantially in terms of size, memory capacity, physical construction, and distance over which they can be operated. Consequently tags are often custom designed to meet the needs of specific applications and are therefore typically proprietary in nature. Readers manage the data communications process with the tags, which often utilize a proprietary protocol based on the unique characteristics of the tag technology employed.

While the commercial sector has been quick to adopt RFID for inventory control, the widespread usage of such technology in the military environment has been hampered by the lack of a common set of RFID standards and communication protocols. Numerous RFID vendors exist who typically use their own proprietary tag and reader technologies including communication protocols. This means that tags from one manufacturer require readers from the same vendor and readers from one manufacturer cannot read tags from other vendors. In addition, while national regulatory bodies allocate different radio transmission frequencies for RFID, there is no one global standard. Compatibility with local, national, and international regulations is a key issue in the management of inventory across borders.

RFID technology uses frequencies within the range of 50 kHz to 5.9 GHz where the type of application determines the frequency that is used for a specific system solution, and, in turn, finds which reader and tag will provide the best solution in terms of range, reliability, data transfer rate, and cost. Asset tracking applications that require read ranges greater than 10 feet are based on either backscatter techniques (where a small portion of the reader's RF energy is reflected from the tag, encoded with information stored in it) or two-way active tags (for higher value asset control) and operate at frequencies in the 433 MHz, 915 MHz, 2.45 GHz, and 5.8 GHz bands.

The form factor and frequency of the RFID tag vary widely depending on the application, and environmental factors also play a significant role. For example, a tag operating at 13.56 MHz has a size of 7.6 cm $\times$ 4.5 cm while one operating at 2.45 GHz has a size of only 6 cm $\times$ 1 cm. Figure 1.3 shows a commercially available active RFID tag developed for conventional corrugated shipping cartons [7]. As shown in this photograph, the integration of RFID transponders in self-adhesive functional components expands their potential uses. When combined with labels, RFID transponders can be fastened to virtually any substrate with either permanent or removable adhesion, visibly or invisibly. For extreme application requirements, RFID labels are coated with highly durable polyurethane resin. Figure 1.3 also shows that the RFID antenna and electronics are bonded on to a flexible acrylic substrate.

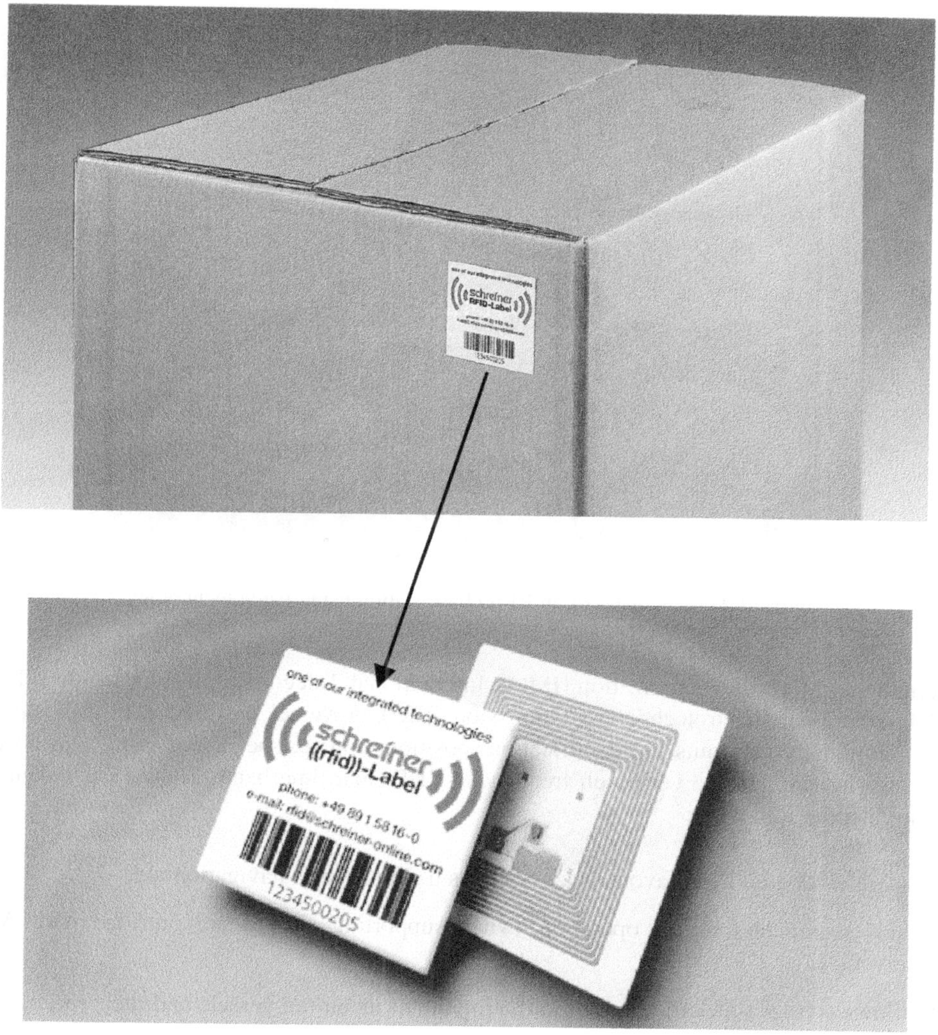

Figure 1.3 Photograph of a commercially available RFID label for shipping cartons [7].

Since RFID systems do not require direct line-of-sight for communications, the tag can also be concealed and embedded directly in the casing of the product.

Figure 1.4 shows a schematic of an RFID base station module for 2.4 GHz utilizing a dual polarized patch antenna for improved performance. The radiator is easily integrated with the required RF and low-frequency electronics and is protected from the environment by a radome.

1.2.2 Automatic Identification System (AIS)

In the last five years, an Automatic Identification System (AIS) shipboard broadcast transponder system has gone from development to deployment in large ocean-going vessels. The AIS is a communication protocol developed, and being extended (for Class B), by the

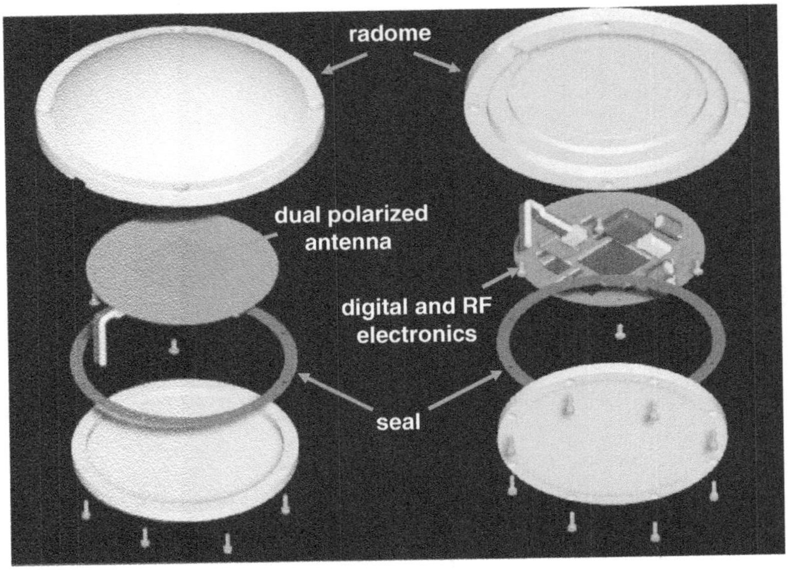

Figure 1.4 Schematic of an RFID base station using patch technology.

International Maritime Organization (IMO), International Telecommunication Union (ITU), and International Electrotechnical Commission (IEC). A vessel outfitted with AIS equipment continually transmits its identification, position, course, speed, and other data to all other nearby ships in the very high frequency (VHF) maritime band. The AIS is meant to accomplish the following:

- It operates as a collision avoidance system as a ship-to-ship navigation aid.
- It aids search and rescue operations when supporting aircraft are outfitted with AIS receivers.
- It allows coastal states to obtain information about incoming vessels and their cargo.
- It allows ports to better manage their waters' traffic when integrated into a Vessel Tracking System (VTS).

The large-scale use of AIS equipment is a key foundation for improving waterways management and building a waterway and coastline border security programme. Requiring all blue water vessels to deploy a common AIS sensor infrastructure will allow countries to maintain persistent surveillance of large and small vessel movements. Integrated with a VTS, this border security system will have a priori data available on the identity of vessels entering certain waters, knowledge of movement within the sensor system, knowledge of previous ports visited, and knowledge of potential at-sea rendezvous of multiple vessels.

In accordance with the International Convention for the Safety of Life at Sea (SOLAS) as amended in 2000, AIS transponders were mainly deployed in vessels larger than 300 tons. Pursuant to new rules published in 2003 by the United States Coast Guard, the Department of Homeland Security's lead agency for maritime homeland security, deployment of AIS was required for considerably smaller commercial vessels, as small as 65 feet or 100 tons, by the end of 2004. Clearly, larger vessels pose the greatest consequence resulting from a collision. However, these larger vessels represent but a fraction of vessels on the water. The preponderance of maritime traffic is smaller commercial vessels and recreational craft, which are not currently required to deploy AIS equipment. The high cost of currently available AIS equipment is an impediment to the broad adoption of AIS equipment for these smaller maritime vessels. In fact, when the US Coast Guard sought comments regarding the expansion of AIS to vessels not previously required, it amended the interim rules such that only a smaller set of SOLAS certified vessels are required to carry an AIS. However, the US Coast Guard still envisions a time when all vessels operating in US waters carry AIS and is seeking a lower cost AIS solution.

A block diagram of the electrical/physical architecture is shown in Figure 1.5. The position data originates from an integrated GPS receiver which obtains position and time data that is then processed to calculate course data. The GPS receiver is connected to an external GPS antenna via a separately supplied cable. The GPS data are transferred to a central processing unit (CPU), likely a microcontroller, where the information is used to construct AIS transmissions and is used as a variable in the collision avoidance algorithms. External AIS and DSC (digital selective calling) transmissions are received via three independent receivers within the AIS Class B Basic device. A single VHF transmitter is used for all transmissions from this device. The transceiver is connected to an external VHF antenna via a separately supplied cable. The AIS/DSC messages are deconstructed and formatted in a programmable logic device. Power is supplied to all the electronics via a supply that converts the vessel's 12 V DC power to the appropriate voltages for the electronic circuits.

A central processing function is provided by a microcontroller. This microcontroller executes collision avoidance algorithms using the position and track data from the host vessel and the AIS information reported by other vessels. The microcontroller manages the user interface, including reading the user input via buttons and driving the display information. The microcontroller works with the programmable logic device to aid in the construction and formatting of AIS transmissions/reception of AIS reports from other vessels.

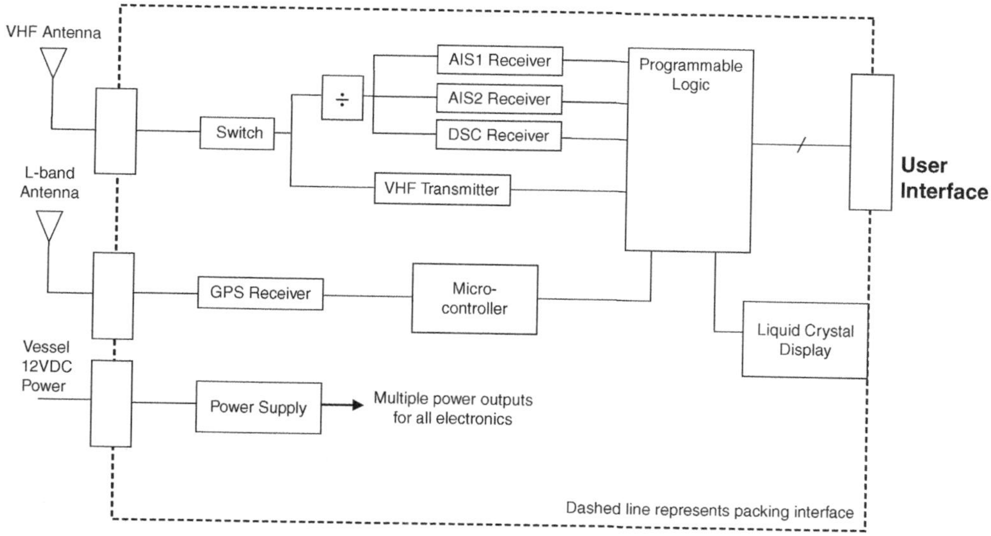

Figure 1.5 Block diagram of the AIS device functional and physical architecture.

1.2.3 *Joint Tactical Radio System (JTRS)*

One of the prime objectives of the Joint Tactical Radio System (JTRS) is to realize a single software radio capable of processing different waveforms from many different RF services over a wide band of frequencies [8]. The JTRS waveforms supported in the 2 MHz–2 GHz frequency range include future Wideband Networking Waveform (WNW), Single Channel Ground Air Radio System (SINCGARS), Ultra High Frequency (UHF) SATCOM, High Frequency (HF) Independent Side Band (ISB) with Automatic Link Establishment (ALE), Link-16, Identification Friend or Foe (IFF), Digital Wideband Transmission System (DWTS), Soldier Radio, Wireless Local Area Network (WLAN), Cellular Radio, Personal Communication Services (PCS), and future expansion of Mobile Satellite Services (MSS) [9]. The implementation of these services in JTRS will require the development of a wideband, airborne antenna platform including a very sophisticated filtering procedure. The radiating structure is the key enabling technology for the proposed software defined radio since wideband amplifier technologies as well as the supporting control software for these communication links are reasonably well advanced. The schematic in Figure 1.6 highlights the multitude of communication levels and therefore the overall complexity of communications in a battlefield environment.

1.2.4 *Mobile ad hoc Network (MANET)*

While ad hoc networks are of great interest for military and tactical applications, other applications have also emerged. These include commercial applications such as an infrastructureless network of notebook computers in temporary offices or a conference/campus

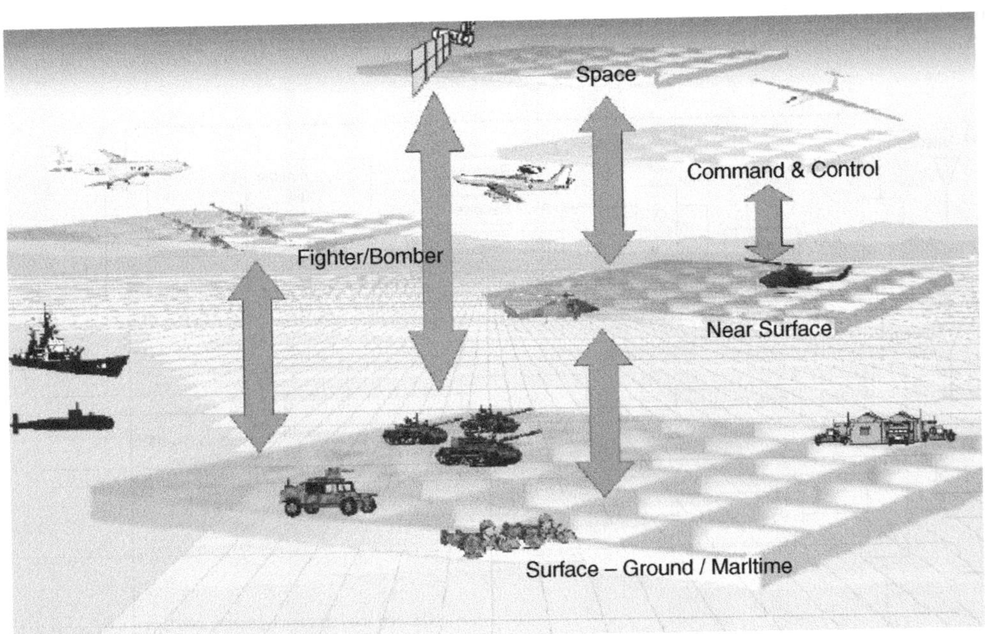

Figure 1.6 Schematic showing the levels of communications in a battlefield environment.

setting, as well as situations where service is to be provided without the visible presence of a wired backbone. There is also a growing interest in large-scale sensor network applications where hundreds or thousands of sensors may be scattered throughout a city for biological detection. Thus, this technology has great potential where it can reduce the cost and time associated with installing a fixed infrastructure.

In the commercial market, the proliferation of wireless local area networks, and wireless personal area networks, such as Bluetooth, pave the way for ad hoc networking as a key enabling technology. Ad hoc networking will have the ability to open up a range of business, gaming, and content-distribution applications that will lead to increased network usage and potential revenue. In addition, a number of companies are already including ad hoc networking concepts in their products based on wireless mesh networks, including MeshNetworks, Flarion, and Cybiko. Ad hoc networks are also being considered for the extension of the cellular airinterface, leading to improvements in cellular capacity and coverage. As a result, ad hoc network technologies have the opportunity to tap into a multibillion industry in the evolution towards next-generation cellular communication systems. A schematic highlighting a typical ad hoc wireless system architecture is shown in Figure 1.7.

A mobile ad hoc network (MANET) is an autonomous system of mobile hosts (also serving as routers) connected by wireless links that form a communication network with no wired backbone infrastructure [10]. This is in contrast to the conventional single-hop cellular network model where fixed prelocated cell sites and base stations (BSs) are installed as access points and communication between two mobile nodes relies completely on the wired backbone and BSs. Since the geographic location of all network nodes can change in a MANET, the network topology as well as the paths between two terminating points are highly dynamic. In addition, there is no predetermined organization of available links between users; thus an ad hoc network forms automatically and adapts to the changing network, with individual nodes responsible for dynamically discovering other nodes that they can directly communicate with. Since not all nodes can directly communicate with each other, they are also required to relay packets on behalf of other nodes to deliver data across the network. A MANET is thus an example of a multihop wireless network.

Ad hoc networks are of particular use where infrastructure is either not available or cannot be relied on in times of emergency. A MANET builds on the concept of packet switched radio networks that date back to the early 1970s with the US Department of Defense (DoD) funded programs to investigate such concepts for the military and enable mobile communication in infrastructureless, hostile battlefield environments where the nodes forming the network could be soldiers, tanks, or aircraft. While MANETs are of still great

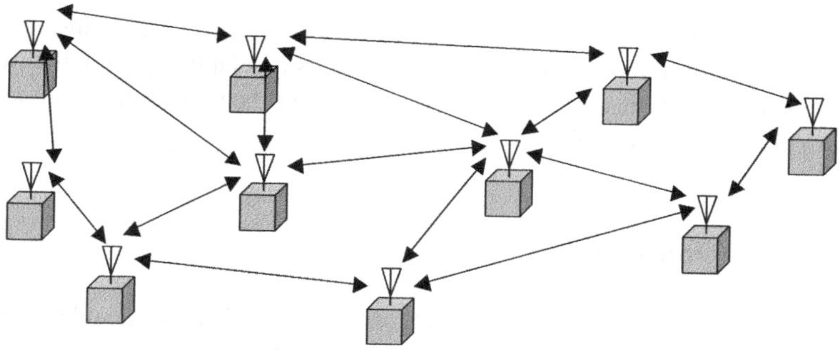

Figure 1.7 A typical architecture for a mobile wireless ad hoc network.

interest for military and tactical applications (e.g. law enforcement, emergency rescue, or exploration missions), other applications have since emerged. These include commercial applications such as an infrastructureless network of notebook computers in temporary offices or a conference/campus setting, as well as situations where service is to be provided without the visible presence of a wired backbone. There is also a growing interest in large-scale sensor network applications where hundreds or thousands of sensors may be scattered throughout a city for biological detection.

Multicasting in MANETs, where messages are sent from one node to multiple recipients, can efficiently support a variety of applications and is typically the most characteristic operation of MANETs used in military and tactical applications. In these scenarios, multicasting is generally used for disseminating important and confidential information. Therefore multicasting routing protocols need to ensure a reliable message delivery, an area that is actively being pursued [11]. Since the routes between nodes in a MANET are typically multihop in nature, a global routing approach that operates across the whole network and maintains and updates the routing structure whenever changes occur in the system is required. Due to the dynamic nature of the mobile ad hoc environment, the routing protocol must be able to adapt dynamically to changes in the network topology as well as optimize the path between any two nodes. A key challenge in ad hoc network routing is achieving such a goal while also being constrained by the nature of the network itself, since nodes in MANETs typically have limited lifetimes and processing capabilities and the wireless medium also has limited bandwidth.

Due to their inherent characteristics there are many challenges in implementing mobile wireless ad hoc networks that are currently being pursued across all layers of the network. The physical layer has to be able to adapt to rapid changes in the characteristics of the link between two nodes. The medium access control (MAC) layer needs to minimize collisions, allow fair access, and transport data reliably over the shared wireless links. The network layer must determine and distribute information used to calculate routing paths in a way that maintains efficiency when links change often and bandwidth is at a premium. Thus ad hoc networking is a cross-layer design challenge. An emerging challenge is the design of MANETs that can exploit the properties of new hardware technologies. One example is that of smart or beamforming antennas, which do not have a fixed antenna pattern but can adapt to the current radio conditions. Similar to cellular networks, the ability to focus or steer RF energy can provide increased throughput in the network and reduced delay via an increase in spatial reuse [12]. These benefits also enable MANETs to be density adaptive. The use of directional antennas on mobile terminals, however, introduces the complex issue of finding the desired direction for transmission or reception. This is most critical in a MANET that does not have centralized control and the terminals may also have limitations with respect to size and the complexity/cost of communication hardware. Transmission range power control can also help to make ad hoc networks density adaptive [13], but new network layer algorithms are needed if the use of power control and beamforming antennas is to be successfully exploited in MANETs.

1.2.5 Satellite Communications

The XTAR system is a satellite communications network that is capable of providing tactical aircraft with over-the-horizon communications. XTAR's two-satellite system is based on Space Systems/Loral's space-proven 1300 satellite platform and has been designed specifically for US and NATO SATCOM systems [14]. XTAR features two very important aspects in its system. Firstly, it offers flexibility since its spot beams can be dynamically controlled. In addition, XTAR provides larger bandwidth and throughput compared to existing X-band terminals as it has 'stackable' spot beams and also dual polarization capability. The dual polarization feature of XTAR, namely right-hand circular polarization (RHCP) and left-hand circular polarization (LHCP) effectively doubles the spectral efficiency of the system. All of these key elements of the XTAR system make it appear extremely useful for defence-specific communication services.

1.2.6 Global Positioning Satellite (GPS) System

The global positioning satellite (GPS) system [15] uses 24 satellites in medium earth orbits to provide accurate position information (latitude, longitude, and elevation) to users on land, in the air, or at sea. Originally developed as the NAVSTAR system by the military at a cost of approximately $12B, the GPS has quickly become one of the most pervasive applications of wireless technology for military, consumers, and businesses throughout the world. Today, GPS receivers are found on commercial and private airplanes, boats, and ships as well as ground vehicles. When integrating GPS equipment into vehicles and systems, it is imperative to take into consideration not only parameters such as weight, size, and number of channels when designing the receiver module, but also the jamming susceptibility of the GPS application.

GPSs transmit low-power signals that must travel great distances to reach mobile GPS receivers. In contrast, GPS jamming devices are usually located in relatively close proximity to these GPS receivers and transmit significantly higher power signals. The simplest airborne GPS antenna structures are small single-element omnidirectional radiators. However, the use of omnidirectional antennas makes these devices particularly susceptible to jamming. Techniques such as beam steering and null steering can reduce the effects of jamming. It is well known that one of the most effective means of mitigating the potential issues of jamming is to utilize a controlled reception pattern antenna (CRPA), or, put more simply, a phased/adaptive array. The CRPA is an array of antenna elements that has the ability to modify its radiation pattern for the purpose of rejecting interference signals and/or emphasizing the GPS satellite signal. Present-day CRPAs employ seven or eight elements and are over 13 inches wide in diameter; there is much interest in reducing the overall size of a CRPA without compromising the antenna's efficiency and antijamming capability.

1.2.7 Wireless Personal Area Network (WPAN)

Very wide unlicensed frequency ranges in the V-band have been allocated by regulatory agencies around the world, including 57–64 GHz in the US and 59–66 GHz in Japan. Operation of wireless networks in such large frequency bands at such high frequencies enables the transmission of very high data rates. In addition, the high atmospheric absorption due to oxygen that occurs at 60 GHz provides other benefits, such as high security, frequency reuse, and immunity to electromagnetic interference, as highlighted in Figure 1.8. The security features of the 60 GHz transmission band have long been exploited in radio communications for military applications, but more recently it is also being actively pursued for commercial wireless telecommunications.

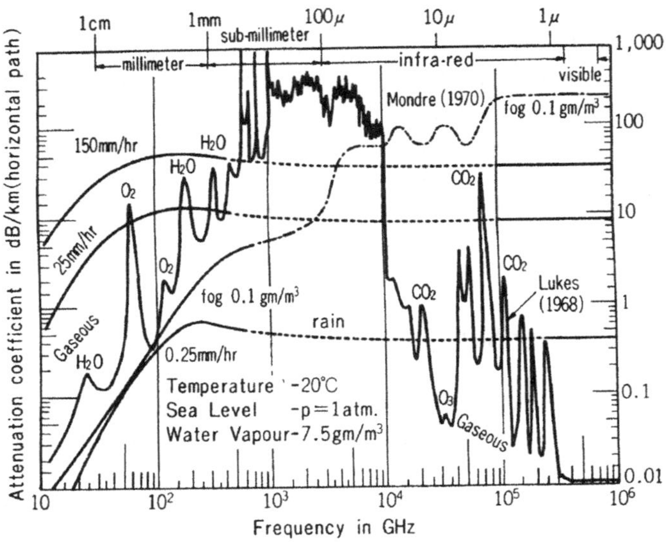

Figure 1.8 Attenuation coefficients of various gases as a function of frequency [19].

While wireless local area network (WLAN) standards such as Bluetooth, IEEE 802.11b, and IEEE 802.11a/g have demonstrated a progression in data rate capability from around 1 Mb/s, 11 Mb/s, and 55 Mb/s [16], respectively, the unlicensed V-band spectrum provides an opportunity to continue the trend for data rate increases, for operation at Gb/s rates. As an example, the IEEE 802.15.3 Study Group 3c (SG3c) was formed in March 2004 to explore the use of the 60 GHz band for wireless personal area networks (WPANs) [17]. SG3c is developing a millimetre-wave-based alternative physical layer (PHY) for the existing high data rate IEEE 802.15.3 WPAN Standard 802.15.3-2003 [18]. The millimetre-wave WPAN will allow very high data rate applications such as high-speed internet access, streaming video download, real-time data streaming as well as the wireless databus for cable replacement.

1.2.8 Combat Identification (CID)

Combat identification (CID), the ability to differentiate friend or foe with high confidence, is a serious and challenging problem for military forces. Recent military engagements have shown the increased need for forces to be able to identify quickly the presence of their own soldiers within an active battle or combat area. Today, the battles being fought by personnel are increasingly moving from remote combat zones such as rural geographies and jungles to urban and populated centres, where the identification of the presence of friend or foe is particularly problematic. In addition, statistics indicate that approximately 60 % of fratricides historically occur in ground-to-ground scenarios.

Improved CID concepts and technologies can not only increase the effectiveness of military operations but also significantly reduce the number of friendly-fire casualties as well as lower the risk of collateral damage. The focus of this proposed effort is the development of an affordable multimode. Presently deployed CID systems are either inherently expensive, not capable of operating over a wide range conditions or environments, or vulnerable to enemy exploitation. The Individual Combat Identification System (ICIDS) is a dismounted soldier CID technology consisting of a weapon-mounted laser interrogator and a helmet-mounted radio transponder.

1.3 NEW WIRELESS PLATFORMS

Not only are innovative wireless systems being considered to provide new and improved services but new platforms are being developed. Two such platforms are unmanned vehicles and body wearable systems. Both technologies can potentially provide significant advances in wireless communication/sensing services and as a result are receiving a lot of attention from both the military and commercial sectors.

1.3.1 Unmanned Vehicles

The role of unmanned vehicles in today's military and first responder sectors is becoming more important. These platforms can be used to provide reconnaissance information without the expense (cost and safety) of a manned mission. An example of this technology is a low-cost, expendable tactical UAV (unmanned aerial vehicle) with interchangeable sensor payloads tailored to the specific operation, which can be launched from another aircraft. There are several types of UAVs and two examples are shown in Figure 1.3.1.

Another example of an unmanned vehicle is the unmanned undersea vehicle (UUV). These self-propelled untethered submersibles operate either autonomously or under minimal supervisory control. Over the last decade, interest in UUVs for commercial applications has increased markedly. These seacraft are receiving wider acceptance in the natural resource and telecommunication industries for monitoring and maintenance of underwater pipes and cables. Oil and gas companies such as Shell now routinely employ UUVs offshore, where they offer particular benefits for offshore oilfield route surveys and pipeline surveillance and repair in deep waters. UUVs are also finding a role in other commercial applications, such as operation in hazardous subsea environments, as well as automated inspections of ship hulls and other infrastructure located underwater. In addition to military applications, other government agencies are considering UUVs for operations such as hazardous waste management (Department of Energy), fisheries research (National Oceanic and Atmospheric Administration), and drug seizures (US Coast Guard). Figure 1.10 shows an example of a UUV.

One of the most critical components of any unmanned vehicle system is the data communication link between the vehicle and the control platform. This data link must be both highly reliable and secure in order to properly control the UAV or UUV and its sensor payload, while also providing real-time jam-resistant transmission of the sensor data. Presently JTRS and even WLAN are being considered for these unmanned vehicles.

1.3.2 Body Wearable Platforms

The fact that the current antennas are hindering the ability of soldiers to accomplish their mission due to their unwieldy size helps to substantiate the critical need for antennas integrated into tactical clothing: body wearable solutions. In addition to the military, textile antennas will also find use in the activities of first responders and law enforcement personnel in applications where communication is critical. To achieve interoperability for emergency

(a)

(b)

Figure 1.9 Examples of UAVs: (a) Northrop Grumman's *Predator* [20] and (b) Aerosonde's *MK IV* [21].

personnel, first responder voice/data communication equipment is moving to the 700/800 MHz frequency band and data communication devices operating within the IEEE 802.11 WLAN 2.4–2.5 GHz band continue to proliferate. The ability to provide firefighters with the capability to carry out hands-free communications without the need for obtrusive antennas helps to realize a safer communication system for such emergency personnel. In general, the elimination of obtrusive antennas either through helmet-mounted antenna technology or integrated uniform antenna solutions, will aid many first responders. An example of a body wearable antenna integrated within a tactical vest is shown in Figure 1.11.

Figure 1.10 A UUV developed by Bluefin [22].

Figure 1.11 Photograph showing body wearable antennas integrated in a combat vest.

1.4 ANTENNA REQUIREMENTS

As should be apparent from the previous examples of wireless systems being considered for both the military and the commercial markets, there is a wide range of antennas needed to interface with these systems; that is a 'one antenna' solution is not realistic. This is why this book comprises more than one chapter! There are, however, some fundamental properties that the antenna engineer must address to ensure the chosen solution operates in the best possible manner, for a given volume. Of course these characteristics are not independent and of course the engineer typically must compromise between the parameters to give the optimum result.

1.4.1 Size

The size (volume) of the antenna and its overall impact on the surrounding environment is extremely important for most wireless communication systems. Not only does the antenna need to be unobtrusive and extremely small but it is typically desired that its performance be equivalent to an antenna developed in isolation. There are of course limits to how small an antenna can be, that is to say if an efficient solution is required. Having said this, there are technologies that can be used to reduce the size of an antenna effectively and some of these are investigated in later chapters. However, associated with most of these procedures is a compromise in performance, typically bandwidth (return loss bandwidth) and gain. Also, as a general rule of thumb, the more narrowband the solution the more susceptible it is to its surrounding environment. Fortunately a lot of communication systems (such as WLAN) do not require large instantaneous bandwidths and therefore this is not an issue. Also there are a lot of sophisticated electromagnetic (EM) software packages available that can accurately model the surrounding environment of an antenna and so variations in the performance due to, say, the packaging can be taken into account.

One point that should be made clear here is that the size of the antenna includes the ground-plane used to create the radiator. Unfortunately there has been a plethora of 'small' antennas that have recently been proposed in the literature that actually have large ground-planes. The fact is that if the antenna requires a large ground-plane to operate, then the size of the antenna is set by the size of that ground-plane.

Probably the biggest issue with utilizing small antennas for wireless communications is the reduction in efficiency. Effectively, the lower the efficiency, the more power is required from the front-end RF amplifiers and therefore the lower the battery life of the wireless terminal (if it is a terminal under consideration). Despite this, advances in efficient RF amplifiers and new battery technology are helping to resolve this potential problem.

1.4.2 Integration

The ease of integration is a very important consideration in antenna technology. Radiators that can be easily integrated with the RF front-end are overall cheaper solutions. Typically, too, the easier the integration the more efficient the solution, which is very important for wireless applications in the high microwave frequency bands. Integration does not only refer to the RF engineering, but also the structural properties. For example, in the UAV applications, it is imperative the antennas used can reside within the surface of the platform such that the aerodynamics of the vehicle is not greatly impacted. Of course the 'integrated' radiator cannot adversely affect the structural integrity of the vehicle. This is why appliqué antennas are receiving considerable attention today.

1.4.3 Efficiency

As mentioned before, efficiency is very important for any radiator; the greater the efficiency the better the link budget. Other than the size of the radiator, another factor that impacts the efficiency of the antenna is the material used to create the structure, or surround the radiator. For printed antennas, the supporting substrate/dielectric laminate needs to be low in loss, particularly for high-frequency applications. Although FR4 is a low-cost readily available laminate, for frequencies above 2 GHz the dielectric losses significantly reduce the overall efficiency of the radiator, making this a cheap but unattractive solution. High dielectric constant ceramics are becoming more abundant and due to the dielectric loading effect of the material on the radiator, small solutions can be developed. Care must be taken to ensure that the high dielectric constant material is low loss, as typically the loss tangents of these materials are higher than their low dielectric constant counterparts and therefore the overall impact on the radiation performance can be severe.

As there are many antenna solutions for wireless applications, there is a common quandary often considered: should I choose a wideband solution or one that just meets the bandwidth requirements? In terms of efficiency, if low-loss material is used typically a narrowband solution will be smaller and also more efficient over the desired wireless band than a wideband radiator. Having said this, a wideband solution is typically more 'forgiving' of fabrication tolerances, although more dependent on its surrounding environment.

1.4.4 Bandwidth

The designed antenna must satisfy the bandwidth requirements for the wireless system. Of course there are two fundamental bandwidth criteria the antenna must typically meet: the impedance bandwidth (return loss bandwidth) and the gain bandwidth. There are no universal definitions of these bandwidths that apply to every wireless system, but typically the return loss bandwidth is the frequency range in which the return loss is better than -10 dB, which means that less than 10 % of the power is reflected into the RF circuitry. There have been cases where -6 dB has been acceptable for an antenna, due to the stringent size constraints placed on the radiator. However, there are other systems where -20 dB is required, in particular if 'ghosting' is an issue. The gain bandwidth is usually less than a 3 dB variation across the range of frequencies, although for some systems a 1 dB variation is more appropriate. For narrowband wireless systems, the gain requirement is typically not as hard to meet compared to the return loss bandwidth for a constrained volume.

There have been some alarming trends noticed in the literature, in particular with the development of wideband printed antennas; researchers tend to focus on the return loss bandwidth of their proposed antenna, without considering the gain bandwidth. This could be a consequence of two factors: history and ease. Historically, when one of the most prevalent printed antennas, the patch antenna, was investigated, because of its inherent narrowband nature, the impedance variation was more of an issue than the gain variation of the antenna. In fact in its basic form the return loss bandwidth of a patch antenna is smaller than the gain bandwidth by a factor of 10 or so. For this reason, much research was dedicated towards accurately modelling this form of radiator and also the enhancement of the return loss bandwidth. Unfortunately in the process it appears that some of the more recent researchers have forgotten about the gain response, which is equally as important, if not more so. This is especially true when investigating radiators with bandwidths in excess of several octaves. It should always be remembered that a matched load has an excellent return loss bandwidth, but does not radiate! The other reason for this trend may be associated with the cost of making the appropriate measurements. Return loss measurements are relatively straightforward to make using a vector network analyser, whereas gain measurements need some form of antenna range, whether it be an anechoic chamber, near-field facility, or outdoor range. These facilities are expensive and perhaps not accessible to most researchers.

1.4.5 Polarization

The polarization of an antenna is an interesting issue and in some wireless system cases its investigation and consequences have become a 'lost art'. Satellite communications have always had well-defined polarization requirements, but terrestrial systems have really been a case-by-case scenario and it really depends on what part of the link is being designed. For example, most base stations in present-day cellular systems use dual linear polarization to take advantage of multipath effects and also the different propagation effects on horizontal and vertical polarization, whereas the handset terminal requirement is typically 'whatever you can get' simply to the restricted size constraints and therefore there is great difficulty in controlling the polarization.

In general most wireless systems prefer some form of polarization diversity, whether it be to reduce the probability of 'drop-outs' or to increase the capacity of the system. The latter requires strict polarization control.

1.4.6 Power Handling

The power handling requirements for an antenna once again really depend on the application, with some radar systems requiring greater than 100 kW capability. Most wireless systems do not need such large power handling requirements and typically maxima in the order of 1–10 W are acceptable. The power handling capability of an antenna is fundamentally determined by the materials used to develop the radiator. Having said this, certain classes of radiators can handle high power levels better than others; for example waveguide-based radiators can operate efficiently at very high power levels. For base station antennas where the power handling requirements are higher, to avoid arcing and other high power level effects, the number of solder joints and other potential current discontinuities should be minimized.

1.4.7 Summary

Looking at these generalized requirements and also the need for low-cost solutions, printed antennas would appear to be the best choice of radiator. Of course printed antennas come in many forms and so the more appropriate choice for a particular application may not be obvious. The reader should keep in mind that having made such a statement, there will always be applications and the need for conventional antennas: waveguide-based, reflectors, and wire-based. In fact a lot of the printed antennas presented in the upcoming chapters have their origins in nonprinted antenna technology or can easily be derived from them. Several obvious examples are the shorted patch/planar inverted-F antenna (PIFA) and its derivation from the F-type wire antenna created in the 1940s. Also, the relationship between the printed reflectarray and the classic reflector, as well as the printed quasi Yagi Uda and its wire predecessor, seem quite obvious. Maybe not such an obvious link is the tapered slot antenna and the horn waveguide radiator, although close examination of the characteristics and design trends associated with both of these radiators do imply a strong connection.

1.5 FUNDAMENTAL PRINTED ANTENNAS

In this section, an attempt will be made to summarize the many forms of printed antennas. Most of these radiators will be discussed in more detail in subsequent chapters. A decision was made to divide printed antennas further into three categories: inherently wideband radiators; resonant style antennas; and small radiators. Of course these radiators are not totally independent; for example it is possible to develop a narrowband version of one of the wideband solutions and conversely by using multiple narrowband radiators a wideband element can be designed. Thus the divisions used here are somewhat grey.

1.5.1 Wideband Printed Antennas

Inherently wideband printed antennas are typically derived from travelling wave antenna concepts. From the IEEE Standard Definitions of Terms for Antennas (IEEE Std 145-1983) [23], travelling wave antennas are defined as 'an antenna whose excitation has a quasi-uniform progressive phase, as a result of a single feeding wave traversing its length in one direction only'. Thus any antenna that relies on this mechanism to radiate falls into this class, and as can be imagined there is a plethora of antennas that do so. Examples include beverage antennas, horn antennas, reflectors, tapered slot antennas, and lens antennas. A more simplified interpretation of the above IEEE standard definition, which is not necessarily valid for all forms of travelling wave antennas, is that this form of radiator converts guided electromagnetic energy into radiated energy in a gradual or smooth transition. An example of a travelling wave antenna, whose form is obviously compliant with this statement, is a horn antenna where the flange of the waveguide is gradually opened to match effectively the impedance/mode of the guided wave to that of free space. As stated before, travelling wave antennas are not necessarily wideband in nature, but because of the smooth transition from the guided medium to the unguided medium they tend to be, at least with careful design.

1.5.1.1 Tapered Slot Antennas

The tapered slot antenna is undoubtedly the most common printed travelling wave antenna (see, for examples, References [24] to [27]). These antennas can operate over multioctave bandwidths and can operate well in a phased array environment. Effectively this form of printed radiator can be thought of in the same manner as a horn antenna, where a

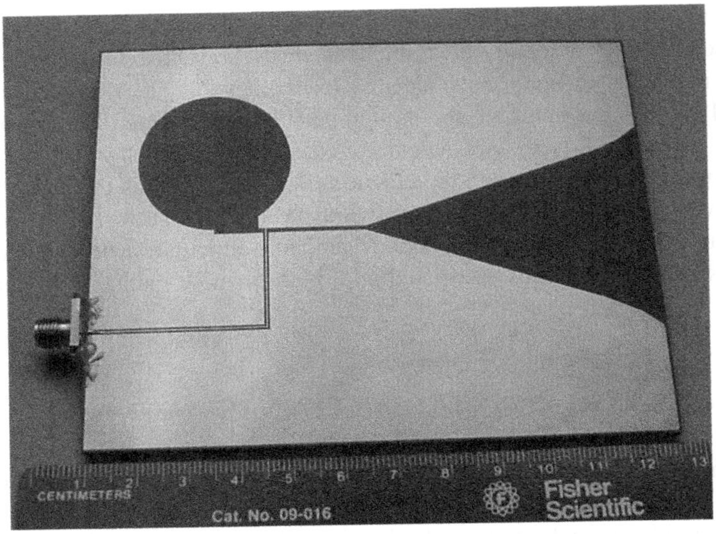

Figure 1.12 Photograph of a tapered slot.

waveguide is flared out to 'match' the electromagnetic wave to free space, although in this case the transmission medium is a slot line. However, unlike a conventional horn antenna the fundamental mode of the slot line is transverse electromagnetic (TEM) in nature and so some of the issues associated with the wave matching of a horn antenna are mitigated here. Having said that, the tapered slot is a relatively complicated radiating element with many degrees of freedom. As is the case for a horn antenna, the profile of the taper (or flare) contributes to the input impedance behaviour as well as the radiation performance of this printed travelling wave antenna. Common profiles include linear, exponential, and piecewise linear. One particular challenge with a tapered slot antenna is the transition from the slot line to a more common transmission media used in microwave circuitry design.

Figure 1.12 shows a photograph of a tapered slot antenna with a coplanar waveguide (CPW) feed [28]. Incorporated in this design is a wideband balun (part of which is the large slot-line cavity in the top left-hand corner of the photograph), which efficiently transfers energy from an unbalanced transmission line (CPW) to the balanced slot-line. Figure 1.13 shows the predicted and measured return loss of the developed travelling antenna, as well as a typical radiation pattern.

1.5.1.2 Printed Spiral Antennas

Spiral-shaped antennas are well known to give very wideband impedance and radiation responses [29]. In their traditional form these antennas radiated equally in the planes both above and below the spiral radiator (whether in the form of a wire/conductor or a slot). To create a unidirectional radiation pattern, spirals were loaded by a lossy cavity. A consequence of such a loading was that the volume of the antenna increased dramatically as the cavity depth had to be greater than the radius of the spiral, and therefore it was more challenging to integrate these radiators into the surfaces of potential platforms.

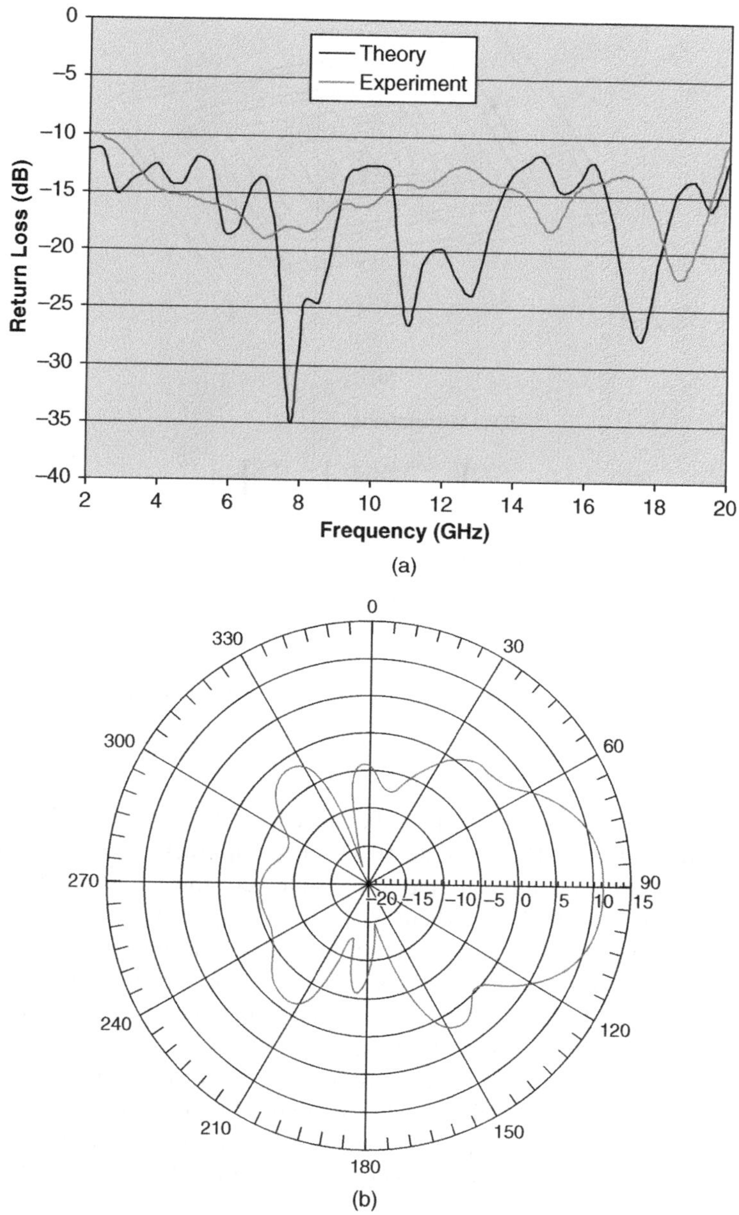

Figure 1.13 Performance of the tapered slot of Figure 1.12: (a) return loss and (b) radiation patterns.

In the early 1990s, a low-profile spiral was developed that significantly reduced the depth of the antenna [30]. Here a wideband printed spiral antenna was developed using a grounded laminate of thickness comparable to that used for microstrip patch antennas over the same frequency range. The key to obtaining a good impedance response and radiation performance from this low-profile spiral was to incorporate an absorbing material around the perimeter of the spiral conductor and also include the design of a wideband balun. A simpler version of this approach is shown schematically in Figure 1.14. Here one arm of the spiral

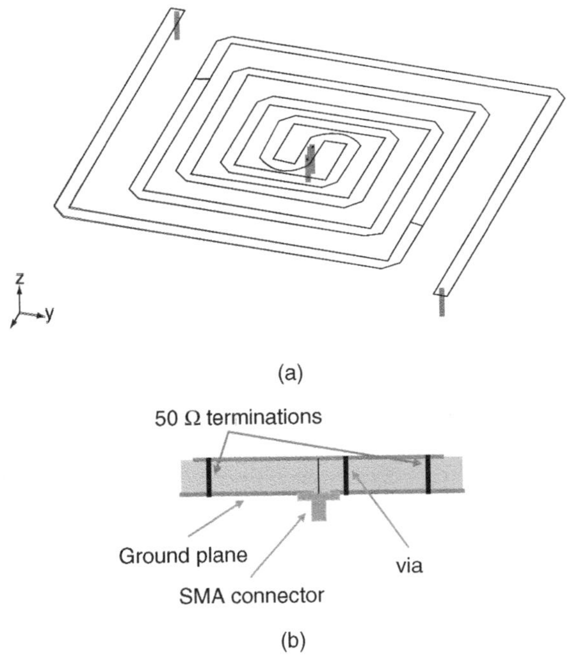

Figure 1.14 Schematic of a simplified printed spiral antenna: (a) three-dimensional view and (b) cross-sectional view.

is probe fed, while a via is used to connect the other arm of the spiral to the ground-plane below the printed conductor. The arms are 50 Ω microstrip tracks and are terminated with 50 Ω resistors. Not using a balun can give responsible performance as the thickness of the dielectric layer is less than 0.05 λ_0 (where λ_0 is the free-space wavelength at the highest frequency of operation). As expected, the return loss performance of this simplified printed spiral antenna degrades at higher frequencies because of the oversimplified feed structure. Of course spiral antennas loaded with resistive elements do have lower radiation efficiencies than nonloaded radiators (such as tapered slots).

Figure 1.15 shows the predicted return loss performance of the printed spiral shown in Figure 1.14. As can be seen from Figure 1.14, the return loss is better than -10 dB over a wideband bandwidth and as the frequency increases the return loss performance degrades. Figure 1.16 shows the predicted radiation patterns of the printed spiral at 1 GHz.

1.5.2 Resonant-Style Printed Antennas

Although the introduction of 'wideband' wireless systems (for example JTRS) would imply that inherently narrowband technologies are no longer relevant, this is not the case. A key factor is the definition of 'wideband', and it seems that some systems have 'stretched' the definition perhaps due to the marketing potential of the phrase or perhaps to differentiate the proposed technology from previous generations. For example, some of the cellular systems near 2 GHz were touted as being wideband because the user bandwidth was more than the 900 MHz systems. This statement is true in a relative sense, relative to previous generations

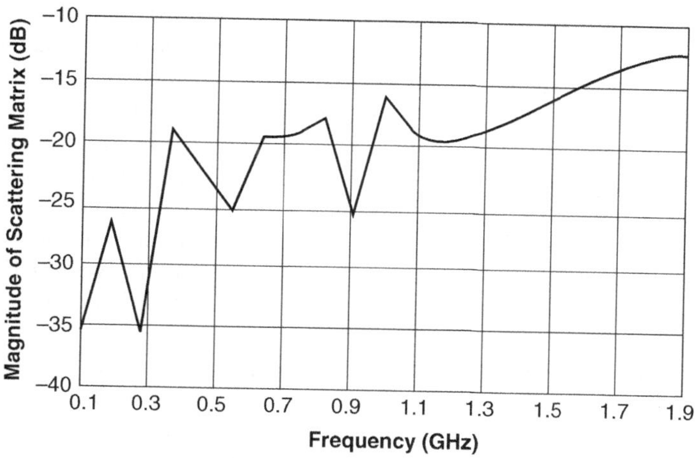

Figure 1.15 Return loss performance of the spiral antenna presented in Figure 1.14.

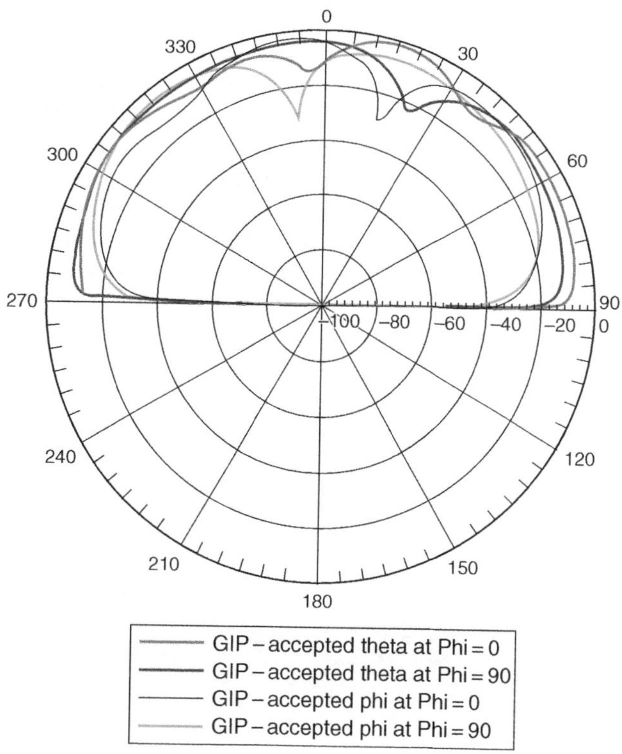

Figure 1.16 Typical radiation patterns of the spiral antenna presented in Figure 1.14.

not frequency; the bandwidth is more than the previous generations, but with respect to the frequency of operation the increase in bandwidth is minor and could hardly be deemed wideband. Having said this, there are some truly wideband applications, such as JTRS or UWB, and there are multiband systems too, of which an option may be just to utilize a wideband radiator.

From a designer's prospective, when would using a resonant-style radiator be considered and why would narrowband technology be considered when wideband radiators can easily accommodate the electrical requirements? Resonant-style radiators do have several antennas over their travelling wave counterparts. If the bandwidth is small enough, resonant-style antennas tend to be more efficient (if low-loss materials are used). Resonant-style antennas tend to be smaller in size (typically less than $\lambda_g/2$, where λ_g is the guided wavelength), which can allow for arrays of these elements to be formed readily with therefore good control of the radiation performance. Resonant-style antennas, because of their inherent narrowband nature, can help with out-of-band rejection of noise and therefore assist the overall RF performance of the link, although the first two reasons are probably the most important advantages of resonant-style antennas.

The three fundamental elements for printed resonant-style antennas are: the dipole, the slot, and the patch. Although there have been many variations of these printed radiators developed over the years, the principles of operation of the 'new' versions can be traced back to these three fundamental radiators. Feeding or coupling power to and from these antennas plays a critical role in the overall performance of the antenna, as is the case for any radiator, but it is probably more correct to say that the feeding process, if done properly, allows advantage to be taken fully of the inherent properties of the antenna.

As mentioned before, there is definitely a 'grey' zone when addressing wideband and narrowband antenna technologies. A good example is the aperture stacked patch [31], of which a photograph of this multilayer/element radiator with the layers separated and sandwiched together is shown in Figure 1.17. Here the antenna is developed by combining three 'narrowband' elements (two patches and one slot). The resulting radiator can have bandwidths in excess of an octave, thereby perhaps qualifying it as a wideband radiator. Another example of a wideband antenna developed from narrowband elements is the log-periodic antenna. Once again, the overall response of the antenna is wideband in nature but the individual elements are not.

Figure 1.17 Photograph of an aperture stacked patch.

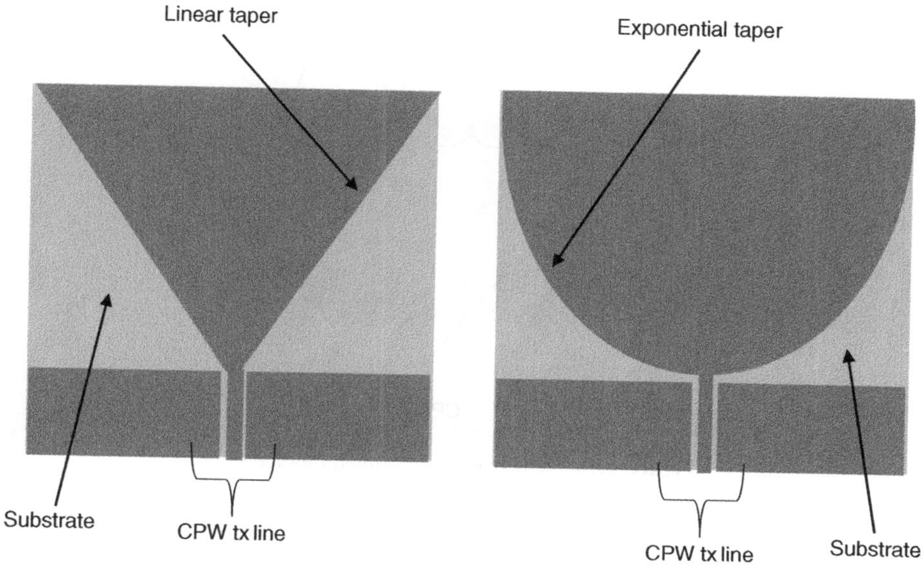

Figure 1.18 Schematics of profile optimized monopoles.

To complicate this issue further, several wideband printed antennas have recently been reported, which can be considered as variations of these narrowband counterparts. In particular, a class of printed dipoles, monopoles, and slots can be considered as 'profile optimized' versions, of which a 'bow-tie' dipole is a well-known example. Figure 1.18 shows schematics of some CPW-fed 'profile optimized' printed monopoles. Here effectively the shape of the conductor (or aperture for the slot case) has been modified to improve the impedance matching response of the antenna, in a similar vain to many travelling wave antennas, such as a tapered slot. These antennas can operate relatively efficiently over multioctaves but it can be argued that their origins stem from narrowband alternatives.

1.5.2.1 Printed Dipoles

Figure 1.19 shows a schematic of a printed dipole antenna, one of the fundamental printed resonant-style radiators. Here the balanced feed, in this case the coplanar stripline (CPS), is used to couple power to and from the resonant antenna. Usually a thin dielectric laminate is used to develop the printed feed and dipole. Figure 1.20 shows the predicted return loss performance and radiation patterns (at the centre of the matched band) of a typical printed dipole. One issue associated with a dipole solution is the sensitivity of the radiator to the presence of a ground-plane, which is hard to avoid in most communication platforms. For this reason, monopoles are a much more attractive radiator solution.

1.5.2.2 Printed Slots

Printed slots are the magnet equivalent of a printed dipole radiator. Because of the presence of the inherent ground-plane these printed antennas are more common than their dipole counterparts. Figure 1.21 shows a schematic of a folded printed slot antenna fed by a

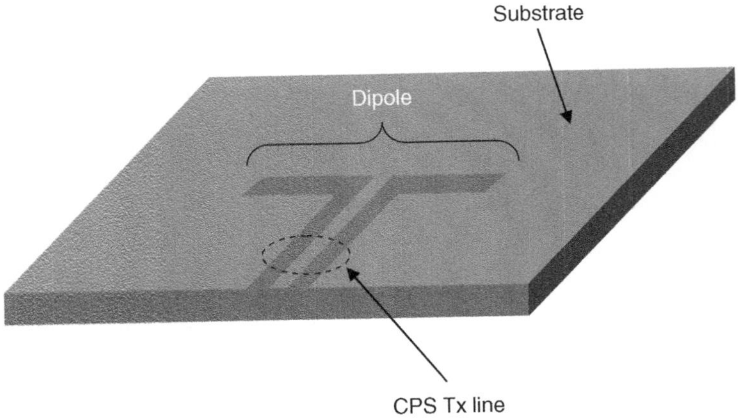

Figure 1.19 Schematic of a printed dipole.

microstrip transmission line. Unlike a dipole, a balanced transmission line is not required to feed this form of printed antenna. The printed slot is folded to ensure a good impedance match at the designed frequency; the number of folds depends on the dielectric constant of the material used to fabricate the antenna [32]. The width of the slot is typically very narrow ($0.01\lambda_g$), although the bandwidth of the antenna can be enhanced by making it larger [33].

Figure 1.22 shows the predicted return loss performance and radiation patterns of a printed folded slot antenna. For this case considered, the ground-plane is assumed infinite in extent.

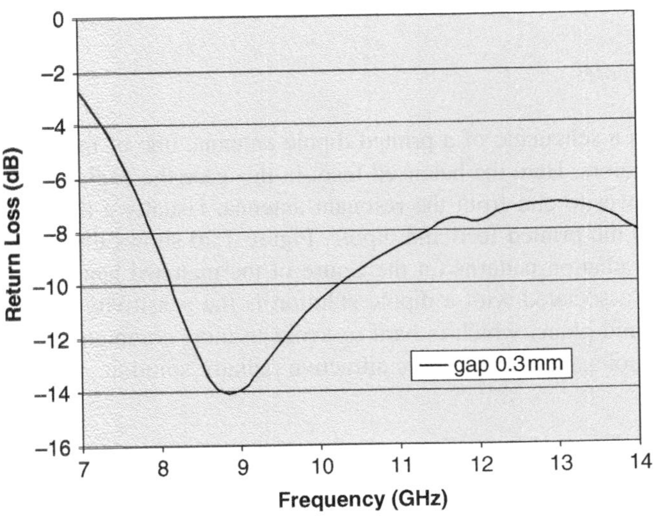

(a)

Figure 1.20 Predicted performance of a printed dipole: (a) return loss and (b) radiation patterns.

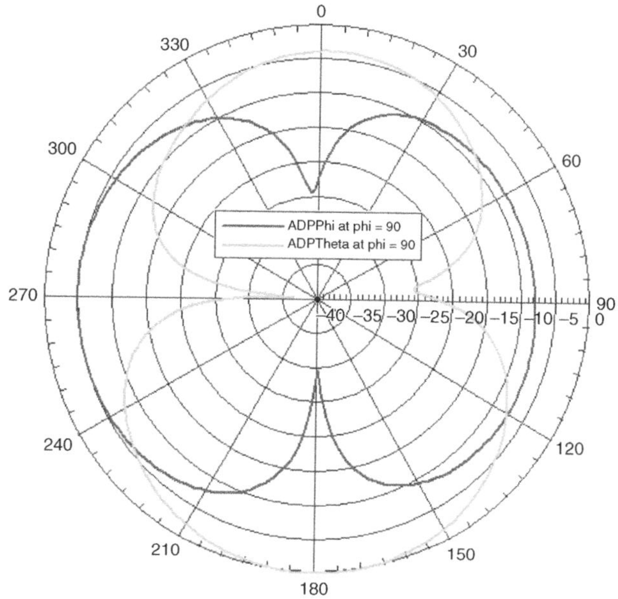

(b)

Figure 1.20 (Continued).

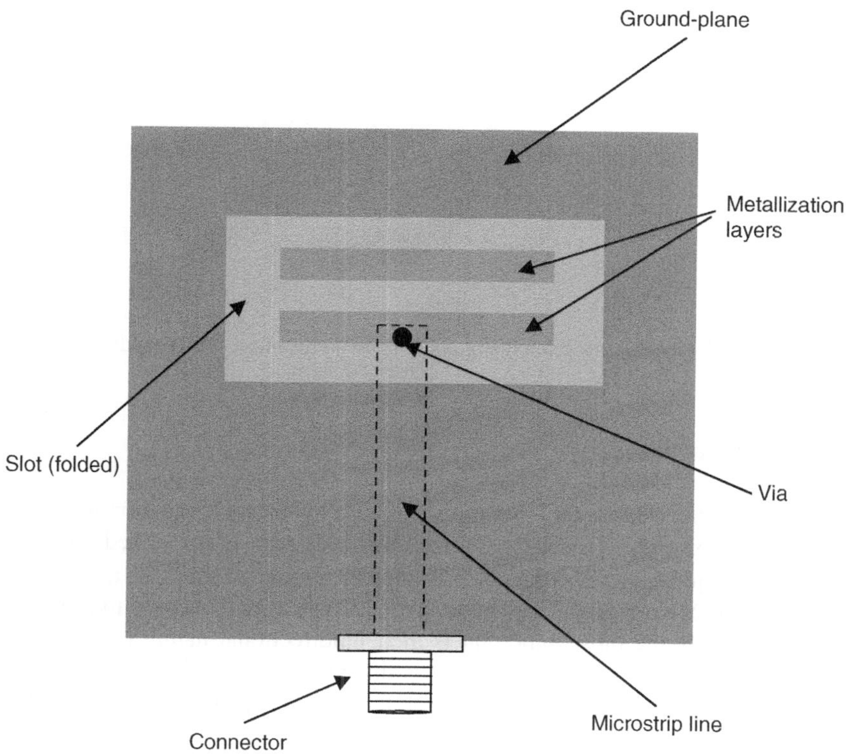

Figure 1.21 Schematic of a microstrip feed folded slot.

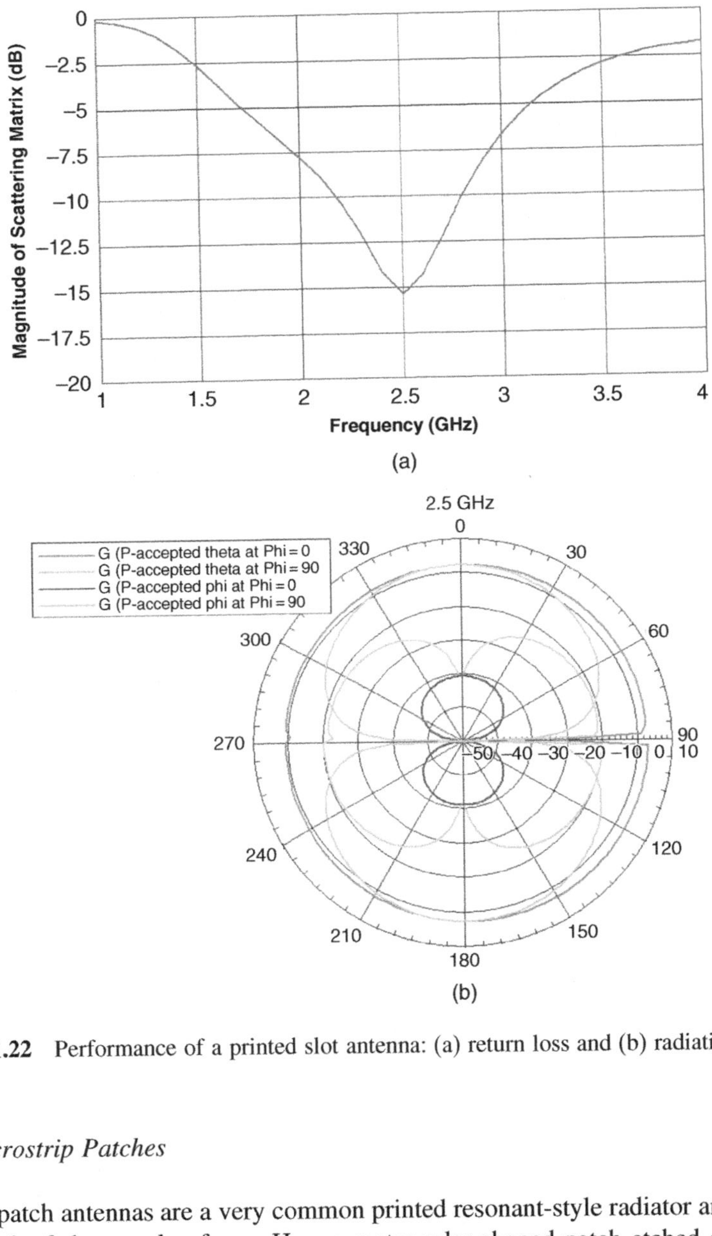

Figure 1.22 Performance of a printed slot antenna: (a) return loss and (b) radiation patterns.

1.5.2.3 Microstrip Patches

Microstrip patch antennas are a very common printed resonant-style radiator and Figure 1.23 shows a probe-fed example of one. Here a rectangular-shaped patch etched on a grounded dielectric material is fed by a probe that rises through the grounded substrate and is soldered to the patch conductor. Microstrip patches have several advantages compared to printed dipoles and slots and probably the most important is their unidirectional nature (and consequently higher gain) because of their ground-plane. It is this ground-plane that unfortunately limits the bandwidth of the antenna; of the three fundamental printed resonant-style radiators, it probably has the smallest bandwidth.

Like the slot and dipole, there have been many versions of patch antennas examined over the years, including different-shaped patches and varieties of feeding procedures. As mentioned

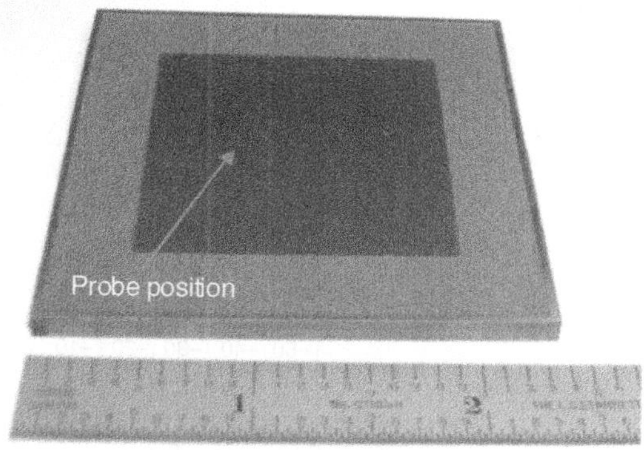

Figure 1.23 Photograph of a probe-fed microstrip patch antenna.

before, the method of feeding an antenna can prove beneficial in terms of optimizing the overall performance, both return loss and radiation patterns, of the radiator. For microstrip patches the two general classes of feeding procedures can be divided into direct contact, where the feed electrode comes into contact with the patch radiator, and noncontact, where the feed line is parasitically coupled to the antenna. Each method has its relative merits and disadvantages and a detailed discussion on these can be found in Reference [34].

Figure 1.24 shows the return loss performance of a typical microstrip patch antenna as well as the predicted radiation patterns at the centre of the −10 dB return loss band. For the case considered here, the ground-plane is assumed infinite.

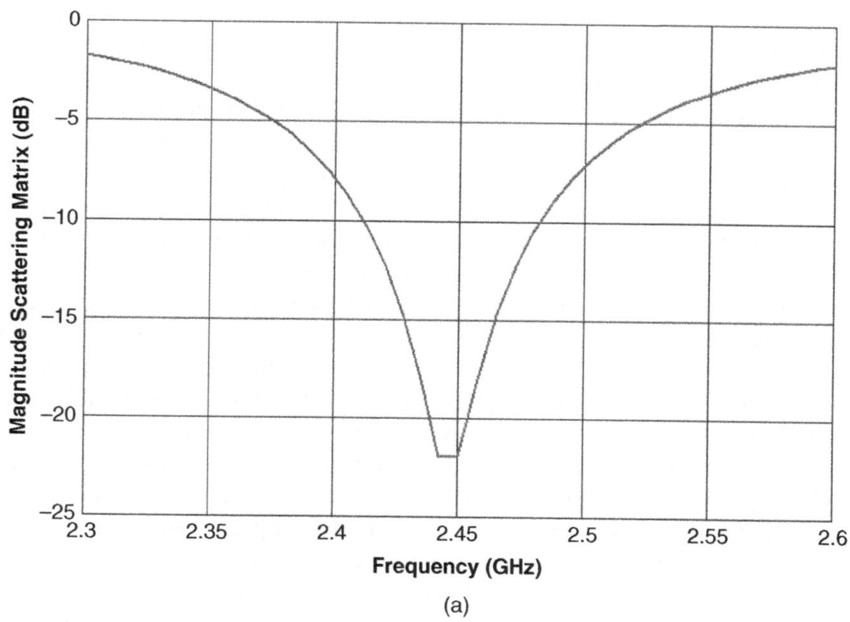

Figure 1.24 Performance of a microstrip patch antenna: (a) return loss and (b) radiation patterns.

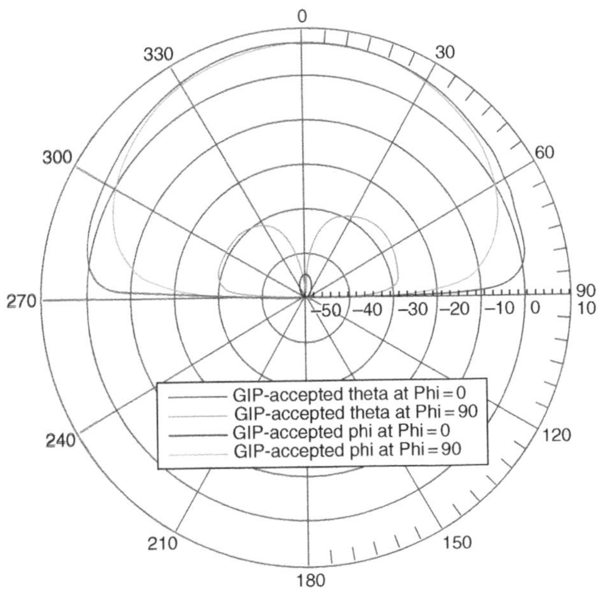

(b)

Figure 1.24 (Continued).

1.5.3 Small Printed Antennas

It is because of the requirements for small antennas in wireless terminals (in particular handsets) that there has been considerable research and development undertaken in the area of small printed antennas over the years. Researchers throughout the world have taken the fundamental resonant-style printed antennas and applied size-reduction procedures to yield antennas that can reside in small volumes. Of course this is one of the most difficult challenges an antenna engineer can face, namely to develop an electrically small radiator that can satisfy the radiation performance requirements of most wireless systems. Fortunately there have been significant advancements made in other EM aspects of engineering that has helped the designer, in particular the development of sophisticated electromagnetic software tools that can model not only the radiator but the surrounding environment. These software tools tend to be computationally intensive and therefore the advancements made in computing power cannot be ignored, which has also helped the antenna engineer to use these software packages as design tools (not just as an analysis engine). Having said this, there are compromises that are made when making an antenna small, and it would be almost irresponsible to ignore them. Bandwidth and gain of any antenna are related to its volume and therefore a smaller antenna will have lower gain and smaller bandwidth. Once again, through advancements made in other aspects of engineering, this impact on the overall performance of the wireless system can be minimized. For example, higher quality RF amplifiers (in terms of efficiency and cost) are being developed that ease the gain and return loss requirements of the small antenna. It is interesting to note that most commercially available antennas for cellular handset terminals quote a −6 dB return loss bandwidth and have gains of less than 0 dBi. It can be extrapolated that this is because the requirements of −10 dB return loss and gains greater than 0 dBi are just too difficult to achieve for such constrained volumes, and other parts in the system (including the amplifiers) can be used to offset the degradation in the system performance.

There are a variety of procedures that have been successfully employed to reduce the overall size of a printed antenna. These include dielectric loading (both permittivity and permability); using shorting posts (both shorted patches and planar inverted-F antennas, PIFAs); fractal engineering; and maximizing the radiator size within the volume allocated for the antenna, or three-dimensional antenna engineering. Some of these topics will be discussed in later chapters.

Figures 1.25 to 1.28 show four examples of how small versions of the fundamental printed resonant-style radiators can be developed. In Figure 1.25 a CPW-fed printed monopole is shown where the monopole conductor meanders back and forth in order to maximize the area for the given radiator effectively [35]. Figure 1.26 shows a three-dimensional version of a flared (or triangular-shaped) monopole, where the conductor is folded over the edge of the supporting substrate to increase the length of the antenna effectively and subsequently reduction of the operation frequency is realized [36]. Figure 1.27 shows a photograph of a shorted patch antenna. Here shorting pins are used to reduce the size of the patch conductor. Two pins are used to increase the separation distance between the shorting points (the pins) and the probe feed [37].

In Figure 1.28 three concepts are combined to yield a small version of a slot radiator: (1) dielectric loading; (2) folding the slot; and (3) using a short-circuit [38]. Of course there are many more procedures that can be used to reduce the size of the antenna, and judging by

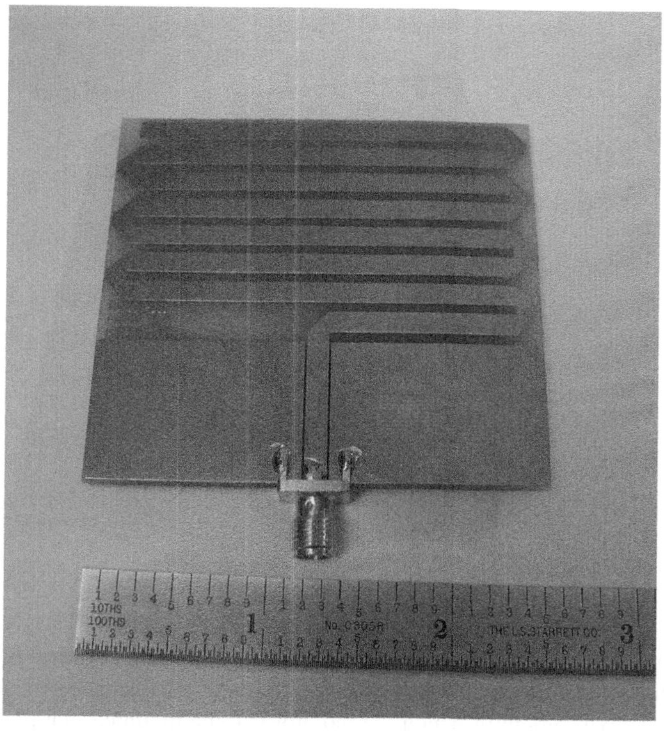

Figure 1.25 Photograph of a printed meander line monopole.

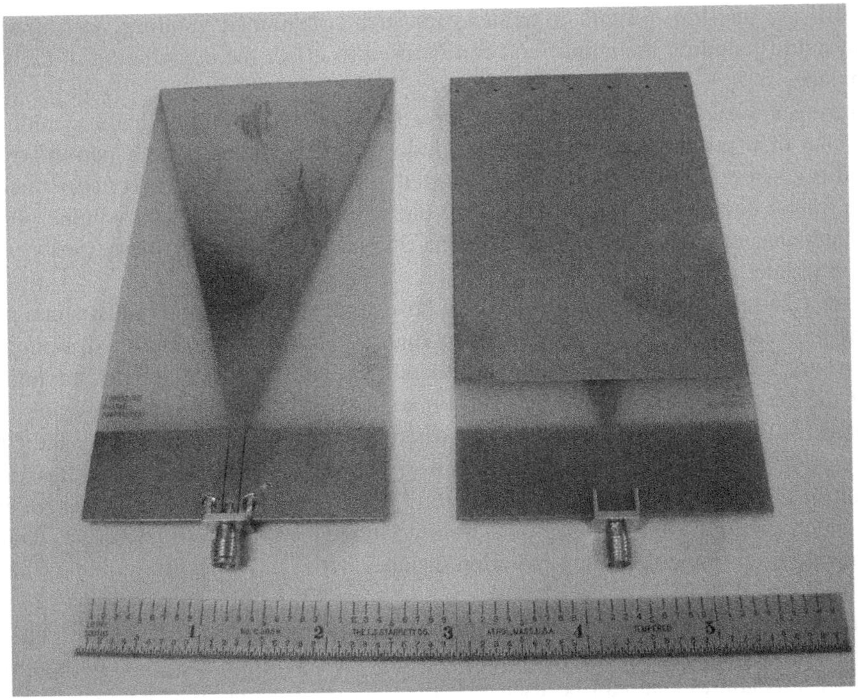

Figure 1.26 Photograph of a three-dimensional flared monopole [36].

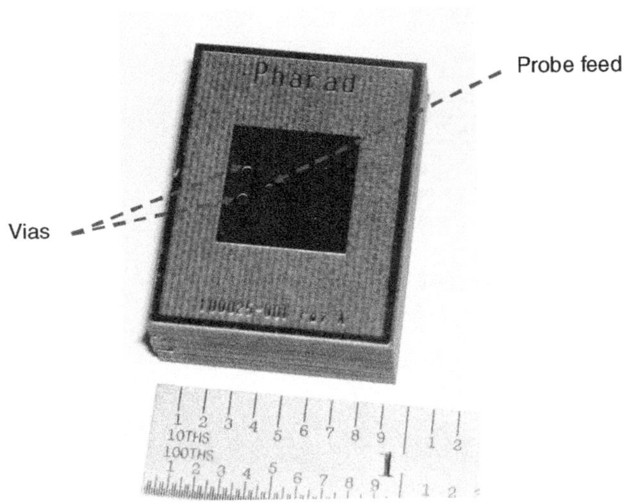

Figure 1.27 Photograph of a shorted patch antenna.

the number of publications in this area of R&D there will be many more to come. It is the authors' opinion that no one technique gives the ultimate performance of a small radiator and really it is the environment in which the radiator is to be mounted that plays a significant role in which is the best size-reduction method.

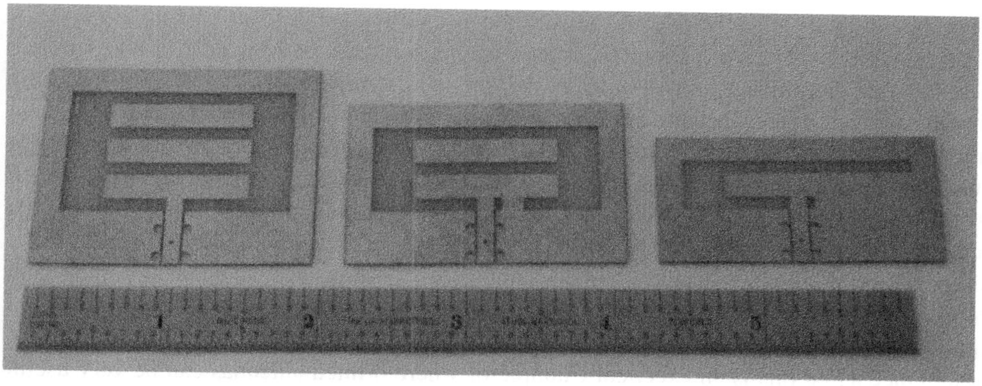

Figure 1.28 Photograph of printed folded slots [38].

1.6 OUTLINE OF THE BOOK

In this book several concepts related to wireless systems and printed antennas are presented. The remainder of the book is divided into three parts: Part I, Fundamental Wideband Printed Antennas for Wireless Systems; Part II, Small Printed Antennas for Wireless applications; and Part III, Advanced Concepts and Applications in Wireless Systems. The first part describes wideband printed antenna radiators suitable for a variety of wireless systems. In Chapter 2 multilayered patch antennas are presented including aperture coupled and proximity coupled patches. Stacked patches are investigated and design strategies and bandwidth optimization procedures are given. In Chapter 3 variations of printed quasi-Yagi antennas are discussed and several designs presented, including a case for operation at 120 GHz. Examples of arrays of these antennas are also given. These two chapters represent wideband variations of resonant-style printed antennas as previously discussed. In the next chapters travelling wave versions of printed antennas are explored. In Chapter 4 printed spiral antennas are presented and design strategies and a parameter study are given. In Chapter 5 a variation of the printed spiral, the printed folded Beverage antenna, is discussed and several examples given, including three-dimensional versions. In Chapter 6 the characteristics of tapered slot antennas are presented and, once again, there is a focus on looking at design trends so that the reader can establish how to optimize the performance of this class of radiator. Chapters 2 to 6 represent most of the wideband fundamental radiating elements that are presently being considered for wireless systems.

The next part of the book presents four chapters focusing on small printed antennas. In Chapter 7 several versions of PIFAs are examined, including multiband versions, and the impact of capacitive and inductive loading and fractal forms are summarized. In Chapter 8 a variety of shorted patches are presented including circular polarized and multiband versions. Techniques to reduce cross-polarization and also procedures to further reduce the conductor size of the antenna are summarized in this chapter. Chapter 9 looks at printed antenna specifically designed for WLAN applications, in particular versions of the printed monopole. Antennas capable of operating over several wireless communication system bandwidths are presented and a surface-mounted spiral printed monopole is examined. Chapter 10 presents how printed antennas can be incorporated into computers and other wireless communication components. In particular, this chapter shows how the performance of the antenna can be impacted by the surrounding environment.

In the third part of the book some advanced concepts and applications in wireless systems are presented. Not only variations of printed antennas but also advanced EM and wireless concepts that can impact printed antenna designs are discussed. In Chapter 11 reflectarray antennas are presented, including the principles of operation and design methods of passive and active printed reflectarray antennas. This chapter also presents designs for automotive radar systems, base stations, and spatial power combiners. Chapter 12 looks at the use of artificial magnetic conductors and other surfaces and how they can be used to improve the performance of printed antennas. Chapter 13 presents an overview of integrated transceivers. Integrated transceivers are a very good example of where printed antennas are extremely useful. Chapter 14 investigates the art of reconfigurable antennas for a software defined radio. Here the radiators are designed to be inherently matched to the active elements feeding them. Once again this is an excellent example of where printed antennas can be very useful radiators, as the control of the input impedance of a printed antenna (especially a patch antenna) is easily controllable. Finally, Chapter 15 examines multiple-input multiple-output (MIMO) systems and how patch antennas can be used in these wireless systems.

REFERENCES

1. R. Prasad and L. Munoz, *WLAN and WPANS towards 4G*, Artech House, Norwood, Massachusetts, 2003.
2. D. M. Pozar, *Microwave and RF Design of Wireless Systems*, John Wiley & Sons, Inc., New York, 2000.
3. E. D. Kaplan and C. Hegarty (eds), *Understanding GPS: Principles and Applications*, 2nd edn, Artech House, Norwood, Massachusetts, 2006.
4. J. D. Kraus and D. A. Fleisch, *Electromagnetics with Applications*, 5th edn, McGraw-Hill, New York, 1999.
5. *Memorandum on Draft Policy for Radio Frequency Identification (RFID)*, DoD Acquisition, Technology and Logistics, 2 October 2003.
6. *Memorandum on Update on Draft Policy for Radio Frequency Identification (RFID)*, DoD Acquisition, Technology and Logistics, 20 February 2004.
7. www.schreiner-online.com/en/html/presse/06_LogiData/06_3_rfid-Label/Transponder_und_Etiketten.html
8. http://jtrs.army.mil/
9. Joint Tactical Radio System Operational Requirements Document (Extract of JROC approved final with waveform Table 4-2 and Annex E), Version 3.2, April 2003.
10. A. J. Goldsmith and S. B. Wicket, 'Design challenges for energy-constrained ad hoc wireless networks', *IEEE Wireless Communications,* 8–27, August 2002.
11. C. de Morais Cordeiro, H. Gossain and D. P. Agrawal, 'Multicast over wireless mobile ad hoc networks: present and future directions', *IEEE Network,* 52–59, January/February 2003.
12. S. Bellofiore, J. Foutz, R. Govindarajula, I. Bahceci, C. A. Balanis, A. S. Spanias, J. M. Capone and T. M. Duman, 'Smart antenna system analysis, integration and performance for mobile ad-hoc networks (MANETs)', *IEEE Transactions on Antennas and Propogation,* **50**, 571–581, May 2002.
13. R. Ramanathan, 'Making ad hoc networks density adaptive', in *Proceedings of MILCOM,* 957–961, October 2001.
14. http://wwww.xtarllc.com/capabilities/indelx.htm
15. http://msl.jpl.nasa.gov/Programs/gps.html
16. http://grouper.ieee.org/groups/802/11/
17. http://www.ieee802.org/15/pub/SG3c.html
18. http://grouper.ieee.org/groups/80215/
19. ITU Recommendation ITU-R P.676-3, 'Attenuation by atmospheric gases', Geneva, 1997.

20. http://www.northropgrumman.com/index.html
21. http://www.aerosonde.com/
22. http://www.bluefinrobotics.com/
23. 'IEEE standard definitions of terms for antennas', IEEE Std 145–1983, *IEEE Transactions on Antennas and Propagation*, **31**, November 1983.
24. K. S. Yngvesson *et al.*, 'The tapered slot antenna – a new integrated element for millimeter-wave applications', *IEEE Transactions on Microwave Theory and Techniques*, **37**, 365–374, February 1989 (invited paper).
25. K. S. Yngvesson *et al.*, 'Endfire tapered slot antennas on dielectric substrates', *IEEE Transactions on Antennas and Propagation*, **33**, 1392–1400, December 1985.
26. C. H. Chio and D. H. Schaubert, 'Parameter study and design of wide-band widescan dual-polarized tapered slot antenna arrays', *IEEE Transactions on Antennas and Propagation*, **48**, 879–886, June 2000.
27. M. C. Greenberg, K. L. Virga and C. L. Hammond, 'Performance characteristics of the dual exponentially tapered slot antenna (DETSA) for wireless communications applications', *IEEE Transactions on Vehicular Technology*, **52**, 305–312, March 2003.
28. R. B. Waterhouse, 'Traveling wave antennas', in *Encyclopedia of RF and Microwave Engineering* (ed. K. C. Chang), Vol. 6, John Wiley and Sons, Inc., New York, pp. 5363–5373, 2005.
29. J. Dyson, 'The equiangular spiral antenna', *IRE Transactions on Antennas and Propagation*, **AP-7**, 181–187, April 1959.
30. J. J. H. Wang and V. K. Tripp, 'Design of multioctave spiral-mode microstrip antennas', *IEEE Transactions on Antennas and Propagation,* **39**, 332–335, March 1991.
31. S. D. Targonski, R. B. Waterhouse and D. M. Pozar, 'Design of wideband aperture-stacked patch microstrip antennas', *IEEE Transactions on Antennas and Propagation*, **46**, 1246–1251, September 1998.
32. H.-S. Tsai and R. A. York, 'FDTD analysis of CPW-fed folded slot and multiple-slot antennas on thin substrates', *IEEE Transactions on Antennas and Propagation,* **44**, 217–226, February 1996.
33. C. A. Balanis, *Antenna Theory: Analysis and Design*, 2nd edn, John Wiley & Sons, Inc., New York, 1997.
34. R. B. Waterhouse, *Microstrip Patch Antennas: A Designer's Guide*, Kluwer, Boston, 2003.
35. L. C. Godara (ed.), *Handbook on Antennas in Wireless Communications*, CRC Press, New York, 2001.
36. R. B. Waterhouse and D. Novak, 'Printed folded flared monopole antenna', *IEE Electronics Letters*, **42**, February 2006.
37. R. B. Waterhouse, S. D. Targonski and D. M. Kokotoff, 'Design and performance of small printed antennas', *IEEE Transactions on Antennas and Propogation*, **46**, 1629–1633, November 1998.
38. R. Waterhouse and D. Novak, 'Small folded CPW fed slot antennas', in *Proceedings of the IEEE Antennas and Propagation Symposium,* Albuquerque, New Mexico, July 2006.

Part I

Fundamental Wideband Printed Antennas for Wireless Systems

2

Multilayered Patch Antennas

Wayne S. T. Rowe
School of Electrical and Computer Engineering
RMIT University, Melbourne, Victoria, Australia

2.1 INTRODUCTION

Microstrip patch antennas are an attractive solution to many wireless communication scenarios due to their low-cost, low-profile, conformable, and easy-to-manufacture architecture. However, the primary obstruction for their application in many systems is their limited bandwidth. The bandwidth of a conventional printed antenna element is only in the order of a few percent. Because of this fundamental drawback, a large body of research has been devoted to overcoming the inherent bandwidth limitation of microstrip antennas.

An established means of enhancing the bandwidth of a conventional microstrip patch is to use electrically thick substrates. Unfortunately this relatively simple method does have its drawbacks. For edge-fed microstrip patch antennas, an increase in the thickness of the microwave substrate results in an expansion in the width of the microstrip feed lines. Consequently, the structure generates additional spurious radiation (or feed radiation). This results in elevated levels of cross-polar radiation from the patch radiator, which can limit the performance of the antenna, especially when it is configured in an array. For example, increasing the thickness by a factor of 2 can increase the bandwidth by a similar factor, but also degrades the cross-polar radiation levels significantly [1]. In the case of probe-fed patch antennas, a thicker substrate requires the length of the probe pin to be extended. This can lead to excessive feed inductance and counteract the bandwidth enhancements achieved with the thick substrate.

Another concern with the use of thick substrates is increased surface wave mode generation [2, 3]. Surface wave power deducts from the radiated power causing a reduction in antenna efficiency, and can also cause pattern degradation when the surface wave diffracts at the edge of the ground-plane or substrate. Various methods have been proposed to overcome the spurious feed radiation [4] and surface wave generation [2, 3, 5] limitations of printed

Printed Antennas for Wireless Communications Edited by R. Waterhouse

antennas on electrically thick substrates. The ideal scenario is to be able to increase the antenna bandwidth without incurring these limitations.

A diverse range of other bandwidth enhancement techniques has also been reported. It is often possible to devise an impedance matching network to manipulate the input impedance of a printed antenna to increase the bandwidth. This matching network may evolve as a stub tuner at the input of the antenna [6], as a direct modification to the patch element [7], or as a tuning element gap coupled to the patch [8]. Parasitically coupled coplanar patch elements can also form a wideband multiresonator subarray [9]. This is by no means an exhaustive list, as it only highlights a few of the basic techniques.

2.2 TYPES OF MULTILAYERED PATCH ANTENNAS

Multilayered printed antennas are possibly the most common method used to enhance bandwidth. The multilayered approach will typically use two methods to achieve this goal:

(a) increase the overall thickness of the substrate materials between the patch radiator and the ground plane and/or

(b) insert multiple coupled resonant elements into the structure.

An example of a multilayered approach that uses the first method listed above is the proximity coupled patch antenna [10].

Figure 2.1 displays a schematic of a proximity coupled microstrip antenna. The patch element resides on a separate substrate to the microstrip feed line. There is no physical electrical contact between the feed line and the patch. Energy is electromagnetically coupled from the feed line to the patch due to their close proximity with each other. This architecture therefore allows a thin feed substrate (and, hence, a thin microstrip feed line) to be employed while the thickness of the material seen by the patch can be increased.

The aperture coupled configuration [11, 12] is another simple method of expanding the bandwidth of a single-element patch antenna. By virtue of the common ground-plane, this configuration also has the ability to isolate the feed line from the patch radiator. This reduces the effects of spurious feed radiation on the antenna pattern.

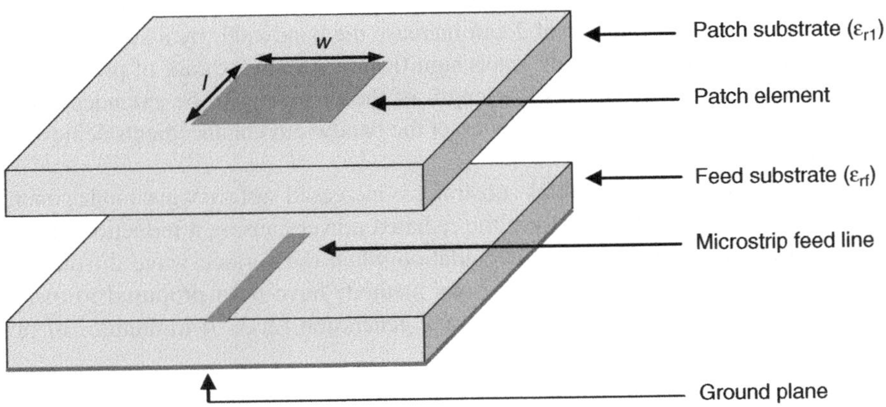

Figure 2.1 Proximity coupled microstrip antenna.

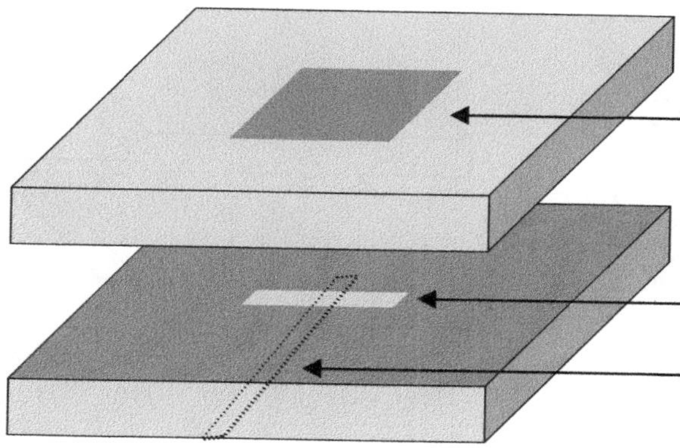

Figure 2.2 Aperture coupled patch antenna.

The structure of an aperture coupled microstrip antenna is shown in Figure 2.2. Energy from the microstrip feed line (located on the underside of the feed substrate in Figure 2.2) is coupled through the ground-plane to the patch element via an aperture in the metallic boundary. The isolation provided by the ground-plane enables the feed and patch characteristics to be optimized almost independently of each other. An additional parameter, the dimensions of the aperture, can also be used to assist in the impedance matching of the antenna. Alternatively, the aperture can be utilized as an additional resonator. This facilitates the creation of a dual band antenna or, in the case where the aperture resonance is mutually coupled to the patch element resonance, a broadband antenna. One consequence of the aperture being a resonant element is that the level of backward-directed radiation is typically quite high, as radiation from the aperture directly contributes to the rear-directed fields.

To expand the bandwidth further, printed antennas with several resonant elements may be used. As mentioned previously, the resonant elements may be placed in the same plane [9] but are commonly assembled on top of each other. These multilayer antennas are regularly referred to as stacked patch antennas, due to the vertical formation of the resonant elements. Impedance bandwidths in excess of 20 % are readily achieved using the stacked patch method. Various stacked patch configurations are shown in Figure 2.3. The most fundamental forms of the stacked patch are displayed in Figures 2.3(a) and (b), consisting of a probe-fed or edge-fed driven patch and a parasitic element placed on top. Figures 2.3(c) and (d) depict proximity coupled and aperture coupled versions of the stacked patch antenna. Note that these two structures contain two parasitic patch radiators as there is no direct connection of the feed.

2.3 DESIGN TRENDS OF MULTILAYERED PATCH ANTENNAS

In microstrip patch antennas, the radiation bandwidth typically has a more wideband characteristic than the impedance bandwidth. It is typically assumed that for relatively symmetric conventional patch antenna designs if you have a good impedance response, the radiation characteristics should be acceptable within this band. For this reason, the design of patch antennas is usually focused on obtaining an acceptable impedance response, and once this is achieved the radiation attributes are validated. Hence, this section examines the design

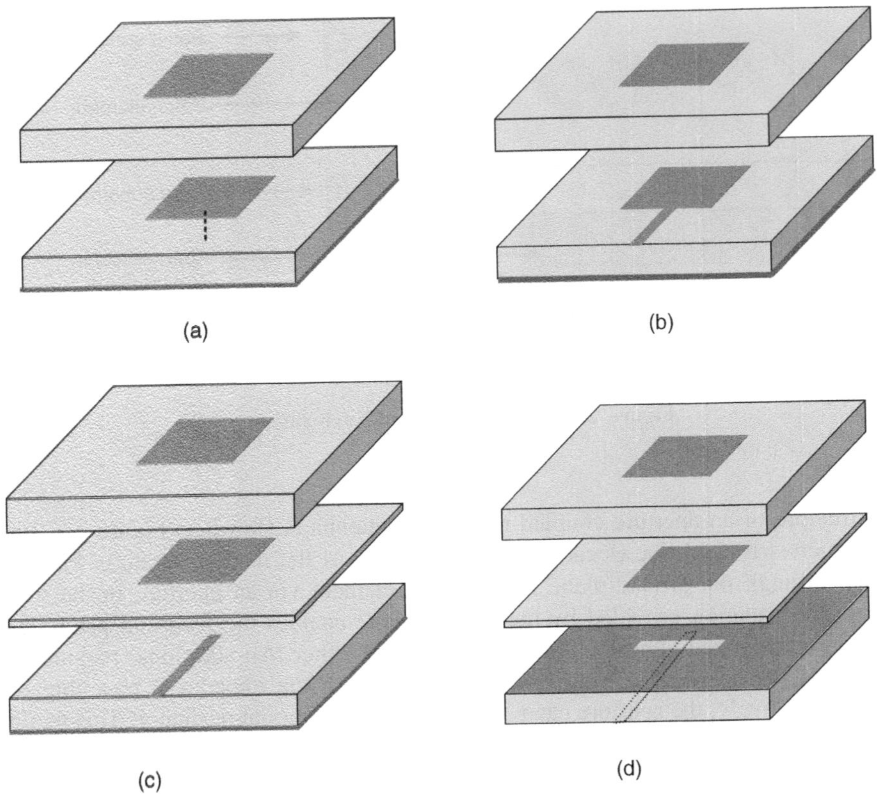

Figure 2.3 Stacked microstrip patch antenna structures: (a) probe-fed; (b) edge-fed; (c) proximity coupled; and (d) aperture coupled.

trends for multilayer patch antennas with a heavy focus on the impedance locus. Substrate selection is generally the initial step in the design of multilayer patch antennas. The choice of material is a critical factor in the performance of the antenna, and is examined in depth in Section 2.4. The design principles for the proximity coupled and aperture coupled configurations are briefly summarized here, followed by a more detailed examination of the parameter-dependent impedance variations for stacked patch antenna architectures. The structures analysed in this section utilize low-permittivity microwave substrates, which typically provide optimal antenna performance.

2.3.1 Parameter Study of a Proximity Coupled Antenna

In Reference [13] a parameter study was performed on a proximity coupled patch antenna similar to that of Figure 2.1, varying each attribute of the design individually while keeping the remainder constant. The parameter variation study verified the anticipated results common to single-element patch radiators. Increasing the length (l) of the patch element decreased the operating frequency and enlarged the size of the resonant loop in the input impedance. The gain of the antenna was also marginally diminished. A shorter patch had the converse effect. Varying the width (w) of the patch element had a much more subtle effect on the operating frequency, while offering control of the resonant loop size. The location of the microstrip line open-circuit termination can influence the position of the resonant loop, with a minimal effect on other characteristics. Extra length on the microstrip line before the open-circuit termination shifts the resonant loop towards the inductive region of the Smith chart. These results confirm the generally accepted design procedure for proximity coupled patches of using the patch length to set the frequency of operation, and fine tune with the patch width and open circuit position to maximize bandwidth.

If the patch substrate is made thinner a slight improvement in gain is observed. However, an enlarged impedance loop and increased operating frequency also result. A thicker patch substrate exhibits the opposite consequences. The operating frequency also increases when a thinner feed substrate is used. However, the gain is noticeably reduced, as is the size of the impedance loop. Once again, the effects are reversed for a thicker feed substrate. Each time the feed substrate thickness is varied, the width of the microstrip line must be altered to maintain a characteristic impedance of 50 Ω.

2.3.2 Aperture Coupled Patch Antenna Design Trends

The principles for designing an aperture coupled patch antenna similar to Figure 2.2 were articulated in Reference [12]. It was found that the resonant frequency of the aperture coupled patch antenna is primarily determined by the length of the patch element, as is the case for most single-patch microstrip antennas. However, there is a secondary influence on the resonant frequency, the aperture (or slot) length. As the length of the aperture increases, the resonant frequency shows a slight decrease.

A decrease in the length of the aperture is also seen to cause a diminished level of coupling between the microstrip feed line and the patch element. This manifests as a reduction of the impedance circle on the Smith chart. The centre of the impedance circle also moves towards the extremity of the chart, closer to the short-circuit position in this case.

If the length of the open-circuit stub on the end of the microstrip feed line is varied, the input impedance at each frequency follows a contour of constant resistance. This indicates that near the resonant frequency of the antenna, the aperture and the patch element resemble a series load along an open-circuit transmission line and can be modelled with an equivalent circuit. Hence, a very simple process can be followed in order to obtain a desired antenna impedance, as the aperture length can be used to tune the resistance, and the reactance can then be adjusted by varying the open-circuit stub length.

The properties of the antenna substrates were also examined. It is interesting to note that for the aperture coupled patch configuration, an increase in the permittivity of the feed layer had minimal impact on the resonant frequency. However, the coupling level from the feed line to the patch element increased. This can be interpreted as being a consequence of the aperture being electrically longer, even though its physical dimensions have not changed. If the thickness of the feed material is increased, which moves the feed line further away from the aperture, the level of coupling to the patch is diminished. As with the permittivity variation, the resonant frequency remains relatively constant. The effect of increasing the thickness of the antenna substrate was similar to that of making the feed material thicker. It is possible to compensate for an increase in thickness of the layers by making the aperture longer, but this can yield an elevated level of backward-directed radiation.

2.3.3 Design Rules for Probe-Fed Stacked Patch Antennas

The design rules for a probe-fed stacked patch antenna (Figure 2.3(a)) should be analogous to that for an edge-fed stacked patch antenna (Figure 2.3(b)), as the two feeding techniques closely resemble one another. Presented here is a strategy for creating probe-fed stacked patch antennas with circular radiating elements [14]. It is stated in Reference [14] that rectangular and annular ring elements exhibit the same trends as the circular radiators.

As is the case with all antennas with mutually coupled resonances, a loop in the input impedance locus is observed for probe-fed stacked patches. If the size of the driven patch element is increased while holding all other parameters constant, a decrease in the size of this loop is seen. The real component of the input impedance at the centre of the loop also moves to a lower value. The upper parasitic patch element exhibits the opposite trends. Hence if the size of the upper patch element is increased in isolation, the loop expands and the centre has a higher real value of impedance.

The position of the probe has only minimal consequences on the impedance behaviour of probe-fed stacked patch antennas. Increasing the distance of the probe relative to the centre of the patch exhibits a similar trend to that of increasing the overall size of the upper patch element. However, the magnitude of the shift in the centre of the impedance loop is far less pronounced for varying the probe position.

The thicknesses of the dielectric layers in the probe-fed stacked patch antenna are critical factors to the performance. For the lower dielectric layer, there is a trade-off between the bandwidth that can be achieved and the level of control over the impedance. Traditional concepts of antenna design would suggest that a thicker substrate would yield a larger bandwidth. However, the lower patch element in isolation must be overcoupled and achieve a wideband response. Hence, a thinner substrate for the lower dielectric layer is required, so when the parasitic patch element is placed on top of the driven patch element, the coupling loop formed is located close to the centre of the Smith chart.

The thickness of the dielectric between the two patch elements primarily governs the level of mutual coupling. The thicker this layer is made the smaller the loop in the impedance locus becomes, and vice versa. The upper substrate layer is also strongly linked to the thickness of the lower substrate. If the material selected for the lower layer is electrically thick, the liberty in selection of the upper dielectric thickness is diminished.

2.3.4 Characteristics of Stacked Proximity Coupled Antennas

The bandwidth of proximity coupled patch antennas can be enhanced by stacking an additional parasitic patch over the conventional structure of Figure 2.1. This configuration is seen in Figure 2.3(c). An antenna of this type was constructed in Reference [13] and displayed a measured 10 dB return loss bandwidth of 20 % and a gain of approximately 9 dBi. To ascertain the design characteristics of the stacked proximity coupled antenna, a parameter study was performed on the architecture in Figure 2.3(c). Each physical attribute of the antenna was varied independently, while holding all others constant. The stacked proximity coupled antenna presented in Reference [13] was used as a baseline.

The parasitic patch elements of the baseline antenna are square. Preserving square patch elements is advantageous as it enables the formation of a dual polarized stacked proximity coupled patch antenna simply with the addition of a second orthogonal feed line. Circular polarization can also be obtained by driving these two feed lines at quadrature. The ensuing parameter study maintains the square geometry of the patch elements, but varies the overall size. Throughout this parameter study, the baseline response is located in the centre of each figure (for comparison) with variations below and above the baseline parameter value on the left and right, respectively.

Figure 2.4(a) displays the effect on the input impedance when the size of the lower patch element is altered by 0.8 mm ($\sim 0.013\lambda_0$). The variation of patch element 1 affects the level of mutual coupling between the two patches, which can be observed as a change in size of the small inner loop in the impedance loci of Figure 2.4(a). An increase in patch size corresponds to a reduction in the magnitude of mutual coupling. The larger outer loop is a manifestation of the coupling from the open-circuit microstrip feed line to the patch elements, which is also altered when changes are made to the size of patch element 1. As patch element 1 is physically located in the middle of the antenna, it is understandable that variation of this element will influence both coupling levels. Patch element 2 is situated close to the top of the antenna structure, and primarily influences the interaction between the two patches, as can be seen in Figure 2.4(b). It should be noted that an increase in the size of patch element 2 has the converse effect to patch element 1 on the magnitude of mutual coupling between the two patches. The input impedance response is less sensitive to changes in the size of patch element 2 than patch element 1. The change in patch element 2 dimensions shown here is 2 mm ($\sim 0.033\lambda_0$).

The input impedance plots in Figure 2.5(a) depict the consequence of varying the height of patch substrate 1. Typical substrate heights for Rogers *RT/Duriod*™ *5880* were used in this case study. As this substrate is situated between the feed line and the lower patch element, a change in the coupling characteristics from feed to patches is predicted to be the primary consequence. This is evident in Figure 2.5(a), although the values of substrate height remain relatively small ($\sim$ 0.0021 to $0.0085\lambda_0$), so strong coupling is still observed for all cases. For the foam substrate variation in Figure 2.5(b), an increase in height diminishes the magnitude of mutual coupling between the patches. However, the feed to patch coupling is also reduced. This could be due to the feed to patch coupling being governed by the characteristics of the mutually coupled patches as a single entity, and disturbing this will therefore transform the way the feed and patches interact. Once again typical commercially available foam thicknesses were used in this study.

A feed substrate height variation predominantly induces a change in the level of coupling between the feed and patch elements, as evident in Figure 2.6(a). This effect is analogous to the conventional single-layer proximity coupled patch antenna case. The microstrip feed line width was adjusted with the alteration of the feed substrate height to maintain a 50 Ω characteristic impedance. The stub offset is defined as the distance of the microstrip feed line open-circuit from directly underneath the geometric centre of the patch element. Modifying the stub offset position had a minimal consequence on the impedance response of the stacked proximity coupled antenna, unless large alterations were considered (e.g. $> 0.02\lambda_0$). This fact can be valuable if a dual or circularly polarized proximity coupled antenna is desired. If a second orthogonal feed line is added to the structure to obtain a dual or circularly polarized antenna (as stated previously), the open circuit terminations on the end of the microstrip feed lines can be moved away from the centre of the antenna without greatly affecting the impedance response. This assists in decoupling the feed lines and improves the isolation between them.

The majority of the parameter variations in this study had only a minor influence on the gain of the antenna. Typically, the gain varied less than 0.2 dB from that of the baseline antenna. As for the single-layer proximity coupled patch antenna case the exception was the

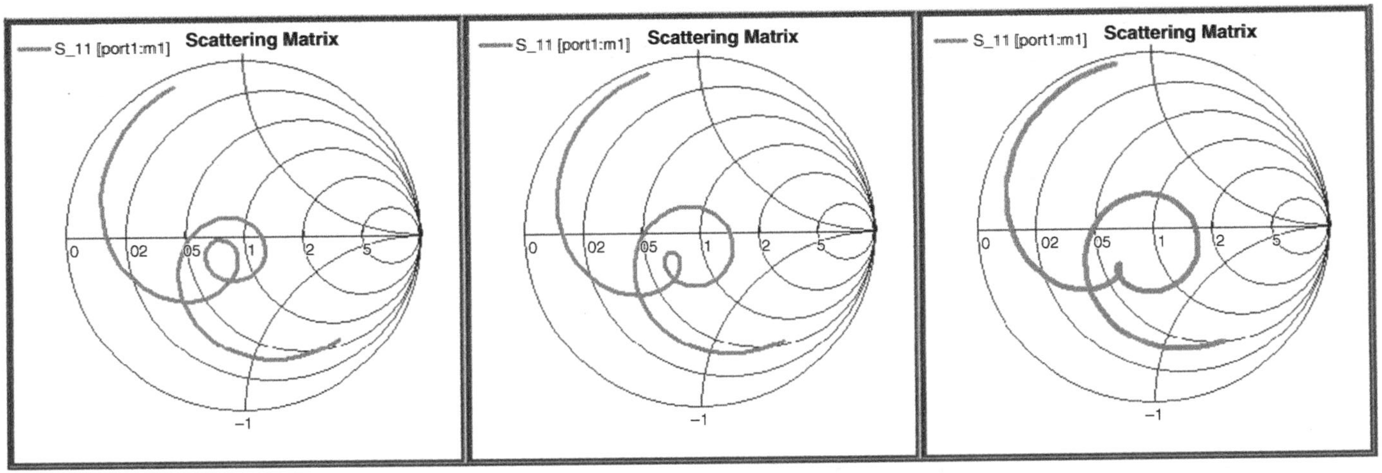

(a)

Figure 2.4 (a) Patch element 1 size variation (19.6 mm^2, 20 mm^2, 20.4 mm^2) and (b) patch element 2 size variation (21 mm^2, 21.6 mm^2, 23 mm^2).

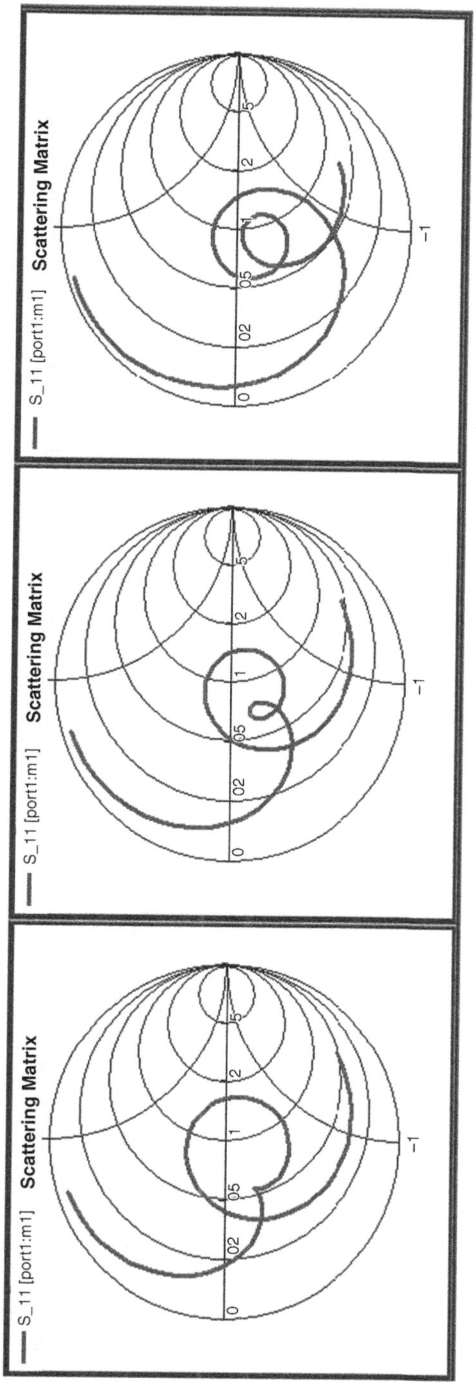

(b)

Figure 2.4 (Continued).

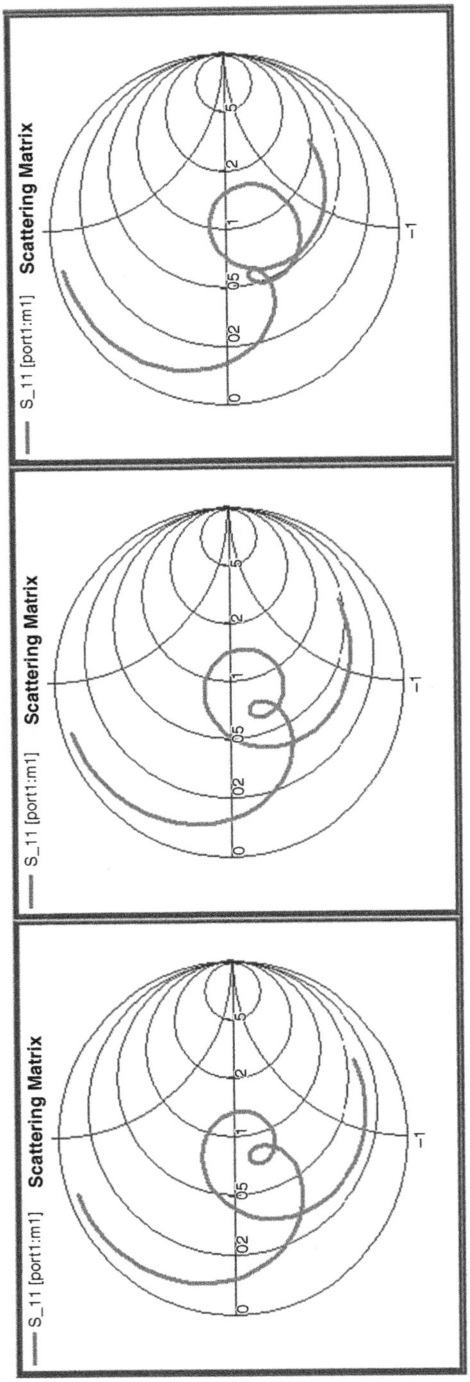

(a)

Figure 2.5 (a) Patch substrate 1 thickness variation (0.127 mm, 0.254 mm, 0.508mm) and (b) foam thickness variation (3 mm, 4 mm, 5 mm).

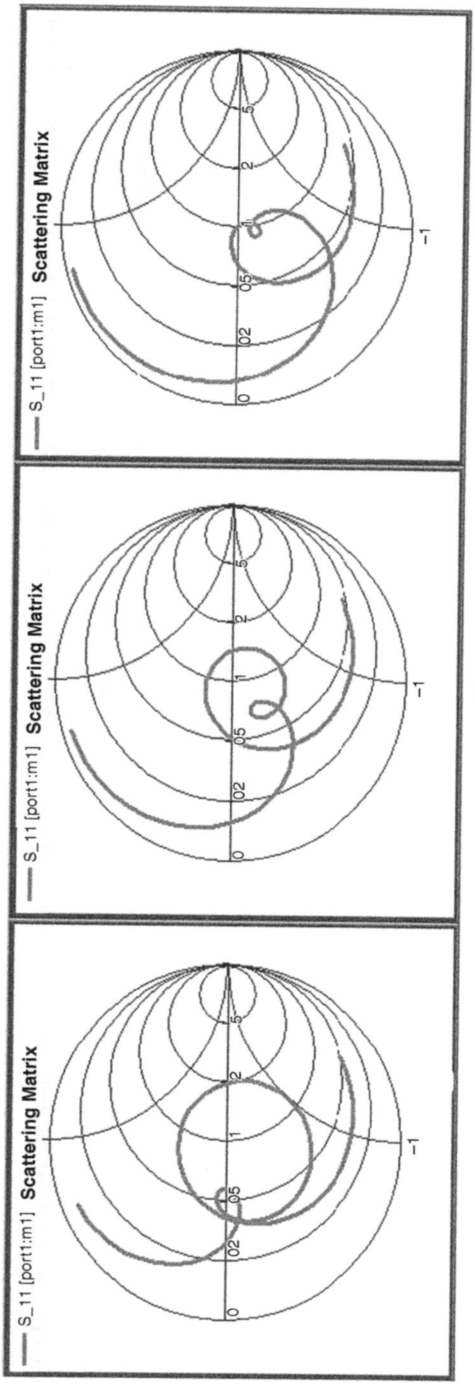

(b)

Figure 2.5 (Continued).

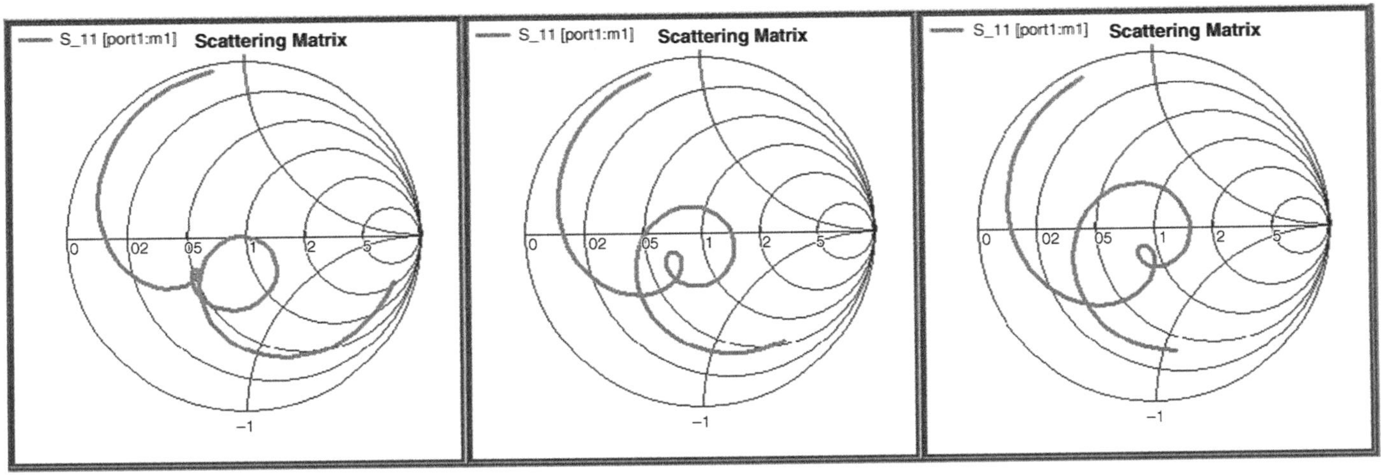

(a)

Figure 2.6 (a) Feed substrate thickness variation (1.016 mm, 1.575 mm, 2.083 mm) and (b) stub offset variation (−2 mm, −3.6 mm, −5.2 mm).

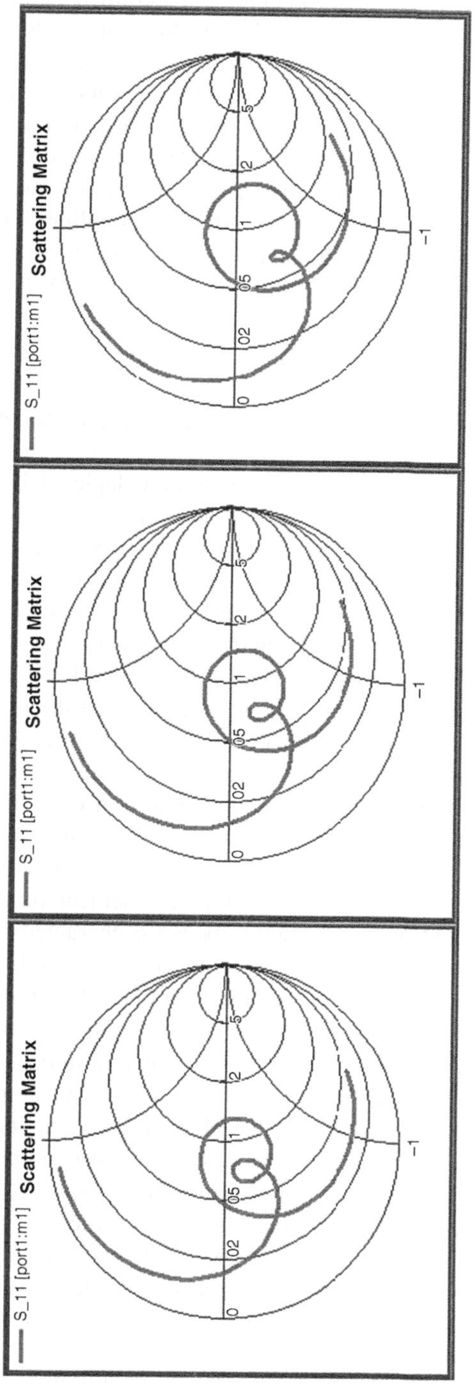

(b)

Figure 2.6 (Continued).

feed substrate height, where the gain showed the most significant rise when this substrate thickness was increased.

As the stacked proximity coupled patch antenna is a multiresonant structure, there are numerous interactions that influence the response. The dominant performance trends have been identified via this parameter study. These can be summarized as follows: the mutual coupling of the two patch resonances primarily corresponds to a loop in the input impedance locus (for this case it is the small inner loop); the second (large outer) loop relates to the interaction between the microstrip feed line and the resonantly coupled patch elements. Thus, a good starting point to achieve a broadband impedance response is to use a stacked structure with an electrically thick feed substrate. To ensure sufficient coupling to the patch elements, a thin substrate layer between the microstrip line and patch element 1 is required. The trends observed in the parameter study may then be employed to optimize the stacked proximity coupled antenna.

2.3.5 An Investigation of Aperture Coupled Stacked Patches

The aperture coupled stacked patch antenna configuration depicted in Figure 2.3(d) can deliver wideband impedance performance as well as providing isolation of the feed circuitry from the radiating elements. The broad bandwidth is primarily a product of the coupled patch resonances. Due to the fact that the aperture in the ground plane is not close to the resonant length, this antenna also exhibits reasonably low levels of back radiation. This section presents a summary of the mutual coupling trends in the parametric study of wideband aperture coupled stacked patch antennas disclosed in Reference [15]. The investigation in Reference [15] also analyses the effects of parameter variation on the individual resonances of the patch elements. However, to maintain consistency with the explorations of other antenna architectures in this chapter, this information is not presented here.

Firstly, to gain an understanding of the relative excitation of each of the parasitic patch elements in an aperture coupled stacked patch antenna, each patch is individually removed from the structure and the input impedance is analysed. If an aperture coupled stacked patch antenna with a broadband response has the upper patch element removed, the resulting structure exhibits an overcoupled or overexcited impedance characteristic. If the top patch is replaced to its original position and the lower patch is not considered in the analysis, the result is an undercoupled resonance. The combination of the two produces a moderately coupled loop in the impedance locus, which can be centred on the Smith chart.

The parametric study investigates the effect on the antenna impedance of varying one characteristic of the antenna, while maintaining all others at the value required to produce a wideband response. The size of the lower patch element was the first to be considered. If the lower patch is varied from it optimal value, the mutual coupling between the two patch elements decreases, and hence the loop in the impedance locus diminishes. It is known from Reference [12] that the coupling to the patch decreases when the patch size increases relative to the aperture size. Increasing the lower patch size also has the effect of blocking the top patch from coupling to the fields from the aperture excitation. When the lower patch is reduced in size the coupling loop becomes inductive, and the impedance tends towards the undercoupled response seen when the lower patch is removed.

For the variation of the upper patch element, similar results were seen on the mutual coupling level of the patch resonators. Any shift away from the optimal value causes the mutual coupling to decrease. The smaller the upper patch becomes, the closer the impedance locus resembles the overcoupled patch response seen when the top patch is totally removed. An increase in the upper patch size makes the response inductive, the opposite to the result for the lower patch.

If the thickness of the dielectric layer between the aperture and the lower patch is increased, it is instinctive to predict that the level of interaction between the aperture and the patches would be reduced as they are physically moved further apart. This postulation held true for the lower patch, but not for the upper patch element. The impedance locus also became more capacitive as the thickness of this layer reduced.

The thickness variation of the substrate that separated the two patch elements displayed a more intuitive result. Increasing the thickness, and hence the distance between the two patch radiators, reduced the level of mutual coupling between the two. This caused the loop in the impedance locus to shrink in size.

2.3.6 Parameter Study of Aperture Stacked Patch Antennas

The design of aperture stacked patch (ASP) antennas shown in Figure 2.3(d) can be a very complex process due to the three mutually coupled resonant elements. Not only must the coupling between the two parasitic patch radiators be considered, but the aperture in the ground-plane is also a resonator that contributes to the response of the antenna. To produce a wide impedance bandwidth the individual resonators must be weakly coupled to produce tight loops in the impedance locus, and these loops need to be alongside each other so that a continuous match over a large frequency range can be achieved Reference [16]. It is postulated in Reference [16] that one of the resonant loops is caused by the interaction of the aperture and lower patch, while the other arises from the mutual coupling of the two patch elements.

This section (taken from Reference [16]) investigates the effect on the impedance locus of modifying the principle dimensions of the ASP antenna. The impact on the response antenna will be assessed in terms of the variation in the size of the impedance loops and their relative position on a Smith chart. The dimensions of the baseline ASP antenna are given in the caption of Fig. 4 in Reference [16]. All parameters are held constant except the one being considered, and the baseline antenna results are depicted as the central figure. This allows the influence of the parameters under observation to be gauged.

The plots in Figure 2.7 highlight the effect of changing the length of the aperture. This variation shows a distinct change in the lower frequency loop in the impedance locus, while having a minimal influence on the higher frequency loop. The increase in the lower frequency loop size transpires from increased coupling between the aperture and lower patch as the aperture gets longer. The two loops are nestled closely together at the optimum aperture length and merge if the aperture length is excessive.

The lower patch size is a crucial parameter on the input impedance response of an ASP antenna. The lower patch is mutually coupled to both the upper patch element and the resonant aperture, contributing to the formation of both loops in the impedance locus. The influence of the lower patch size is clearly evident in Figure 2.8. The effect on the coupling level of each impedance loop is opposite as the lower patch size is varied. The higher frequency loop shrinks, while the lower frequency loop expands at the size of the lower patch increases. Interestingly, the opposite effect on the coupling levels occurs if the size of the upper patch element is the parameter that is varied. This can be seen in the input impedance plots of Figure 2.9. The inverse relationship leads to the conclusion that 'it is not

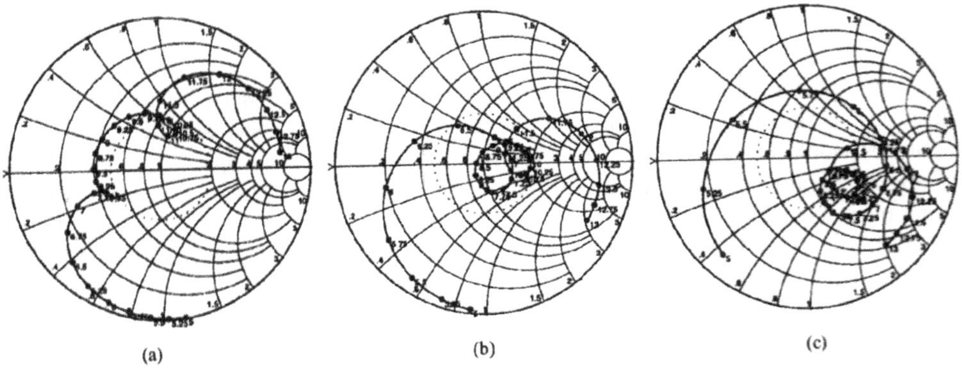

Figure 2.7 Input impedance as a function of aperture length: (a) SL (slot/aperture length) = 8 mm; (b) SL = 10 mm; and (c) SL = 12 mm.

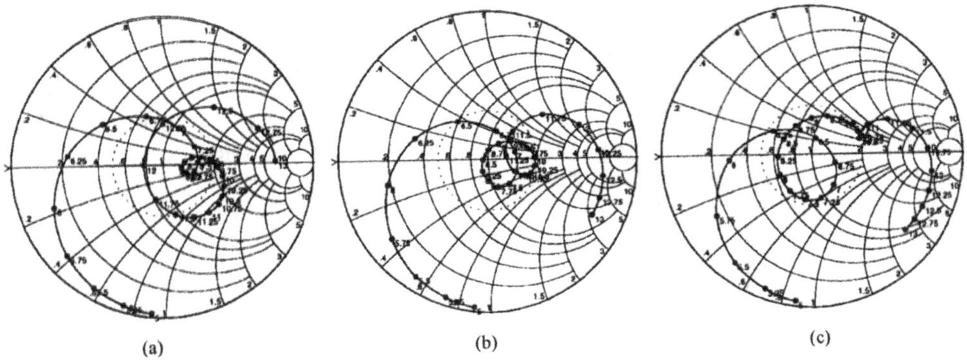

Figure 2.8 Input impedance as a function of the lower patch size: (a) PL_1 (patch length) = PW_1 (patch width) = 8 mm; (b) $PL_1 = PW_1 = 9$ mm; and (c) $PL_1 = PW_1 = 10$ mm.

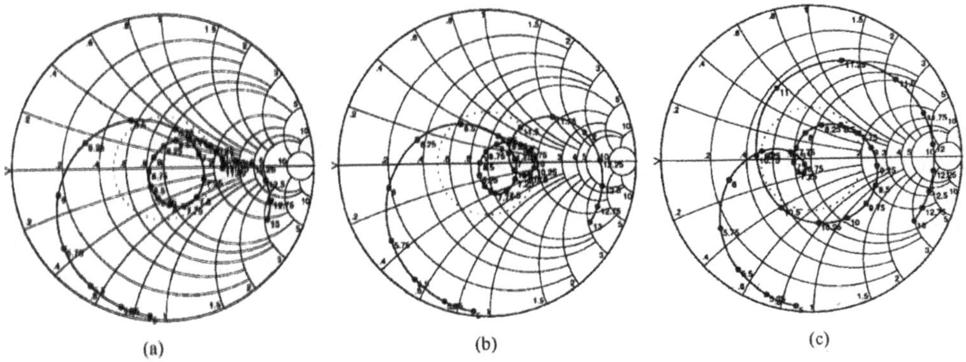

Figure 2.9 Input impedance as a function of the upper patch size: (a) $PL_2 = PW_2 = 9$ mm; (b) $PL_2 = PW_2 = 10$ mm; and (c) $PL_2 = PW_2 = 12$ mm.

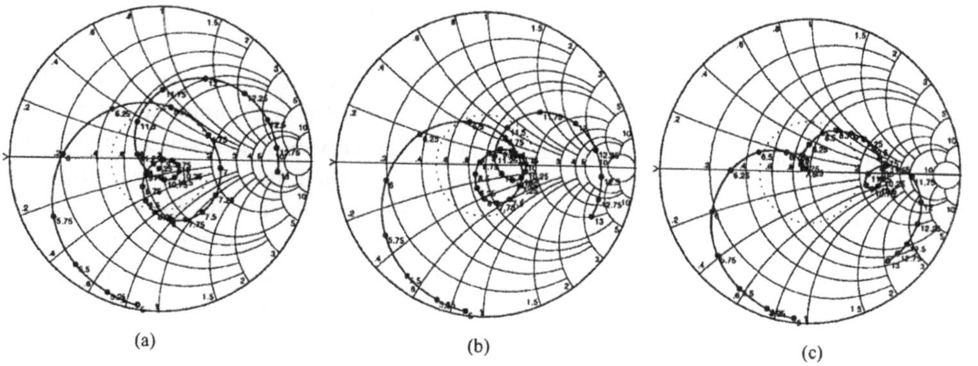

Figure 2.10 Input impedance as a function of the layer 1 dielectric thickness: (a) $d_1 = 2.5$ mm; (b) $d_1 = 3$ mm; and (c) $d_1 = 4$ mm.

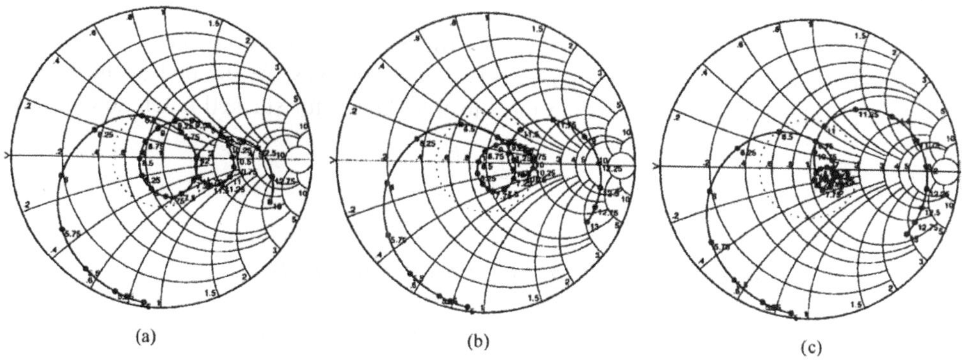

Figure 2.11 Input impedance as a function of the layer 2 dielectric thickness: (a) $d_2 = 2$ mm; (b) $d_2 = 3$ mm; and (c) $d_2 = 4$ mm.

the *absolute* dimensions of each patch that govern the impedance behavior, but the *relative* size of each patch to the other that is important' [16].

The effect of varying the lower substrate thickness is shown in Figure 2.10. As this parameter controls the physical distance between the aperture and lower patch, it is intuitive to predict that increasing the lower substrate thickness would reduce the coupling between them, and hence the size of the coupling loop in the impedance locus would diminish. A similar effect occurs when using a smaller aperture, as both reduce the aperture to the lower patch coupling. This can be a very useful link when designing and fabricating ASP antennas. If a certain thickness material required for the lower substrate is unavailable, a thicker material may be used if the aperture length is also increased to compensate.

In Figure 2.11, the impact of the upper substrate thickness on the impedance locus is displayed. Altering this parameter directly affects the coupling between the two patches, which is manifested as a reduction in the upper frequency loop with increased substrate thickness. However, the thickness of the upper substrate also influences the interaction between the two mutual resonances, decreasing the size of the lower frequency loop when a thicker substrate is used.

2.4 ANALYSIS OF LAYERED STRUCTURES

The most integral step in the design of multilayered antennas is to choose the layer configuration. Several general 'rules of thumb' have been proposed for determining the dimensions of microstrip patches (for example see Reference [17]). However, generalized approaches to the selection of the materials for use in a multilayer patch antenna are less prevalent. The dielectric material selection must take into consideration the bandwidth and efficiency performance required for the particular application, as well as the requirements/configuration of the system the antenna is to interface with. This selection process is of great importance as it is the dielectric substrates that primarily control the printed antenna performance.

In Reference [18] a figure of merit is proposed that equips an antenna designer with a method for comparison of different substrate configurations, with respect to radiation efficiency and impedance bandwidth performance. The quantity has relevance to single-layer patches and stacked configurations and can be easily determined. The figure of merit is defined as η_{sw}/Q, where the quality factor (Q) and surface wave efficiency (η_{sw}) are two imperative characteristics of a patch antenna. The figure of merit applies for any particular dielectric material.

This section analyses the substrate selection, influence of the material characteristics, and effects of layer alignment offsets in multilayer printed antennas. At the end of the section, some general selection tips for choosing the substrate layers for stacked patch antennas are given.

2.4.1 Substrate Selection for Stacked Patch Antennas

The conventional school of thought indicates that thick microwave laminates with a low dielectric constant enable the design of stacked patch antennas with wide impedance bandwidths and low surface wave generation. This philosophy has generally held true. However, stacked patches that employ high-permittivity materials in combination with low-permittivity substrates can also yield broadband, highly efficient results [19–23]. The primary reason for considering antenna elements printed on high-permittivity material is to enable direct integration with monolithic microwave integrated circuit (MMIC) or optoelectronic integrated circuit (OEIC) devices and systems.

To achieve the broadest possible impedance response, a low-permittivity microwave laminate (i.e. $\varepsilon_r = 2.2$) in combination with dielectric foam is typically used. This combination also has the simplest design procedure for stacked patch antennas. The use of two foam dielectric layers would improve the surface wave efficiency at the expense of the impedance bandwidth due to more tightly coupled patch resonators [14].

The thicknesses of the dielectric substrates are a critical parameter in the impedance bandwidth and can be achieved from probe-fed and edge-fed stacked patch antennas. The lower substrate must be sufficiently thick to maximize the bandwidth, but not overly so, which would result in a driven patch that is excessively inductive [14]. The lower patch in isolation should display only a slightly overcoupled response. This allows the addition of the parasitic patch element to 'pull' the impedance response closer to a matched position, as mentioned in Section 2.3.3. In Reference [14], a good starting point for the thickness of the dielectric layers was quoted as being $0.04\lambda_0$ for the lower substrate and $0.06\lambda_0$ for the upper layer.

2.4.2 *The Effect of Substrate Layer Permittivity in Multilayered Patch Antennas*

There are very few investigations into the effects of substrate layer permittivity in multi-layered printed antennas in the literature. For aperture coupled patch antennas, Sullivan and Schaubert [12] reported an escalation in the coupling between the aperture and the patch element as the permittivity of the feed material rises. This was attributed to the electrical length of the aperture increasing. This effect is also covered in Section 3.2.

2.4.2.1 Effect of Substrate Layer Permittivity in Proximity Coupled Patch Antennas

The proximity coupled patch antenna is a promising candidate for direct integration with MMIC/OEIC devices. The primary advantage of the proximity coupled structure is that a minimal area on the expensive semiconductor wafer is consumed, with only the formation of the open-circuit microstrip feed line required. However, MMIC wafers are typically of high permittivity; hence the proximity coupled patch antenna must be capable of exhibiting reasonable performance on such a material. A theoretical investigation (using Ensemble 6.0) into the proximity coupled antennas capabilities for this application was conducted in Reference [23].

Figure 2.12 illustrates the effect on the bandwidth and gain of the conventional proximity coupled patch antenna seen in Figure 2.1 as the feed substrate permittivity is raised from 2.2 to 10.2. At each value of permittivity, the feed was altered to maintain a 50 Ω characteristic impedance and the patch element was optimized for a maximum return loss. As expected, a decline in the antenna bandwidth is observed as the dielectric constant of the feed substrate is increased. The antenna gain diminished rapidly up to a permittivity of approximately 5 and then levels off to oscillate in the vicinity of 5.6 dBi. This effect can be attributed to the decrease in energy coupling from the feed to the patch. As the ε_r of the feed layer is increased, the energy will be more confined between the microstrip track and the ground-plane. To compensate for this, the height of the patch substrate should be decreased.

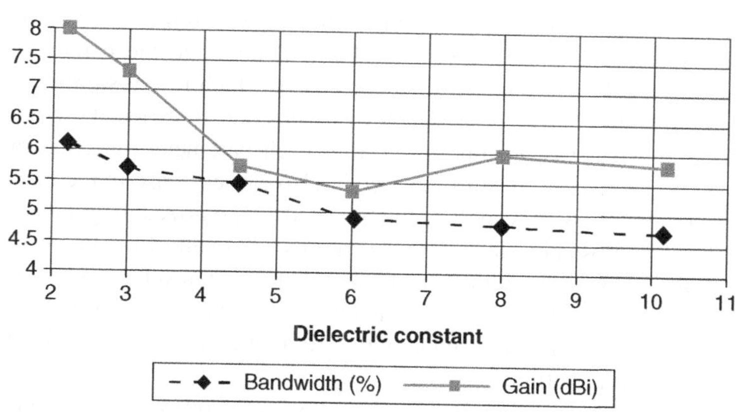

Figure 2.12 Variation of the feed substrate permittivity on the bandwidth and gain of a proximity coupled patch antenna.

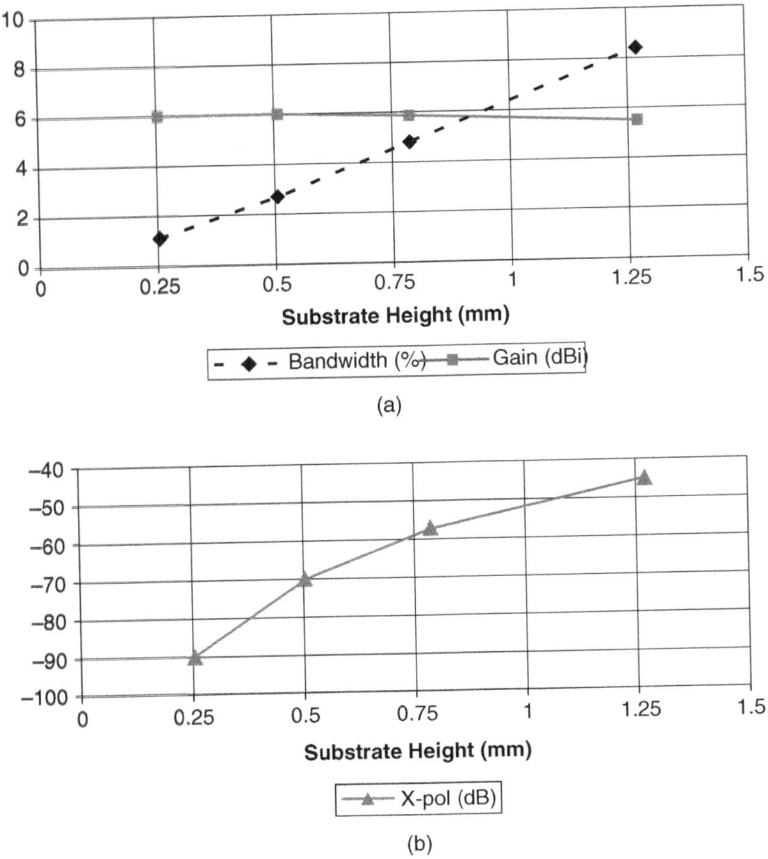

Figure 2.13 Effect of altering the high-permitivity feed substrate height on: (a) the bandwidth and gain; and (b) the cross-polarization level.

Figure 2.13 examines the performance of the conventional proximity coupled patch antenna with a high-permittivity feed substrate due to changes in the substrate height. As seen in Figure 2.13(a), the antenna bandwidth is proportional to the substrate height, while the antenna gain remains relatively constant. This is contrary to the parameter studies in Reference [13], where a low ε_r feed substrate was used, which reveal significant variations in gain when the height of a low-permittivity feed substrate is varied. It should be noted, however, that the antenna was not optimized after each parameter variation in Reference [13], as it was here. An increase in substrate height also caused a rise in the maximum level of cross-polarized radiation with respect to the broadside copolarized magnitude, as depicted in Figure 2.13(b). The cross-polarization values are exceptionally low due to the lack of a current discontinuity associated with the feed and radiator and also because of the low level of cross-polar radiated fields from a printed antenna mounted on a high dielectric constant material [19]. It is expected that an antenna with a finite ground-plane would exhibit higher levels of cross-polarized radiation.

Figure 2.3(c) showed the structure of a stacked proximity coupled patch antenna. As with any stacked patch antenna configuration, there are several degrees of freedom in the design of stacked proximity coupled patch antennas. However, as was shown in Reference [24] for a direct contact stacked patch, the dielectric layers govern the overall performance of the antenna.

In Reference [25], an investigation into the composition of the dielectric layers to achieve good impedance responses is undertaken for proximity coupled stacked patch antennas. Before launching into this investigation it should be noted that the permittivity of the material between the patch elements, ε_{r2}, should be as close to unity as possible. This has been shown to be true for all stacked patch configurations, independent of the feeding technique. This is to ensure that the antenna efficiency is as high as possible [24].

In Figure 2.14, the characteristics of a stacked proximity coupled patch antenna are tracked as the permittivity of both the feed substrate and the lower patch substrate are incremented, so that they retain the same value. The microstrip feed line width is altered at each step to maintain a 50 Ω characteristic impedance. The structure was then optimized for the maximum impedance bandwidth. Figure 2.14(a) displays the impedance bandwidth as a function of the permittivity of the feed and patch substrates. Interestingly, the maximum bandwidth was not achieved using the lowest permittivity value of 2.2. The bandwidth remains relatively constant for permittivity values between 3 and 7, and then begins to reduce in the region from 7 to 10. For all of the permittivity values analysed, the bandwidth remained in excess of 20 %.

The centre frequency gain for the stacked proximity patch antenna versus the feed and patch substrate permittivity is depicted in Figure 2.14(b). The maximum gain of 8.9 dBi was registered for the case when both the feed substrate and lower patch substrate have a dielectric constant of 2.2. The gain remained high as the two substrate permittivities were sequentially incremented, oscillating around a value of 8 dBi.

The proximity coupled patch structure inherently quells spurious feed radiation by avoiding the current discontinuity associated with contacting feed methods. This results in proximity coupled patch architectures exhibiting very low levels of cross-polarization. Figure 2.14(c) plots the maximum level of cross-polarized radiation (with respect to the broadside copolarized level) in the forward hemisphere of the stacked proximity coupled patch antenna versus the feed and patch substrate permittivity. The cross-polarization level decreases steadily with an increase in the permittivity. A 14 dB difference is observed between the permittivity values of 2.2 and 10.2.

It is shown in Reference [22] that good bandwidth and efficient radiation could be achieved for the case of a proximity coupled stacked patch mounted on a high feed dielectric constant if the next layer was of a similar dielectric constant value. The design strategy in Reference [22] is based on that for conventional proximity stacked patches (where low dielectric constant materials are used) [13] and also for 'hi–lo' edge-fed stacked patches [19]. Simplistically, the objective of the lower patch in an element is to couple power from the feed network. The upper patch must then efficiently radiate this power into free space. To couple power from a microstrip line developed on a high dielectric constant laminate effectively, the first patch must be located in very close proximity to the transmission line. This can be deduced from the observation that as the dielectric constant of the feed substrate increases more energy is trapped within this layer. To assist in the extraction of this energy, the dielectric constant of the layer used between the feed and the lower patch may be made of comparable value to that of the feed layer. As is the case for all stacked patches when an efficient solution is required, the dielectric constant of the substrate between the upper and lower patch elements must be as low as possible (i.e. close to unity, for example foam). The printed antenna incorporates a thin layer of copper clad with a low dielectric material on top of the layer of foam to accommodate the etching of the patch conductor.

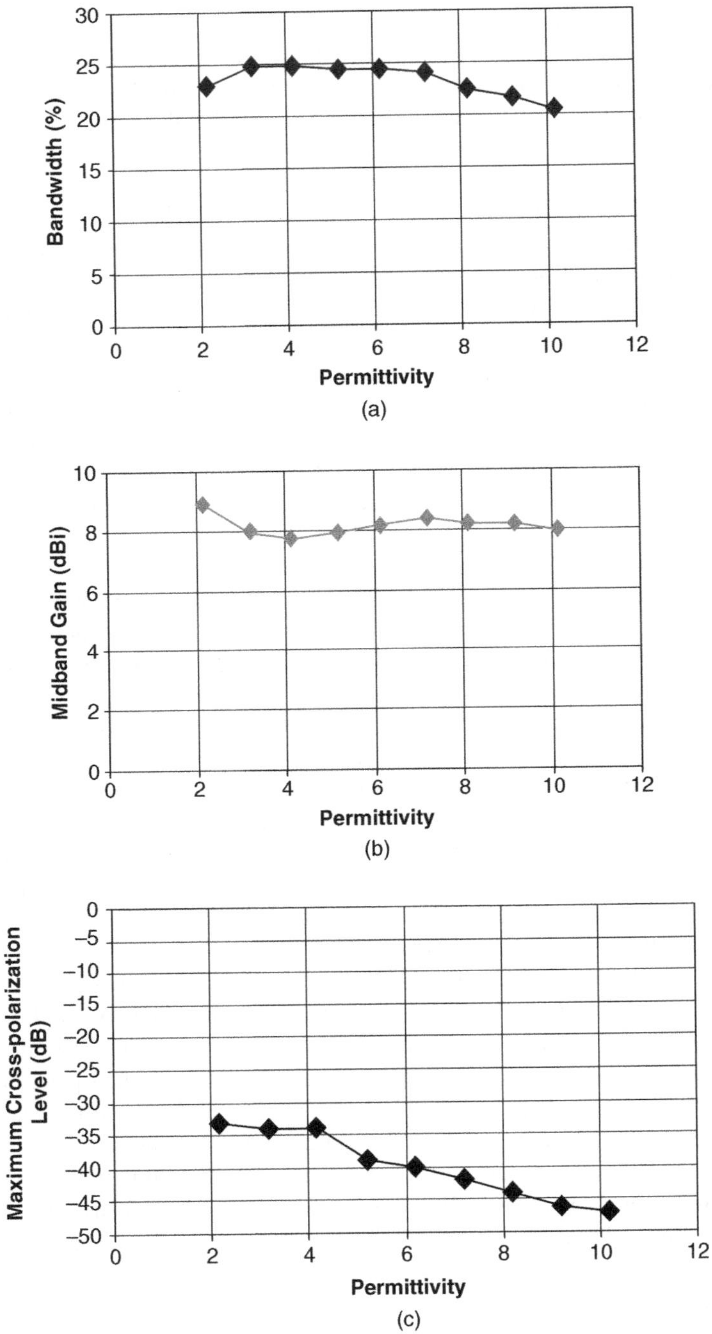

Figure 2.14 The result of varying the feed and patch substrate permittivity simultaneously on: (a) the bandwidth; (b) the gain at the centre frequency; and (c) the maximum cross-polarization level.

The predicted 10 dB return loss bandwidth for the antenna presented in Reference [22] was 25 % and the measured value was 21 %. At the midband frequency of 4.8 GHz, the gain of the antenna was measured as 8.0 dBi, compared to a predicted value of 8.1 dBi. The radiation patterns resemble those of other highly efficient printed antennas, and were consistent across the impedance bandwidth. The measured cross-polarization radiation levels are more than 20 dB below the copolar levels in both principal planes for all observation angles.

For the configuration above, expensive MMIC/OEIC wafers are still required to ensure that the dielectric constants of the feed substrate (ε_{rf}) and the layer between the feed and the lower patch (ε_{rl}) correspond. In Reference [23] a theoretical investigation assessed whether a different dielectric constant material can be substituted for the high dielectric constant first layer, thereby reducing the cost of the directly integrated patch antenna.

The results in Reference [22] were used as a baseline case. The dielectric constant ε_{rl} was reduced by decrements of approximately 2. The antenna was optimized for each case to ensure that the best performance was achieved. Table 2.1 summarizes the bandwidth and gain results. Interestingly, for ε_{rl} values greater than 4.5 the bandwidth and gain retained similar values, and the dimensions of the patches, the thickness of all the materials, and the feed location did not have to be changed to optimize the performance. These results are particularly important for integrating proximity coupled stacked patches with MMIC/OEIC devices. This demonstrates that the material layer between the feed and the lower patch does not need to be matched exactly to the feed substrate, and therefore a more cost-effective material (such as Rogers *RT/Duroid 6010* with $\varepsilon_r = 10.4$) can be used. As long as there is not a dramatic difference between ε_{rf} and ε_{rl} broad bandwidths should still be achievable. The impedance response of the proximity coupled stacked patch starts to degrade for ε_{rl} values less than or equal to 4.5, when all the other parameters have been fixed to those given in Reference [22]. The reason for this is that as ε_{rl} is decreased, less of the fields are coupled from the feed line to the first patch as the high dielectric constant feed material confines most of the energy. To resolve this, the thickness of the layer between the feed and the lower patch can be reduced. The bandwidths and gains shown in Table 2.1 for $\varepsilon_{rl} = 4.5$ are for when the thickness of this layer has been halved to 0.125 mm. For $\varepsilon_{rl} = 2.2$, an ultra thin layer (0.05 mm) is required to give the results in Table 2.1.

One of the interesting findings from the previous study is that as the dielectric constant of the first layer is decreased, the thickness of the material must also decrease to achieve a good impedance bandwidth. It is worth investigating whether this relationship can be used to remove some of the design restrictions for a proximity coupled stacked patch mounted on low dielectric constant materials.

An investigation in Reference [25] uses the proximity coupled stacked patch in Reference [13] as the baseline case. The materials used were a combination of Rogers *RT/Duriod 5880* ($\varepsilon_r = 2.2$) and foam ($\varepsilon_r = 1.07$) with the feed substrate and lower patch substrate utilizing

Table 2.1 Relationship between ε_{rl} and bandwidth when ε_{rf} is high.

ε_{rl}	10.4	8.8	6.8	4.4	2.2
2:1 VSWR bandwidth (%)	23.7	25.6	26.2	21.4	22.0
Gain (dBi)	8.1	8.0	8.0	8.2	8.3

the Rogers *RT/Duroid 5880*. The thickness of the lower patch substrate was 0.254 mm and a feed offset (from the centre of the patch) of 3.6 mm was used to achieve the broadband impedance response. The value of ε_{r1} was increased to 3.5, and then 4.5, and for each case the other parameters of this printed antenna were varied to maintain the original bandwidth. Two important consequences of increasing ε_{r1} were observed. Firstly, to achieve the original bandwidth, there was a need to increase the thickness of the lower patch substrate to 0.508 mm. It was also not necessary to offset the feed line from the centre of the patch. Thus using a higher dielectric constant material for a lower patch substrate to a certain extent removes the restriction of requiring a very thin layer between the feed line and the first patch. It also allows for maximum coupling between the feed and the patch and therefore an increased gain. For the three cases considered, the gains were 8.9 dBi ($\varepsilon_{r1} = 2.2$), 9.2 dBi ($\varepsilon_{r1} = 3.5$), and 9.3 dBi ($\varepsilon_{r1} = 4.5$).

2.4.2.2 Feed Substrate Variation of the ASP Configuration

Exceptionally broad impedance bandwidths in the range of 50–70 % can be achieved from the aperture stacked patch (ASP) configuration [16], as shown in Section 2.3.6. Due to the mutual coupling between the aperture resonator and the parasitic patches, it is imperative that a study be undertaken to see what effect the characteristics of the feed substrate have on the aperture and hence the overall performance of the ASP.

In Reference [26], the microstrip-fed ASP configuration is examined to determine the effect of decreasing the thickness and increasing the permittivity of the feed substrate on the behaviour of the antenna. The development of broadband MMICs/OEICs incorporating ASP antennas relies on knowing the influence of such modifications.

The ASP configuration as given in Reference [16] is not an optimal structure for integration into OEIC/MMIC modules. The primary hindrance is the *RT/Duroid 5880* feed substrate, which is of a very low permittivity ($\varepsilon_r = 2.2$), which is not compatible with the high dielectric constant OEIC/MMIC wafer materials. This antenna will be used as the baseline for this investigation.

The effect of gradually increasing the permittivity while maintaining the thickness of the feed substrate was investigated using Ensemble 6.0. At each increment of permittivity the structure was optimized for maximum bandwidth, but the basic geometry of the ASP antenna given in Reference [16] was maintained. The microstrip line widths were also adjusted at each stage to preserve their characteristic impedance and ensure the efficient operation of the reactive power divider.

As the permittivity of the feed substrate is increased, there is a decline in the optimal bandwidth obtainable from the ASP configuration, as evident in Figure 2.15. The reduction in the bandwidth is primarily due to an increase in the quality factor (Q) of the resonant coupling aperture. To illustrate this, the apertures and microstrip feed network from the ASP antennas with feed substrate permittivities of 2.2 and 10.2 were analysed with the patch elements of the ASP removed. The input impedances of the two single-substrate aperture antennas are compared in Figure 2.16. Distinctly narrower resonant peaks are observed for the aperture antenna with the high dielectric constant substrate, compared to those of the lower permittivity material. Thus the bandwidth of the aperture is very dependent on the surrounding material. When the aperture mounted on the high dielectric constant material is mutually coupled to parasitic patch elements, the overall ASP antenna bandwidth is diminished.

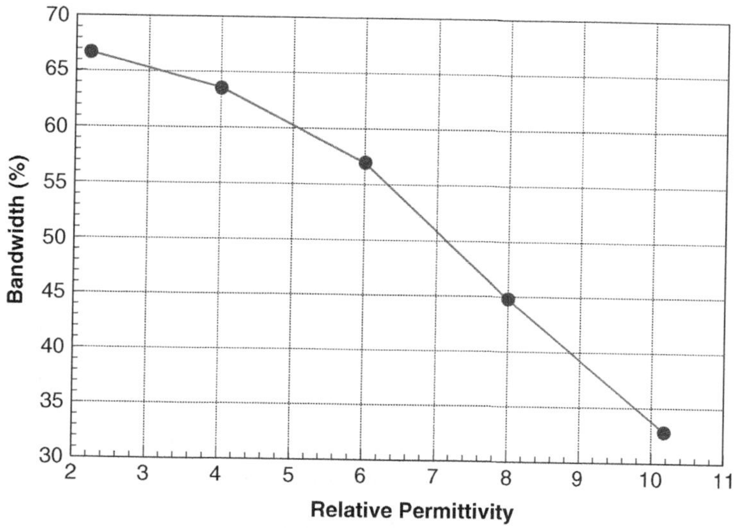

Figure 2.15 Optimized percentage bandwidth versus permittivity of the feed substrate.

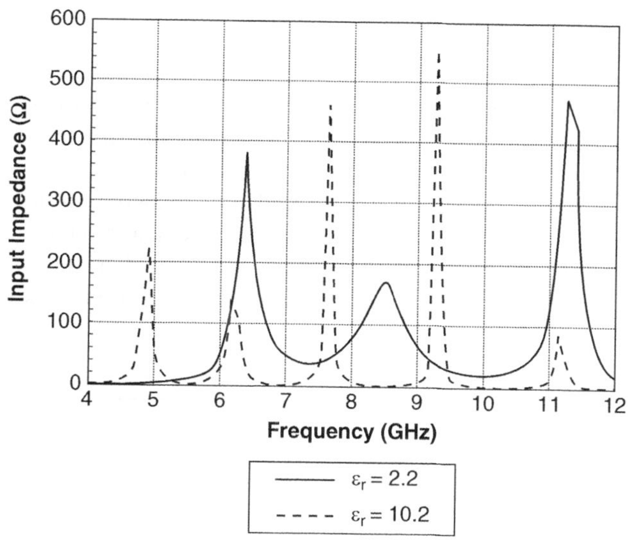

Figure 2.16 Input impedance comparison for high and low feed substrate permittivity aperture antennas.

As depicted in Figure 2.17, the peak gain within the impedance bandwidth of the antenna is also diminished with increasing permittivity, although it still remains above 7 dBi. The minimum gain maintains a consistent level. It should also be noted that the width of the 100 Ω sections of microstrip line in the reactive power divider became extremely thin (80 μm) when the permittivity of the feed material reached 10.2. The level of back radiation remained relatively constant, being approximately 7 dB below the broadside level.

Due to the very thin microstrip lines required to realize the reactive power divider of the ASP antenna structure when a high-permittivity feed substrate is employed, a single feed

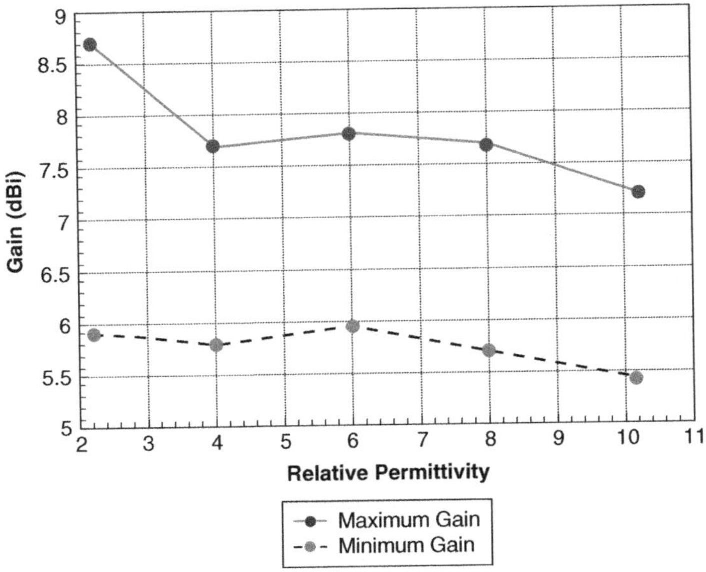

Figure 2.17 Maximum and minimum antenna gains versus permittivity of the feed substrate.

line ASP antenna was developed [26]. For MMIC/OEIC integration purposes, the thickness of the high-permittivity feed substrate ($h_1 = 1.6$ mm) is excessive. A theoretical study of the consequences of decreasing h_1 was undertaken. Once again, the structure was optimized for a maximum operational bandwidth after each iteration of h_1.

Figure 2.18 shows that a decrease in h_1 also leads to a reduction in the achievable bandwidth of the ASP structure. This is due to the aperture element having a lower bandwidth with smaller values of h_1. In general, the bandwidth of a printed antenna element is proportional to the volume it occupies [27], explaining the observed trend. As before, when the

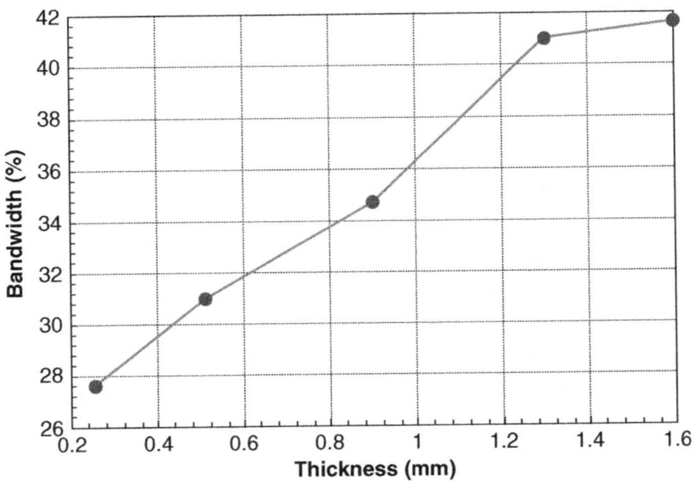

Figure 2.18 Optimized percentage bandwidth versus thickness of the feed substrate.

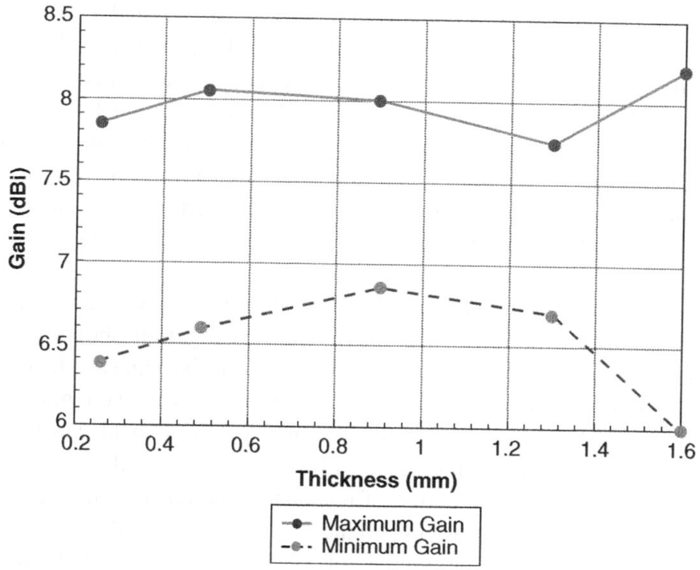

Figure 2.19 Maximum and minimum antenna gains versus thickness of the feed substrate.

aperture is mutually coupled to parasitic patch elements, the overall ASP antenna bandwidth is diminished with decreasing h_1.

From Figure 2.19, the peak gain within the impedance bandwidth of the antenna remained relatively consistent as h_1 is varied, with slight fluctuations around 8 dBi. The minimum gain displayed some minor variations, but maintained a level above 6 dBi. The front-to-back ratio exhibited improvement for lower values of h_1, attaining a value of 15 dB for $h_1 = 0.254$ mm.

An impedance bandwidth in excess of 27 % is obtainable from an ASP antenna that incorporates a thin, high-permittivity feed substrate. This bandwidth is not as broad as the ASP antennas in Reference [16] due to the effect of the feed substrate on the resonant coupling aperture. However, the bandwidth is sufficient for a wide range of communications system applications. The gain maintains a level similar to the initial ASP structure that utilizes a low-permittivity feed substrate, implying that the surface wave activity remains low.

The impedance bandwidth of the ASP antenna with a high-permittivity feed substrate is comparable to that of a frequency scaled version of the MMIC/OEIC compatible 'hi–lo–lo' antenna geometry [21]. The 'hi–lo–lo' structure has the advantage of a ground-plane to reduce the level of back radiation, enabling a superior front-to-back ratio to that of the ASP configuration. However, the ASP ground-plane provides isolation between the antenna elements and the feed arrangement, reducing spurious radiation and coupling to other devices in the circuit.

2.4.3 Substrate Offsets in Multilayered Patch Antennas

As stacked patch antenna architectures require multiple substrate layers to be situated on top of each other, the effect of misalignment due to fabrication inaccuracy is a pertinent issue. The input impedance of an aperture coupled patch antenna is relatively insensitive to small variations in patch position over the aperture, but can be significantly disrupted for larger value patch offsets [12]. The coupling factor between the aperture and patch resonators is defined by the size of the loop in the impedance locus, and is greatest when the patch is centred directly over the aperture. The coupling factor decreases significantly as the patch is offset in the direction of resonance. However, small lateral movements of the patch orthogonal to the resonant direction causes only minor changes in the coupling factor if the entire aperture remains under the confines of the patch element. A slight increase in coupling is observed when the edge of the aperture is in the vicinity of the edge of the patch, but drops sharply as more of the aperture emerges from underneath the patch.

An analysis of layer offsets in a conventional stacked patch antenna, as depicted in Figure 2.3(a), is conducted in Reference [28]. This investigation shows that deliberately adding offsets into the stacked patch structure can broaden the bandwidth by manipulating the fringing fields on the lower patch, and hence changing the effective length of the patches. There are limitations in the extent to which the upper patch element can be offset, as the asymmetry introduced can lead to increased cross-polarization and a skewed main beam direction. For patch offsets in the direction of resonance, the results show an increase in the size of the mutual coupling loop in the impedance locus. Offsets orthogonal to the resonant direction have a minimal effect on the mutual coupling loop as they do not alter the effective resonant length of the patch elements. However, large offsets can generate a secondary loop in the impedance locus. Diagonal offsets are shown to have the same properties as the sum of the individual contributions from the offsets in the resonant and orthogonal directions.

Typically, offsets due to manufacturing tolerances will not incur severe degradation for most stacked patch antennas. The mutual coupling of the patch elements tolerates minor inaccuracies in alignment. However, at higher frequencies, such as in the millimetre-wave region, offsets may play a more significant role in the performance of the stacked patch antenna. At these frequencies, the magnitudes of the fabrication offsets are generally a more significant proportion of a wavelength compared to lower microwave frequencies. Intentionally adding offsets into a stacked patch architecture to extract extra bandwidth has minimal application, as the enhanced cross-polarization and beam skew will preclude its use in many situations.

2.4.4 General Substrate Selection Tips for Wideband High-Performance Stacked Patch Antennas

The following delineates some general guidelines for the material selection in stacked patch antennas to obtain optimal performance:

1. Use all low-permittivity/loss tangent materials. The conventional theory of having low dielectric constant microwave laminates for the entire stacked patch structure is still the simplest way of obtaining large-bandwidth, high-performance antennas.
2. Select substrates to be as thick as possible for the broadest bandwidth. In particular, the upper substrate(s), which supports the parasitic patch elements, should be as thick as possible. The feed material needs to provide appropriate balance between the goals of obtaining wideband performance and avoiding the generation of surface waves and spurious feed radiation.

3. If a high-permittivity substrate is required for the feed layer, ensure that the uppermost patch element resides on a thick low dielectric constant material. This ensures a wide bandwidth and high efficiency.

These design guidelines come with a caveat; they are limited by the system environment in which the antenna resides. If the antenna is a stand-alone element, then there are minimal limitations imposed on the selection of the antenna materials. However, if the antenna is to be directly integrated on to an MMIC wafer, the thickness and permittivity of the feed substrate is governed by the material system used. There may also be practical restrictions on the materials and thicknesses that can be used for the upper substrate layers in this case. Factors such as how the upper substrate layers are attached, how they are formed, if they interfere with surrounding circuitry or devices, and packaging must be considered.

2.5 SUMMARY

In this chapter multilayered patch antennas have been investigated and a method was found to optimize the performance of these radiators. A variety of multilayered patch structures have been discussed and design trends for these radiator have been investigated. An investigation into the material composition of these printed antennas was presented as well as a guideline for substrate selections. It is hoped that the information provided can give readers a solid foundation to enable them to design patch-based antennas for a variety of wireless solutions.

REFERENCES

1. R. B. Waterhouse (ed.), *Microstrip Patch Antennas: A Designers Guide*, Kluwer Academic Publishers, Boston, Massachusetts, 2003.
2. D. R. Jackson, J. T. Williams, A. K. Bhattacharyya, R. L. Smith, S. J. Buchheit and S. A. Long, 'Microstrip patch designs that do not excite surface waves', *IEEE Transactions on Antennas and Propagation*, **41**(8), 1026–1037, 1993.
3. C. S. Lee, V. Nalbandian and F. Schwering, 'Gain enhancement of a thick microstrip antenna by suppressing surface waves', *IEEE Antennas and Propagation Society International Symposium Digest*, **1**, 460–463, June 1994.
4. W. S. T. Rowe and R. B. Waterhouse, 'Edge-fed patch antennas with reduced spurious radiation', *IEEE Transactions on Antennas and Propagation*, **53**(5), 1785–1790, 2005.
5. D. M. Kokotoff, R. B. Waterhouse, C. R. Birtcher and J. T. Aberle, 'Annular ring coupled circular patch with enhanced performance', *Electronics Letters*, **33**(24), 2000–2001, 1997.
6. H. F. Pues and A. R Van de Capelle, 'An impedance-matching technique for increasing the bandwidth of microstrip antennas', *IEEE Transactions on Antennas and Propagation*, **37**(11), 1345–1354, 1989.
7. T. Huynh and K.-F. Lee, 'Single-layer single-patch wideband microstrip antenna', *Electronics Letters*, **31**(16), 1310–1312, 1995.
8. Y.-K. Sohn, Y. Oh and Y.-G. Koo, 'Analysis of a parasitic-tuned microstrip antenna using the FDTD method', *IEEE Antennas and Propagation Society International Symposium Digest*, **1**, 404–407, July 1996.
9. G. Kumar and K. C. Gupta, 'Broad-band microstrip antennas using additional resonators gap-coupled to the radiating edges', *IEEE Transactions on Antennas and Propagation*, **32**(12), 1375–1379, 1984.

10. D. M. Pozar and B. Kaufman, 'Increasing the bandwidth of a microstrip antenna by proximity coupling', *Electronics Letters*, **23**, 368–369, 1987.
11. D. M. Pozar, 'A microstrip antenna aperture coupled to a microstrip line', *Electronics Letters*, **21**, 49–50, 1985.
12. P. L. Sullivan and D. H. Schaubert, 'Analysis of an aperture coupled microstrip antenna', *IEEE Transactions on Antennas and Propagation*, **34**(8), 977–984, 1986.
13. W. S. T. Rowe and R. B. Waterhouse, 'Investigation of proximity coupled antenna structures', *IEEE Antennas and Propagation Society International Symposium Digest*, **2**, 904–907, June 2003.
14. R. B. Waterhouse, 'Design of probe-fed stacked patches', *IEEE Transactions on Antennas and Propagation*, **47**(12), 1780–1784, 1999.
15. F. Croq and D. M. Pozar, 'Millimeter wave design of wide-band aperture coupled stacked microstrip antennas', *IEEE Transactions on Antennas and Propagation*, **39**(12), 1770–1776, 1991.
16. S. D. Targonski, R. B. Waterhouse and D. M. Pozar, 'Design of wideband aperture-stacked patch microstrip antennas', *IEEE Transactions on Antennas and Propagation*, **46**(9), 1246–1251, 1998.
17. J. R James and P. S. Hall, *Handbook of Microstrip Antennas*, Peter Peregrinus, London, 1989.
18. D. M. Kokotoff, R. B. Waterhouse and J. T. Aberle, 'Figure of merit for the design of microstrip patch antennas', *IEEE Antennas and Propagation Society International Symposium Digest*, **4**, 2478–2481, July 1999.
19. R. B. Waterhouse, 'Stacked patches using high and low dielectric constant material combination', *IEEE Transactions on Antennas and Propagation*, **47**(12), 1767–1771, 1999.
20. W. S. T. Rowe and R. B. Waterhouse, 'Broadband CPW fed stacked patch antenna', *Electronics Letters*, **35**(9), 681–682, 1999.
21. W. S. T. Rowe and R. B. Waterhouse, 'Broadband microstrip patch antennas for MMICs', *Electronics Letters*, **36**(7), 597–599, 2000.
22. R. B. Waterhouse and W. S. T. Rowe, 'MMIC compatible printed antennas', *Electronics Letters*, **39**(21), 1493–1495, 2003.
23. W. S. T. Rowe and R. B. Waterhouse, 'Investigation of proximity coupled patch antennas suitable for MMIC integration', *IEEE Antennas and Propagation Society International Symposium Digest*, **2**, 1591–1594, June 2004.
24. A. Mitchell, M. Lech, D. M. Kokotoff and R. B. Waterhouse, 'Search for high-performance probe-fed stacked patches using optimization', *IEEE Transactions on Antennas and Propagation*, **51**(2), 249–255.
25. W. S. T. Rowe and R. B. Waterhouse, 'Investigation into the performance of stacked proximity coupled patches', *IEEE Transactions on Antennas and Propagation*, **54**, 1693–1698, June 2006.
26. W. S. T. Rowe and R. B. Waterhouse, 'Theoretical investigation on the use of high permittivity materials in microstrip aperture stacked patch antennas', *IEEE Transactions on Antennas and Propagation*, **51**(9), 2484–2486, 2003.
27. R. B. Waterhouse, 'Microstrip antennas', Chapter 6 in *Handbook on Antennas in Wireless Communications*, CRC Press, New York, 2001.
28. E. Rajo-Iglesias, G. Villaseca-Sanchezand and C. Martin-Pascual, 'Input impedance behavior in offset stacked patches', *IEEE Antennas and Wireless Propagation Letters*, **1**(1), 28–30, 2002.

3

Printed Quasi-Yagi Antennas

Kevin M. K. H. Leong and Tatsuo Itoh
Department of Electronic Engineering
University of California Los Angeles, Los Angeles, California, USA

3.1 INTRODUCTION

The configuration of the printed 'quasi-Yagi antenna' can be compared that of the classic Yagi–Uda dipole type antenna. In both antennas the structures acting as the driver, director, and reflector can be identified. However, major differences arise when discussing the fundamental operation of the two types of antennas. The planar quasi-Yagi antenna is able to produce an endfire radiation pattern by using the grounded portion of the antenna substrate as the reflecting element rather than relying on an additional reflector dipole or metal plate, making the overall structure compact in size. In this chapter the basis of the operation of the novel quasi-Yagi antenna will be discussed. Performance metrics such as measurements of return loss, gain, radiation efficiency, and full antenna radiation patterns of quasi-Yagi antennas designed with microstrip and coplanar waveguide feeding will be shown. The demonstration of the quasi-Yagi antenna in passive and active antenna arrays will also be described. Furthermore, application of the antenna for integration with millimetre-wave circuitry along with its use in active integrated antennas will be discussed.

3.2 QUASI-YAGI ANTENNA ELEMENT

The classic Yagi–Uda dipole antenna was first presented in an English language journal in 1928 [1]. Yagi–Uda antennas are frequently used for low-frequency applications (UHF) such as for television receiver antennas that are placed on the tops of homes and buildings. They consist of a linear array of parallel dipoles forming a parasitically coupled array as shown in Figure 3.1.

Printed Antennas for Wireless Communications Edited by R. Waterhouse

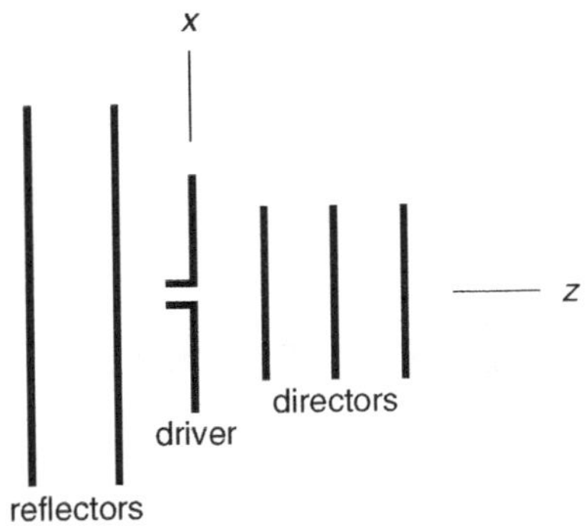

Figure 3.1 Schematic diagram of a classic Yagi–Uda antenna.

This type of antenna is known to exhibit a high-gain endfire (z direction) radiation pattern with moderate bandwidth. The dipole array consists of three main components' namely a driver dipole, a reflector dipole, and one or, more commonly, several parasitic director dipoles. The driver dipole is used to excite the array structure; subsequently the field radiated by this dipole is reflected by the long reflector dipole positioned at the reverse endfire direction. The additional parasitic director dipoles serve to enhance further the radiation in the forward endfire direction.

The quasi-Yagi antenna [2–6] shares certain similarities in configuration and operation to the classic Yagi–Uda antenna. However, definite distinct differences between the two antennas do also exist. For this reason, the name 'quasi-Yagi' antenna was coined. A schematic of the recently developed microstrip-fed quasi-Yagi antenna is shown in Figure 3.2. The antenna configuration is very similar to the configuration of the Yagi–Uda dipole array in such a way that all three dipole components can be readily identified. However, the role of the driver dipole has become the generator of surface wave power in the high

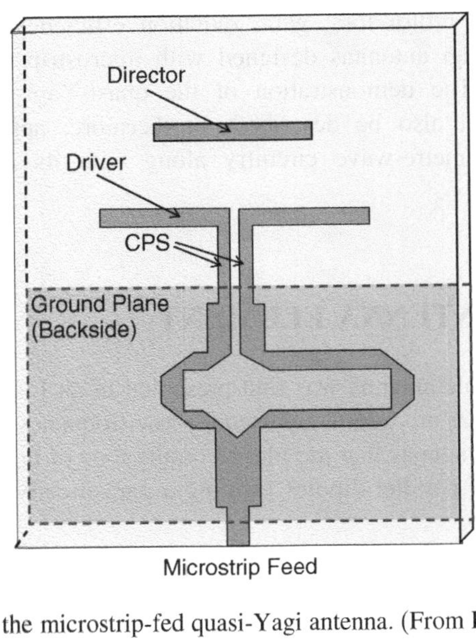

Figure 3.2 Schematic of the microstrip-fed quasi-Yagi antenna. (From Reference [4]. © 1999 IEEE. Reproduced by permission of the IEEE.)

dielectric constant substrate on which the antenna is printed. Antennas printed on high dielectric constant substrates typically suffer from undesired effects due to generation of surface waves, including low-radiation efficiency, high cross-polarization radiation, and strong mutual coupling [7]. On the contrary, the presented novel printed antenna in fact takes advantage of the generation of surface waves. Specifically, the driver dipole is designed to excite TE_0 (transverse electric) surface wave, within the base substrate. The generated TE_0 surface wave energy directly contributes to free-space radiation. The endfire nature of the antenna comes in part due to the truncated microstrip ground-plane on the backside of the substrate that acts as an ideal reflector for the TE_0 mode surface wave. The phenomenon occurs since in the grounded dielectric slab region (microstrip feed line region), the TE_0 mode is completely cut off until 39 GHz in the X-band antenna prototype, which uses a substrate with a permittivity of $\varepsilon_r = 10.2$ and thickness of 0.635 mm. Furthermore, because they share the same field polarization, the TE_0 surface wave is strongly coupled to the parasitic director dipole element of the quasi-Yagi antenna. For this reason, the position of the parasitic director element greatly influences the optimization of the antenna's front-to-back ratio as well as its impedance bandwidth.

The efficient generation of surface wave energy depends greatly on the choice of dielectric substrate. To be precise, a high-permittivity substrate with moderate thickness is required for proper excitation of the desired surface wave. This can be observed in the plot shown Figure 3.3, which plots the calculated radiating TE_0 surface wave power generated by a current element for various types of substrate materials using the formulation described in Reference [8]. It can be observed that surface wave power increases with an increasing material dielectric constant and for a given material an optimum material thickness can be determined. Furthermore, by increasing the substrate thickness, the TM_0 (transverse magnetic) mode is also excited along with the dominant TE_0 mode. In the case of the quasi-Yagi antenna, this will result in an increase in cross-polarization radiation and a reduction in radiation gain and front-to-back ratio.

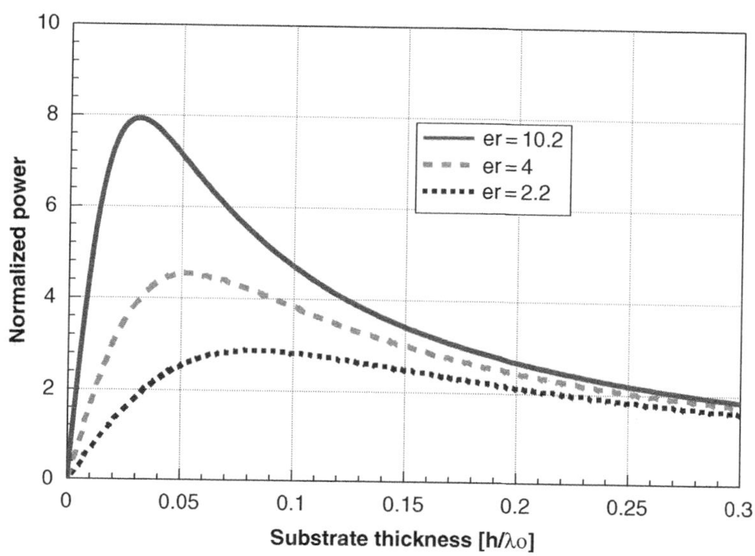

Figure 3.3 Normalized TE_0 surface wave power generated by an electric current element as a function of substrate thickness.

The first demonstration of the quasi-Yagi antenna was done using a substrate material with a permittivity of $\varepsilon_r = 10.2$ and thickness of 0.635 mm at X-band frequencies, corresponding to a thickness of $0.02\lambda_0$. The calculated optimum TE_0 surface wave excitation for the $\varepsilon_r = 10.2$ material is $0.03\lambda_0$. The thickness ratio must be maintained when scaling the antenna to different frequency bands. Evidence of this will be shown in later sections. A photograph of the antenna fabricated on 0.635 mm *RT/Duroid*™ with a permittivity of $\varepsilon_r = 10.2$ is shown in Figure 3.4.

The printed microstrip-fed quasi-Yagi antenna consists of a driver element, which is fed via a coplanar stripline (CPS). The balanced CPS mode is transformed from the input microstrip mode using a broadband transition [9]. The transition itself is implemented by first branching off the single microstrip line into two lines. Then by introducing a 180° phase difference between the signal two paths using an optimized delay line, a high-purity odd microstrip mode is induced at the edge of the microstrip ground-plane truncation point. The odd mode is converted to the CPS mode to drive the driver dipole. The backside metallization acts as the surface wave reflector element for the antenna. Finally, the parasitic director element serves to direct the antenna's electromagnetic fields towards the endfire direction while simultaneously serving as a parasitic impedance matching element.

A finite difference time domain (FDTD) simulation software [10] was used to simulate and optimize the antenna prototype. Simulated and measured return loss of the X-band prototype demonstrates broad impedance bandwidths of 43 and 48 %, respectively (Figure 3.5). The broadband performance offered by the quasi-Yagi antenna is comparable with waveguide-based horn antennas at a fraction of the size and weight.

The antenna's broadband nature can also be seen in the radiation characteristics of the antenna. Simulated radiation patterns of the X-band antenna at 9.5 GHz are shown in Figure 3.6. The FDTD simulations reveal a well-defined endfire radiation pattern in both the E- and H-planes. The broad pattern suggests the potential of using the quasi-Yagi antenna in an array environment. This application will be demonstrated in later sections of this chapter. Additionally, a front-to-back ratio (FTBR) of 17 dB and maximum cross-polarization of −16 dB are demonstrated. Further FDTD simulations reveal that the radiation patterns are very similar across the entire operating band.

For comparison, the measured radiation patterns of the quasi-Yagi antenna at 7.5, 9.5, and 11.3 GHz are shown in Figures 3.7, 3.8, and 3.9, respectively. These frequencies

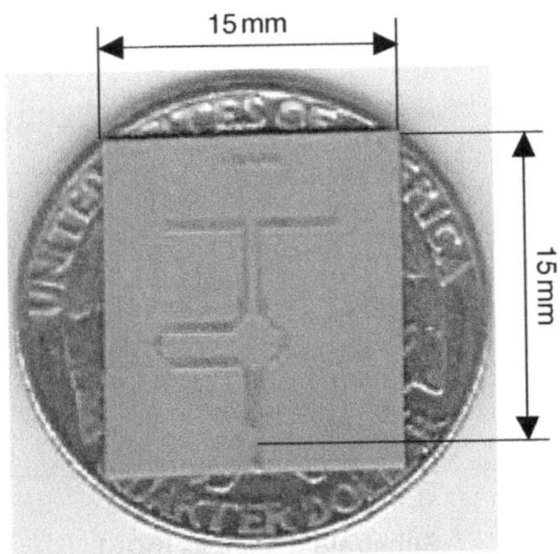

Figure 3.4 Fabricated X-band quasi-Yagi antenna. (From Reference [4]. © 1999 IEEE. Reproduced by permission of the IEEE.)

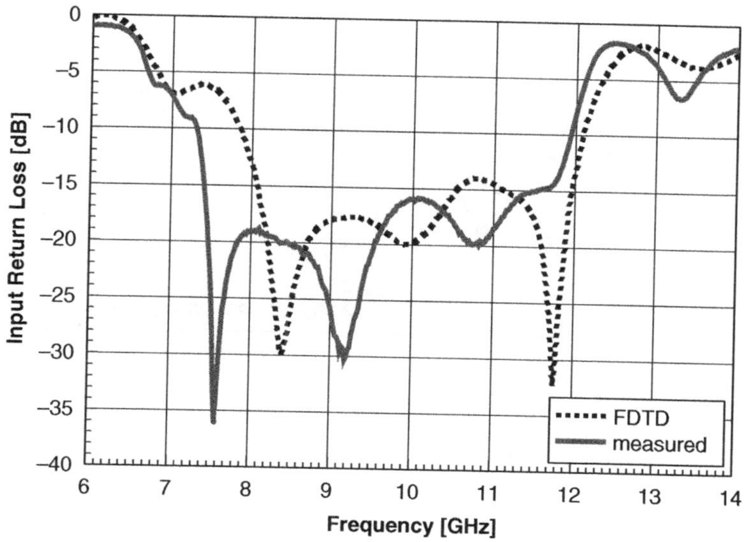

Figure 3.5 Simulated and measured return loss of the X-band quasi-Yagi antenna. (From Reference [4]. © 1999 IEEE. Reproduced by permission of the IEEE.)

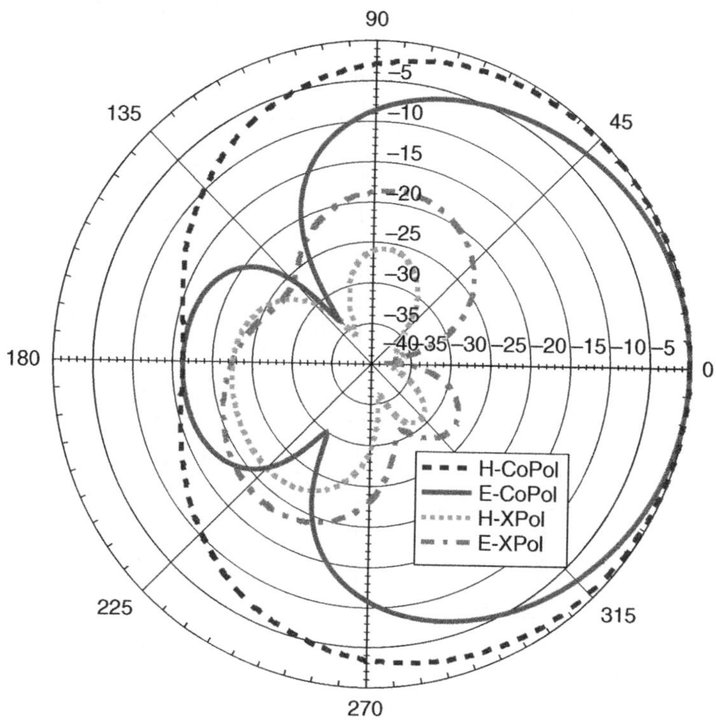

Figure 3.6 Simulated FDTD radiation patterns of the quasi-Yagi antenna at 9.5 GHz. (From Reference [26]. © 2000 IEEE. Reproduced by permission of the IEEE.)

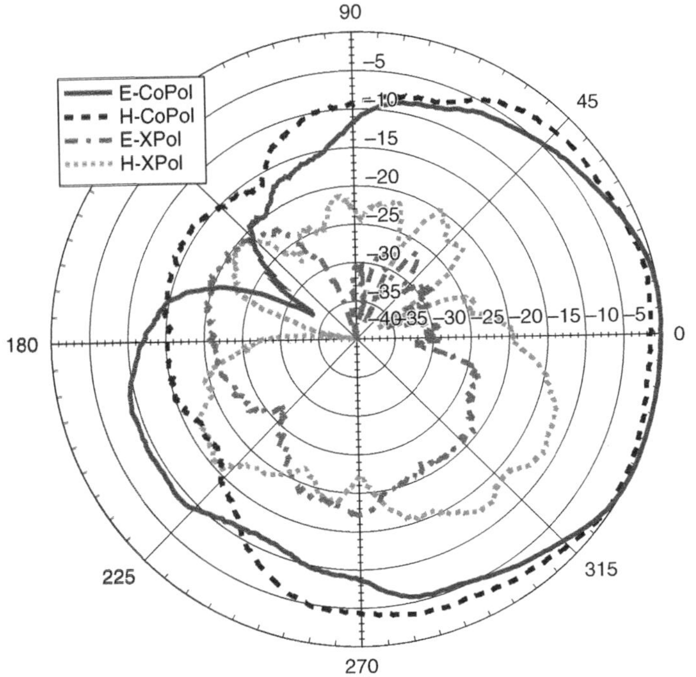

Figure 3.7 Measured patterns of the quasi-Yagi antenna at 7.5 GHz.

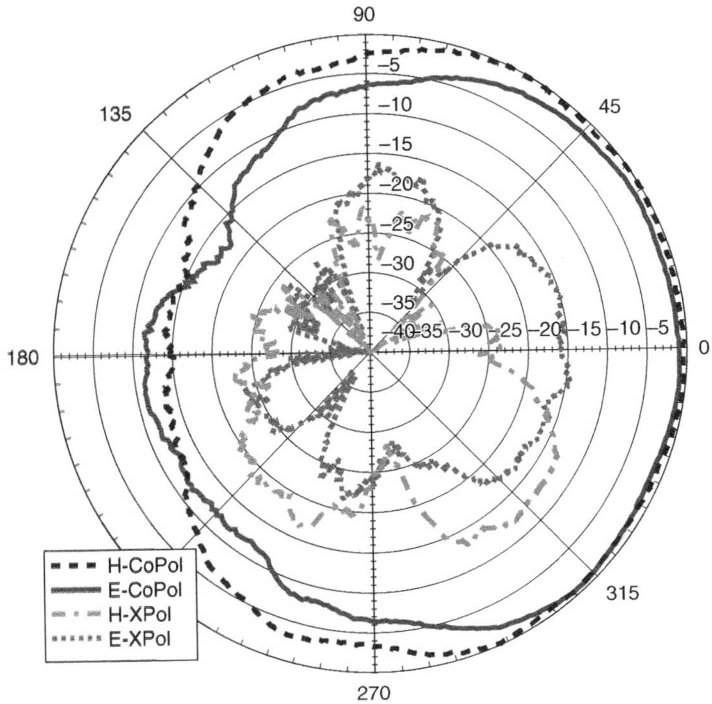

Figure 3.8 Measured radiation patterns of the quasi-Yagi antenna at 9.5 GHz. (From Reference [26]. Reproduced by permission of the IEEE.)

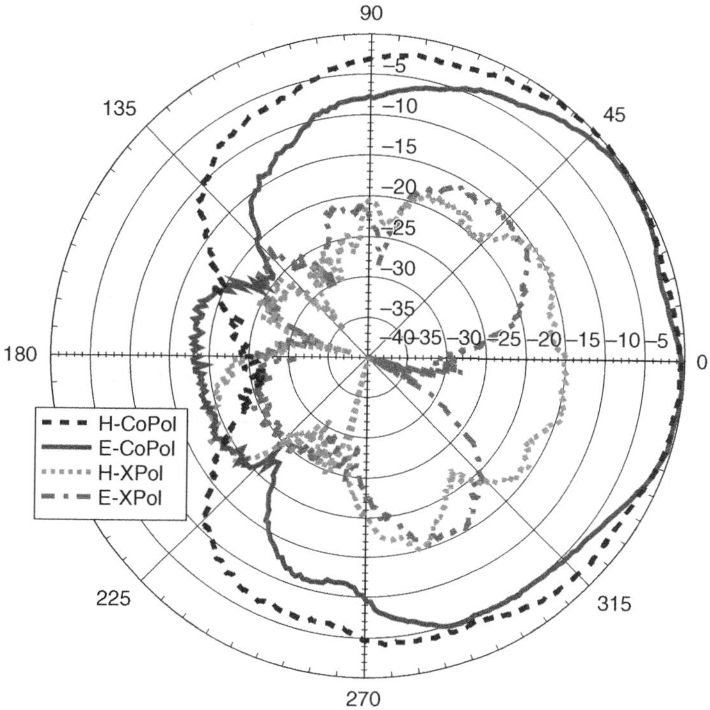

Figure 3.9 Measured radiation patterns of the quasi-Yagi antenna at 11.3 GHz.

approximately correspond to the lower end, middle, and upper end of the antenna's operating band. The results show fair agreement between measurement and simulation. A 12 dB front-to-back ratio with cross-polarization better than −12 dB is observed for all three patterns. The antenna can be optimized to achieve better FTBR (20 dB) and cross-polarization radiation (−20 dB) by sacrificing bandwidth.

It is interesting to note that a three-element traditional Yagi–Uda antenna can only achieve a front-to-back ratio of 5.6 dB. However, in the case of the quasi-Yagi antenna where surface wave energy is radiated, the truncated microstrip ground-plane acts as a very good reflector, directing the radiation to the endfire direction. This is done without the use of the reflecting metal plate used in other endfire dipole antennas [11], allowing the overall radiating structure to be compact and lightweight.

Another metric of an antenna's radiation performance is its gain. Antenna gain was measured using a pair of identical quasi-Yagi antennas and deriving the gain based on the Friis transmission formula. The input return losses of the antennas as well as cable losses are calibrated out. Connector losses and microstrip feed losses have not been accounted for, but their effects are assumed to be minimal. The results are shown in Figure 3.10. The simulated gain is obtained using the Agilent HFSS. Fair agreement is observed between measurement and simulation. Simulations predict gains from 4.5 to 5.1 dB while the measured gain ranges from 3.4 to 5.1 dB. Moreover, the variation in gain is relatively small across the wide operating bandwidth of the antenna.

Radiation efficiency of the antenna was also evaluated. Since the quasi-Yagi antenna relies on surface waves for its basic functionality, it should not exhibit reduced efficiency

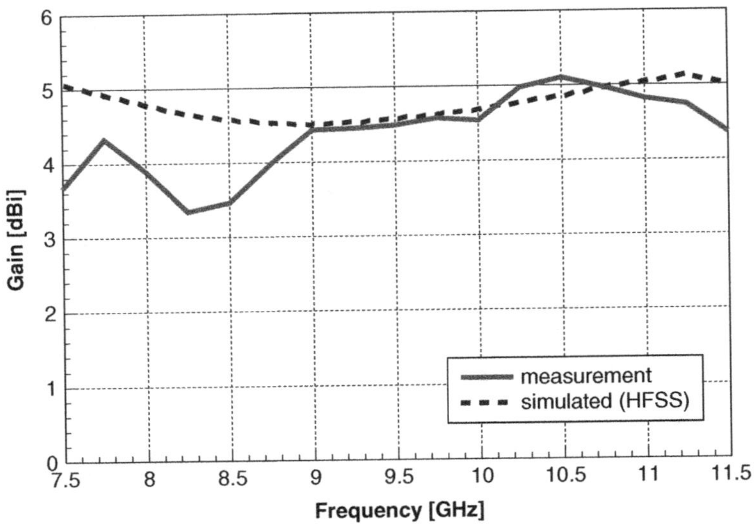

Figure 3.10 Simulated and measured curves of the quasi-Yagi antenna gain. (From Reference [26]. © 2000 IEEE. Reproduced by permission of the IEEE.)

due to the surface wave leakage phenomena, which is present in other antennas fabricated on high dielectric constant substrates. Additionally, the quasi-Yagi is not strongly resonant, so standing wave type cavity losses should be minimal as well.

To measure the efficiency of the quasi-Yagi antenna, the reflection method was used as described in References [12] and [13]. In this procedure, the antenna is placed in a waveguide-sliding short structure and the input return loss is then measured at 20 positions.

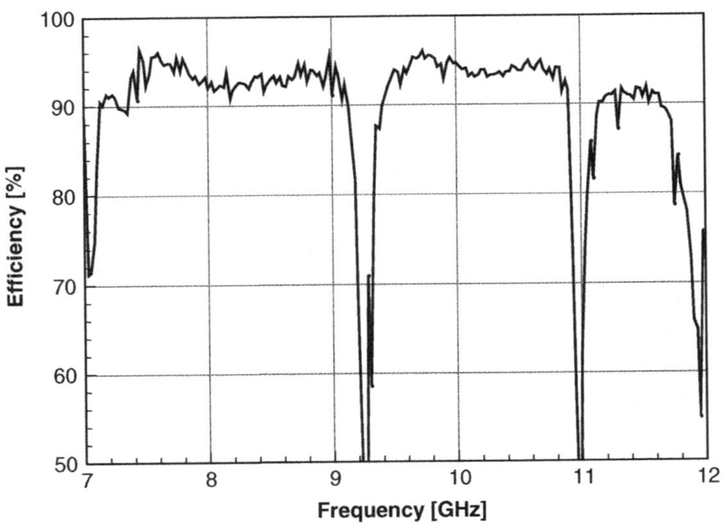

Figure 3.11 Measured radiation efficiency of the quasi-Yagi antenna.

These data and the input return loss for the antenna radiating in free space were used to calculate the antenna's efficiency. The results are shown in Figure 3.11. The radiation efficiency of the antenna was nominally 93 % across the operating band of the antenna. The error in measurement is estimated from the ripple to be roughly ±3 %. Two dips in the efficiency plot are observed at 9.2 and 11.0 GHz. These dips correspond to resonances in the waveguide-sliding short structure that were observed for all sets of sliding short data. Furthermore, since this phenomenon is not observed for the antenna radiating in free space, it can be concluded that these dips in the radiation efficiency are unphysical and represent a limitation of the measurement technique at the resonant frequencies.

The applicability of the quasi-Yagi antenna to any type of microwave or millimetre-wave system is truly realized by the ability to scale the antenna easily to other frequencies bands. This can be done by simply scaling all of the physical dimensions of the antenna, including the substrate thickness, while maintaining the dielectric constant of the material on which it is printed. By doubling both the thickness of the dielectric substrate and all other antenna dimensions of the X-band antenna (centre frequency, 11 GHz), a C-band prototype (centre frequency, 5.5 GHz) is realized. Like the X-band antenna, the C-band quasi-Yagi yields a very broad bandwidth, both in simulation (51 %) and in measurement (50 %). The simulated and measured return loss curves of the C-band antenna are shown in Figure 3.12 while measured radiation patterns taken at 5 GHz are shown in Figure 3.13. The radiation patterns show a front-to-back ratio of 15 dB and cross-polarization better than −14 dB. All of the traits of the C-band quasi-Yagi antenna are consistent with those previously demonstrated in the X-band version. This method of frequency scaling has been proven at millimetre-wave frequencies [14] (V-band) as well as lower frequencies [15] (L-band). In all cases the broad bandwidth and radiation pattern characteristics are retained.

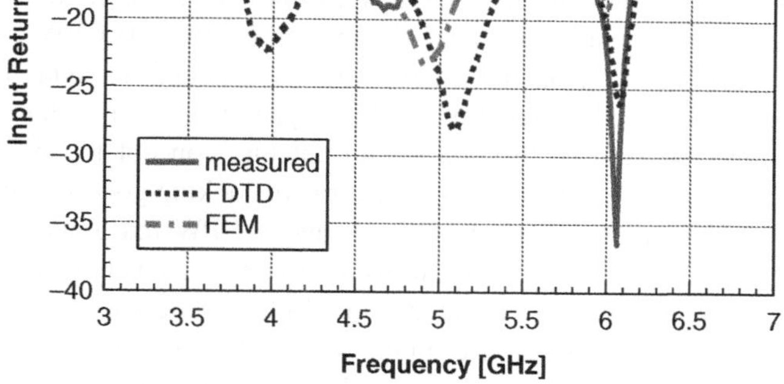

Figure 3.12 Simulated and measured return loss curves of the C-band quasi-Yagi antenna.

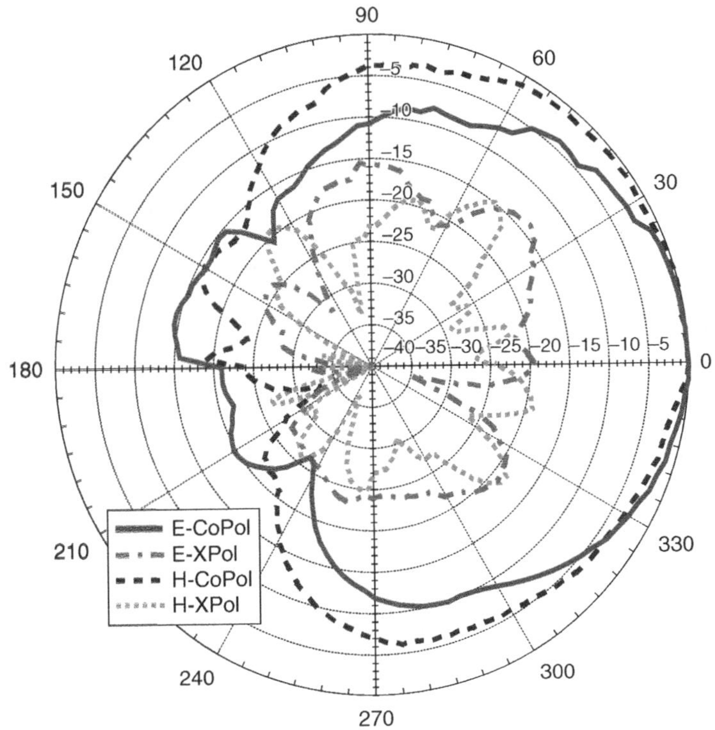

Figure 3.13 Measured radiation patterns of the C-band quasi-Yagi antenna at 5 GHz.

3.3 CPW-FED AND CB-CPW-FED QUASI-YAGI ANTENNAS

Use of single-layer transmission lines such as the coplanar waveguide (CPW) and conductor-backed coplanar waveguide (CB-CPW) [16] are quite common in the implementation of microwave and millimetre-wave integrated circuits. The CPW offers the advantage of lower dispersion characteristics than microstrip lines, especially at higher frequencies. Additionally, no grounding vias are required for integration of active devices, eliminating the parasitic grounding effect, which often limits the frequency range of microstrip circuits. The CB-CPW is an attractive option because it provides the benefit of added mechanical support and heat sinking ability compared to the conventional CPW. In practice, circuits fabricated using the CPW are placed on a metal carrier for heat sinking, emulating CB-CPW lines. There is a strong interest in the integration of millimetre-wave circuits and antennas for applications such as power combining and development of compact front-end technology. The next section describes the extension of the quasi-Yagi antenna concept to be compatible with CPW transmission line feeding developed for the forward-looking goal of circuit and antenna integration. Two similar antennas are developed, one a CPW-fed antenna and another a CB-CPW-fed antenna [17, 18]. Although the transmission lines are very similar, each presents a slightly different geometry of the truncated ground-plane reflector element, affecting the performance of the antenna. The differences in input return loss and antenna radiation patterns are presented and discussed. The antenna design done first at X-band frequencies was used as a model for the design of millimetre-wave frequency antennas at 120 GHz.

3.3.1 CPW-Fed Quasi-Yagi Antenna

The schematic of the CPW-fed quasi-Yagi antenna is shown in Figure 3.14. The hollow patch pictured at the end of the CPW line acts as a wideband transition between the CPW and the coplanar strip mode [19]. The patch represents a wideband open-circuit, which forces the current to flow between the two conductors of the CPS lines, which are used to feed the driver dipole. The antenna, the second dipole, behaves as a parasitic director element, which directs the antenna propagation towards the endfire direction. As in the microstrip quasi-Yagi, the truncated CPW ground-plane serves as the antenna's reflector element.

The proposed X-band prototype is fabricated on a relatively high permittivity substrate (0.635 mm thick *RT/Duroid* $\varepsilon_r = 10.2$). Figure 3.15 shows a photograph of the fabricated antenna. Measurement of the return loss (see Figure 3.16) of the antenna shows a broad bandwidth (S11< -10 dB) of 39 % from 8.0 to 11.9 GHz. Radiation patterns taken at 8, 9, and 11.5 GHz corresponding to the lower, middle, and upper band edges are shown in Figures 3.17, 3.18, and 3.19, respectively. The antenna radiates an endfire beam with reasonable pattern characteristics. The front-to-back ratio was measured to be 9.5 dB with cross-polarization levels better than -15 dB in all directions, having a 3.84 dB gain at 9 GHz. Measurements also show similar radiation characteristics across the entire frequency band, with only slight deterioration at the upper band edge. The plot of antenna gain shown in Figure 3.20 shows a minimal variation from 4.8 to 2.8 dB across the band.

3.3.2 CB-CPW-Fed Quasi-Yagi Antenna

In the design of the CB-CPW-fed antenna the exact dimensions of the CPW-fed antenna were used, the only difference being that metallization was added to the backside of the circuit. Backside metal was truncated at the same point as the top CPW ground-plane. As

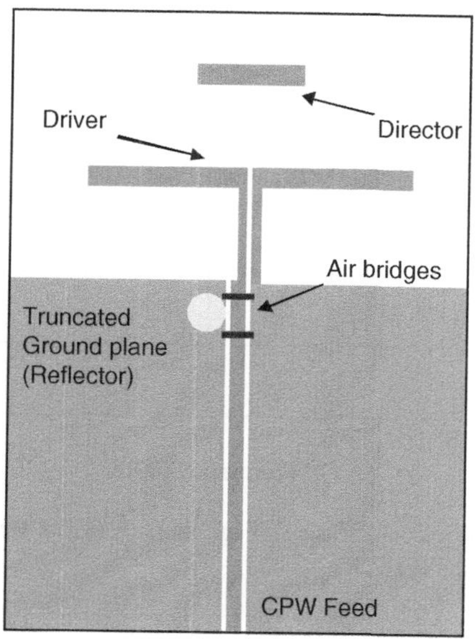

Figure 3.14 Schematic of the quasi-Yagi antenna with CPW feeding. (From Reference [18]. © 2000 IEEE. Reproduced by permission of the IEEE.)

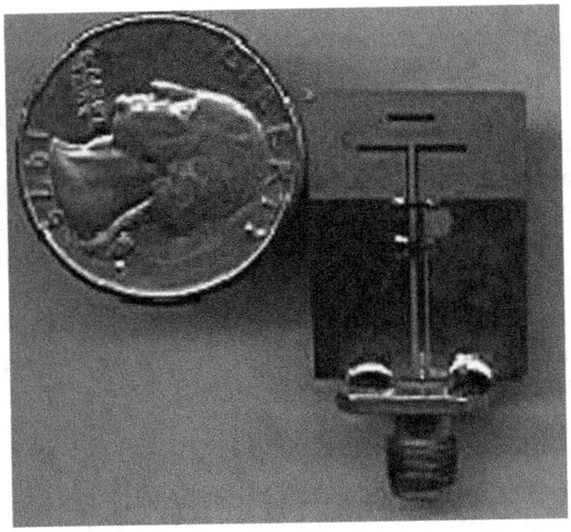

Figure 3.15 Photograph of the X-band prototype quasi-Yagi antenna. (From Reference [18]. © 2000 IEEE. Reproduced by permission of the IEEE.)

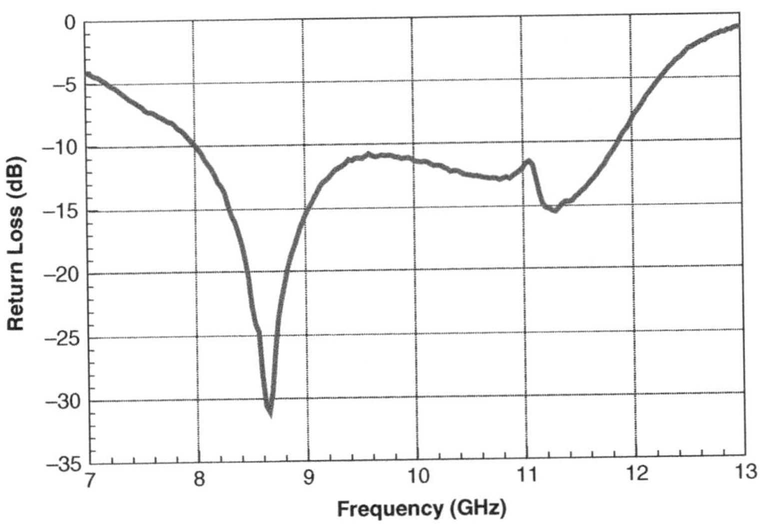

Figure 3.16 Measured return loss of the quasi-Yagi antenna. (From Reference [18]. © 2000 IEEE. Reproduced by permission of the IEEE.)

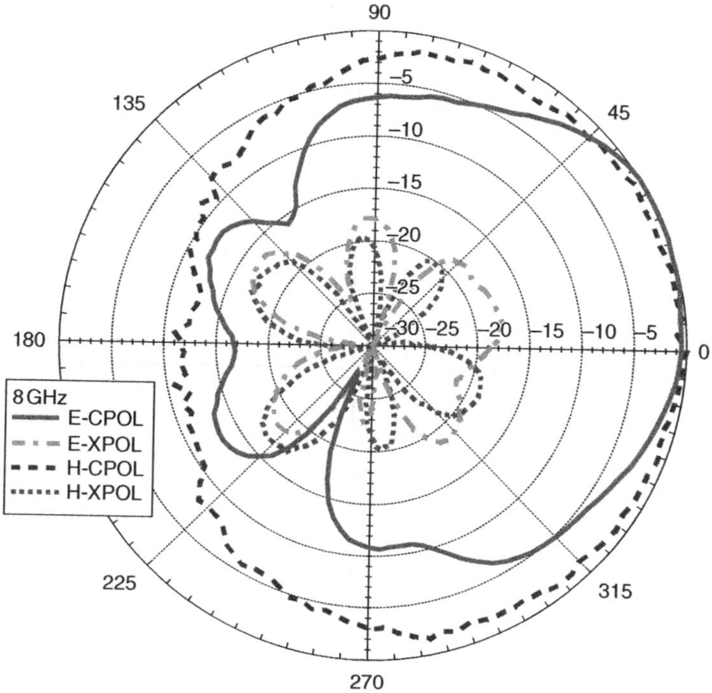

Figure 3.17 Measured radiation patterns of the quasi-Yagi antenna at 8 GHz.

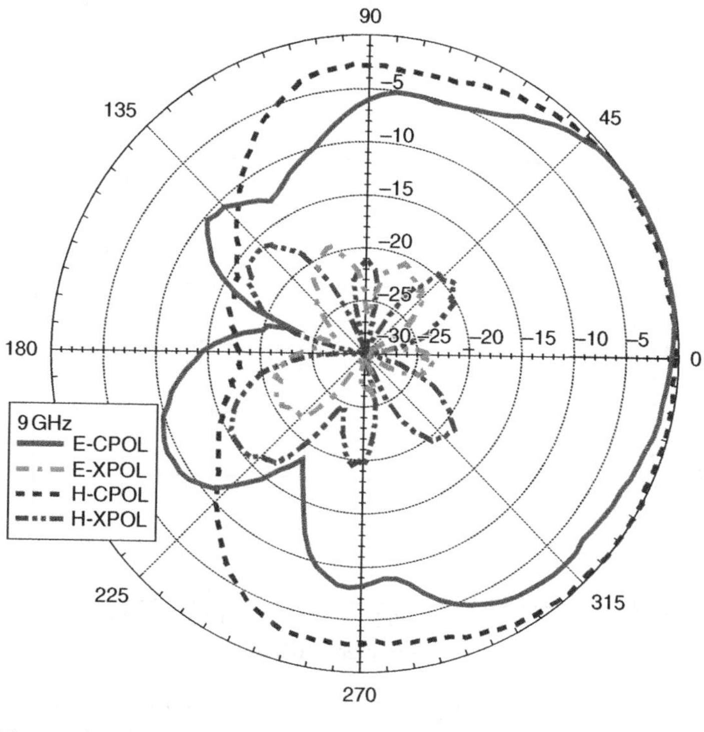

Figure 3.18 Measured radiation patterns of the quasi-Yagi antenna at 9 GHz. (From Reference [18]. © 2000 IEEE. Reproduced by permission of the IEEE.)

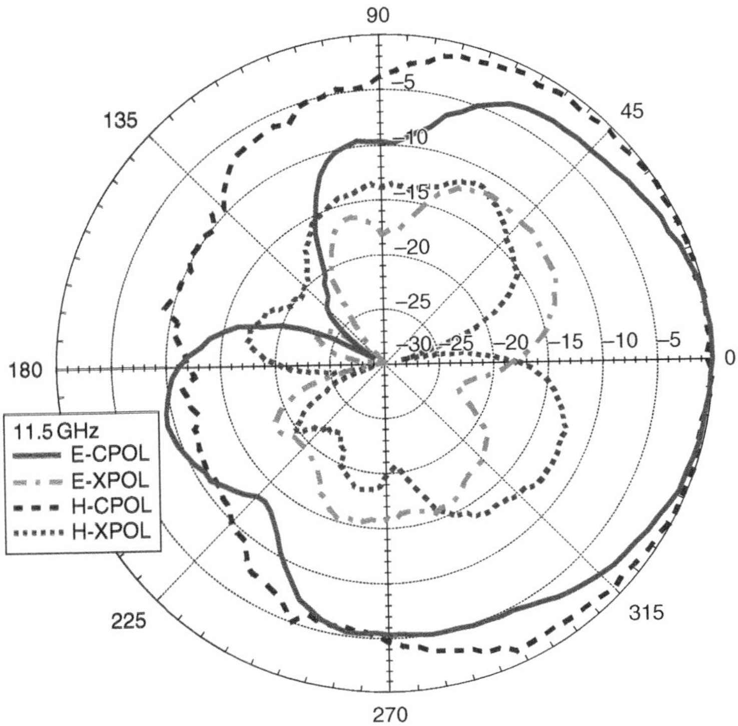

Figure 3.19 Measured radiation patterns of the quasi-Yagi antenna at 11.5 GHz.

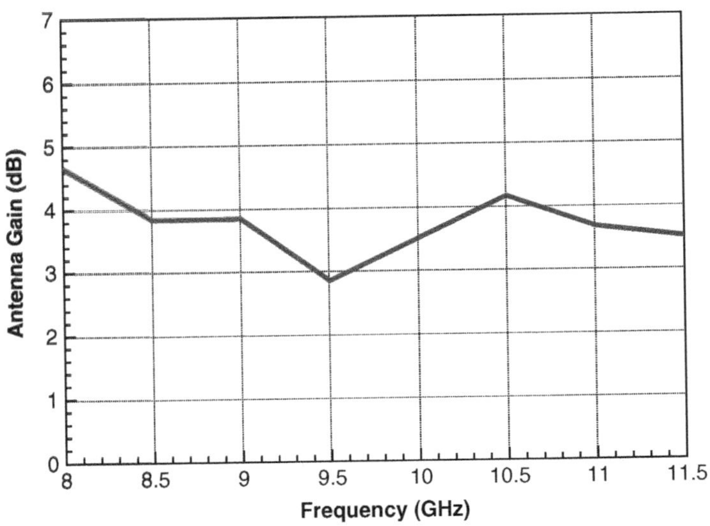

Figure 3.20 Measured antenna gain of the quasi-Yagi antenna.

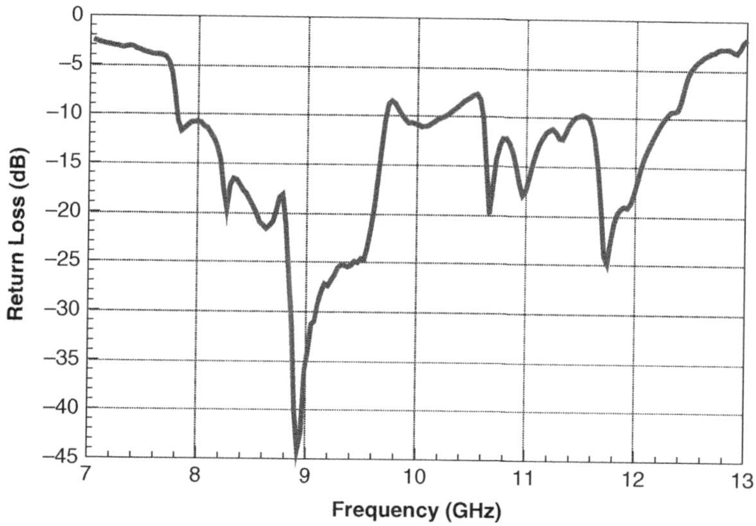

Figure 3.21 Measured return loss of the CB-CPW-fed quasi-Yagi antenna. (From Reference [18]. © 2000 IEEE. Reproduced by permission of the IEEE.)

in all CB-CPW circuits, the parasitic parallel-plate mode is excited, as well as the CPW mode, which presents a problem of power leakage. However, this was kept at a minimum by providing an electrical connection between the top ground-planes and the backside metal using metal vias. Because nothing was done to optimize the dimensions of the transmission line or the transition it was expected that the bandwidth would be reduced compared to the antenna with no conductor backing. Figure 3.21 shows the measurement results of the return loss. A reduced bandwidth ($S_{11} < -10$ dB) of 27 % from 7.8 to 10.2 GHz is observed.

Far-field measurements at 9 GHz show an improved radiation pattern compared to the CPW-fed antenna pattern at the same frequency (Figure 3.22). A front-to-back ratio of 14 dB and a cross-polarization level of less than −16 dB were recorded with an increased gain of 4.2 dB. Patterns measured at 8.5 and 10 GHz were also observed as having similar characteristics. Comparing the front-to-back ratios of the two antennas, it can be concluded that the added backside metallization improves the effectiveness of the antenna's reflector element. It is expected that further optimization of the CB-CPW antenna will result in a further increase in bandwidth.

3.3.3 Millimetre-Wave CB-CPW-Fed Quasi-Yagi Antenna

The power of several devices is often combined, either by using waveguides or using free-space power-combining techniques [20–22]. Due to its intrinsically high efficiency and excellent MMIC compatibility, the quasi-Yagi antenna provides a perfect solution for extracting and radiating power generated by MMIC circuitry into either free space or a waveguide. Because of the antenna's reliance on the generation of TE_0 surface waves as its primary source of free-space radiation, it is well suited for fabrication on materials with high dielectric constants, making it ideal for integration with MMICs that are fabricated on InP or GaAs substrates.

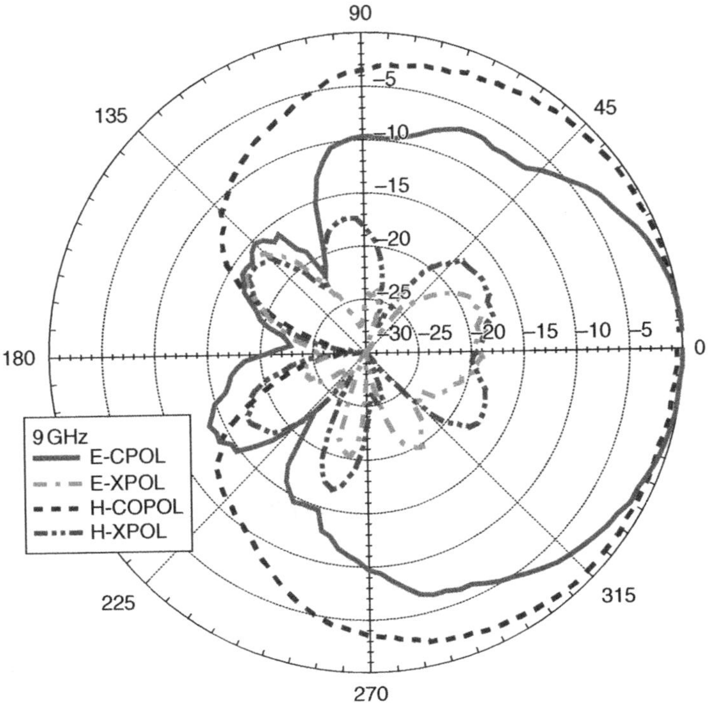

Figure 3.22 Measured radiation patterns of the quasi-Yagi antenna at 9 GHz. (From Reference [18]. © 2000 IEEE. Reproduced by permission of the IEEE.)

An F-band (90–140 GHz) version of the quasi-Yagi antenna with CB-CPW feeding was designed and fabricated on InP substrate [23]. In order to obtain a first-order approximation of the F-band (90–140 GHz) design, a linearly scaled version of the X-band antenna discussed in the previous section was used. The FDTD was again used to reoptimize the antenna on 50 μm InP substrate, with an estimated $\varepsilon_r = 12.6$, while neglecting metal loss.

After optimization, simulation predicts a 40 % bandwidth from 96 to 144 GHz, as shown in Figure 3.23. Also from FDTD simulation results a 16 dB front-to-back ratio and −18 dB cross-polarization level at 120 GHz (Figure 3.24) are expected. These results are very similar to the measured antenna pattern of the X-band prototype antenna.

Monolithic fabrication of the antenna was done using standard MMIC processing. Although special attention was given to the precise removal of the backside metallization, it simply required refinement of an already standard part of the fabrication process, which is used to make dicing of the circuit wafer easier. A photograph of the monolithically fabricated CB-CPW-fed antenna is shown in Figure 3.25.

On-wafer probe measurements from 90 to 140 GHz were performed to determine the antenna's return loss. Special considerations were taken to ensure that the dipole portion of the antenna was not placed directly above any metal plane, including the probe station chuck. This would effectively short-out the antenna and prevent it from radiating. A 'dummy' wafer was placed under the test wafer for added mechanical support and the antenna was allowed to hang off the edge of the metal chuck. Measurement indicates adequate 50 Ω impedance matching ($S_{11} < -10$ dB) for the entire band (Figure 3.23). The measurement compares well with the FDTD simulation having a similar curve shape with an approximately 5 % shift in

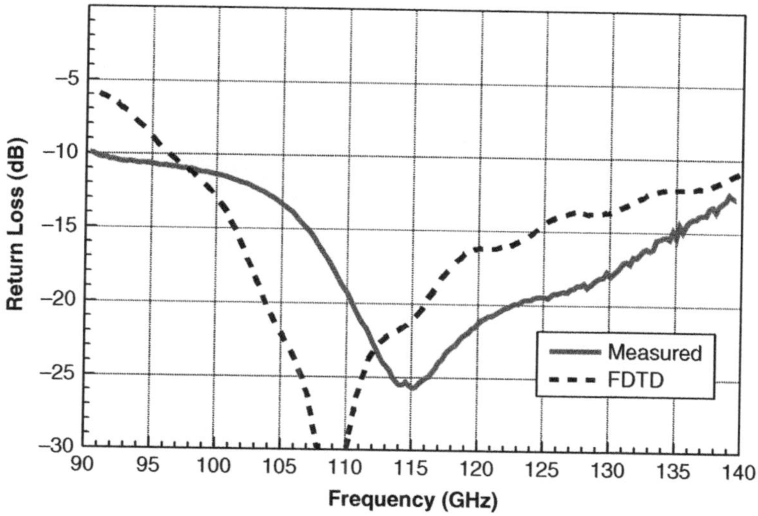

Figure 3.23 Measured and simulated quasi-Yagi antenna return loss. (From Reference [23]. © 2000 IEEE. Reproduced by permission of the IEEE.)

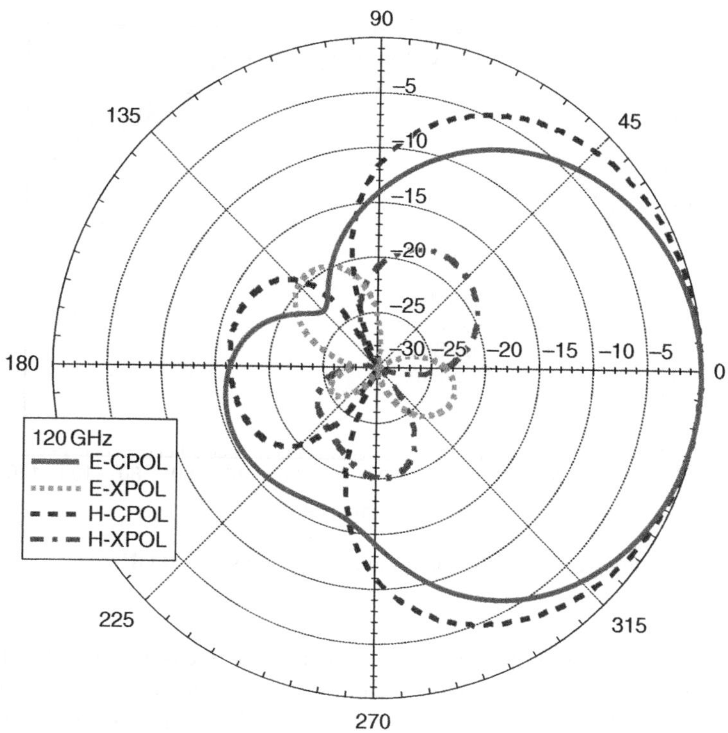

Figure 3.24 Simulated radiation patterns at 120 GHz using the FDTD. (From Reference [23]. © 2000 IEEE. Reproduced by permission of the IEEE.)

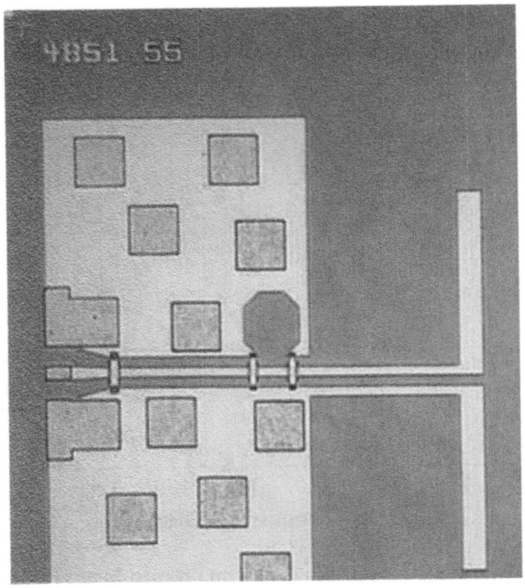

Figure 3.25 Photograph of the fabricated F-band quasi-Yagi antenna (chip size = 1.6mm^2). (From Reference [23]. © 2000 IEEE. Reproduced by permission of the IEEE.)

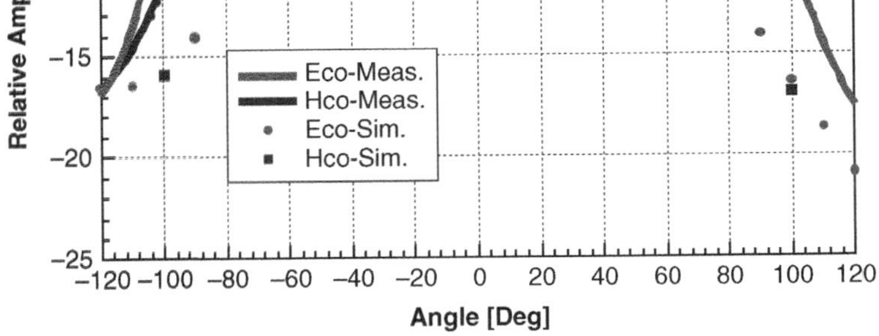

Figure 3.26 Measured radiation patterns at 120 GHz.

centre frequency. The radiation pattern of the antenna was measured at 120 GHz using an MMIC oscillator driving the planar antenna and a harmonic mixer fed by a horn antenna as the receiving antenna. The results are shown in Figure 3.26. A front-to-back ratio of 15 dB along with a cross-polarization level of −20 dB was measured for the broadbeam radiation pattern. An approximate gain of 4 dB was measured using a direct power measurement at 120 GHz. These results are consistent with measurements of the X-band version of the antenna.

3.4 QUASI-YAGI ANTENNA ARRAYS

In this section, the application of the microstrip-fed quasi-Yagi antenna as an array antenna is explored. The antenna's broad radiation pattern, broad bandwidth, and compact size make it ideal for array integration. Furthermore, the endfire antenna can be easily configured into a two-dimensional array by simply stacking multiple cards of one-dimensional subarrays. The conceptual architecture of an active two-dimensional quasi-Yagi array is shown in Figure 3.27. Because the antenna is completely compatible with microstrip circuitry, components such as amplifiers and phase-shifters can be easily incorporated into each card. The problem of heat dissipation, which is often a problem in large planar two-dimensional arrays, can be addressed by adding additional heat sinking to the microstrip ground-plane.

In this section the mutual coupling in both horizontal and vertical planes are quantified through rigorous measurements. Two implementations of passive quasi-Yagi arrays are presented [24–27]. The first array has an endfire pattern and will be used as a building block in larger two-dimensional arrays [28]. The second array uses delay lines to realize a progressive phase shift. The result is a tilted fan beam pattern with possible applications as a base station antenna. The realization of a two-dimensional 8×8 passive quasi-Yagi array is also presented. In the final portion, an active eight-element array is demonstrated.

3.4.1 Mutual Coupling

One important parameter in antenna array design is the mutual coupling between elements within the array. Along with the antenna's physical size, mutual coupling restricts the element spacing within the array. Strong mutual coupling may cause scan blindness, which limits the actual beam scanning range of the antenna array. Furthermore, since mutual coupling is frequency dependent, it will also pose a restriction of maximum usable bandwidth of the array. Two types of coupling were measured. The first case, horizontal coupling (E-plane), is the coupling between two adjacent elements on the same substrate cards. The second case, vertical coupling (H-plane), is the coupling between two elements on adjacent cards coupling

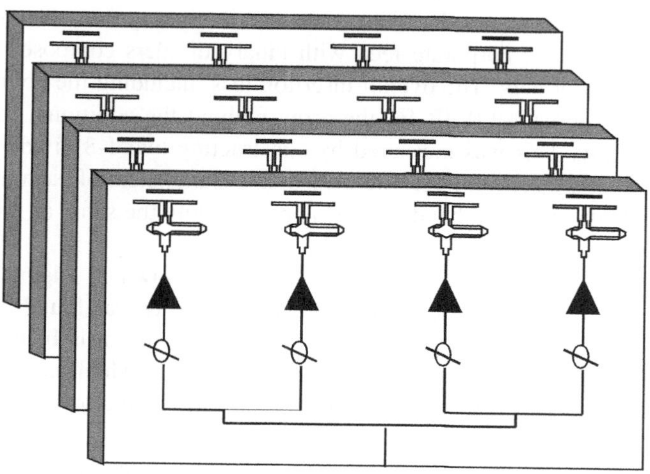

Figure 3.27 Proposed architecture of a two-dimensional array using quasi-Yagi antennas. (From Reference [26]. Reproduced by permission of the IEEE.)

through the air. The mutual coupling is determined by the measured direct transmission coefficient S_{21} between elements.

Horizontal coupling was experimentally determined as a function of array spacing by fabricating 15 antennas, each with a different spacing to its neighbour on a single substrate. The minimum centre-to-centre spacing was extremely close, 10.0 mm, while the maximum centre-to-centre spacing was 25.0 mm. The results are shown in Figures 3.28 (a) to (c), where coupling at specific frequencies in the lower, middle, and upper operating bands of the antenna are shown as a function of frequency. A horizontal centre-to-centre spacing greater than 15.0 mm, which corresponds to $\lambda_0/2$ at the centre frequency of the antenna, will insure that coupling will be below −18.0 dB and in most cases below −20.0 dB.

Vertical coupling was also experimentally determined. In this case, four cards, each with three elements horizontally spaced by 15.0 mm, were tested with centre-to-centre vertical spacings of 15.0 and 20.0 mm. The results are shown in Figure 3.29. The vertical coupling is somewhat stronger than horizontal coupling for comparable centre-to-centre spacing. This is due to the broader H-plane pattern of the quasi-Yagi element. Maximum coupling for 15.0 mm spacing is −17.5 dB at 8.5 GHz and −16.5 dB at 10.0 GHz for the 20.0 mm case. However, mutual coupling is below −20 dB for two-thirds of the bandwidth for both cases.

3.4.2 *Passive Quasi-Yagi Arrays*

A simple equal-amplitude eight-element linear array is used to demonstrate the viability of the quasi-Yagi as an array antenna. Two arrays are built, the first with the main beam at the endfire direction of the quasi-Yagi elements and the second using microstrip delay lines so that the main beam is tilted to approximately 12° from the endfire array. The low mutual coupling between adjacent quasi-Yagi elements and the narrow physical width of the antenna gives great flexibility in array spacing not available with other endfire antennas. Each array is fabricated on a single piece of *RT/Duroid* with $\varepsilon_r = 10.2$ and a substrate thickness of 0.635 mm. A photograph of the tilted beam array is shown in Figure 3.30.

The width of each board is approximately 13.5 cm. The only difference between the two arrays is the inclusion of microstrip delay lines on the tilted beam array to realize a progressive phase delay between elements, which produces a beam tilted away from the endfire direction. Each array utilizes a simple corporate feed with binary dividers composed of T-junctions and quarter-wave transformers. The overall insertion loss, including the connector loss, has been measured to be about −1.0 dB for the array without the additional microstrip delay lines of the tilted array. This was measured by constructing two 1-8 dividers back-to-back and taking half the total insertion loss. The measured *S*-parameters of each array are shown in Figure 3.31. In both cases, the bandwidth is approximately the same as the bandwidth of the quasi-Yagi element, about 50 %.

Radiation patterns for the endfire array at 9 GHz is shown in Figure 3.32. For this particular frequency, the front-to-back ratio of the endfire array was found to be better than 20 dB. Additionally, the cross-polarization level in the main beam is better than −15 dB. About a 2 dB discrepancy in the expected first-sidelobe level, which is −13.5 dB for this type of array, is present. This is due to slight amplitude and phase errors in the feed network present after fabrication.

The radiation patterns of the tilted beam array were also measured. The E-plane of the tilted beam antenna is shown in Figure 3.33 at 8, 10, and 11.7 GHz. The measured 3 dB beamwidth ranges from 10 to 17° in this frequency range. Additionally, the sidelobe levels

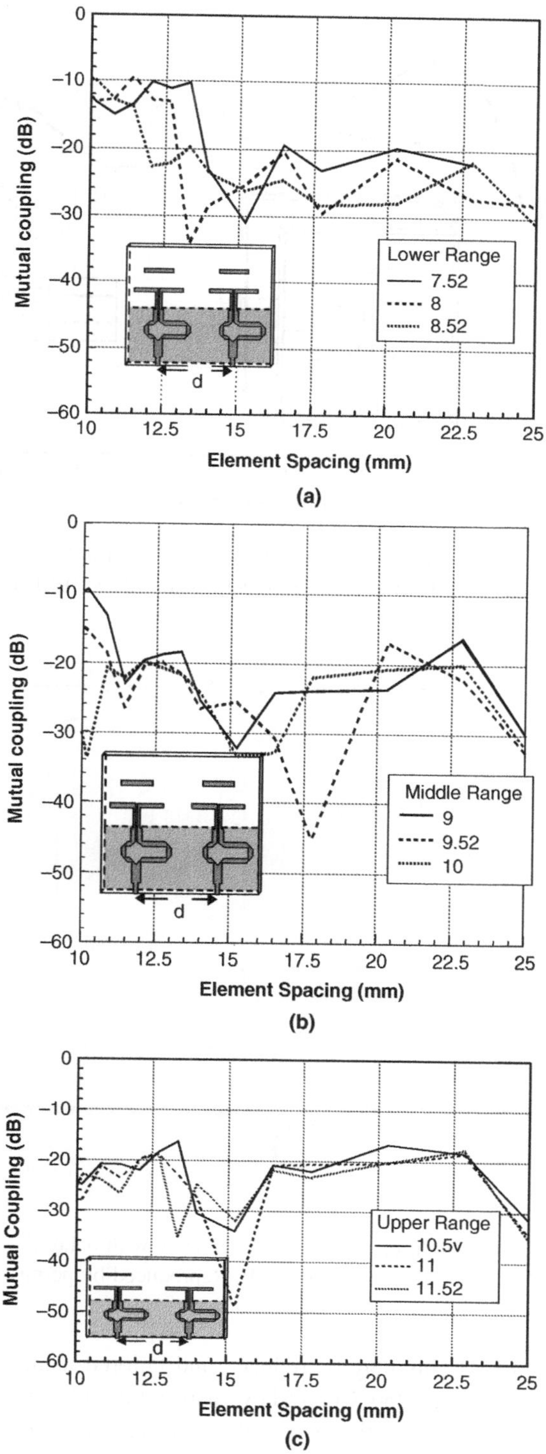

Figure 3.28 Mutual coupling as a function of element spacing *d* for: (a) lower; (b) middle; and (c) upper frequency ranges. (From Reference [26]. Reproduced by permission of the IEEE.)

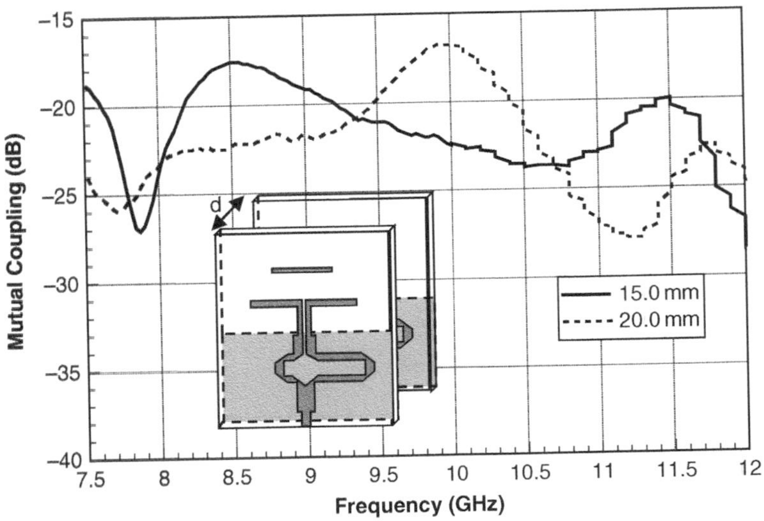

Figure 3.29 Vertical mutual coupling as a function of frequency for 15.0 and 20.0 mm spacings. (From Reference [26]. © 2000 IEEE. Reproduced by permission of the IEEE.)

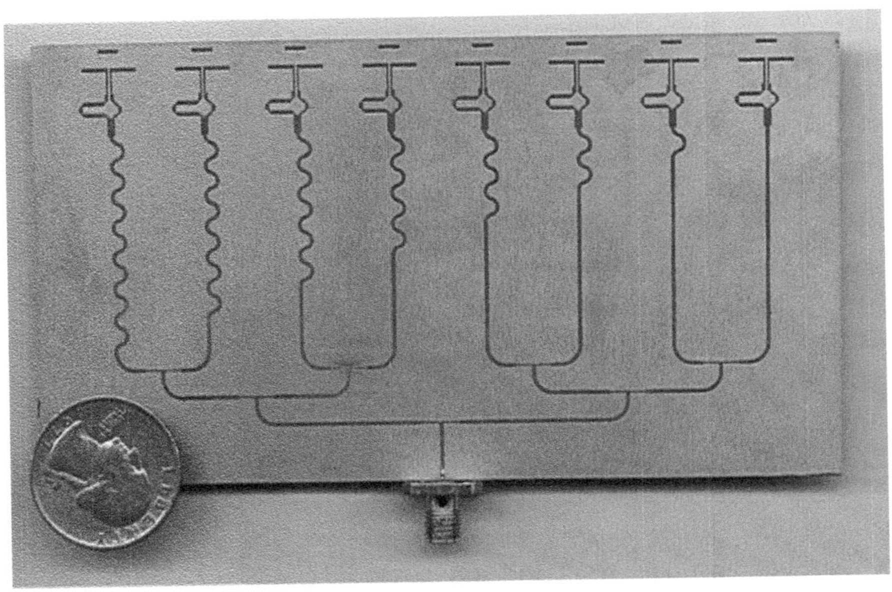

Figure 3.30 Photograph of an eight-element quasi-Yagi array with delay lines for a tilted fan-beam radiation pattern. (From Reference [26]. © 2000 IEEE. Reproduced by permission of the IEEE.)

range from −11 to −14 dB. This may be reduced and tailored using a more sophisticated beamforming technique such as the Chebyshev distribution [29] rather than the simple equal-amplitude distribution used with this proof-of-concept array at the cost of increased beamwidth.

Demonstrating the quasi-Yagi applicability to realization of two-dimensional antenna arrays, a 64-element two-dimensional antenna array was constructed and characterized. An

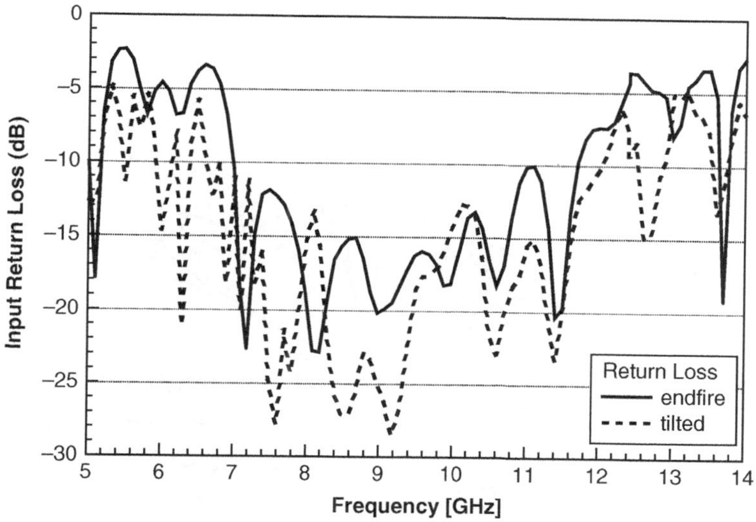

Figure 3.31 Measured return loss for endfire and tilted fan-beam quasi-Yagi arrays. (From Reference [26]. © 2000 IEEE. Reproduced by permission of the IEEE.)

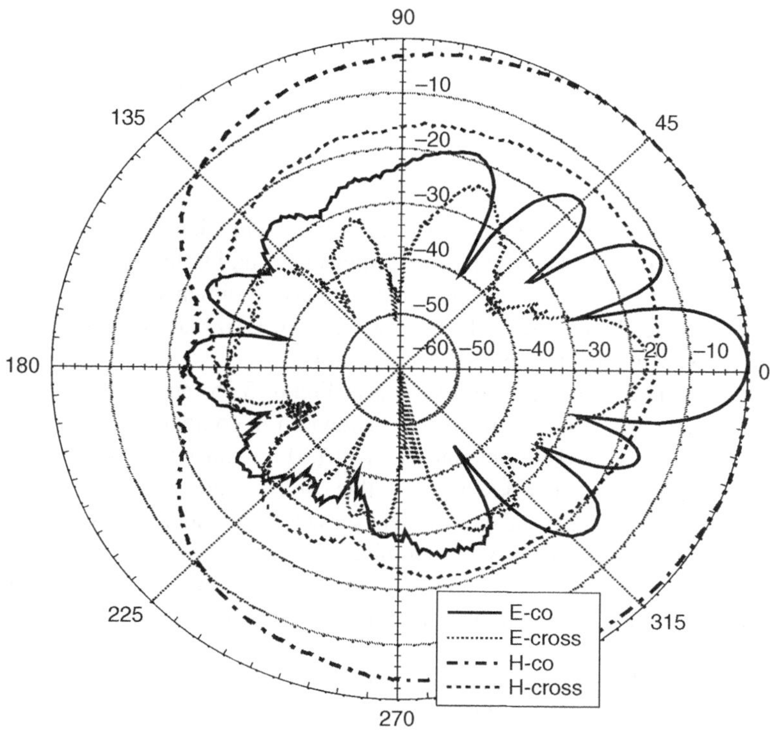

Figure 3.32 E-plane radiation patterns of an eight-element endfire quasi-Yagi array. (From Reference [26]. © 2000 IEEE. Reproduced by permission of the IEEE.)

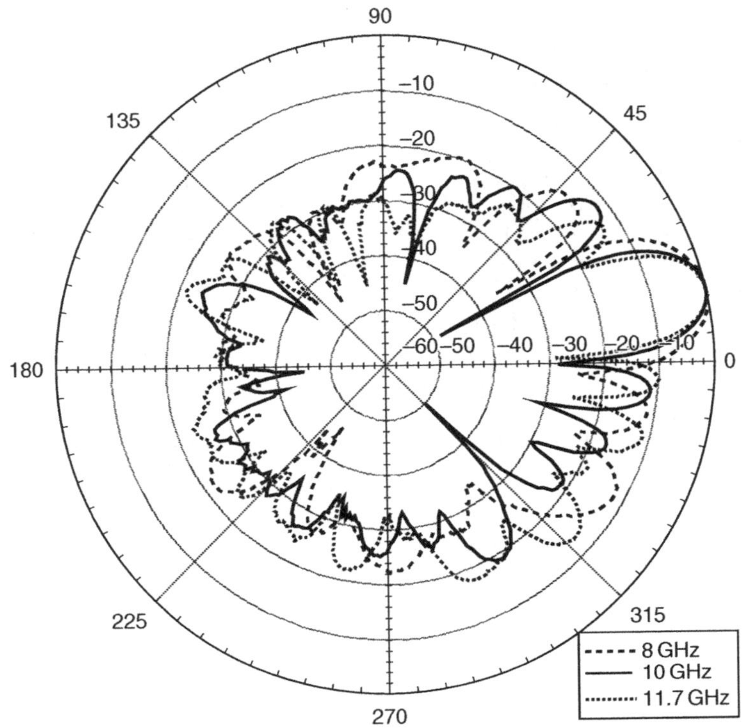

Figure 3.33 E-plane radiation patterns of a tilted fan-beam quasi Yagi array at different operating frequencies. (From Reference [26]. © 2000 IEEE. Reproduced by permission of the IEEE.)

X-band two-dimensional array was constructed using eight identical eight-element cards and is shown in Figure 3.34. The same one-dimensional eight-element array card layouts presented earlier were used in the construction of the two-dimensional array. Each of the eight cards were vertically spaced by 15 mm ($\lambda_0/2$ at 10 GHz) and fed using an eight-way printed circuit power divider through SMA connectors.

Measurement of the input return loss shown in Figure 3.35 indicates a 40 % bandwidth from 7.8 to 11.7 GHz. The slight reduction in bandwidth in comparison with a single eight-element card (50 %) and a single antenna element (50 %) can be attributed to fabrication discrepancies in the feeding network. However, this can be improved by using a more robust feeding system. Antenna patterns taken at 9 GHz show a 12 dB sidelobe level for both E- and H-planes, which is typical for linearly excited phased arrays, as well as a 3 dB beamwidth of 12° and cross-polarization level better than −19 dB (Figure 3.36). A measured front-to-back ratio of 17 dB corresponds well with the radiation characteristics of a single-antenna element, as expected.

The main features of the array radiation performance are summarized over the operating band in Figure 3.37. The array maintains cross-polarization radiation less than −12 dB and a front-to-back ratio of better than 12 dB, which match the single-element characteristics exactly. The level of the sidelobes is ultimately determined by the relative amplitude and phase provided by the feeding network as well as the array spacing, but the shape of the element pattern also has an effect.

Figure 3.34 Photograph of an 8 × 8-element two-dimensional quasi-Yagi antenna array. (From Reference [28]. © 2000 IEEE. Reproduced by permission of the IEEE.)

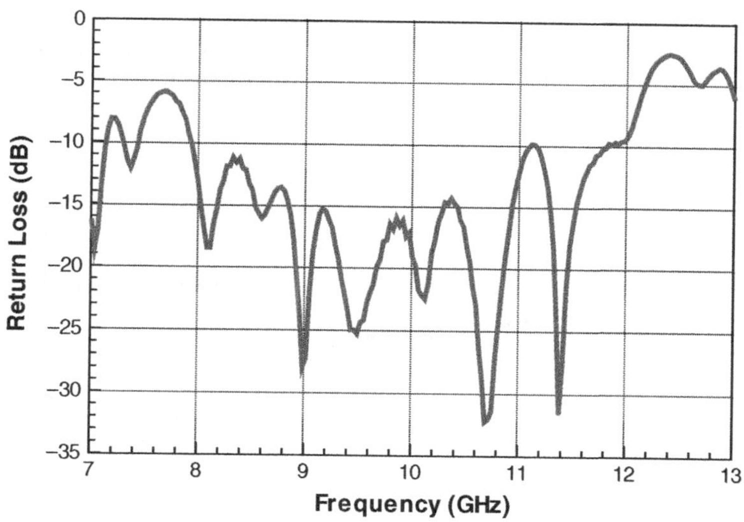

Figure 3.35 Measured return loss of a two-dimensional quasi-Yagi antenna array. (From Reference [28]. © 2000 IEEE. Reproduced by permission of the IEEE.)

Antenna gain measurements shown in Figure 3.38 reveal a gain variation between 20 and 14 dB over the wide frequency band. The directivity of the array was also approximated by considering the 3 dB beamwidth measured in the E- and H-planes. Radiation efficiency can be approximated by noting the difference between the antenna gain and directivity. Notice that at 9.5 GHz the efficiency of the two-dimensional array is estimated to be 50 %. This figure includes all connector and antenna feed losses. Previous results indicate that the

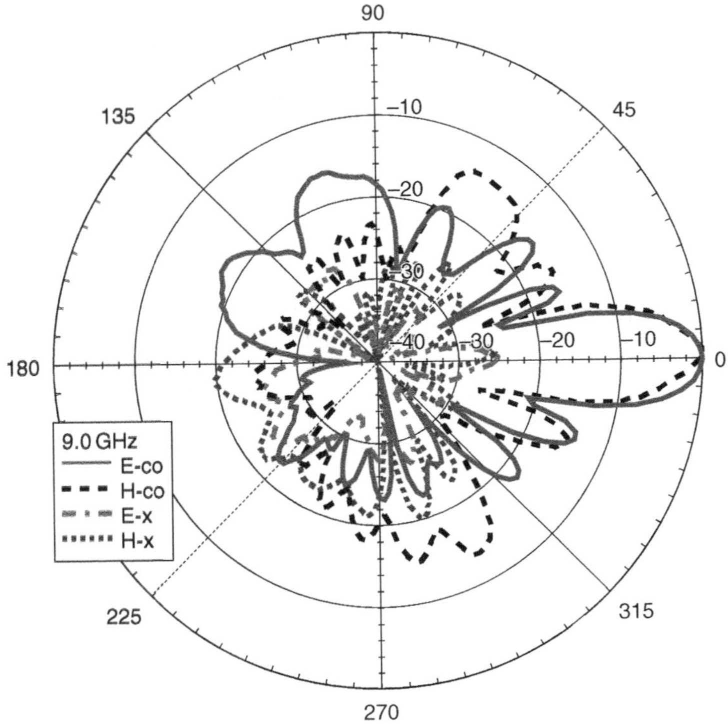

Figure 3.36 Measured radiation patterns of a two-dimensional quasi-Yagi antenna array at 9 GHz. (From Reference [28]. © 2000 IEEE. Reproduced by permission of the IEEE.)

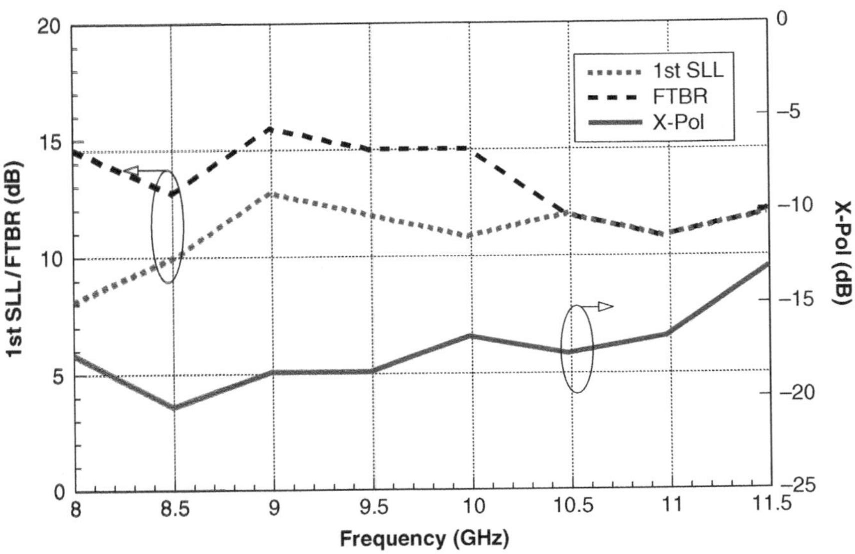

Figure 3.37 Summary of two-dimensional array radiation characteristics.

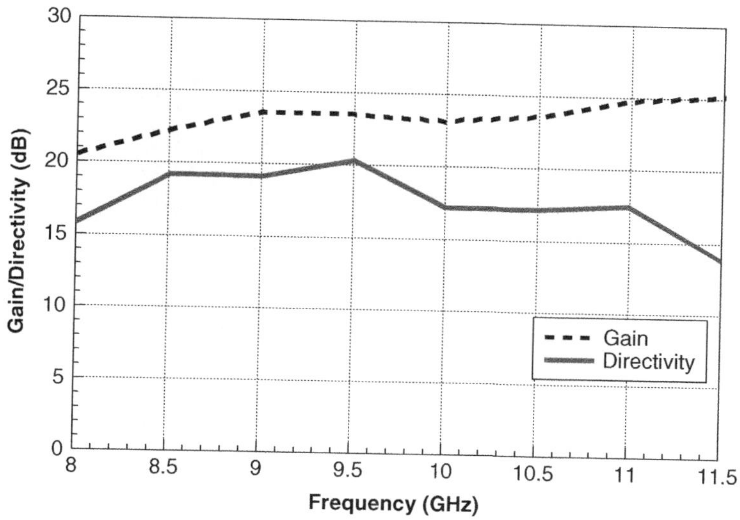

Figure 3.38 Measured antenna gain and directivity of a two-dimensional array.

efficiency of a one-dimensional eight-element array is nearly 65 % and the efficiency of a single element is 93 %. It is believed that the high efficiency performance of the quasi-Yagi antenna can be further highlighted in array applications, especially when a large number of elements must be used.

3.4.3 Active Quasi-Yagi Antenna Array

Antenna arrays with integrated amplifiers are commonly known as active arrays. In the case of a transmit array, the power amplifiers (PAs) are placed directly at each antenna to avoid the losses associated with the feed network or other components, such as phase-shifters. The increase in system efficiency can be dramatic, thus allowing smaller and lighter heat-sinking and batteries. LNAs can be incorporated directly at the antenna platform of a receive array for increased sensitivity. In this section, amplifiers are incorporated into the linear eight-element array to form an active transmit array.

The fabricated transmitting eight-element array is shown in Figure 3.39. For simplicity, matched GaAs field effect transistor (FET) gain block amplifiers are used as the active device. By placing the amplifier output directly at the antenna input, the power-combining losses are reduced.

The measured input return loss of the completed active array is shown in Figure 3.40. Due to the internal matching of the GaAs FET amplifiers, the bandwidth of the active array is 60 %, which is larger than that of the passive eight-element array and, is quite large for a planar active antenna array. However, the increased bandwidth is added at the lower frequency end where the quasi-Yagi cross-polarization and front-to-back ratio deteriorates. Therefore, the increased input match does not directly translate to an increased usable bandwidth in this case.

The radiation patterns were measured in the operating range of the quasi-Yagi antenna and were seen to be identical to those of the passive array, as shown in Figure 3.41. The gains of the passive and active one-dimensional arrays have been measured from 8 to 11.7 GHz

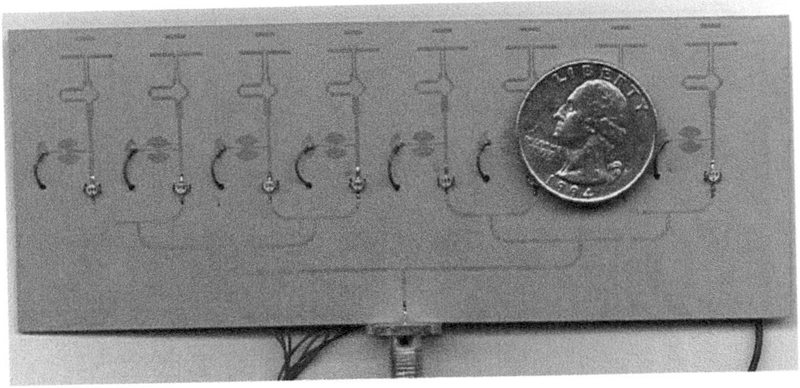

Figure 3.39 Photograph of an active quasi-Yagi antenna array. (From Reference [26]. © 2000 IEEE. Reproduced by permission of the IEEE.)

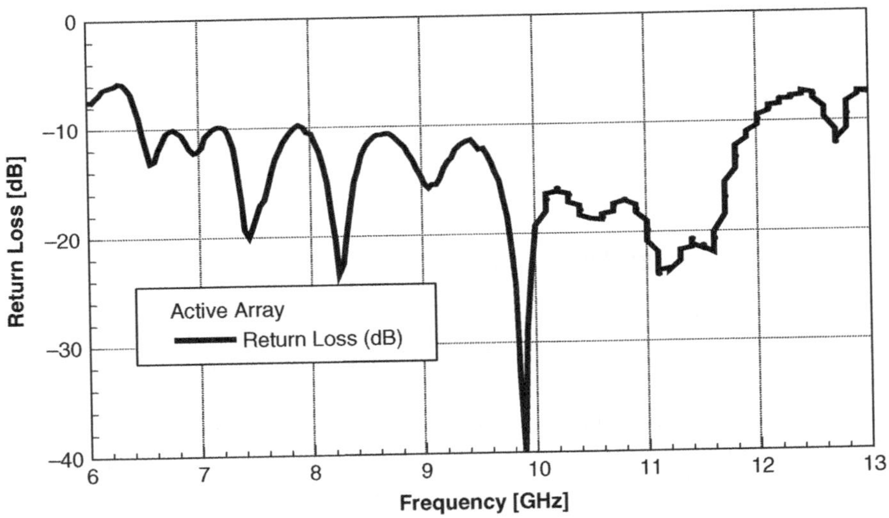

Figure 3.40 Measured return loss of the active eight-element quasi-Yagi array. (From Reference [26]. © 2000 IEEE. Reproduced by permission of the IEEE.)

and are shown in Figure 3.42. The gain was measured using the gain substitution method with a standard gain horn. The difference in the gain between the two plots corresponds to the gain of the amplifier. The gain changes by about 3 dB in this bandwidth. The average gain is about 12 dB for the passive array and 23 dB for the active array. The gain difference between the passive and active arrays can be approximated by summing the positive gain of the gain block amplifier and the associated feed network loss, which was determined to be about −1.0 dB, including the connector loss. The discrepancy between the expected gain and measured gain is 0.6 dB at 9.0 GHz, 0.3 dB at 9.5 GHz, and 0.5 dB at 10.0 GHz.

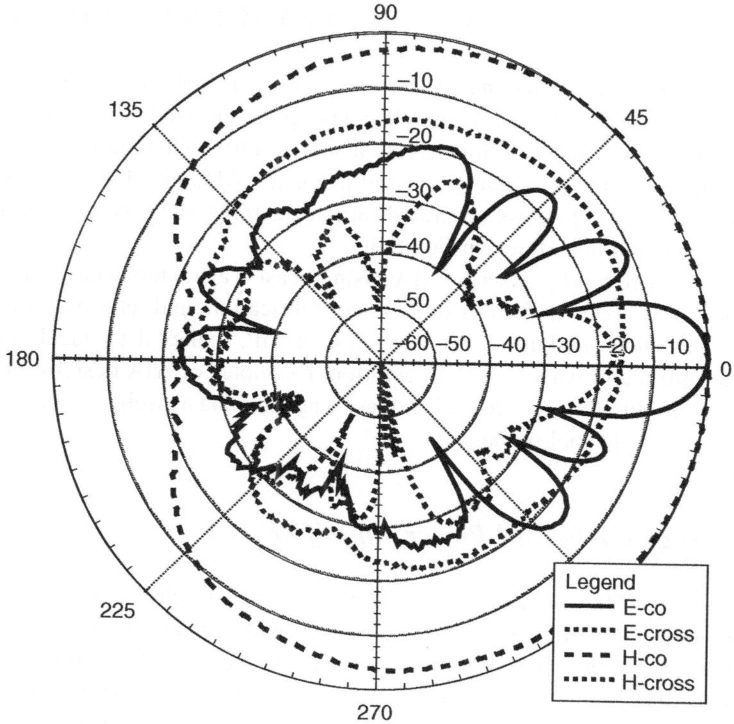

Figure 3.41 E-plane radiation patterns of an eight-element active array.

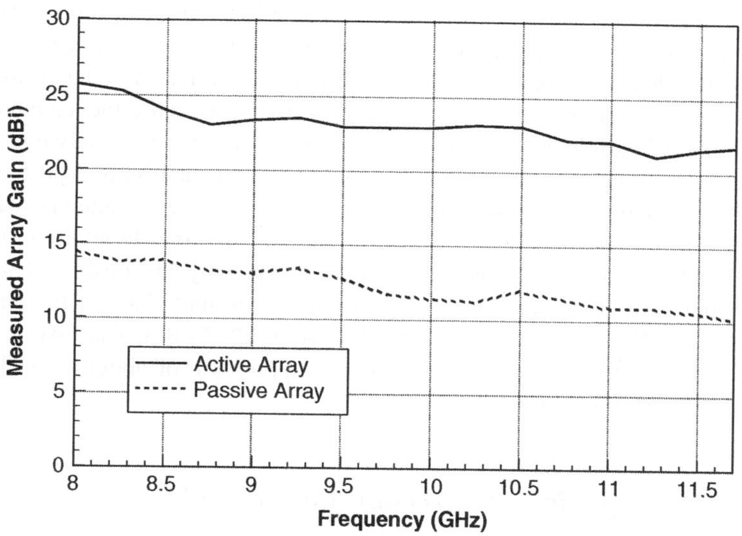

Figure 3.42 Measured gains of active and passive endfire quasi-Yagi antenna arrays. (From Reference [26]. © 2000 IEEE. Reproduced by permission of the IEEE.)

3.5 ACTIVE ANTENNAS WITH QUASI-YAGI RADIATORS

The idea of circuit integrated antennas was conceived by first observing the existing codependence of the antenna and circuit and then taking it one step further to achieve full integration of the antenna and circuit to create a single entity. This merging of the antenna and circuit, often referred to as active integrated antennas (AIAs) [30], has led to innovative RF front-end designs that possess several desirable features such as compactness, lower power consumption, and added design flexibility.

This next section will briefly describe AIAs using quasi-Yagi antenna radiators. The quasi-Yagi's compact size, broad bandwidth single-layer fabrication, and its use of high dielectric constant substrates such as alumina, or Si, GaAs, or InP, makes it an ideal candidate for microwave and millimetre-wave circuit integration. Examples of AIA designs will include a push–pull power amplifier integrated with a quasi-Yagi antenna as well as a direct conversion receiver AIA with IQ channel outputs.

3.5.1 Quasi-Yagi Push–Pull Power Amplifier AIA

Push–pull amplifiers offer twice as much output power as single class-B amplifiers, high theoretical power added efficiency (PAE) (78.5 %), and are able to suppress even-order harmonic generation intrinsically by virtue of their balanced configuration. Conventional microwave frequency push–pull amplifiers use a pair of Class B amplifiers along with input and output baluns to split and combine the power. The minimization of combining loss is crucial to the performance of this type of amplifier. Even the slightest addition of insertion loss at the amplifier output leads to significant reductions in PAE. For example, a power-combining loss of 0.5 dB reduces an amplifier's efficiency from 60 to 53 %.

The AIA design methodology of this type of amplifier makes use of the antenna to play the role of an output balun and harmonic tuner, while still serving its primary function as a radiator forming an intimately integrated design of circuit and antenna. Because no external circuits (i.e. balun, harmonic tuner) or interconnects are required, the related insertion loss is minimized. The type of printed antenna utilized is crucial. Because the antenna is used to replace the output balun, it must perform the same function of accepting only odd excitations and rejecting even excitations. Figure 3.43 shows the implementation of a push–pull power amplifier AIA [31] using the balanced CPS feeds of the quasi-Yagi antenna, which act as the odd excitation power combiner for efficient power combining. In addition, to suppress the undesired radiation of the second harmonic generated by the push–pull amplifier, a corrugation pattern has been introduced at the truncated ground-plane of the antenna. The maximum measured PAE is 60.9 % at an output power of 28.2 dBm. The PAE is better than 50 % from 4.08 to 4.29 GHz. A second harmonic suppression of about −30 dB has been measured in both the E- and H-planes.

3.5.2 Quasi-Yagi IQ Direct Conversion Receiver AIA

Direct conversion receivers convert the received RF signal directly down to baseband. This reduces receiver complexity by eliminating the need for intermediate frequency (IF) circuitry and filters. Furthermore, they reduce system concerns of image frequency rejection. The

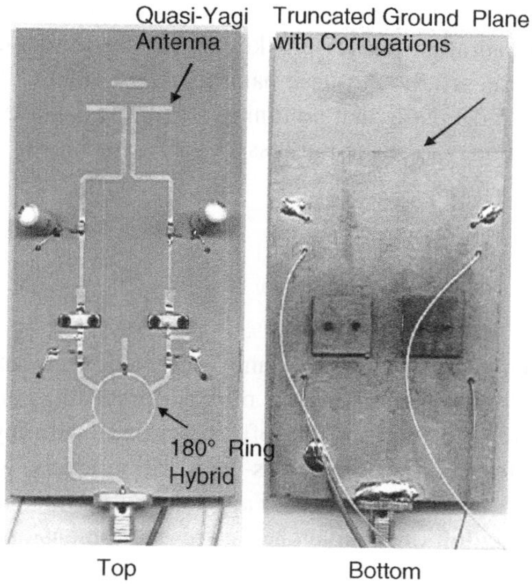

Figure 3.43 Photograph of a push–pull PA AIA. (From Reference [31]. © 2000 IEEE. Reproduced by permission of the IEEE.)

integration of a C-band quasi-Yagi and diode mixers is used to implement a subharmonic direct conversion mixer [32]. Furthermore, the receiver provides quadrature outputs (I/Q). In this case the antenna serves to receive the RF signal, as well as to split the signal into I and Q components. Figure 3.44 shows the AIA prototype. The two CPS feeds of the quasi-Yagi antenna split the RF input power to each subharmonic Schottky diode mixer. The LO signal is split with a Wilkinson divider that has a 45° delay line on one branch, so that there is a 90° phase difference at the second harmonic of the LO. The subharmonic design

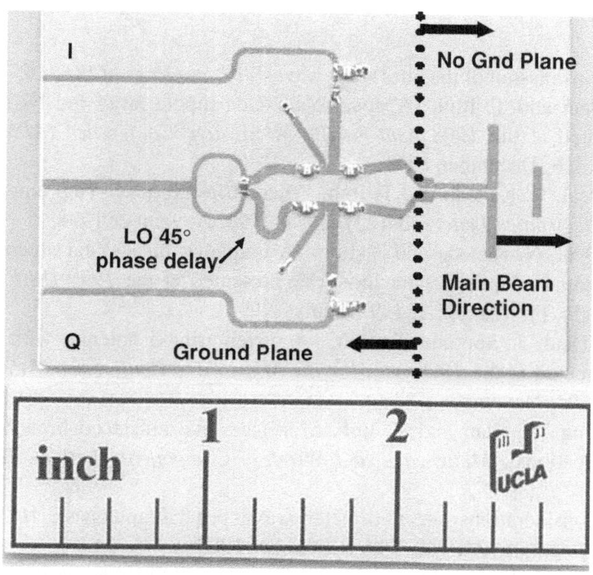

Figure 3.44 Photograph of a direct downconversion mixer AIA with IQ outputs. (From Reference [32]. © 2000 IEEE. Reproduced by permission of the IEEE.)

cancels even-mixing products, thereby improving even-order distortion. Downconversion and demodulation of a quadrature phase shift keying (QPSK)-modulated RF carrier signal has been demonstrated successfully. A phase balance of less than 6° was measured with an amplitude imbalance of 0.5 dB. Note that no tuning attempt was made to improve the circuit performance after fabrication. A conversion loss of 9 dB is achieved for a LO pump power of 4 dBm.

3.6 SUMMARY

This chapter has summarized the operation and presented some implementations of the quasi-Yagi antenna. Prototype antennas with microstrip feeding and CPW feeding have been discussed. This type of planar antenna offers a broad operating bandwidth, compact antenna size, good front-to-back ratio and cross-polarization levels, with moderate gain and high efficiency. The application of the quasi-Yagi antenna in antenna arrays has also been investigated by characterization of passive and active one-dimensional arrays as well as an 8×8 two-dimensional array. Finally, use of the planar antenna as part of an active integrated antenna power amplifier and mixer has been described.

ACKNOWLEDGEMENTS

The authors would like to acknowledge the research efforts of Dr Y. Qian, Dr W. R. Deal, Dr N. Kaneda, Dr J. Sor, Dr M. Sironen, Dr M. DeVincentis, Dr S. M. Lin, and Dr C. Y. Hang, which have contributed to the contents of this chapter.

REFERENCES

1. H. Yagi, 'Beam transmission of the ultra short waves', *Proceedings of IRE*, **16**, 715–741, June 1928.
2. N. Kaneda, Y. Qian and T. Itoh, 'A novel Yagi–Uda dipole array fed by a microstrip-to-CPS transition', presented at the *1998 Asia-Pacific Microwave Conference (APMC'98)*, Yokohama, Japan, pp. 1413–1416, December 1998.
3. Y. Qian, W. R. Deal, N. Kaneda and T. Itoh, 'Microstrip-fed quasi-Yagi antenna with broadband characteristics', *Electronics Letters*, **34**(23), 2194–2196, November 1998.
4. Y. Qian, W. R. Deal, N. Kaneda and T. Itoh, 'A uniplanar quasi-Yagi antenna with wide bandwidth and low mutual coupling characteristics', presented at the *1999 IEEE AP-S International Symposium*, Orlando, Florida, pp. 924–927, July 1999.
5. Y. Qian, W. R. Deal, J. Sor and T. Itoh, 'A novel printed antenna with broadband circular polarization', presented at the *1999 Asia-Pacific Microwave Conference (APMC '99)*, Singapore, pp. A5_1P1–A5_1P4, November 1999.
6. K. M. K. H. Leong, Y. Qian and T. Itoh, 'Surface wave enhanced broadband planar antenna for wireless applications', *Microwave and Wireless Components Letters, IEEE*, **11**(2), 62–64, 2001.
7. D. M. Pozar, 'Considerations for millimeter wave printed antennas', *IEEE Transactions on Antennas and Propagation*, **31**, 740–747, September 1983.
8. N. G. Alexopoulos, P. B. Katehi and D. B. Rutledge, 'Substrate optimization for integrated circuit antennas', *IEEE Transactions on Microwave Theory and Techniques*, **31**, 550–557, July 1983.

9. Y. Qian and T. Itoh, 'A broadband uniplanar microstrip-to-CPS transition', in the *1997 Asia-Pacific Microwave Conference (AMPC'97)*, Hong Kong, pp. 609–612, December 1997.
10. E. Yamashita and Y. Qian, *Analysis of Microwave Circuits and Planar Antennas Using the FDTD Method*, Realize Inc., Tokyo, May 1996.
11. B. Edward and D. Rees, 'A broadband printed dipole with integrated balun', *Microwave Journal*, 339–344, May 1987.
12. R. H. Johnston and J. G. McRory, 'An improved small antenna radiation-efficiency measurement method', *IEEE Antennas and Propagation Magazine*, **40**(5), 40–48, October 1998.
13. H. A. Wheeler, 'The Radiansphere Around a Small Antenna', *Proceedings of IRE.*, 1325–1331, August 1959.
14. M. Sironen, Y. Qian and T. Itoh, 'A 60 GHz conical horn antenna excited with quasi-Yagi antenna', in *2001 IEEE MTT-S International Microwave Symposium*, Vol. 1, pp. 547–550, 2001.
15. K. M. K. H. Leong, Y. Qian and T. Itoh, 'Compact printed surface wave assisted antenna with multiple system compatibility', in *IEEE AP-S International Symposium*, Vol. 3, pp. 448–451, 2001.
16. Y. Shih and T. Itoh, 'Analysis of conductor-backed coplanar waveguide', *Electronics Letters*, **18**, 538–540, June 1982.
17. J. Sor, Y. Qian and T. Itoh, 'A coplanar waveguide fed quasi-Yagi antenna', *Electronics Letters*, **36**(1), 1–2, January 2000.
18. K. M. K. H. Leong, Y. Qian and T. Itoh, 'First demonstration of a conductor backed coplanar waveguide fed quasi-Yagi antenna', *2000 IEEE AP-S International Symposium Digest*, **2**, 924–927; **3**, 1432–1435, July 2000.
19. T. Q. Ho and S. M. Hart, 'A broad-band coplanar waveguide to slotline transition', *IEEE Microwave and Guided Wave Letters*, **2**, 415–416, October 1992.
20. N.-S. Cheng, P. Jia, D. B. Rensch and R. A. York, 'A 120-W X-band spatially combined solid-state amplifier', *IEEE Transactions on Microwave Theory and Techniques*, **47**(12), 2557–2561, December 1999.
21. R. Bashirullah and A. Mortazawi, 'A slotted-waveguide power amplifier for spatial power-combining applications', *IEEE Transactions on Microwave Theory and Techniques*, **48**(7, part 1), 1142–1147, July 2000.
22. B. Deckman, D. S. Deakin Jr, E. Sovero and D. Rutledge, 'A 5-watt, 37-GHz monolithic grid amplifier', *2000 IEEE MTT-S International Microwave Symposium Digest*, **2**, 805–808, June 2000.
23. K. M. K. H. Leong, V. Radisic, L. Samoska, Y. Qian and T. Itoh, 'Monolithic millimeter-wave source incorporating planar surface wave assisted antenna', *European Microwave Conference Digest*, **1**, 73–76, September 2001.
24. W. R. Deal, J. Sor, Y. Qian and T. Itoh, 'A broadband uniplanar quasi-Yagi active array for power combining', in *1999 IEEE Radio and Wireless Conference (RAWCON'99)*, pp. 231–234, August 1999.
25. J. Sor, W. R. Deal, Y. Qian and T. Itoh, 'A broadband quasi-Yagi antenna array', in *29th European Microwave Conference*, Munich, Germany, pp. 255–258, October 1999.
26. W. R. Deal, N. Kaneda, J. Sor, Y. Qian and T. Itoh, 'A new quasi-Yagi antenna for planar active antenna arrays', *IEEE Transactions on Microwave Theory and Techniques*, **48**(6), 910–918, June 2000.
27. F.-C. E. Tsai and M. E. Bialkowski, 'An X-band tray-type spatial power combiner using uniplanar quasi-Yagi antennas', *2004 IEEE Antenna and Propagation Symposium Digest*, **4**, 4143–4146, June 2004.
28. K. M. K. H. Leong, J. Sor, W. R. Deal, Y. Qian and T. Itoh, 'A Broadband 64-element 2-D quasi-Yagi antenna array', in *RAWCON'00*, Denver, Colorado, pp. 67–70, September 2000.
29. W. R. Deal, J. Sor, R. Y. Miyamoto, Y. Qian and T. Itoh, 'A sixteen element quasi-Yagi antenna array with Chebyshev excitation coefficients', in *Millennium Conference on Antennas and Propagation (AP2000)*, Davos, Switzerland, April 2000.
30. Y. Qian and T. Itoh, 'Progress in active integrated antennas and their applications', *IEEE Transactions on Microwave Theory and Techniques,* **46**(11), 1891–1900, November 1998.

31. C. Y. Hang, W. R. Deal, Y. Qian and T. Itoh, 'High-efficiency push–pull power amplifier integrated with quasi-Yagi antenna', *IEEE Transactions on Microwave Theory and Techniques*, **49**, 1155–1161, June 2001.
32. S. Lin, Y. Qian and T. Itoh, 'Quadrature direct conversion receiver integrated with planar quasi-Yagi antenna', in *2000 IEEE MTT-S International Microwave Symposium*, Boston, Massachusutts, 1285–1288, June 2000.

4

Printed Spiral Antennas

Chi-Chih Chen and John Volakis

ElectroScience Laboratory, Electrical and Computer Engineering Department, The Ohio State University, Columbus, Ohio, USA

4.1 INTRODUCTION AND HISTORICAL OVERVIEW

Spiral antennas were introduced in the 1950s by Edwin Turner [1] who demonstrated experimentally that an Archimedean spiral resulted in constant input impedance and circular polarization over a wide range of frequencies. His work ignited great interest in spiral antennas [2, 3] and frequency-independent antennas [2]. Some of the noteworthy early works include those of Dyson for planar and equiangular antennas [4–7] (see Figure 4.1 for typical spiral shapes). For works on two-wire Archimedean spirals, the reader is referred to References [2] and [8]. The second-mode printed spiral is described in Reference [9] and examples of array applications are given in References [3] and [10] to [15]. Four arm spirals have been found to be particularly attractive for direction-finding applications due to their polarization agility [16], whereas resonant [17–22] and other forms of spirals (spiral-mode microstrip antenna [23]) have been considered for various wireless communication needs. Over the past 10 years, conformal broadband antennas have been of interest [24, 25] with efforts focused on recovering the gain and bandwidth lost in cavity-backed configurations. The works of Nurnberger and Volakis [25, 26], Volakis *et al.* [27], Filipovic and Volakis [28, 29], Kramer *et al.* [30], Wang and Tripp [31], and Wang [32] have focused in this direction, demonstrating bandwidths greater than 25:1 for conformal, cavity-backed spirals.

Analysis has played an important role in the understanding of spiral antennas. As early as 1960, Curtis [8] gave an explanation of the spiral's radiation using a series of semicircles, a concept also employed by Wheeler [33] to generate the pattern of the equiangular spiral. A more general explanation of the spiral antenna radiation was given by Cheo *et al.* in 1961 [34]. More recently, Lee *et al.* [35] employed the concept of concatenated concentric loops with a transmission line model to describe the curved wire element spiral's operation below its radiation band. By and large, recent spiral characterizations have been carried out using

Printed Antennas for Wireless Communications Edited by R. Waterhouse

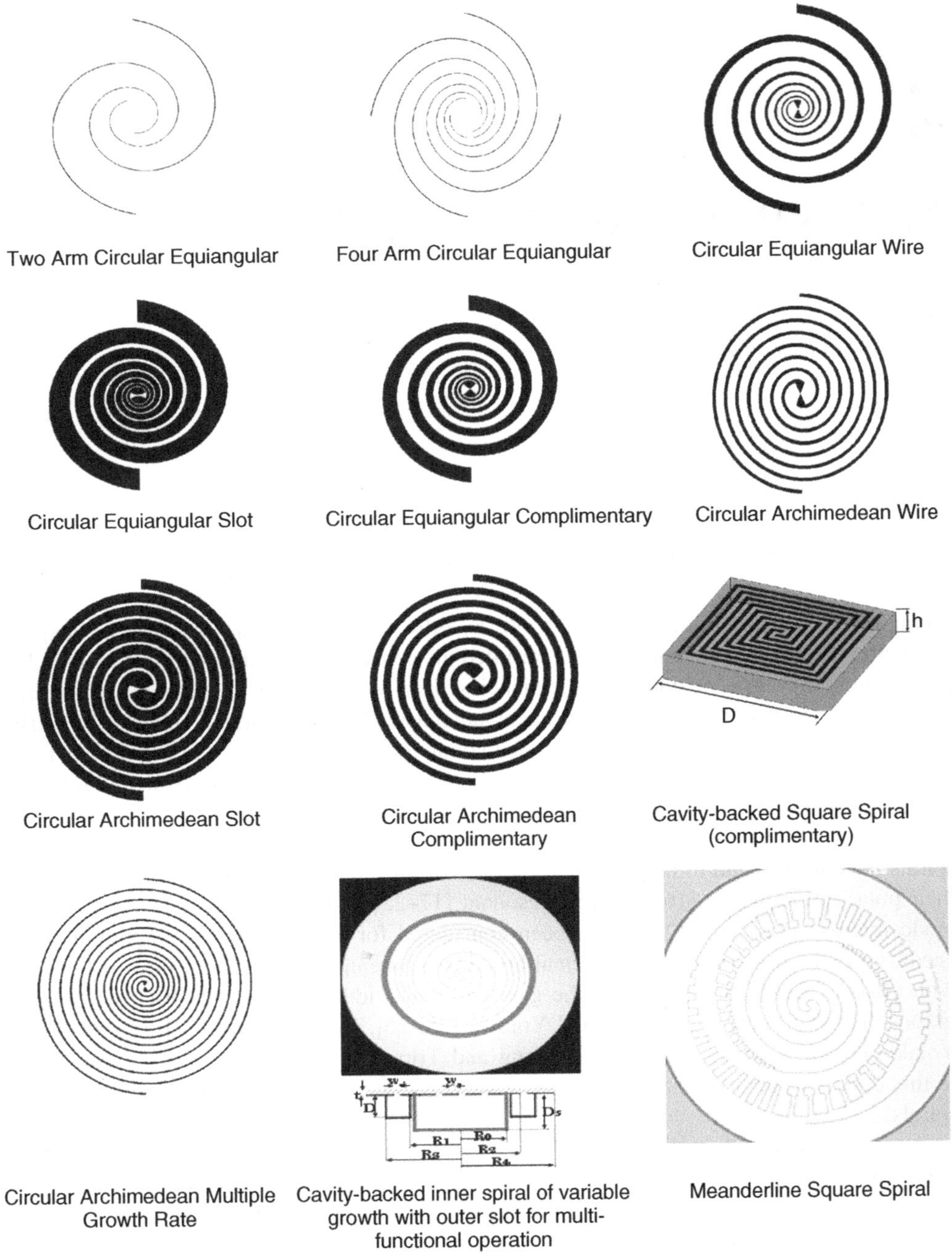

Figure 4.1 Various wire, slot, and cavity-backed (conformal) shapes of spiral antennas used in the literature.

any of the popular numerical techniques, such as the moment method [36–38], the finite difference time domain (FDTD) method [39], and the finite element method (FEM) [40, 41] (see also References [42] and [43]).

This chapter is intended to provide a brief overview of spiral antennas with particular emphasis on recent work relating to conformal and cavity-backed antennas. The chapter begins with a brief review of the spiral's operation principles (radiation modes) and analysis models. Both free-standing, cavity-backed, normal (first-mode), and second-mode operations are described and typical ideal patterns are provided. This is followed by a description of pertinent design issues and more specifically cavity-backed antennas. Dielectric loading and its impact on gain is considered and cavity depth, spiral growth, and antenna loading terminator schemes are described. Subsequently miniaturization issues and reactive loading for impedance correction are discussed aimed at enhancing antenna performance down to UHF and VHF frequencies as size is reduced.

4.2 SPIRAL ANTENNA FUNDAMENTALS

4.2.1 Frequency Independence

Spirals belong to a class of frequency-independent antennas [27, 44–46] and are therefore often described from this traditional point of view. Specifically, the single arm spiral shown in Figure 4.2 will be considered.

The arm geometry follows the relationship:

$$r(\theta, \phi) = g(\theta)\rho(\phi) = g_0\rho(\phi) = e^{a(\phi+\phi_s)} \tag{4.1}$$

in which ρ, θ, and ϕ refer to the usual spherical coordinates and $g(\theta) = g_0 = 1$ for the typical flat spiral. Thus, the entire spiral curve can be described by the cylindrical coordinate $\rho(\phi)$,

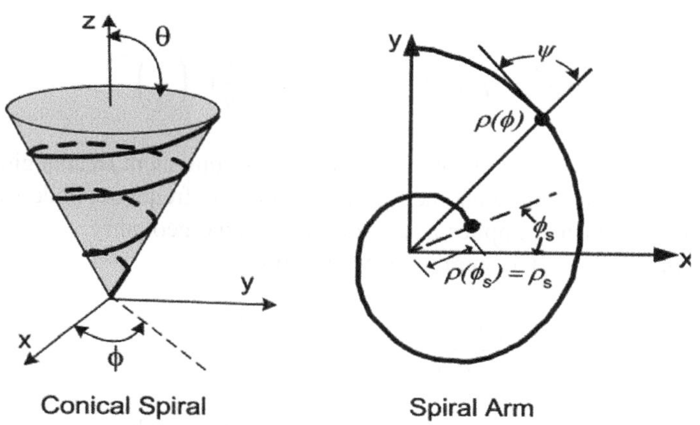

Figure 4.2 Coordinates for the spiral geometries.

viz. the radius in the xy plane measured from the origin to a point on the spiral curve. The choices of

$$\rho(\phi) = e^{a(\phi+\phi_s)} = \rho_0 e^{a\phi} \tag{4.2}$$

and

$$\rho(\phi) = \rho_0 e^{\ln\phi} = \rho_0 \phi \tag{4.3}$$

refer to the equiangular and Archimedean spirals, respectively (see Figure 4.1). Clearly, the constants a and ρ_0 in Equations (4.2) and (4.3) control the expansion of the spiral (how quickly it grows out of the origin) as the angle ϕ increases. Thus, they are often referred to as growth factors.

Mathematically, a frequency-independent antenna must satisfy the relation

$$k\rho(\phi) = \rho(\phi + c), \quad c = \text{constant} \tag{4.4}$$

so that $k\rho(\phi + c)$ will generate the same pattern at the frequency $f = kf_0$, where f_0 is the operational frequency for the shape $\rho(\phi), 0 < \phi < 2\pi$. This scaling holds true for the input impedances as well, and it can be remarked that c depends on the scaling factor k but not on (θ, ϕ). Thus, from Equation (4.4),

$$\frac{d}{dc}[k\rho(\phi)] = \rho(\phi)\frac{dk}{dc}; \qquad \frac{d\rho(\phi + c)}{dc} = k\frac{d\rho}{d\phi} \tag{4.5}$$

implying that

$$\rho\frac{dk}{dc} = k\frac{d\rho}{dc} \qquad \text{or} \qquad \frac{1}{\rho}\frac{d\rho}{d\phi} = \frac{1}{k}\frac{dk}{dc} = \text{constant} = a \tag{4.6}$$

It follows that $d\rho/d\phi = a\rho$, giving the solution $\rho(\phi) = \rho_0 e^{a\phi}$, stated in Equation (4.2) and corresponding to the equiangular spiral. It is interesting to note that Equation (4.2) can be rewritten as

$$\frac{\rho(\phi)}{\lambda} = \rho_0 e^{+a(\phi+\phi_1)}; \qquad \phi_1 = \frac{1}{a}\ln\left(\frac{1}{\lambda}\right) \tag{4.7}$$

In other words, changing the wavelength or frequency is equivalent to changing the starting point of the spiral. This is simply another verification of the frequency independence of the spiral since frequency change implies a rotation of the same geometry.

A parameter of interest is the spiral expansion ratio:

$$\varepsilon = \frac{\rho(\phi + 2\pi)}{\rho(\phi)} = \frac{\rho_0 e^{a(\phi+2\pi)}}{\rho_0 e^{a\phi}} = e^{2\pi a} \tag{4.8}$$

Typically, $\varepsilon \approx 4$, which is similar to the expansion factor for log-periodic dipole arrays (LPDAs).

4.2.2 Radiation Occurrence

Radiation from the spiral occurs when the currents from nearby arms are in phase for constructive radiation, which occurs (normal or first mode) at the radius corresponding to a circumference length of $c = \lambda$. A more general condition that can be applied to antenna reactive loading for varying the current phase velocities is $\beta_e l \sim 2\pi$, where β_e is the effective phase velocity of the current along spiral arms. Figure 4.3 illustrates the current phases along a two-arm log spiral with growth rates of $a = 1/100$ and $a = 1/10$, respectively, with a as defined in Equation (4.6). At the feed point, the two arms are excited with a 180° phase difference and each arm is assumed to vary along its length (free-space phase velocity). The dark greyscale indicates the positive phase and the light greyscale refers to the negative phase.

If the spiral is large enough for a chosen frequency, a 'band' formed by several adjacent arms with similar phases is seen to be located around the one-wavelength circumference (shown as a thick dark area within the spiral). The amplitude of the current within this band is also quite uniform. Therefore, the radiation of a spiral antenna at each frequency is similar to that of a one-wavelength loop, except for the rotating excitation point in the spiral causing circular polarization.

The slower the growth rate, the better the axial ratio due to better rotational symmetry with the band. Usually, the amplitude of the current becomes highly attenuated outside the band due to loss from radiation. If the growth rate is slow, there is no need for a termination (resistor) at the spiral ends. As the operational frequency varies, the size of the band varies accordingly. Therefore, the lower limit of the operational frequency is determined by the spiral diameter. On the other hand, the upper frequency limit is determined by the arm details (tighter arms) near the centre region.

Spiral radiation can be best described by considering the phase along each spiral arm for a two-arm spiral, as shown in Figure 4.4. Therefore

$$\Psi^{1,2} = \Psi_0^{1,2} - \beta_e s = \Psi_0^{1,2} - \frac{2\pi}{\lambda_e} s \tag{4.9}$$

where $\Psi^{1,2}$ refers to the phase along each of the arms (arm1 and arm2), $\beta_e = 2\pi/\lambda_e$ is the effective propagation constant along the spiral, and s is the distance from the centre. For a balanced two-arm spiral feeding, $\Psi_0^1 = 0$ and $\Psi_0^2 = \pi$ must be set, as shown in Figure 4.4.

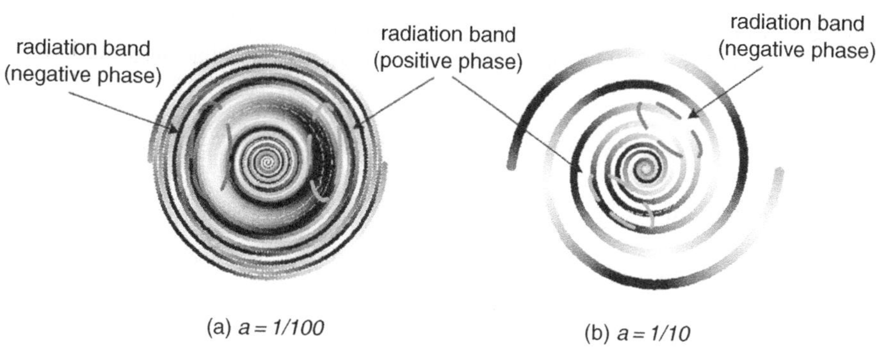

Figure 4.3 Phase of currents flowing on an infinite two-arm equiangular spiral antenna with tight winding. The greyscale indicates the absolute value of phase from 0° (black) to 180° (white).

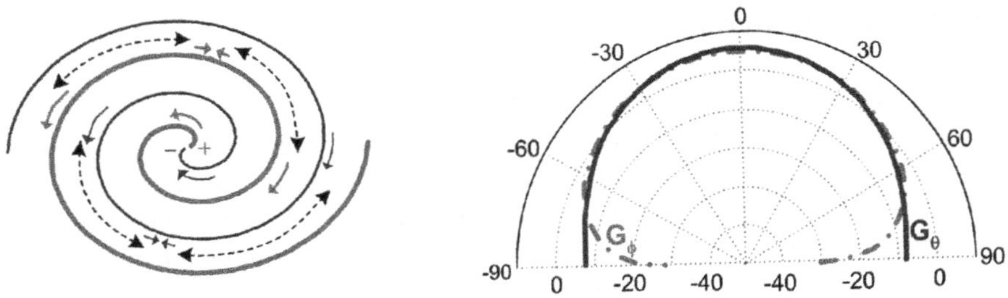

Figure 4.4 Typical normal mode radiation pattern of a two-arm spiral antenna with out-of-phase excitation.

Referring to Equation (2) and substituting for s into Equation (4.8) gives

$$\begin{aligned} \Psi^1(\phi) &= \Psi_0^1 - 2\pi \frac{e^{a\phi} - 1}{a\lambda_e} \\ \Psi^2(\phi) &= \Psi_0^2 - 2\pi \frac{e^{a(\phi-\pi)} - 1}{a\lambda_e} \end{aligned} \tag{4.10}$$

where the $\phi - \pi$ exponent in Equation (4.10) is due to the half-winding difference between the two arms. To have constructive phases from the two arms, it is required that

$$\Psi^2(\phi) - \Psi^1(\phi) = 2m\pi, \quad m = 0, \pm 1, \pm 2, K \tag{4.11}$$

or

$$2\pi e^{a\phi} \frac{1 - e^{-a\pi}}{a} = (2m - 1)\lambda_e \tag{4.12}$$

However, if the growth rate is slow (small a), this reduces to

$$2\pi\rho = (2m - 1)\lambda_e \quad \text{for } |a\pi| << 1, m = 1, 2, 3, \ldots \tag{4.13}$$

since $\rho = e^{a\phi}$ from Equation (4.2). When $m = 1$, the normal or first spiral mode is obtained, associated with the currents and pattern given in Figure 4.4.

When the feeding for the two-arm spiral is changed so that each arm is fed in phase, that is $\Psi_0^2 = \Psi_0^1$, the condition (4.11) for constructive radiation from adjacent arms leads to

$$2e^{a\varphi} \frac{1 - e^{-a\pi}}{a} = 2n\lambda_e \tag{4.14}$$

and for a slow growth rate

$$2\pi\rho = 2n\lambda_e \quad \text{for } |a\pi| << 1, n = 1, 2, 3, \ldots \tag{4.15}$$

is obtained

It is clear that for in-phase feeding, radiation occurs when the circumference is even multiples of λ_e. The resulting radiation pattern has a null at zenith (see Figure 4.5), which

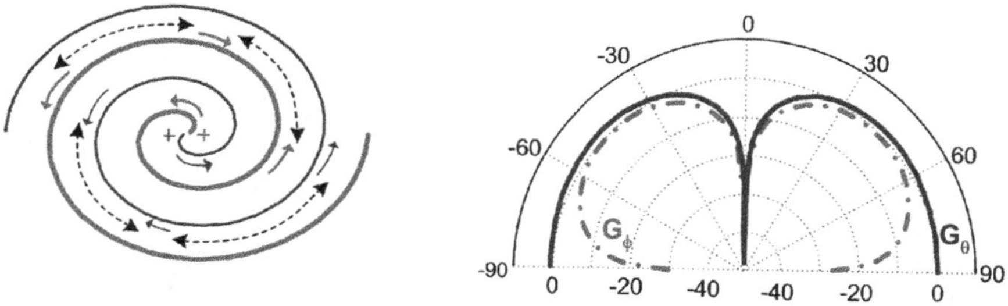

Figure 4.5 Typical even mode (null at zenith, $\theta = 0°$) radiation pattern of a two-arm spiral antenna with in-phase excitation.

is referred to as the second spiral mode. It can be seen that this pattern is particularly useful for applications where good horizon radiation is needed.

Although radiation from adjacent arms within the band are constructive, the total radiation along the normal direction has a null due to cancellations from the opposite side of the currents, which have similar phase but flowing in opposite directions. (In the normal mode, the currents have opposite phase and flow in opposite directions.) The even mode can also be excited with a single-arm spiral (shown in Figure 4.2) provided the phase relations (4.10) of adjacent arms are modified as

$$\begin{aligned} \Psi^1(\phi) &= \Psi_0^1 - 2\pi\frac{e^{a\phi}-1}{a\lambda_e} \\ \Psi^1(\phi-2\pi) &= \Psi_0^1 - 2\pi\frac{e^{a(\phi-2\pi)}-1}{a\lambda_e} \end{aligned} \tag{4.16}$$

Note that $\phi - \pi$ in the exponent has been replaced by $\phi - 2\pi$.

Expressions for the radiation patterns (see Figures 4.4 and 4.5) can be readily obtained by using the radiation integral once the arm currents are available from a numerical analysis tool [42, 43]. However, closed-form expressions for $\rho(\phi) = e^{a(\phi-\phi_0)}$ are available from Cheo *et al.* [34]. Specifically, from Reference [34] the E_ϕ (electric field) component (see Figure 4.2) is given by

$$E_\phi \approx E_0 k_0^3 A(\theta)\, e^{j[n(\phi+\pi/2)-\psi(\theta)]}\frac{e^{-jk_0 r}}{r} \tag{4.17}$$

where $k_0 = 2\pi/\lambda_0$ is the free-space propagation constant with λ_0 being the corresponding wavelength, E_0 is a constant, and m is associated with the radiation mode (first or second mode, see Equations (4.13) and (4.15)). The pattern and phase parameters, $A(\theta)$ and $\psi(\theta)$, are given by

$$\begin{aligned} A(\theta) &= \frac{\cos\theta\,(\tan\theta/2)^m\, e^{m/a\ \tan^{-1}(a\cos\theta)}}{\sin\theta\sqrt{1+a^2\cos^2\theta}} \\ \psi(\theta) &= \frac{m}{2a}\ln\left|1+a^2\cos^2\theta\right| + \tan^{-1}(a\cos\theta) \end{aligned} \tag{4.18}$$

The computed $A(\theta)$ patterns for different growth rates ($a = 0.1, 1/3, 1, 10$) are given in Figure 4.6. As expected, the patterns broaden with increasing a to the point where it loses its typical single-lobe form.

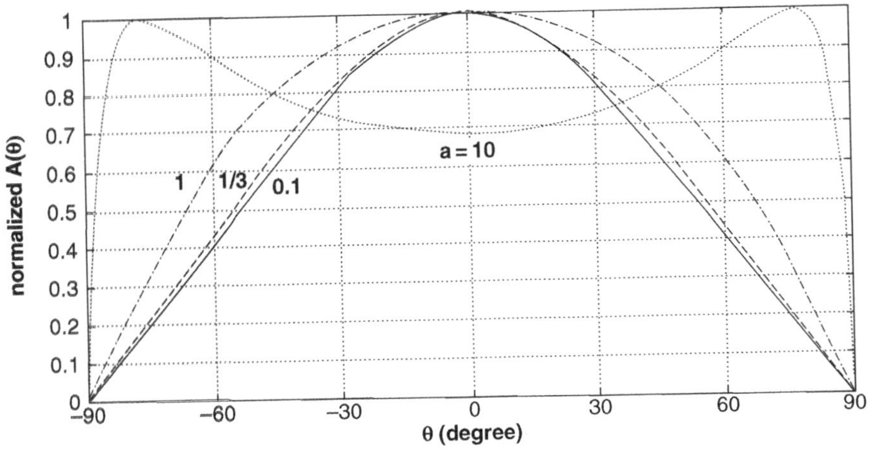

Figure 4.6 Equiangular spiral radiation patterns for different growth rates: $a = 0.1, 1/3, 1$, and 10.

This section will close by noting that the above analysis assumes a nonvarying current propagation constant along the spiral arms. However, given that the spiral behaves as a leaky antenna at certain frequencies, k_{eff} is not always constant over the radiation section of the spiral. In that case, the phase expression in Equation (4.9) needs to be revised to read

$$\Psi^i = \Psi_0^i - \int_{\substack{\text{spiral}\\ \text{length}}} k_{\text{eff}}\,(s')\,\mathrm{d}s' \tag{4.19}$$

4.2.3 Input Impedance

The input impedance of an infinite equiangular spiral antenna with tight windings in free space is purely real and frequency independent. Ideally, currents are excited from the centre of a spiral antenna and flow outwards along the spiral arms. While travelling, they radiate at the resonance ring without reflections. The latter implies that the input impedance at the excitation port is the same as that of the equivalent transmission lines formed by the spiral arms [47] at high frequencies, where the impedance depends on the geometry layout including the conductor-to-air ratio and winding shape. For instance, a square spiral antenna tends to produce more impedance variation due to reflections and diffractions from corners.

Practical spiral antennas are truncated at some radius and thus the input impedance will also be a function of the growth rate and frequency. However, since the currents decay rapidly beyond the resonant rings in a tightly wound spiral due to radiations when the spiral aperture is 5/4 times the radius of the resonant ring, the input impedance is not affected by reflections from the truncation. This suggests a constant resistance, particularly when the wavelength is shorter than 4/5 of the maximum circumference length. At lower frequencies, where the spiral size is not large enough to support a resonant ring (and convert all currents to radiations), a significant proportion of the arm currents are reflected from the spiral ends. Such reflections introduce opposite polarization radiation as they flow inwards towards the centre of the spiral and cause a frequency-dependent impedance with both resistive and

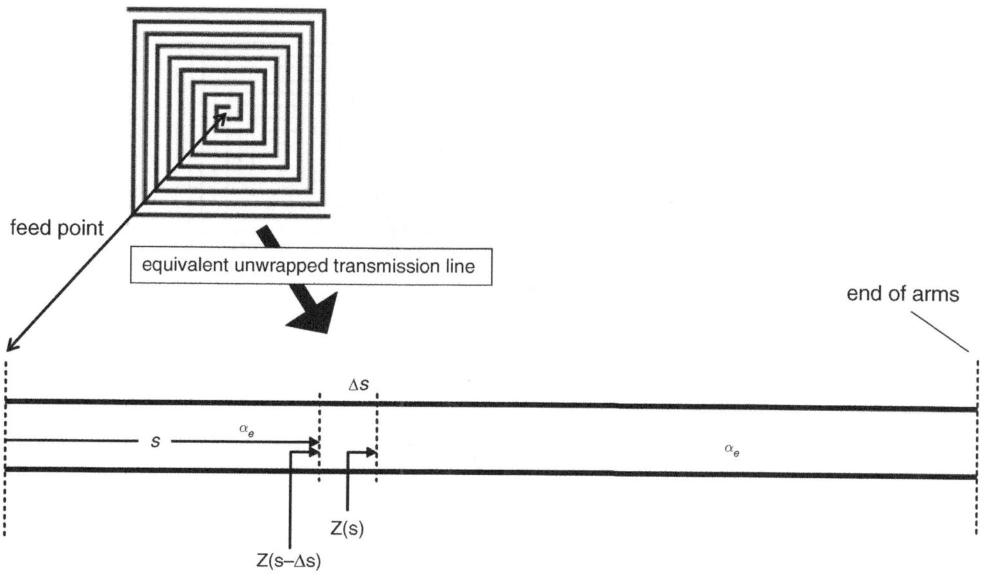

Figure 4.7 Equivalent transmission-line model of spiral antennas.

reactive components. An equivalent transmission-line model can be used to model this situation, as discussed later (see Figure 4.7). The equivalent phase velocity and impedance needed for this model can be obtained from the line capacitance and inductance, given by

$$Z = G\sqrt{\frac{\mu}{\varepsilon}} = \sqrt{\frac{L_{\text{eff}}}{C_{\text{eff}}}} \tag{4.20}$$

$$v = \frac{1}{\sqrt{\mu\varepsilon}} = \frac{1}{\sqrt{L_{\text{eff}}C_{\text{eff}}}} \tag{4.21}$$

Figure 4.8 plots the input impedance for a square (unterminated) Archimedean spiral antenna 5 cm × 5 cm in size for three near-complementary conductor-to-air ratios calculated using HFSS (the dark regions on the spirals indicate the perfect electrical conductor (PEC) section of the spiral). The high-frequency region is indicated in the shadowed region. As expected, above 1.85 GHz when the aperture circumference (20 cm) is approximately 5/4 times of the wavelength (16.2 cm), the impedance becomes mainly resistive and constant. It is also noted that the resistance increases with the conductor-to-slot ratio. The impedance among complementary geometries is of course expected to satisfy the following well-known relationship [48], $Z_{\text{sc}}^2 \approx Z_{\text{s}}Z_{\text{w}}$, where the subscripts 'sc', 's', and 'w' stand for 'self-complementary', 'slot', and 'wire', respectively. However, $Z_{\text{sc}}^2 \neq \eta_0^2/4$ due to spiral geometry truncation. Clearly, the rapid impedance variations at low frequencies are caused by the strong spiral end-reflections and impedance transformation over a long transmission line from the spiral end to its centre.

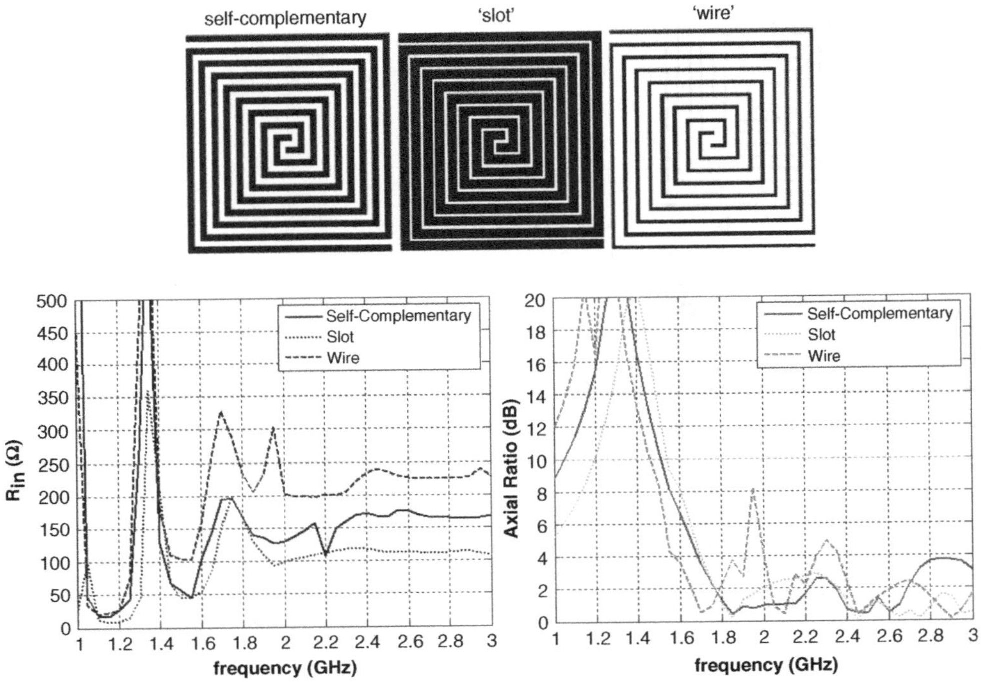

Figure 4.8 Antenna impedance for square Archimedean spiral antennas.

4.2.4 Radiation Centre

In many applications such as arrays and reflector feeds, it is important to know the location of the spiral's radiation centre. This is particularly so for structures placed on platforms, where a ground-plane is present. For the spiral antenna, it is even more important because the phase behaviour described in Equation (4.18) indicates that the phase centre varies with the pattern angle and growth rate. Focusing on the $m = 1$ mode (normal mode), the antenna's rotational symmetry implies that the phase centre is below the z (vertical) axis, as shown in Figure 4.9. Accordingly, it can be assumed that the phase of the far-zone radiated field will be of the form

$$E \approx B(\theta)\mathrm{e}^{-\mathrm{j}\psi(\theta)} \frac{\mathrm{e}^{-\mathrm{j}k_0 r}}{r} \tag{4.22}$$

in which

$$\psi(\theta) = k_0 d \cos\theta \tag{4.23}$$

The phase centre's distance from the aperture is then found by equating (4.18) and (4.23), giving

$$\frac{d(\theta)}{\lambda} = \frac{[1/(2a)]\ln\left|1 + a^2\cos^2\theta\right| + \tan^{-1}(a\cos\theta)}{2\pi\cos\theta} \tag{4.24}$$

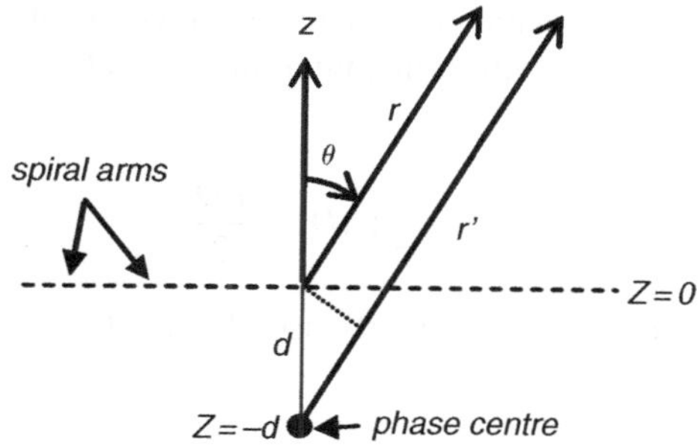

Figure 4.9 Determination of the equivalent phase centre location.

From Equation (4.24), it is clear that the radiation centre is located behind the spiral plane and, in general, is a function of the pattern angle, θ, and growth rate, a. Figure 4.10 plots the normalized phase centre location as a function of θ for four different growth rates ($a = 0.01, 0.1, 1, 2$). It is observed that d becomes independent of the pattern angle when the growth rate is less than unity. Specifically,

$$\frac{d}{\lambda} \approx \frac{a}{2\pi} \quad \text{for } a << 1 \tag{4.25}$$

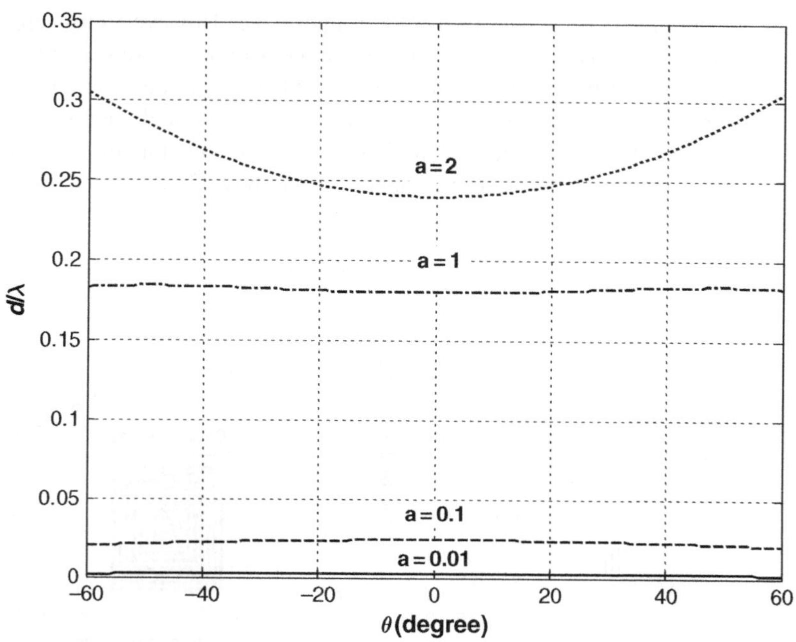

Figure 4.10 Normalized distance of the radiation phase centre located behind an infinite equiangular spiral.

Using resonant band theory and Equation (4.25), the maximal radiation centre movement for a finite equiangular spiral antenna with a maximal diameter D can be estimated using the equation

$$d_{\max} = \frac{a}{2\pi}(\pi D) = \frac{aD}{2} \tag{4.26}$$

Knowledge of the maximal offset is particularly useful when the spiral antenna is used as a reflector feed.

4.3 EFFECT OF GROUND-PLANE BACKING

Many applications require antennas to be of low profile and mounted on electrically conducting surfaces such as aircraft, automobiles, or unmanned vehicles. In these cases, the separation between the antenna and the metallic backing is very small (less than $\lambda/20$). Such an electrically conducting backing greatly affects the pattern, impedance, and bandwidth of the spiral. Of course, the spacing between the antenna and the ground-plane plays a crucial role. At high frequencies, where the antenna size is large enough to establish resonances and the electrical height above a PEC ground-plane is not too small (i.e. $C > 4\lambda_0/3$ and $h > 0.3\lambda_0$, with C being the antenna's circumference and λ_0 the operational wavelength), the ground-plane effect can be described by the interference between the spiral and its image (assuming negligible mutual coupling). The resultant boresight gain varies according to

$$G_{\text{on ground plane}} = G_{\text{free space}} \left|1 - e^{-j2\pi\, 2h/\lambda}\right|^2 \tag{4.27}$$

where h refers to the antenna distance from the ground-plane. However, most applications require low height profiles, making such a large height undesirable. Therefore, it is important to understand the impact of the ground-plane. Figure 4.11 serves to demonstrate the impact of a nearby PEC ground-plane on antenna radiation performance, namely boresight gain. The shown spiral antennas refer to two nearly complementary wire and slot spiral antennas each 2 inch × 2 inch in size. For such a size, resonances exist above 2 GHz (since $\lambda_0 \sim 2$ inch)

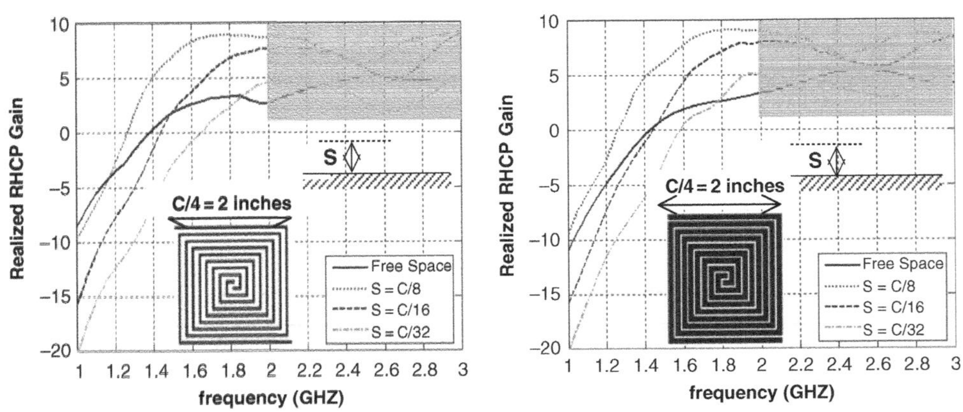

Figure 4.11 Effects of an infinite PEC ground-plane on the realized gain (including mismatch loss) of a square Archimedean wire spiral (left) and a slot spiral (right).

and thus the gain levels off after 2 GHz (see the shaded region in the graph). Since the lowest resonant wavelength is associated with the circumference (i.e. $\lambda_{min} \cong 3C/4$), it is convenient to normalize the height h with respect to C. For the chosen cases, the heights correspond to $\lambda_{min}/6$, $\lambda_{min}/12$, and $\lambda_{min}/24$ at 2 GHz. The characteristic impedance of the feeding lines is set to 225 Ω for the wire spiral case and 110 Ω for the slot spiral case. Thus, in each case, the strip line and the slot line impedances are matched to the antenna resistance at high frequencies. The gain comparisons for the slot and wire spirals are given in Figure 4.11. As seen, the gain is below 0 dBic for frequencies lower than 1.5 GHz and decreases significantly, particularly when the distance from the ground-plane is $S = C/16(\lambda_{min}/16$ at 1.5 GHz). The gain also reduces at higher frequencies as the distance from the ground-plane drops below $\lambda_{min}/24(S = C/32)$.

The gain reduction discussed above is primarily due to the excitation of nonradiating transmission-line modes between the spiral conductors and the ground-plane. This can also be explained by using image theory as illustrated in Figure 4.12. When the separation becomes less than $\lambda_{min}/16$, the strong coupling between the spiral and its image excites a nonradiating transmission-line mode and as the height continues to decrease (in terms of λ_0) the antenna structure becomes an open-ended transmission line that ceases to radiate. This also leads to significant mismatches caused by strong reflections from the spiral ends, as can be seen by the oscillatory behaviour of the input impedance curves depicted in Figure 4.13 (as compared to a spiral without a backing). Clearly, the ground-plane's presence leads to

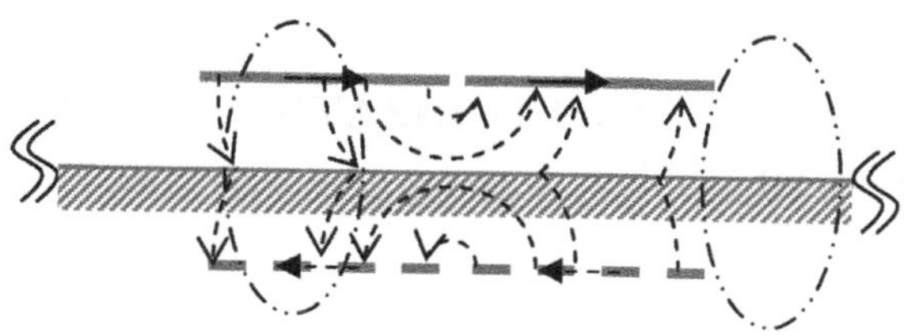

Figure 4.12 Graphical display of the undesired transmission-line mode excited between the spiral antenna and the nearby PEC ground-plane.

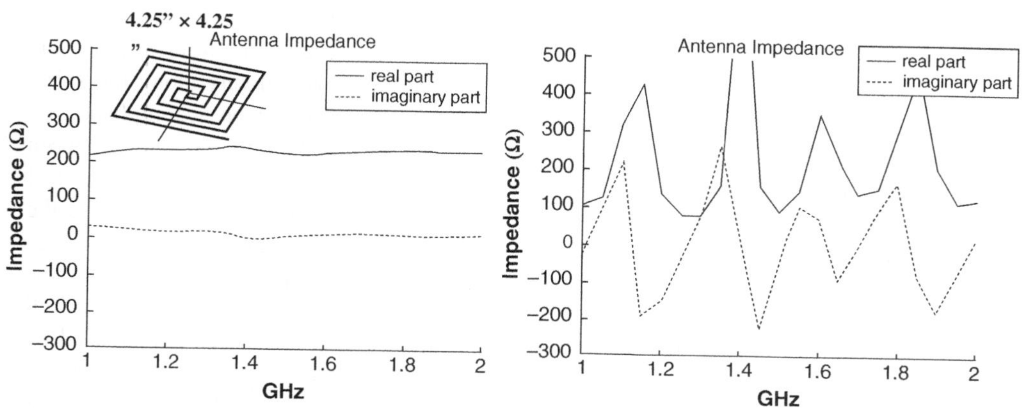

Figure 4.13 Input impedance of a spiral antenna in free space (left) and 1 inch above an infinite PEC ground-plane (right).

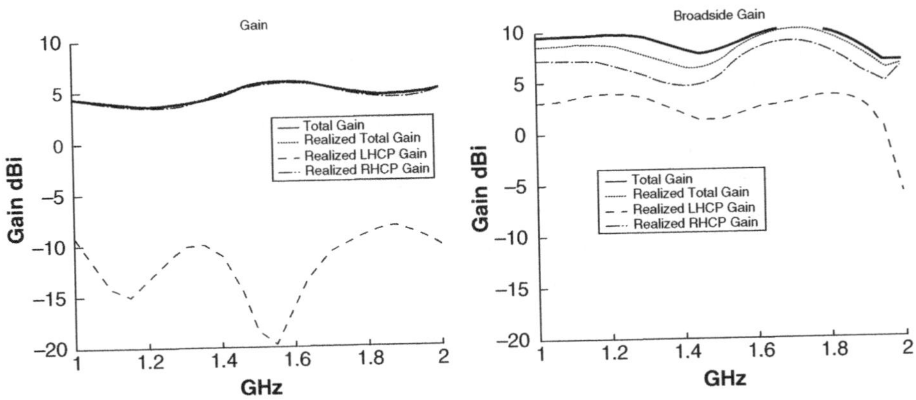

Figure 4.14 LHCP gain for an otherwise RHCP spiral antenna in free space (left) and 1 inch above an infinite PEC ground-plane (right).

oscillations similar to those obtained from a shorted or open-ended transmission line. The strong reflected currents also generate radiation of the opposite polarization sense, resulting in a high level of LHCP gain (see Figure 4.14). Termination of the spiral arm with a resistor of value equal to the characteristic impedance of the spiral will reduce cross-polarization at the expense of efficiency.

4.4 SPIRAL ANTENNA MINIATURIZATION USING SLOW WAVE TREATMENTS

4.4.1 Introduction

Antenna miniaturization techniques using dielectric or reactive loading [49–56] are commonly used to increase the antenna's electrical size without a corresponding increase in its physical size. However, each of these miniaturization techniques by itself faces important performance trade-offs, especially when large miniaturization factors are pursued. In this section, a hybrid approach is presented that involves both dielectric and reactive loading to modify the effective inductance and capacitance of the equivalent transmission line comprising the spiral. Specifically, the aim is to reduce the wave velocity along the spiral arms while retaining a (constant) desirable line impedance. This can be accomplished by increasing the effective capacitance and inductance while maintaining the ratio between them so that the impedance remains unchanged. The present approach involves the use of dielectric material on both sides of the antenna (substrate and superstrate) to increase the effective capacitance of the antenna structure while maximizing the effect of the dielectric material for a given dielectric constant and thickness. Concurrently, the thickness of the dielectric material is tapered to suppress dielectric resonances and surface waves as well as maintain high-frequency performance [47, 53]. The effective inductance is increased by converting the spiral arms into inductive arms [35] using a two-dimensional meander line or three-dimensional conical coils. The former is easier to fabricate but the latter achieves greater miniaturization. To improve performance at low frequencies further, a combination of PEC and ferrite is used as a ground-plane (ferrite in the outer region and PEC in the

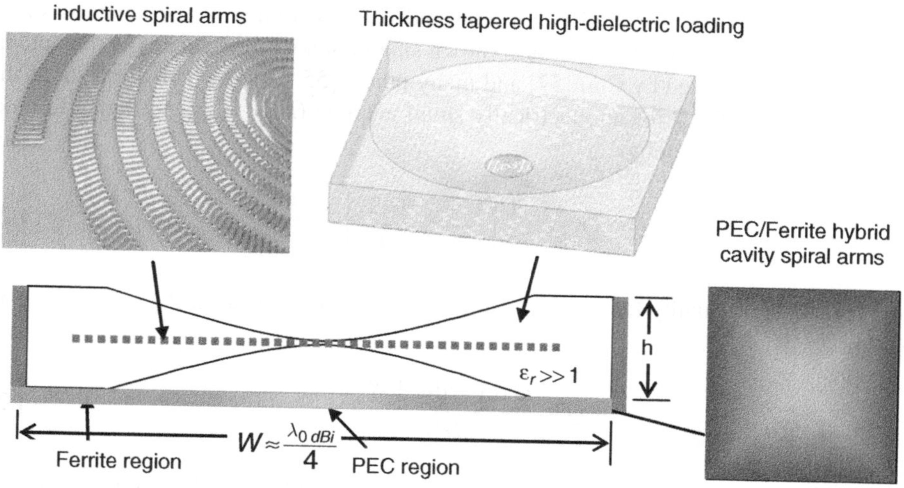

Figure 4.15 Novel miniature spiral antenna design concept.

inner). This hybrid backing serves to suppress the transmission-line mode between the spiral arms and the PEC ground. Figure 4.15 displays the conceptual configuration of the proposed miniature spiral antenna design.

The gain performance of the miniaturized spiral antenna shown in Figure 4.15 can be estimated from three specific points corresponding to the total realized gain of −15, 0, and 5 dBi, using the empirical formulae shown in Table 4.1. In these formulas, W is the square spiral antenna diameter in inches, M_{ATL} is the phase velocity reduction factor due to reactive loading (or inductive arm), and ε_{e} is the effective dielectric constant which is approximately equal to the dielectric constant for double-sided loading. It can be remarked that the product of $\varepsilon_{\mathrm{e}}^{0.2}$ and $M_{\mathrm{ATL}}^{0.4}$ is defined as the miniaturization factor. More details about each component in this design will be addressed later and a design example under a size constraint of 6 inches will be discussed.

4.4.2 Limit of Realized Gain at Low Frequencies (Below Resonance)

Before miniaturization is discussed, it is important to understand some fundamental limitations related to miniature antennas. In doing so the usual size parameter, ka, will be used,

Table 4.1 Predicted gain performance for the proposed miniaturization design.

Total gain (dBi)	Frequency (MHz)
−15	$\approx \dfrac{800}{\varepsilon_{\mathrm{e}}^{0.2} \times M_{\mathrm{ATL}}^{0.4}} \times \dfrac{2}{W(\text{inches})}$
0	$\approx \dfrac{c}{4W}$
+5	$> \dfrac{c}{2W}$

where k is the free-space propagation constant and a is the radius of the smallest sphere enclosing the entire antenna. The first well-known small antenna limit relates to the antenna quality factor, Q, introduced by Chu [57] and many others [55, 56, 58]. This limit states that the minimum achievable Q for an electrically small antenna ($ka << 1$) with a single mode (TE or TM) excited is given by

$$Q_{\min} = \frac{1}{2}\left[\frac{1}{(ka)^3} + \frac{2}{ka}\right] \tag{4.28}$$

However, the antenna quality factor is related to the antenna impedance, $Z_A = R_A + jX_A$, via the relation

$$Q \approx \frac{\omega \, dX_A/d\omega \pm X_A}{2R_A} \tag{4.29}$$

By treating the reactance of a small antenna like a capacitor or an inductor, this can be further simplified to give

$$Q \approx \frac{|X_A|}{R_A} \tag{4.30}$$

Therefore, the antenna impedance here becomes

$$Z_A = R_A + jQR_A \tag{4.31}$$

This implies that the imaginary part is more dominant in miniature antennas since Q is large for small antennas. As will be shown, the magnitude of the real part decreases as the degree of miniaturization increases. It is therefore difficult to attain complex impedance matching (4.31) over a wide bandwidth. However, R_A is typically small and slowly varying. Therefore, it is possible to match the real part of Z_A and, if so, the power mismatch loss is given by

$$|\Gamma|^2 = \frac{Q^2}{Q^2 + 4} \tag{4.32}$$

Combining Equations (4.28) and (4.32) leads to

$$|\Gamma_{\min}|^2 = \frac{4(ka)^4 + 4(ka)^2 + 1}{16(ka)^6 + 4(ka)^4 + 4(ka)^2 + 1} \tag{4.33}$$

Equation (33) can be used to determine the maximum achievable realized gain from

$$\max(G_{\text{realized}}) = G\left(1 - |\Gamma_{\min}|^2\right) \tag{4.34}$$

where G is the directive gain. Using Equations (4.33) and (4.34), the maximum achievable realized gain is plotted in Figure 4.16 along with the curve for a typical spiral matched to the characteristic impedance of the equivalent transmission line. These results indicate a maximum size reduction of 64 % if the feeding line is matched to the radiation resistance.

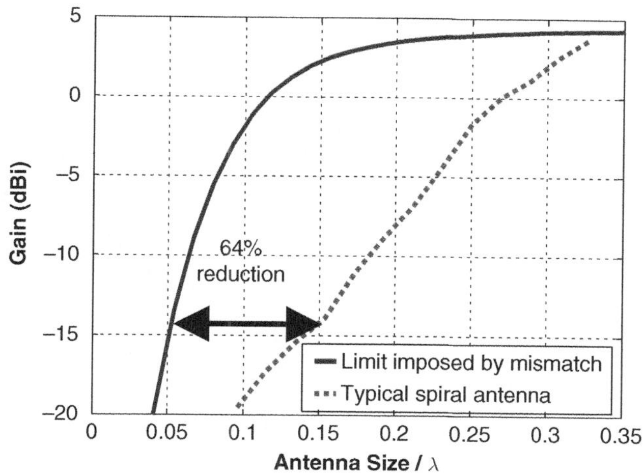

Figure 4.16 Maximum achievable realized gain for a miniature antenna with the feeding network matched to the real part of the antenna impedance.

4.4.3 Limit of Realized Gain Limits at High Frequencies (Above Resonance)

Now consider the effects of miniaturization on gain at higher frequencies where the antenna is resonant when unloaded. As an example, consider a circular loop whose directivity is first evaluated (see Figure 4.17) when its circumference is one wavelength in free space.

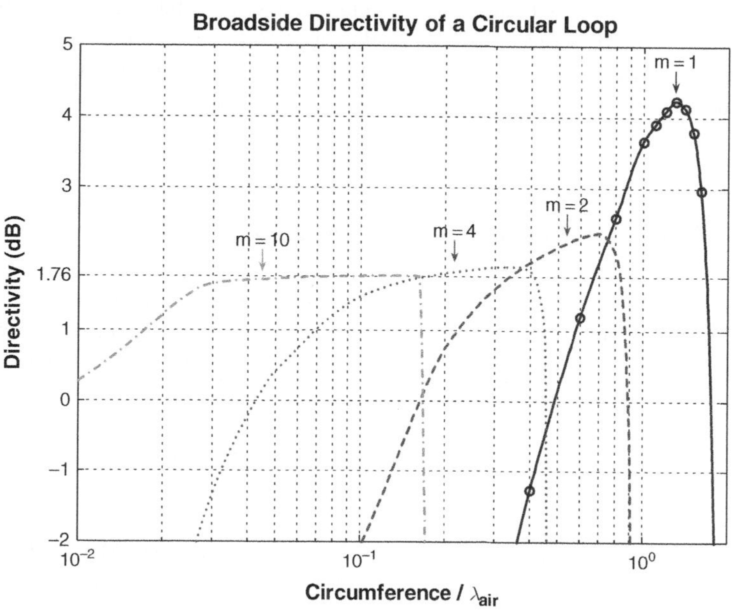

Figure 4.17 Directivity of a miniaturized one-wavelength circular ring antenna.

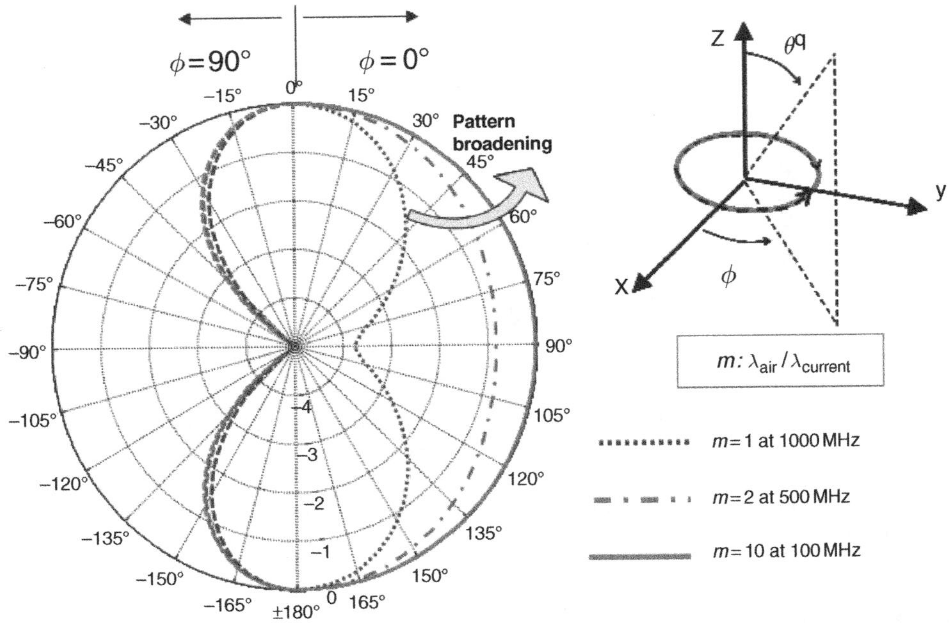

Figure 4.18 Patterns corresponding to the miniaturized circular loop antenna.

This corresponds to the case of $m = 1$ in Figure 4.17. Subsequently, the wave velocity is artificially slowed down by a factor m while concurrently scaling the circumference of the loop by the same factor. Thus, the loop can be considered at resonance while its size is reduced, a situation that occurs during miniaturization. Figure 4.17 shows the directivity of the loop miniaturized by a factor of 2, 4, and 10. As seen, the directivity approaches a limiting value of 1.76 dB, corresponding to the directivity of an ideal dipole. Looking at the corresponding patterns shown in Figure 4.18, it can be concluded that the directivity decrease is due to pattern broadening as the antenna size approaches that of an electrically small loop.

4.4.4 High Refractive Index Material Loading

Antenna miniaturization can be achieved by introducing high-contrast dielectric material in the near-field region. The effectiveness of the dielectric loading depends on how the fields are trapped/contained within the dielectric material. If the antenna was loaded with an infinite half-space or full-space, the effective dielectric constant is given by

$$\varepsilon_e = \begin{cases} (\varepsilon_r + 1)/2, & \text{for an infinite half-space} \\ \varepsilon_r, & \text{for an infinite full-space} \end{cases} \tag{4.35}$$

These values provide the maximum achievable miniaturization (MAM) for single-side or double-side loading cases. In practice, only finite thickness loading is applied (see Figure 4.19). Therefore only a fraction of the MAM is obtainable depending upon the configuration of the loading such as the thickness, width, shape, and so forth. Specifically,

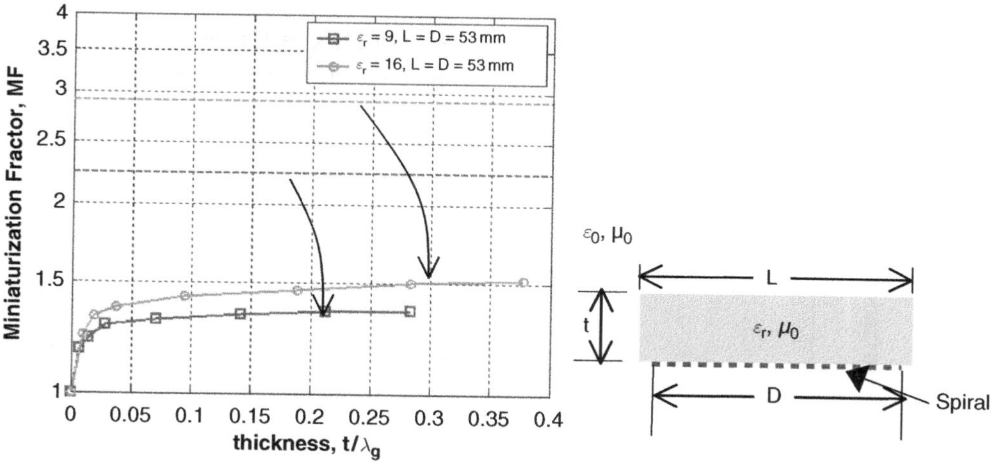

Figure 4.19 Impact of superstrate loading on miniaturization. Plot of MF versus thickness for the superstrate for $\varepsilon_r = 9$ and $\varepsilon_r = 16$.

the combination of the dielectric loading and the distribution of the antenna near-fields determines the effective dielectric constant and, therefore, the equivalent miniaturization. The following discussion focuses on the miniaturization effect of dielectric loading on spiral antennas. The -15 dBi realized gain point will be chosen as a reference point in evaluating the amount of frequency shifting and the miniaturization factor (MF) will therefore be defined as

$$\mathrm{MF} = \frac{f_{\text{unloaded},-15\text{dBi}}}{f_{\text{loaded},-15\text{dBi}}} \tag{4.36}$$

Note that this MF is different from the phase velocity reduction (i.e. v/c). Rather, this MF definition includes all impedance mismatches and pattern broadening effects, and is thus a more practical and useful parameter.

Consider a square printed Archimedean spiral antenna loaded with a finite dielectric slab as a superstrate (see Figure 4.19). The width of the spiral is designated as $D = 53$ mm (or 2 inches) and the superstrate width and thickness are denoted as L and t respectively. The dielectric slab is also assumed to be lossless and has a relative permittivity ε_r. Full-wave simulations were carried out to compute the realized gain with the feeding-line impedance matched to the antenna resistance at high frequencies, where the gain is almost constant. The superstrate width L is chosen to be equal to the antenna diameter D. The thickness of the slab is then varied for two cases with $\varepsilon_r = 9$ and $\varepsilon_r = 16$. The resulting MF curves are shown in Figure 4.19 as a function of normalized thickness with respect to the guided wavelength defined as $\lambda_g \equiv \lambda_0/\sqrt{\varepsilon_r}$. The horizontal dashed lines represent the MAM for a dielectric constant of $\varepsilon_r = 9$ and $\varepsilon_r = 16$ if half-space loading is applied. Initially ($t < 0.02\lambda_g$, where λ_g is the guided wavelength in the dielectric), the MF increases rapidly for both cases. However, this increase slows down significantly as the thickness is increased further ($t > 0.1\lambda_g$) and stops well short of the MAM in each case. It is apparent from this figure that a further increase in thickness will result in diminishing returns. This is primarily because the current superstrate width is unable to capture the near-fields surrounding the sides of the spiral antenna.

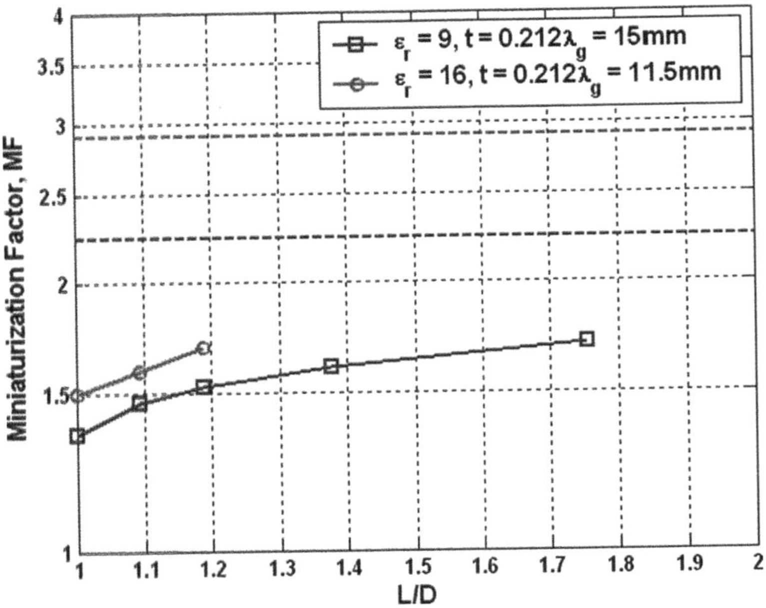

Figure 4.20 Impact of superstrate width on miniaturization.

The effect of the superstrate width L is examined next. Figure 4.20 shows the resultant MF curves for different width ratios, L/D, between the superstrate and the spiral antenna superstrate for a fixed thickness of $0.212\lambda_g$. These results clearly show that the MF increases with the width of the superstrate, thereby confirming the assumption that low-frequency fields extend outside the aperture of the antenna. It can therefore be inferred that the maximum MF for any dielectric constant is achievable as long as the fields of the antenna are loaded sufficiently. However, sufficient loading of the antenna fields is not practical since it equivalently increases the overall antenna size. Therefore, for practical implementation, the width of the slab must be constrained by the diameter of the antenna.

It is of course expected that the combined superstrate and substrate dielectric loading will result in larger MF values. The issue is whether a distribution of the dielectric (equally) between the superstrate and substrate will result in a larger MF than a superstrate of equal total thickness. To answer this question, both sides of the spiral antenna were loaded with a square slab of thickness $t/2(L=D)$ with $\varepsilon_r=9$. The results were then compared to those obtained from loading only a single side with a slab thickness of t, shown in Figure 4.21. It is evident that for the same total thickness a larger or equivalent MF is achieved when both sides of the antenna are loaded. Therefore, it can be concluded that loading both sides of the antenna will always produce the same or more miniaturization for the same total thickness.

Dielectric material loading for the purpose of spiral miniaturization has its limits. Specifically, while the low-frequency gain is usually improved by dielectric material loading the high-frequency gain tends to decrease if high contrast material is used. To demonstrate this, the choice was made to simulate a four-arm spiral antenna (2 inches wide and 0.5 inch high above an infinite ground-plane) with the dielectric material sandwiched between the antenna and the ground-plane (the latter being the same size as the antenna).

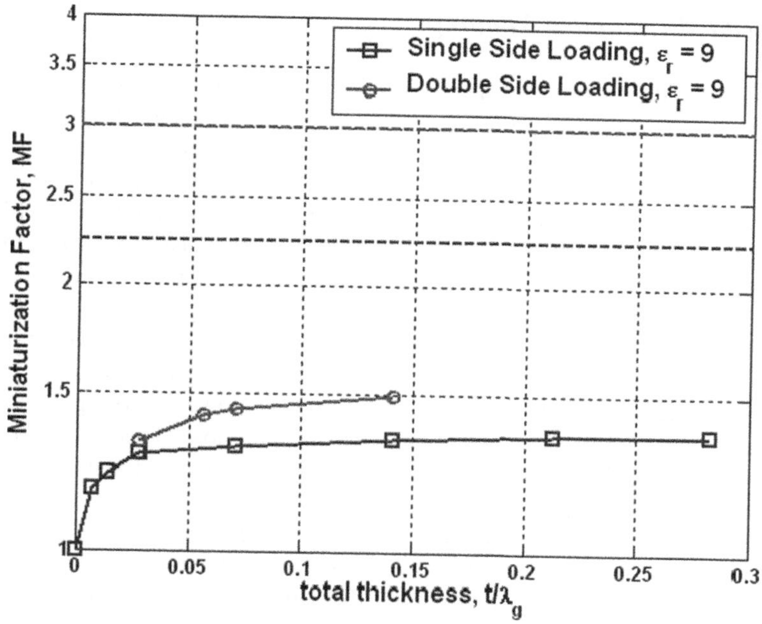

Figure 4.21 Comparison of the MF corresponding to single-sided and double-sided loadings for the same total thickness.

To examine the performance, the broadside circular-polarized gain was extracted at two different frequencies and plotted as a function of the dielectric constant (Figure 4.22). It can be seen that an optimum value of the dielectric constant exists for the material, above which the high-frequency gain begins to decrease.

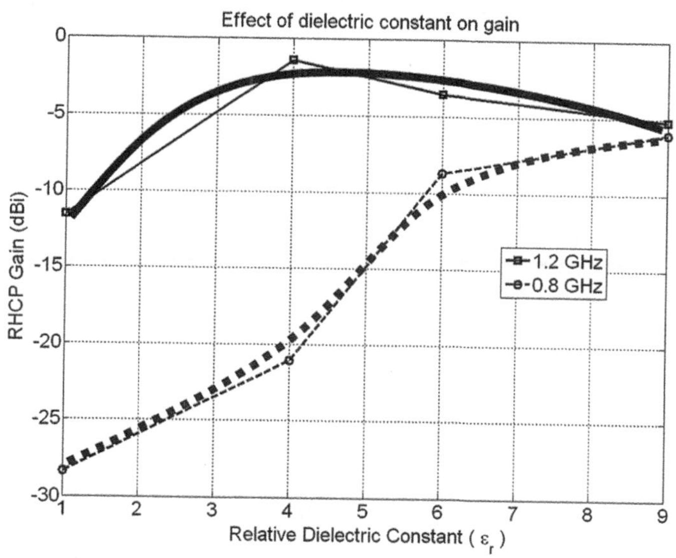

Figure 4.22 Effect of the dielectric constant on spiral antenna gain.

4.4.5 Embedded Inductive Loading

Having discussed miniaturization using dielectric loading, inductive loading will now be considered. Based on the transmission-line model for the spiral antenna, it will now be miniaturized by means of reactive loading, as demonstrated in Reference [35]; that is the phase velocity of the guided wave will be reduced by enhancing the inherent inductance and capacitance of the spiral antenna structure. This allows for an additional wave slow-down with a negligible increase in weight. In Reference [35], this was accomplished using lumped surface-mount inductors placed in series with the spiral arm and lumped surface-mount capacitors placed in parallel with the adjacent spiral arm. This approach works well as long as the number of lumped elements remains small so that the losses associated with soldering and the inductor's inherent resistance is kept to a minimum (there is no need for lumped capacitors since dielectric loading can serve this purpose). However, for aggressive loading the lumped elements lead to an excessive amount of ohmic loss. Therefore, implementing reactive or inductive loading using lumped elements is not practical and some alternative implementation is needed.

Another way to achieve inductive loading is through meandering of the spiral arm [48]. However, the meandering approach does not result in purely inductive loading. A better way to achieve inductive loading is by coiling the spiral arm such that the arm resembles a solenoid. This is accomplished by using standard PCB (printed circuit board) fabrication to create a coil by alternating a series of traces on the top and bottom layers of a board, as illustrated in Figure 4.23. In practice these traces are connected using vias to form square loops. The amount of inductive loading is then primarily a function of the number of turns (pitch) and the loop area (arm width × board thickness), which can be varied to achieve the desired miniaturization. Figure 4.23 shows an example of this type of loading for a 6 inch diameter square spiral. For this design the pitch is 60 mil, the trace width is 30 mils, and the via diameter is 20 mils. The board material is FR4 ($\varepsilon_r = 4$, $\tan\delta = 0.015$) with a thickness of 92 mils. This antenna also features multiple growth rates and a centre section that does not have the inductive treatment. Both of these features help to provide a transition from the untreated section to the heavily inductive section. Figure 4.23 shows a comparison of the measured gain

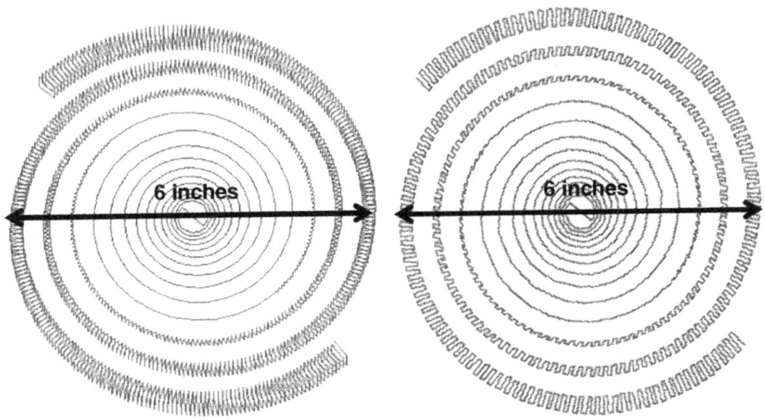

(a)

Figure 4.23 (a) Inductive treatments on a 6 inch diameter circular log wire spiral antenna using either three-dimensional conical-coil arms (left) or planar square meander-line arms (right). (b) The numerical electromagnet code (NEC) calculated RHCP gain comparison.

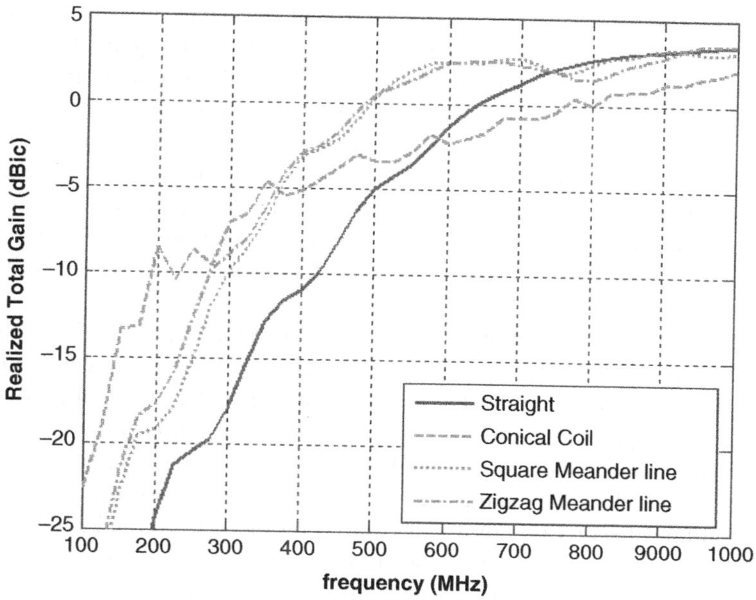

(b)

Figure 4.23 (Continued).

for the untreated and treated (inductive loading) 6 inch spirals. The spiral with the inductive treatment was able to achieve a gain of -15 dBic at 146 MHz compared to only 266 MHz for the untreated spiral. This is a considerable improvement in miniaturization and was achieved with a minimal increase in weight and profile. It should also be noted that the transition can be made more gradual to reduce reflections and thus improve gain between 250 and 500 MHz.

4.4.6 Hybrid PEC/Ferrite Backing

It is well known that the presence of a ground-plane behind an antenna affects the antenna radiation pattern and impedance. The extent of the effect is of course a function of the spacing between the antenna and the ground-plane, which is easily understood from image theory. As the distance between the ground-plane and antenna decreases, the coupling between them increases and eventually, as noted earlier, the antenna becomes a section of an open-ended transmission line and ceases to radiate. This leads to the unavoidable realized gain reduction at lower frequencies due to the increasing mismatch loss. To minimize these effects a large separation is required, usually $\lambda/4$ at the lowest operating frequency. However, such a large separation is impractical for most applications. Also, most applications require unidirectional radiation. Thus, eliminating the ground-plane or cavity is not possible. The ideal solution is to use a perfect magnetic conductor, which can be placed as close to the antenna as desired without negatively impacting on its performance. However, such a material does not exist and alternatives must be considered with the goal being to maintain the free-space antenna gain from the 0 dBi gain point and below.

As an alternative to the PEC ground-plane, a commercially available ferrite (TT2-111) from TransTech was used. The initial permeability at 5 MHz is $\mu_r = 270$, which decreases to unity near 1 GHz. Also, the dielectric constant and loss tangent measured at 9.4 GHz are

12.5 and less than 0.001, respectively. The ferrite backing was made by using seven 1 inch × 6.5 inch × 0.25 inch bars to from a 6.5 inch × 7 inch ferrite tile. The gain of a 6 inch square spiral was then measured using the ferrite tile as a backing with separations ranging from 0.5 to 2 inches. Figure 4.24 shows the measured gain for a separation of 1 inch, which provided the best trade-off between a low profile and a good low-frequency gain. For this spacing the ferrite tile allows recovery of the free-space gain up to 0 dB. However, the ferrite tile did not provide an adequate performance at higher frequencies when compared to the free-space and metallic ground-plane PEC backed spirals. Of course, this can be improved by increasing the separation but is not an option. Therefore, it is desired to modify the ferrite backing to increase the gain above 700 MHz.

Since the ferrite backing performed well below 500 MHz and the PEC ground-plane above 800 MHz, the logical way to improve the ferrite backing is by combining it with the ground-plane; that is the centre of the ferrite should be coated with a highly conductive material while the periphery remains unchanged. In addition, a transition region is needed between the ferrite and conductive material to minimize scattering that might occur at the transition. The conductive coating was designed by first choosing a shape that was subsequently optimized based on constraints placed on the −15, 0, and 5 dB gain points. The shape of the conductive patch was chosen to be a circular patch with serrated edges. The inner and outer radii as well as the number of serrations were obtained via the optimization process using genetic algorithms and pattern search methods [59]. An illustration of the final star patch design is shown in Figure 4.25. It has an inner radius of 1.5 inches and an outer radius of 5.2 inches with 27 serrations. The actual patch was fabricated using copper tape.

Figure 4.26 shows the measured gain for the 6 inch spiral with the hybrid backing. It is apparent that the hybrid surface provides a good compromise between the ground-plane

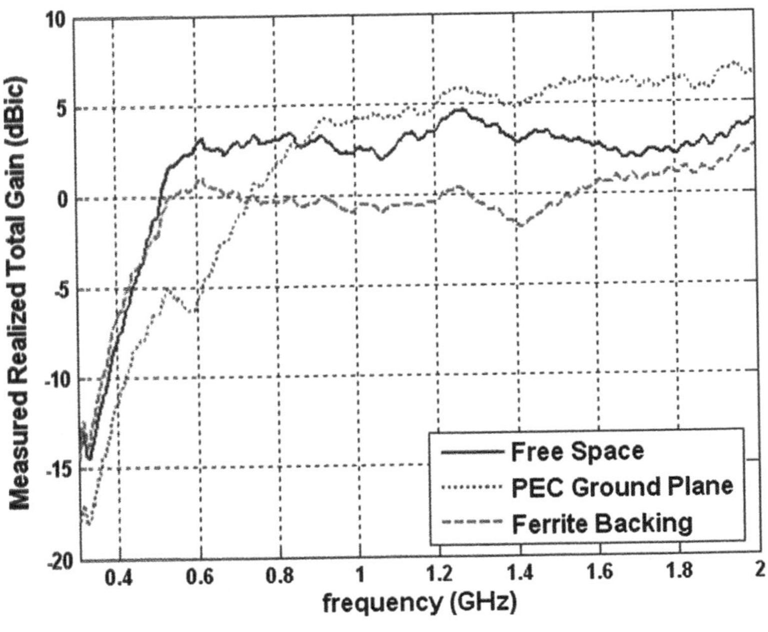

Figure 4.24 Measured realized gain for the 6 inch spiral antenna with ferrite and PEC backing (1 inch separation in all cases).

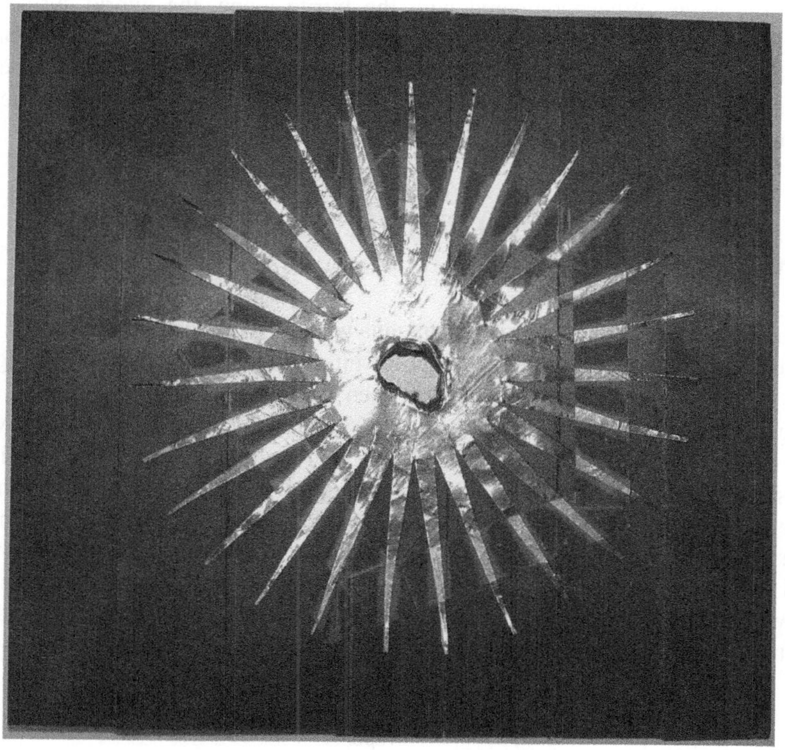

Figure 4.25 Illustration of the hybrid ferrite/PEC backing.

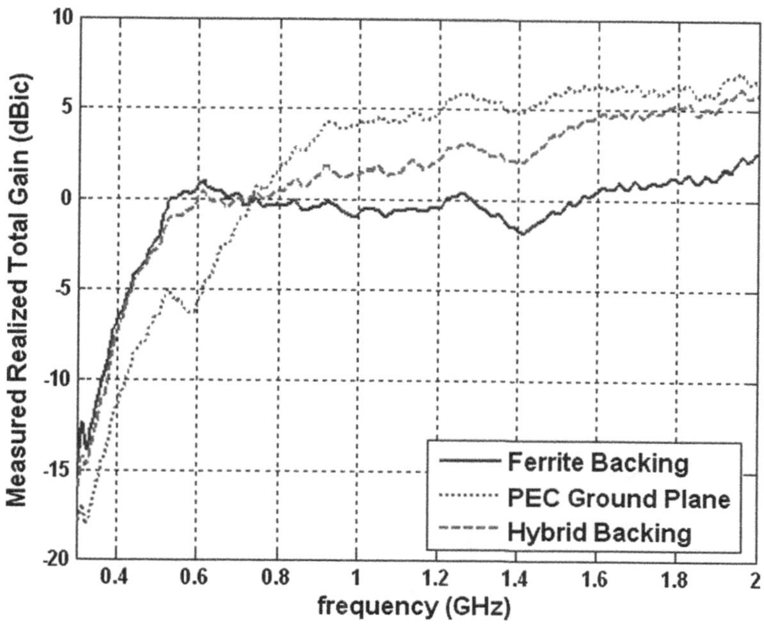

Figure 4.26 Measured realized gain for the 6 inch spiral antenna with hybrid backing (1 inch separation in all cases).

and ferrite tile alone. When compared with the ferrite alone, the hybrid surface is able to recover the free-space gain almost up to the 0 dBi gain point, but most importantly it significantly improved the gain above 800 MHz. Side walls could be added and the spacing and patch shape must be optimized when dielectric material is present between the antenna and backing.

4.5 INTEGRATION OF DIELECTRIC AND INDUCTIVE LOADINGS

The real challenge in realizing a miniature antenna using both dielectric and reactive loadings is their integration into a single antenna. Such an integrated antenna is shown in Figure 4.27 consisting of a square spiral sandwiched between two tapered dielectric pieces. Both the superstrate and substrate have $\varepsilon_r = 9$ and are 0.625 and 0.75 inches thick, respectively. The measured realized gain of the antenna is shown in Figure 4.27 and is -15 dBic at 142 MHz. This means that the untreated cavity-backed spiral would need to be 2.35 times larger to achieve the same gain at this frequency. However, the dielectric is rather ineffective for further miniaturization. This is most likely a result of the coiled spiral arms not being embedded in the $\varepsilon_r = 9$ dielectric material. Since the inductive treatment confines more of the field within the FR4 material, it is important to embed the spiral arms within the $\varepsilon_r = 9$ dielectric material to maximize their effect. Figure 4.28 highlights these results and indicates

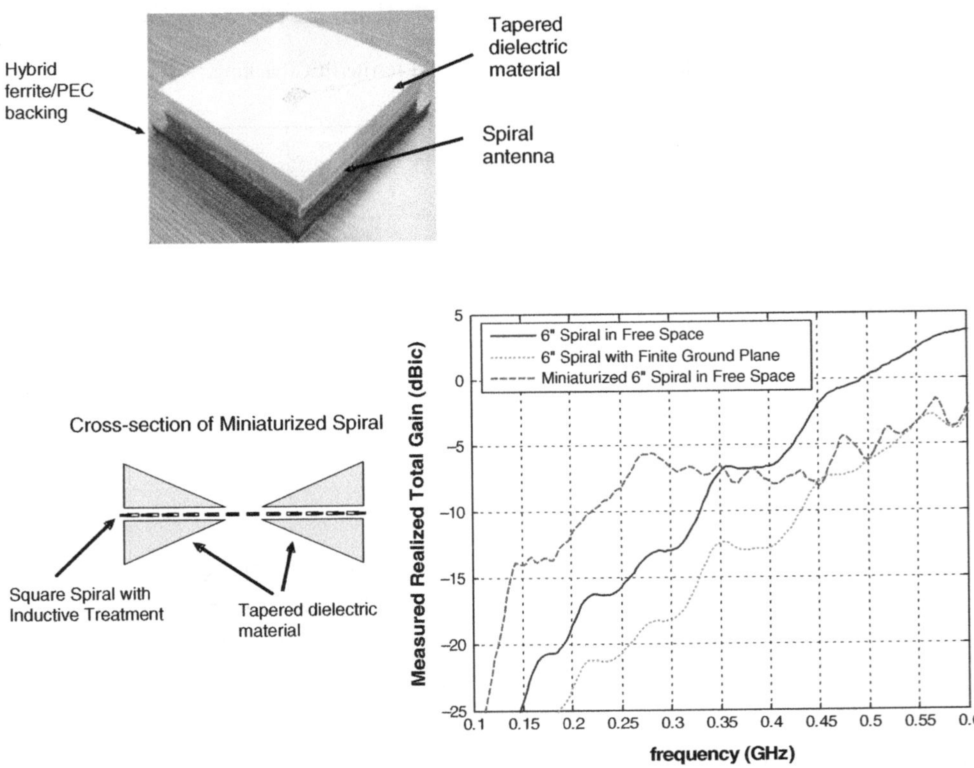

Figure 4.27 Initial results for the integrated 6 inch miniature spiral antenna.

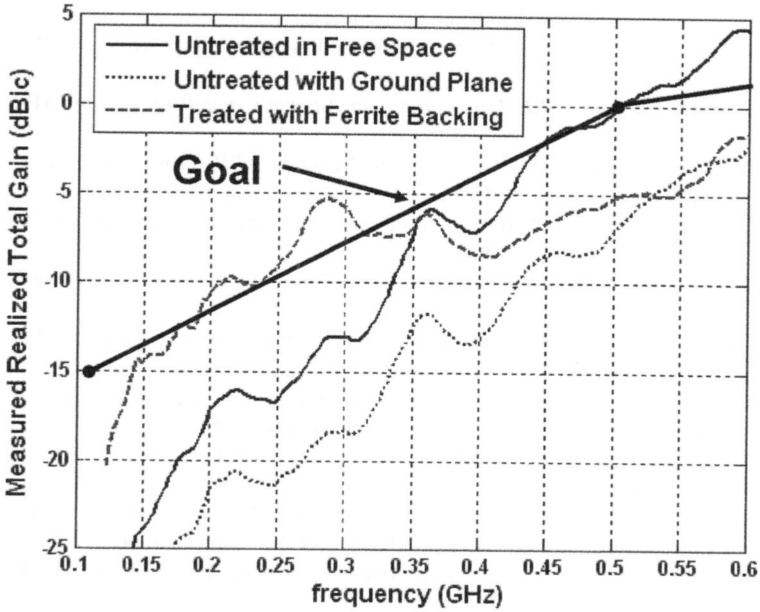

Figure 4.28 Initial results for the integrated miniature 6 inch spiral antenna.

the design goal. Further improvements along these lines are currently being pursued with the goal of best utilizing the volumetric antenna geometry.

4.6 SUMMARY

In this chapter a thorough review of printed spiral antennas has been presented. A historical overview of the technology has been given and the fundamental EM equations that govern the performance of this class of wideband antenna have been presented. Several techniques to reduce the overall size of the spiral and the impact on the EM performance of the antenna when it is loaded with a variety of materials have also been given.

REFERENCES

1. E. M. Turner, 'Spiral slot antenna', US Patent 2863145, October 1955.
2. V. Rumsey, 'Frequency independent antennas', *IRE National Convention Record*, **1**, 114–118, 1957.
3. J. Donnellan and R. Close, 'A spiral-grating array', *IRE Transactions on Antennas and Propagation*, **AP-9**, 291–294, May 1961.
4. J. Dyson, 'The equiangular spiral antenna', *IRE Transactions on Antennas and Propagation*, **AP-7**, 181–187, April 1959.
5. J. Dyson, 'The unidirectional equiangular spiral antenna', *IRE Transactions on Antennas and Propagation*, **AP-7**, 239–334, October 1959.
6. J. D. Dyson, R. Bawer, P. E. Mayes and J. I. Wolfe, 'A note on the difference between equiangular and Archimedes spiral antennas', *IEEE Transactions on Microwave Theory and Techniques*, **9**(2), 203–205, March 1961.

7. J. Dyson and P. Mayes, 'New circular-polarized frequency-independent antennas with conical beam or omnidirectional patterns', *IEEE Transactions on Antennas and Propagation,* **9**(4), 334–342, July 1961.
8. W. Curtis, 'Spiral antennas', *IRE Transactions on Antennas and Propagation*, **AP-8**, 293–306, May 1960.
9. J. Donnellan, 'Second-mode operation of the spiral antenna', *IRE Transactions on Antennas and Propagation*, **AP-8**, 637, November 1960.
10. J. Kaiser, 'The Archimedean two-wire spiral antenna', *IRE Transactions on Antennas and Propagation*, **AP-8**, 312–323, May 1960.
11. J. Donnellan, 'A spiral-doublet scanning array', *IRE Transactions on Antennas and Propagation*, **AP-9**, 276–279, May 1961.
12. W. L. Stutzman, 'Wide bandwidth antenna array design', in *Proceedings of the IEEE Southeastern Regional Meeting*, Raleigh, North Carolina, pp. 92–96, April 1985.
13. A. A. Chatzipetros, 'Planar wideband arrays with variable element size', *Independent Study Report*, Virginia Technical University, Blacksburg, Virginia, March 1993.
14. D. G. Shively, 'Wideband planar array antennas: theory and measurements', Master's Thesis, Virginia Technical University, Blacksburg, Virginia, May 1988.
15. D. B. Shively and W. L. Stutzman, 'Wideband arrays with variable element sizes', *IEE Proceedings H, Microwaves, Antennas and Propagation*, **137**(4), 238–240, August 1990.
16. R. Corzine and J. Mosko, *Four-Arm Spiral Antennas*, Artech House, Norwood, Massachusetts, 1990.
17. H. Nakano, K. Nogami, S. Arai, H. Mimaki and J. Yamauchi, 'A spiral antenna backed by a conducting plane reflector', *IEEE Transactions on Antennas and Propagation*, **34**(6), 791–796, June 1986.
18. H. Nakano, H. Soga, T. Honma and J. Yamauchi, 'Effects of a small disk to a spiral antenna backed by a conducting plane reflector', *IEE Proceedings H, Microwaves, Antennas and Propagation*, **138**(4), 475–477, August 1991.
19. H. Nakano and K. Nakayama, 'A curved spiral antenna above a conducting cylinder', *IEEE Transactions on Antennas and Propagation*, **47**(1), 3–8, January 1999.
20. H. Nakano, J. Eto, Y. Okabe and J. Yamauchi 'Tilted- and axial-beam formation by a single-arm rectangular spiral antenna with compact dielectric substrate and conducting plane', *IEEE Transactions on Antennas and Propagation*, **50**(1), 17–24, January 2002.
21. E. D. Caswell, 'Design and analysis of star spiral with application to wideband arrays with variable elements sizes', PhD Dissertation, Virginia Technical University, Blacksburg, Virginia, 2001.
22. S.-C. Wu, 'Analysis and design of conductor-backed square Archimedean spiral antennas', *Electromagnetics*, **14**, 305–318, 1994.
23. T. Talty, Y. Dai and L. Lanctot, 'Automotive antennas trends and future requirements', in *Proceedings of IEEE APS International Symposium*, Boston, Massachusetts, Vol. I, pp. 430–433, July 2001.
24. T. Iwasaki, A. Freundorfer and K. Izuko, 'A unidirectional semi-circle spiral antenna for subsurface radars', *IEEE Transactions on Electronics Compatability*, **36**(1), 1–6, February 1994.
25. M. Nurnberger and J. Volakis, 'A new planar feed for slot spiral antennas', *IEEE Transactions on Antennas and Propagation*, **44**(1), 130–131, January 1996; see also *IEEE Transactions on Antennas and Propagation*, **50**, 82–85, January 2002.
26. M. W. Nurnberger and J. L. Volakis, 'Extremely broadband slot spiral antennas with shallow reflecting cavities', *Electromagnetics*, **20**(4), 357–376, April 2000.
27. J. L. Volakis, M. W. Nurnberger and D. S. Filipovic, 'A broadband cavity-backed slot spiral antenna', *IEEE Antennas and Propagation Magazine*, **43**(6), 15–26, December 2001.
28. D. S. Filipovic and J. L. Volakis, 'Novel slot spiral antenna designs for dual-band/multi-band operation', *IEEE Antennas and Propagation Transactions*, **51**, 430–440, March 2003.
29. D. Filipovic and J. L. Volakis, 'A flush mounted multifunctional slot aperture (Combo-Antenna) for automotive applications', *IEEE Antenna Propagation Transactions*, **52**(2), 563–571, February 2004.

30. B. A. Kramer, M. Lee, Chi-Chih Chen and J. L. Volakis, 'Design and performance of an ultra wide-band ceramic-loaded slot spiral', *IEEE Antennas and Propagation Transactions*, **53**, 2193–2199, July 2005.
31. J. J. H. Wang and V. K. Tripp, 'Design of multioctave spiral-mode microstrip antennas', *IEEE Transactions on Antennas and Propagation*, **39**(3), 332–335, March 1991.
32. J. J. H. Wang, 'Spiral-mode microstrip (SMM) antennas and associated methods for exciting, extracting and multiplexing the various spiral modes', US Patent, April 1997.
33. M. S. Wheeler, 'On the radiation from several regions in spiral antennas', *IRE Transactions on Antennas and Propagation*, **AP-9**, 100–102, January 1961.
34. B. R.-S. Cheo, V. H. Rumsey and W. J. Welch, 'A solution to the frequency-independent antenna problem', *IRE Transactions on Antennas and Propagation*, **AP-9**, 527–534, November 1961.
35. M. Lee, C.-C. Chen and J. L. Volakis, 'Ultra-wideband antenna miniaturization using distributed lumped element loading', in *2005 IEEE AP-S International Symposium and Usnc/Ursi National Radio Science Meeting*, 2005.
36. Y. Yeh and K. Mei, 'Theory of conical equiangular-spiral antennas: part I – numerical technique', *IEEE Transactions on Antennas and Propagation*, **AP-15**, 634–639, September 1967.
37. Y. Yeh and K. Mei, 'Theory of conical equiangular-spiral antennas: part II – current distributions and input impedance', *IEEE Transactions on Antennas and Propagation*, **AP-16**, 14–21, January 1968.
38. N. J. Champagne II, J. T. Williams, R. M. Sharpe, S.-U. Hwu and D. R. Wilton, 'Numerical modeling of impedance loaded multi-arm Archimedean spiral antennas', *IEEE Transactions on Antennas and Propagation*, **40**(1), 102–108, January 1992.
39. C. Penney and R. Leubbers, 'Input impedance, radiation pattern, and radar cross section of spiral antennas using FDTD', *IEEE Transactions on Antennas and Propagation*, **42**(9), 1329–1332, September 1994.
40. L. Ronglin and N. Guangzheng, 'Numerical analayasis of 4-arm Archimedean printed spiral antenna', *IEEE Transactions on Magnetics*, **33**(2), 1512–1515, March 1997.
41. T. Ozdemir, J. L. Volakis and M. Nurnberger, 'Analysis of thin multioctave cavity-backed slot spiral antennas', *IEE Proceedings on Microwave Antennas and Propagation*, **146**(6), 447–454, December 1999.
42. A. F. Peterson, S. L. Ray and R. Mittra, *Computational Methods for Electromagnetics*, The IEEE/OUP Series on Electromagnetic Wave Theory, Oxford University Press, 1997.
43. J. L. Volakis, A. Chatterjee and L. C. Kempel, *Finite Element Method for Electromagnetics*, IEEE Press–John Wiley & Sons, Inc., New York, 1998.
44. R. C. Johnson and H. Jasik (eds), *Antenna Engineering Handbook*, 3rd edn., McGraw-Hill, New York, 1993.
45. D. S. Filipovic and J. L. Volakis, 'Broadband meanderline slot spiral antenna', *IEE Proceedings on Microwaves, Antennas and Propagation*, **149**(2), 98–105, April 2002.
46. Y. P. Zhang, T. K. Lo and Y. Hwang, 'A dielectric-loaded miniature antenna for microcellular and personal communications', in *Proceedings of the IEEE APS Symposium*, Newport Beach, California, pp. 1152–1155, June 1995.
47. E. A. Mariani, C. P. Heinzman, J. P. Agrios and S. B. Cohn, 'Slot line characteristics', MTT-17, No. 12, pp. 1091–1096, December 1969.
48. R. F. Harrington, *Time-Harmonic Electromagnetic Fields*, 2nd revised edn, IEEE Press–John Wiley & Sons, Inc., New York, 2001.
49. S. Hong and G. Rassweiler, 'Size reduction of a conical log-spiral antenna by loading with magneto-dielectric material', *IEEE Transactions on Antennas and Propagation*, 650–651, September 1966.
50. D. Psychoudakis, S. K. C. Pillai, J. H. Halloran and J. L. Volakis, 'Miniaturization of a bow-tie antenna with textured dielectric superstrates', in *IEEE Antennas and Propagation Society International Symposium*, pp. 2572–2575, 2004.
51. H. Mosallaei and K Sarabandi, 'Antenna miniaturization and bandwidth enhancement using a reactive impedance substrate', *IEEE Transactions on Antennas and Propagation*, **52**(9), 2403–2414, September 2004.

52. J. L. Volakis, C. C. Chen, M. Lee, B. Kramer and D. Psychoudakis, 'Miniaturization methods for narrowband and ultra-wide band antennas', in *IEEE International Workshop on Antenna Technology (IWAT)*, pp. 119–121, 2005.
53. B. A. Kramer, M. Lee, C.-C. Chen and J. L. Volakis, 'Design and performance of an ultra wideband ceramic-loaded slot spiral', *IEEE Transactions on Antennas and Propagation*, **53**(7), 2193–2199, June 2004.
54. D. P. Tran, M. Tian and L. P. Ligthart, 'Antenna element miniaturization and the use of airgap matching techniques', in *Antennas and Propagation Society International Symposium, AP-S. Digest*, Vol. 1, pp. 286–287, 1994.
55. S. R. Best, 'A discussion on the properties of electrically small self-resonant wire antennas', *IEEE Antennas and Propagation Magazine*, **46**(6), 9–22, December 2004.
56. J. S. McLean, 'A re-examination of the fundamental limits on the radiation Q of electrically small antennas', *IEEE Transactions on Antennas and Propagation*, **44**(5), 672, May 1996.
57. L. J. Chu, 'Physical limitations of omni-directional antennas', *Journal of Applied Physics*, **19**(12), 1163–1175, December 1948.
58. A. D. Yaghjian and S. R. Best, 'Impedance, bandwidth, and Q of antennas', *IEEE Transactions on Antennas and Propagation*, **53**(4), 1298–1324, April 2005.
59. B. Kramer, S. Koulouridis, C.-C. Chen and J. Volakis, 'A low profile UWB antenna with imbedded inductive loading', in *2005 Antenna Measurement Techniques Association Symposium (AMTA)*, Rhode Island, 2005.

5

Printed Folded Beverage Antennas

Rod Waterhouse and Dalma Novak
Pharad, Maryland, USA

5.1 INTRODUCTION

In the lower microwave frequency spectrum (less than 6 GHz) there is an abundance of wireless communication systems currently in service. With the multiple cellular communication bands already in use and also all the wireless local area network (WLAN) and wireless personal area network (WPAN) systems, the need for a universal access point, in particular for in-building systems, is paramount. A key element of this access point is the antenna that must be able to operate over the entire frequency range (800 MHz to 5.8 GHz) in terms of impedance match and gain. In addition, such a universal antenna must also exhibit all the other aesthetic traits required for a base station antenna, namely being low in profile, lightweight, and low in cost. A similar quandary is facing military sectors, where a small, conformal antenna is required for the Joint Tactical Radio System (JTRS) [1] platform. In this application there are also size constraints and in some instances a volumetric approach for the antenna solution may be the most appropriate.

There is a variety of travelling-wave-based antennas that may appear suitable for the wireless access point application, or other wideband applications. Several of these solutions are presented in other chapters within this book. Spiral antennas are the most common solution for wideband, omnidirectional applications (for example see Reference [2]). Although the cavity-backed spiral can give the required radiation performance, the size and weight associated with the cavity make this alternative not particularly attractive, especially for low-cost applications in the commercial market. The printed spiral [3] resolves the height problem, but the surface area necessary for the conductors covers many wavelengths and a balanced feed is also required to ensure good matching to the feed network.

Tapered slot antennas can provide excellent bandwidth [4] and are also very efficient as no power is wasted and dumped into a load element as for the case of spiral-based

Printed Antennas for Wireless Communications Edited by R. Waterhouse

antennas. However, tapered slots are usually quite large, especially when required to operate in the lower microwave frequency spectrum. In addition, the radiation pattern of a typical tapered slot antenna is directional, which prevents the use of this form of radiator for applications (such as an indoor wireless) where omnidirectional coverage is required. The broadband transition from the tapered slot to a standard unbalanced transmission medium can also be inefficient, with losses greater than 5 dB. Another issue with a tapered slot solution is that it is principally linearly polarized and the creation of a tapered slot that can accept multiple polarizations dramatically increases the complexity and size of the antenna platform. Bow-tie antennas (for example see Reference [5]) can provide a physically small radiating solution, but there are two drawbacks associated with this technology: the feed can be complicated, especially when requiring operation over a large bandwidth, and the antenna needs to be located away from any ground-planes to operate efficiently (in a similar manner to the tapered slot). Printed antenna solutions based on aperture stacked patches [6], quasi-Yagi–Uda antennas [7], and L-shaped probe excitation [8] simply do not have the inherent wideband nature required here.

Probably one of the most well-known travelling-wave antennas is the Beverage, or long wire or wave antenna [9]. The antenna was invented in the early 1920s by Harold Beverage and variations of this form of radiator are very common, in particular in the MF (300 kHz–3 MHz) and HF (3–30 MHz) ranges. Here a long wire mounted above the ground is excited at one end and terminated at the other port. The antenna is designed to have uniform patterns in both current and voltage. To achieve this, the wire antenna must be appropriately terminated to ensure that no reflections occur. The length of this form of antenna ranges from one to many wavelengths. It is also possible to realize a printed version of a Beverage antenna. In this configuration a probe soldered to the microstrip line is used to excite the antenna, and the characteristic impedance of the transmission line is designed as 50 Ω and so the termination resistance is also this value. The antenna can be well matched over a very wide bandwidth (in excess of a decade) and its typical radiation pattern (including gain) is directed towards the direction of wave propagation, namely endfire. In general, the printed Beverage antenna radiates better when the height of the substrate is increased, but doing so can compromise the input impedance response due to the discontinuity associated with the feed mechanism as well as the load.

In this chapter a novel variation of the printed Beverage antenna is explored, namely a folded version. Here the long conductor is wrapped back on to itself in a spiral fashion. The new antenna has all the wideband traits of the conventional Beverage antenna but is significantly smaller due to the folding effect. Another consequence of folding the conductor is that the radiation response of the antenna is near omnidirectional due to the conductor not being aligned in one direction. In its present form the folded Beverage antenna is similar to a printed spiral antenna, but there are two important distinctions. Firstly, the folded Beverage antenna does not require a balun to feed the radiator as the 'currents' are excited along the one conductor and, in addition, the overall area should be smaller as two spiral-shaped conductors do not need to be accommodated.

The outline of this chapter is as follows. In Section 5.2 a generic form is presented of the proposed wideband printed antenna configuration. The basic design methodology for the radiator is also discussed and the measured performance of the antenna over a large bandwidth is presented, including the return loss response and the radiation patterns. In Section 5.3 the design trends associated with the printed antenna are investigated, including the trade-offs in material selection, the shape of the spiral, and the load impedance. In the next two sections the printed folded Beverage antenna concept is applied to two size-constrained scenarios related to vehicle-mounted JTRS applications. In Section 5.4 the folded Beverage radiator is applied to a height-constrained problem where the height of the antenna can be no greater than $0.05\lambda_0$ (where λ_0 is the free-space wavelength at the lower edge of the frequency band. The measured wideband performance of the radiator is presented in this section. In Section 5.5 a different scenario is investigated; here the height constraint is lifted to $0.25\lambda_0$, but the cross-sectional area has been restricted. Once again the measured wideband performance of the antenna will be presented. Finally, in Section 5.6 the findings on this form of printed antenna will be summarized.

5.2 PROPOSED ANTENNA CONFIGURATION AND GENERAL CHARACTERISTICS

Figure 5.1 shows a schematic of the proposed printed folded Beverage antenna. In this particular configuration the antenna is a rectangular-based spiral with mitred bends (other spiral shapes will be presented in Section 5.3). In Figure 5.1 the radiator is fed at the centre point with a small probe soldered to the etched conductor residing on a grounded substrate of thickness, d, and with a dielectric constant, ε_r. The conductor trace of width, w_B, is then expanded in a spiral-like manner over the area dedicated to the radiator; the larger the area ($L_s \times W_s$) or the greater the length of the antenna (L_B), the more efficient the antenna will be at lower frequencies. The conductor track is terminated at its end with a load resistor (R_L), as for the case of a conventional Beverage antenna or spiral.

There are several design parameters associated with the Beverage antenna that impact its response: the characteristics of the material used to support the track conductor, the track width, the height of the conductor above the ground-plane, and the termination of the antenna. To achieve an 'ideal' travelling-wave response, this termination must be matched to

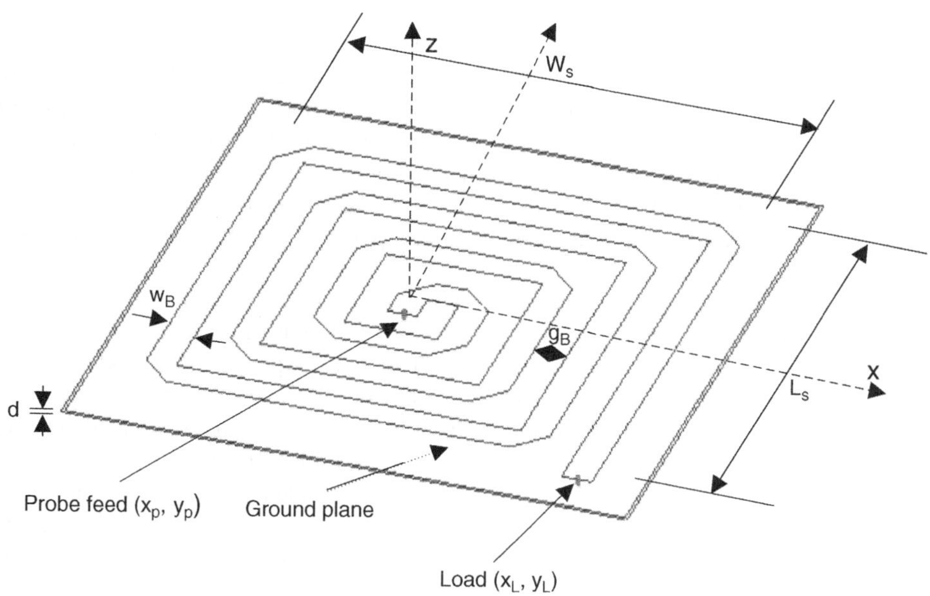

Figure 5.1 Schematic of a printed folded Beverage antenna.

that of the impedance of the transmission line. Doing so will prevent reflections at the end of the antenna and thereby reduce the likelihood of standing wave effects in both the radiation patterns and the impedance response of the antenna. A thorough investigation of the antenna was undertaken using a full-wave planar analysis tool and some of the performance trends that were found will be presented in the next section. However, there are some general characteristics that are pertinent to mention here. The printed Beverage antenna radiates better when the height of the substrate is increased, but doing so can compromise the input impedance response due to the discontinuity associated with the feeding mechanism as well as that associated with the load. Another consequence of using a thicker material for the substrate is that the track width becomes wider, which impacts on the overall size of the radiator since the greater the thickness, the larger the area required to accommodate bends in the folded antenna. This can be circumvented somewhat by using higher dielectric constant substrates, at the expense of radiation efficiency.

The authors set about using these general design strategies to develop a printed folded Beverage antenna that was impedance matched (a VSWR (voltage standing wave ratio) $< 2{:}1$) from 800 MHz to 6 GHz. They used 3 mm *RT/Duroid™ 5880* as the substrate for the antenna as this material has a relatively low dielectric constant ($\varepsilon_r = 2.2$) and the thickness is neither too thin nor too thick over the frequency range to compromise the trade-off between the size of the antenna and the impedance match. Over the required frequency range the electrical thickness of the material varies from $0.008\lambda_0$ to $0.06\lambda_0$, where λ_0 is the free-space wavelength at the corresponding frequency. Using standard microstrip transmission line formulas (refer to Reference [10] for details), the width of the track was determined to be 9.5 mm. The overall area of the antenna was limited to 150 mm × 150 mm (including the ground-plane). Such a size was deemed tolerable for wireless equipment manufacturers [11]. Given this size constraint they then set about laying out the folded beverage antenna. Here the gap was set between the conductors of the antenna (g_B) as 9.5 mm. Standard mitring techniques were used (refer to Reference [10] for details) to minimize reflections from the current discontinuity associated with the bend in the microstrip transmission line. The structure developed was 145 mm × 145 mm in area. The impact on the overall shape of the folded Beverage antenna will be examined in more detail in Section 5.3.

The antenna was fabricated using standard etching procedures. An SMA connector was incorporated to feed the folded antenna. The feed and load ports were located 2 mm from the beginning and end, respectively, of the transmission line. To realize the load for the antenna a simple technique was utilized. Near the end of the transmission line of the antenna the track was cut and a 50 Ω resistor was soldered across the gap. A via to the remaining small section of track was then soldered to the ground-plane. A photograph of the developed antenna is shown in Figure 5.2.

The return loss response of the antenna was measured and is shown in Figure 5.3. As can be seen from this plot, the antenna easily satisfies the return loss requirements of less than −10 dB over the specified frequency range. This is as expected for a loaded antenna and is consistent with other forms of this class of radiator. The periodic ripples in the return loss are due to finite reflections from the connector (which was not time-gated out) and interactions with reflections from the mitred bends of the folded antenna.

The radiation patterns (including gain) of the printed folded Beverage antenna in the principal planes (E- and H-planes) are shown in Figures 5.4 to 5.7 across the specified frequency band. Here $\theta = 90°$ corresponds to the broadside. There are two important trends to observe in Figures 5.4 to 5.7. As suggested earlier, the size constraint limits the gain, particularly at the lower frequencies, and in addition there can be a large undulation in the

Figure 5.2 Fabricated folded Beverage antenna developed on *RT/Duroid 5880*.

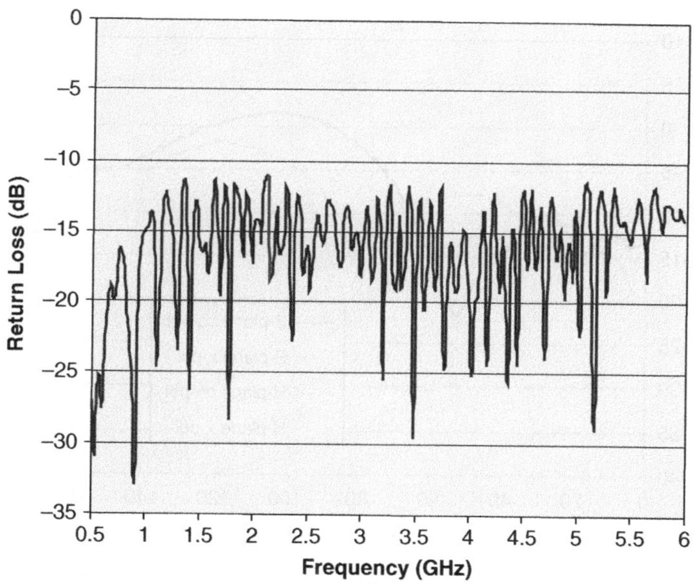

Figure 5.3 Measured return loss of a printed folded Beverage antenna.

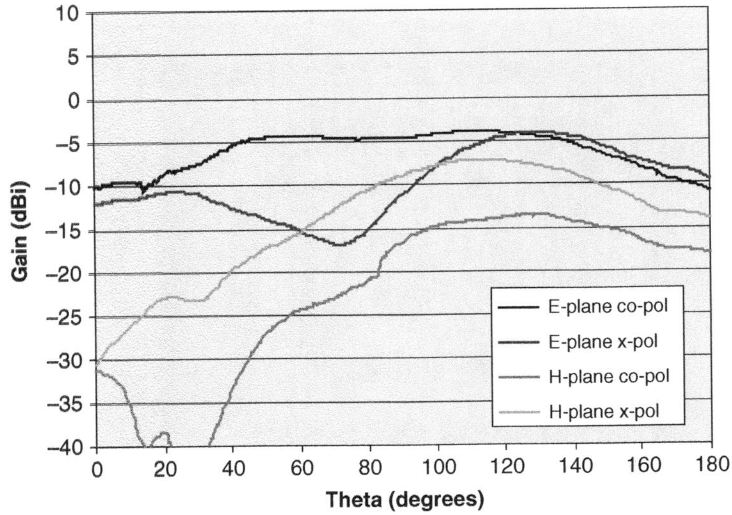

Figure 5.4 Measured radiation patterns and gain at 900 MHz of a printed folded Beverage antenna.

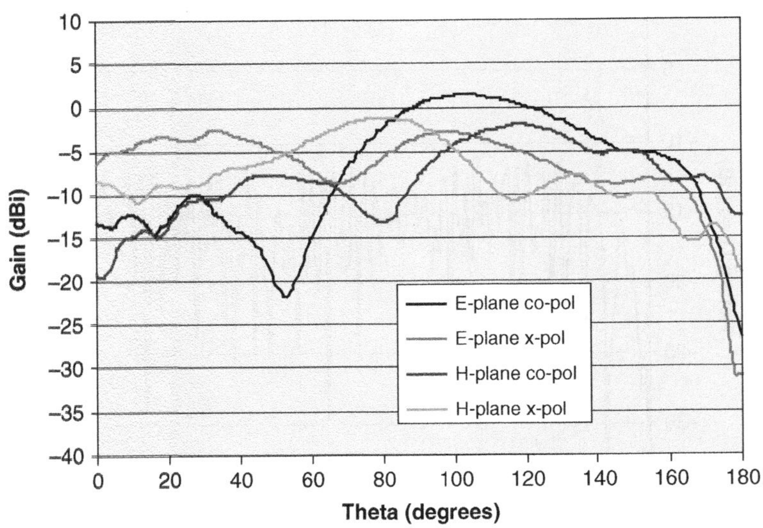

Figure 5.5 Measured radiation patterns and gain at 1.84 GHz of a printed folded Beverage antenna.

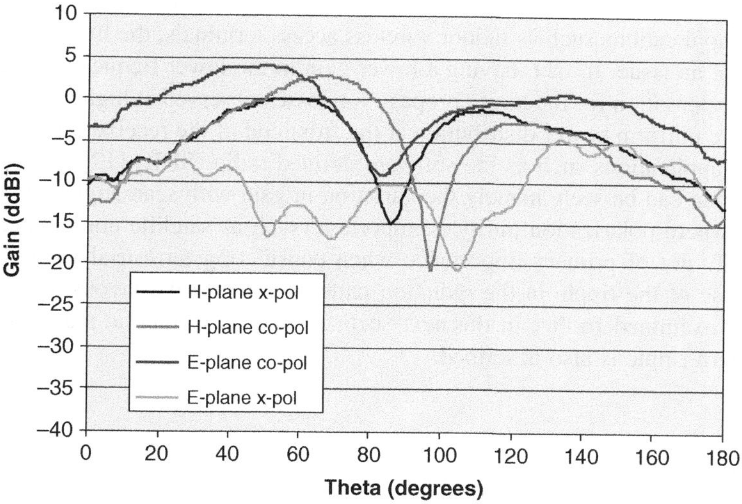

Figure 5.6 Measured radiation patterns and gain at 2.42 GHz of a printed folded Beverage antenna.

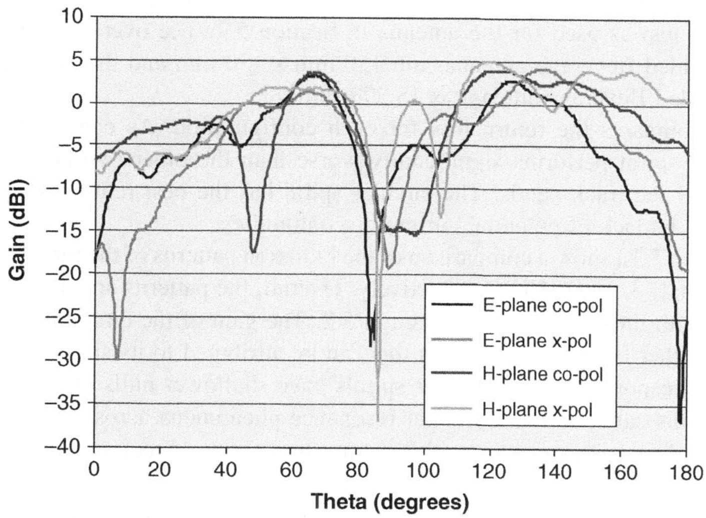

Figure 5.7 Measured radiation patterns and gain at 5.8 GHz of a printed folded Beverage antenna.

gain across the scan angle. The first trend can be easily resolved by making the antenna larger, but in applications such as indoor wireless access terminals, the limited gain may not be too much of an issue. In fact, having a lower gain in the lower frequency range can help to provide compensation for the larger propagation losses observed at higher frequencies and produce a more uniform power distribution at the front end of the receiver. This is important for wideband applications such as the software defined radio (SDR) [12] or JTRS [1]. The second trend that can be seen, namely the variation in gain with scan angle, can be an issue for a system where polarization purity is important (such as satellite communications), but this is typically not of primary importance when considering terrestrial wireless networks [13]. One cause of the ripple in the radiation pattern is due to the layout of the conductor track. This is examined further in the next section where a technique that can reduce the radiation pattern ripple is also described.

5.3 DESIGN TRENDS

In Section 5.2 the general properties of the printed folded Beverage antenna were established. The objective in this section is to investigate a means of achieving the optimum performance of the wideband antenna. Throughout this section a full-wave planar analysis tool is used. The simulation rigorously accounts for the probe feed and load but it does assume an infinite ground-plane. Although the ground-plane can impact on the radiation patterns in terms of gain undulation, the general design trends can still be observed with this design tool.

The first parameter investigated was in the shape of a spiral. Figure 5.8 shows three configurations investigated: a circular spiral, a square spiral, and a square spiral with mitred bends. Each of these folded Beverage antennas has the same number of turns (three), is mounted on the same material used in Section 5.2, and also incorporates the same track width and track gap as used for the antenna in Section 5.2. The overall dimensions of the square-based folded Beverage antennas are 150 mm × 150 mm and the area of the circular spiral-based folded Beverage antenna is 13 273 mm^2.

Figure 5.9 compares the return loss for each configuration. As can be seen from this plot, the square spiral performs significantly worse than the other two configurations due to reflections off the track bends. The circular spiral has the best return loss performance, which is due to the lack of potential current discontinuities.

Figures 5.10 to 5.12 show a comparison of the radiation patterns of the three configurations at frequencies of 1, 3, and 6 GHz, respectively. Overall, the patterns are similar in form and are consistent with those presented in Section 5.2. The gain of the circular spiral is slightly lower than the other two cases, although this can be attributed to its smaller overall length. In addition, the responses of the circular spirals have shallower nulls than the rectangular-based spirals. This can be due to stronger resonance phenomena across the structure of the rectangular-based spirals associated with the transitions in the direction of the printed spiral conductor.

The relationship between the conductor track spacing (g_B) and the performance of the antenna was also explored. A circular spiral was considered similar to that presented in Figure 5.8 with the outer dimension of the spiral set to be the same, but this time the track spacing was reduced to half of the track width. Figure 5.13 shows a comparison of the return loss between the two cases. As can be seen from this result, the return loss performance is similar, although there is one very important advantage of the modified circular spiral,

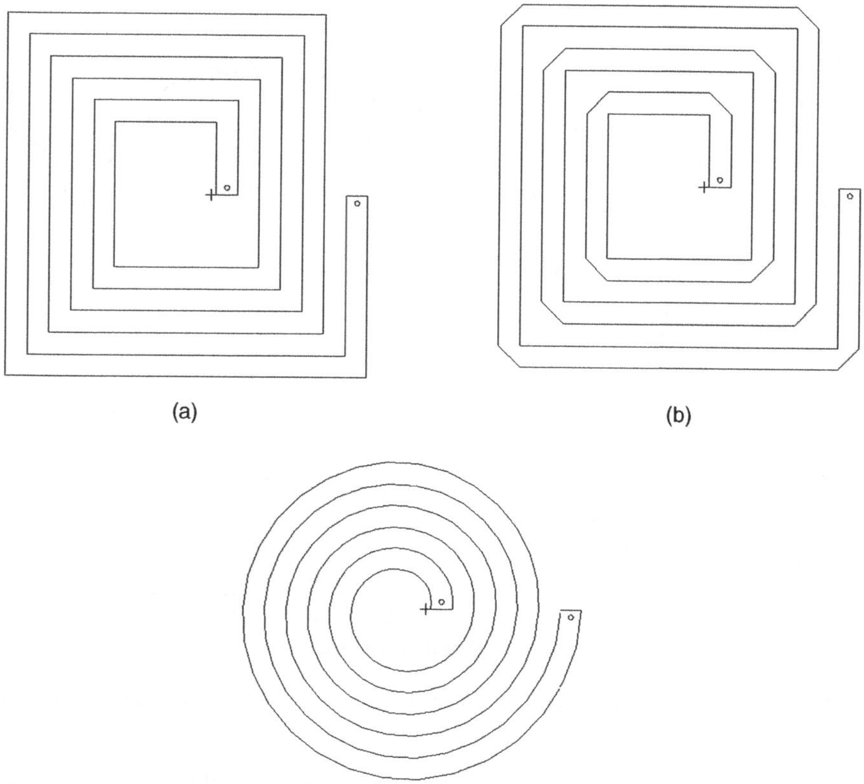

(a) (b) (c)

Figure 5.8 Spiral geometry variations of the investigated folded Beverage antennas: (a) square; (b) square with mitred bends; and (c) circular.

namely that for the same outer dimension, the overall track length is longer. As a result, the gain of the antenna was approximately 2 dB larger across the frequency band.

Obviously the load resistor is a critical component of the radiator and directly impacts the performance of the antenna since the lower the resistor value, the better the efficiency. A parameter study was made of the return loss performance as a function of the load resistor value and no unusual findings were found; as the load was decreased from 50 Ω down to 10 Ω the impedance matching was compromised as expected. Therefore, for the sake of brevity a plot of this trend has not been included here.

The optimal thickness for the folded Beverage antenna is difficult to determine. Generally as the substrate thickness is increased, the antenna becomes more efficient, which can be simplistically attributed to the fact that as the width of a microstrip line increases, more energy is radiated. Thus as the energy travels along the antenna, more of it is efficiently coupled to free space. One issue associated with folded Beverage antennas mounted on thick substrates is that as the thickness increases, the impedance mismatch associated with the probe feed becomes more pronounced, in a similar manner to a probe-fed microstrip patch antenna [14]. This effect will be most dramatic at the higher frequencies of operation where

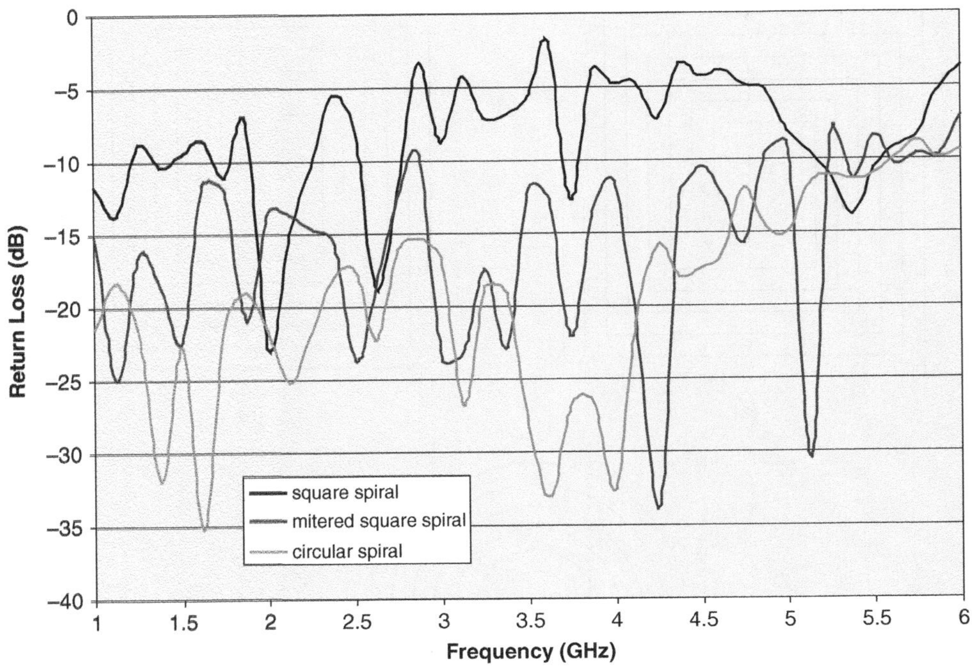

Figure 5.9 Comparison of return loss performance of various spiral-shaped folded Beverage antennas.

the material is electrically thick. A technique that will alleviate this problem is currently being explored.

5.4 LOW-PROFILE, CONSTRAINED AREA DESIGN

Now that the general design methodologies for a folded Beverage antenna have been established this technology will be applied to a constrained area, in this case a nonsymmetrical area. Once again wideband performance is required and for the case in hand the operating frequency range is from 225 MHz to 2.5 GHz. The antenna is to be located on a terrestrial vehicle and the surface area constraints are: 90 cm × 20 cm. This corresponds to an electrical area of $3\lambda_0 \times 0.5\lambda_0$, where λ_0 is the free-space wavelength at the lower frequency edge of the band. As should be evident from this physical limitation, the antenna will not be efficient at the lower frequency limit. The height of the antenna is constrained to be less than 1.5 cm.

From the previous sections it is clear that to make the folded Beverage radiator as efficient as possible, low dielectric constant material should be used as the substrate of the antenna. A setback associated with this approach is that the width of the conductor of the radiator becomes large. For a thickness of 1 cm, the width of the track is 5 cm. Another means of ensuring that the efficiency of the antenna is maximized is to make sure that the radiating conductor should occupy as much of the space available. Doing so is slightly compounded by the difficulty in making bends in the radiating conductor that do not consume a lot of real estate due to the width of the transmission line. The best geometry (ensuring the maximum length of the radiating conductor) was found to be the shape shown in Figure 5.14. Due to limited real estate only three folds in the conductor could be managed.

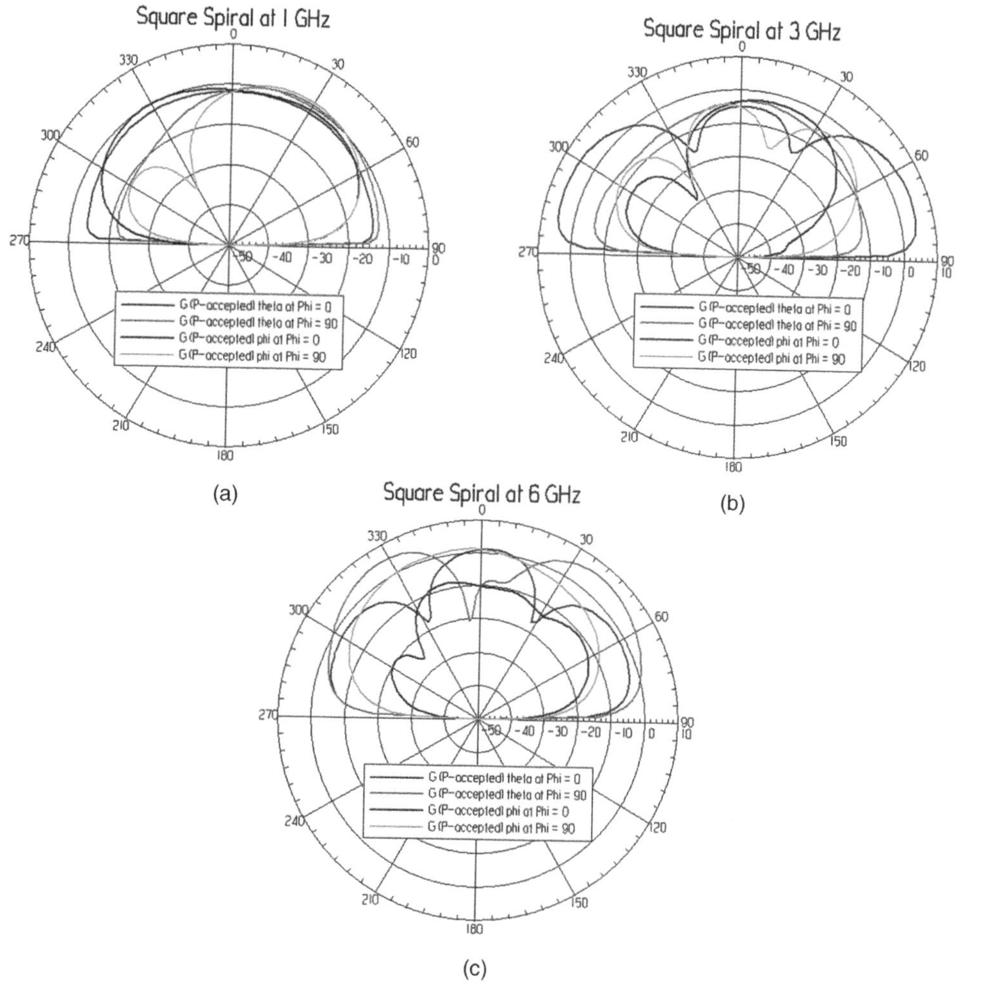

Figure 5.10 Radiation performance of a square spiral-based folded Beverage antenna: (a) 1 GHz; (b) 3 GHz; and (c) 6 GHz.

As can be seen from Figure 5.14, the mounting screws also had to be accommodated within the size constraints, further limiting the available space for the radiator. To realize the conductor of the antenna in Figure 5.14, a very thin layer of FR4 (0.254 mm) was used and the conductor was etched on one side. There is a layer of 1 cm Rohacell foam between the bottom cover and the radiator layer of FR4 (not shown in Figure 5.14). The bottom cover is made from aluminium and acts as the ground-plane for the radiator. The same procedure summarized in Section 5.2 was used to realize the 50 Ω load for the antenna presented in Figure 5.14. A photograph of the developed antenna with its radome above it is shown in Figure 5.15.

The return loss performance of the antenna was measured using a vector network analyser. The response is shown in Figure 5.16. As can be seen from this plot, the radiator has a better than −10 dB return loss over the entire frequency band of interest.

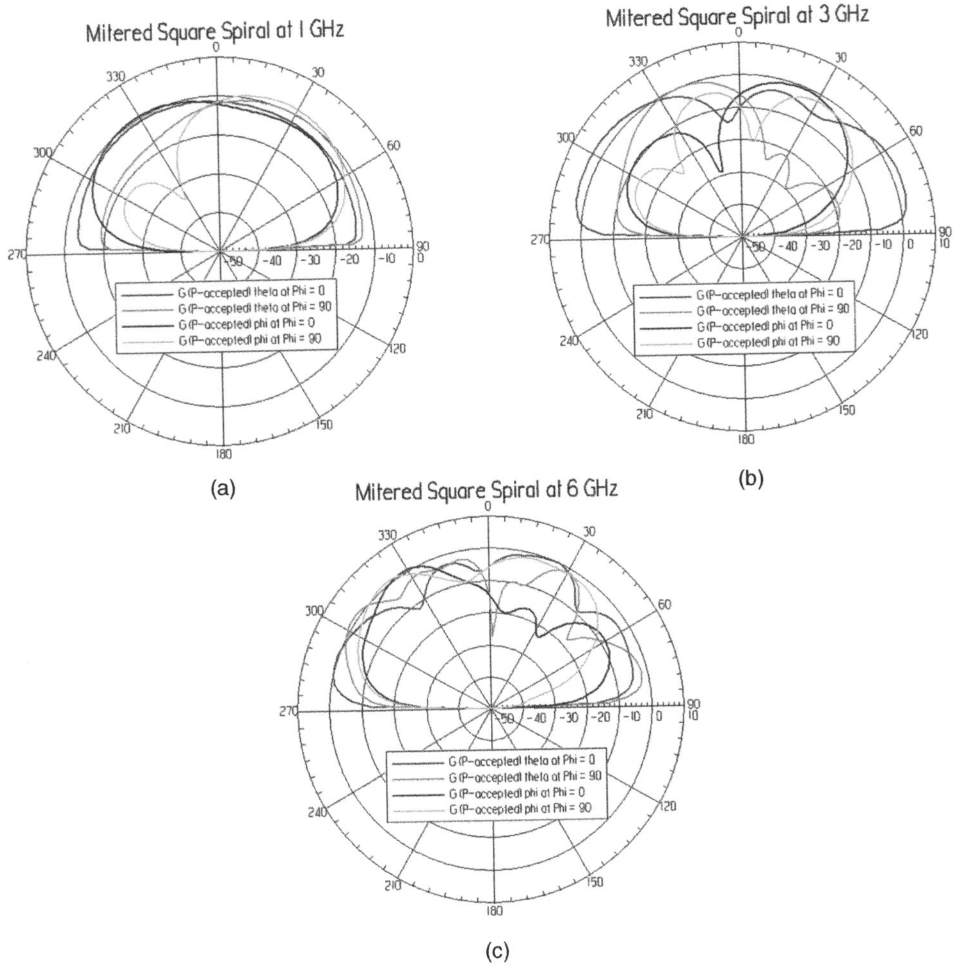

Figure 5.11 Radiation performance of a square spiral with mitred bends based folded Beverage antenna: (a) 1 GHz; (b) 3 GHz; and (c) 6 GHz.

The radiation performance of the antenna was measured across the 225–2500 MHz band and a summary of the patterns including the gain is presented here. The measurements were conducted at 225, 512, 960, 1200, 2000, and 2500 MHz. The radiation patterns were measured at an outdoor antenna testing facility at NAVAIR. Horn and Yagi–Uda gain standards were used to measure the gain of the antennas at each frequency measured: the Yagi–Uda for frequencies below 900 MHz and the TEM horn for frequencies greater than 900 MHz.

For the measurements of a low-profile folded Beverage antenna the following convention was used. The middle of the top of the antenna was aligned to zero degrees. The orthogonal axes (0 and 90°, or x and y) on the antenna were then marked (see Figure 5.17 for the marking convention) and the patterns in these two cuts were rotated and measured, for both vertical and horizontal excitations. These two planes are defined as pitch (P) and roll (R). All radiation patterns in this section use this terminology.

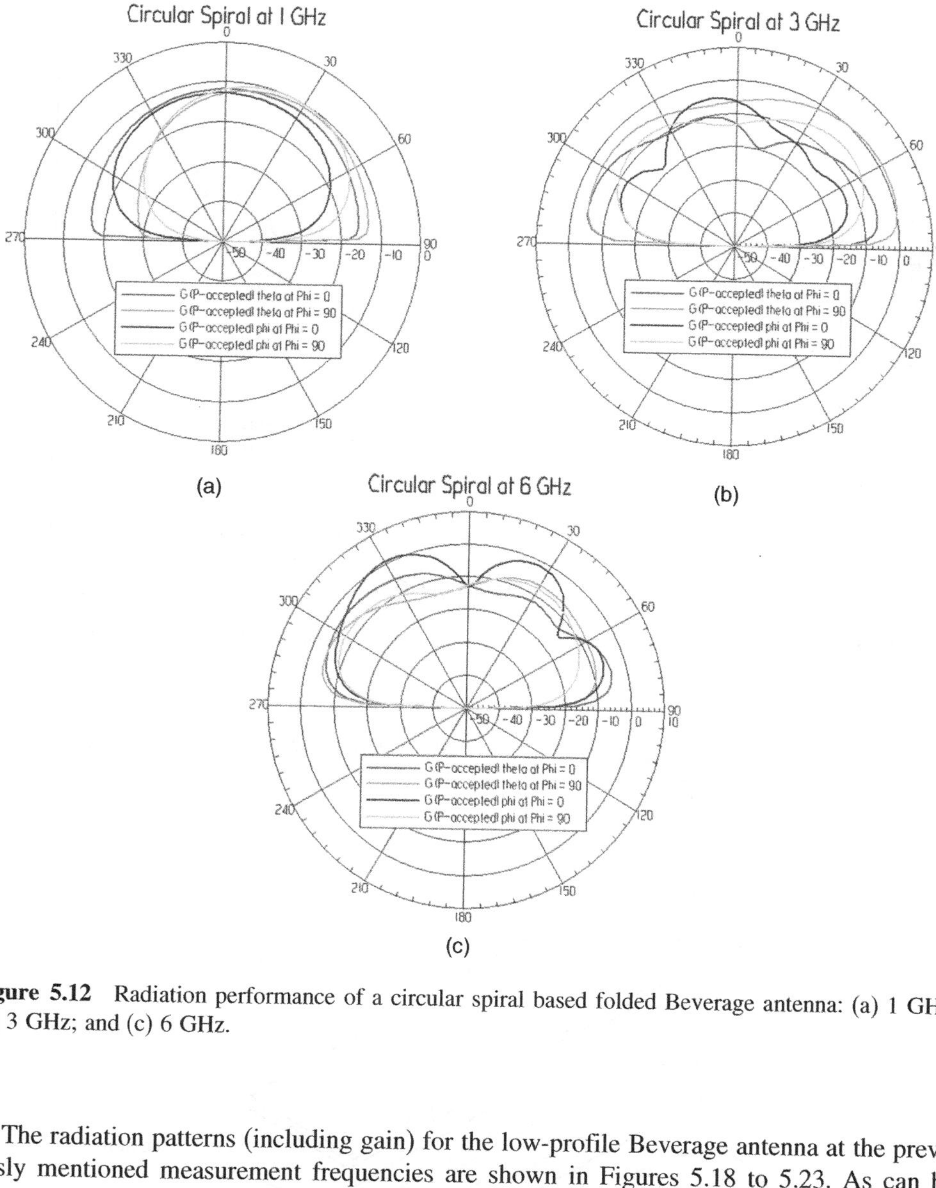

Figure 5.12 Radiation performance of a circular spiral based folded Beverage antenna: (a) 1 GHz; (b) 3 GHz; and (c) 6 GHz.

The radiation patterns (including gain) for the low-profile Beverage antenna at the previously mentioned measurement frequencies are shown in Figures 5.18 to 5.23. As can be seen from these plots, the radiator shows a reasonable coverage area and good polarization diversity. The antenna does, however, suffer from low gain; at 225 MHz the gain is approximately −10 dBi and increases with frequency. This is a consequence of the limited area and height restrictions placed on the antenna. There is also noticeable scalping in the radiation patterns for frequencies above 960 MHz. This is due to the nonsymmetric shape of the antenna. The scalping is evident in the patterns measured along the longer dimension of the radiator, which is consistent with long straight wire radiators. The final observation from the plots shown in Figures 5.18 to 5.23 is that the gain decreases from 3 dBi at 2000 MHz to 0 dBi at 2500 MHz. This could be due to the impedance mismatch discussed previously and the losses in the materials used to develop the antenna since the terminating resistor is only rated to 1 GHz.

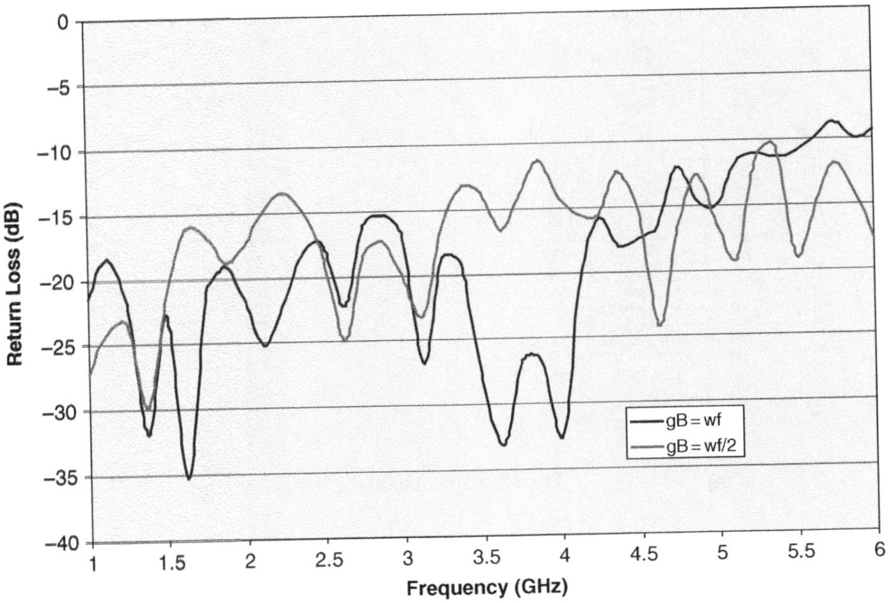

Figure 5.13 Comparison of the return loss performance when the gap between the conductors is halved.

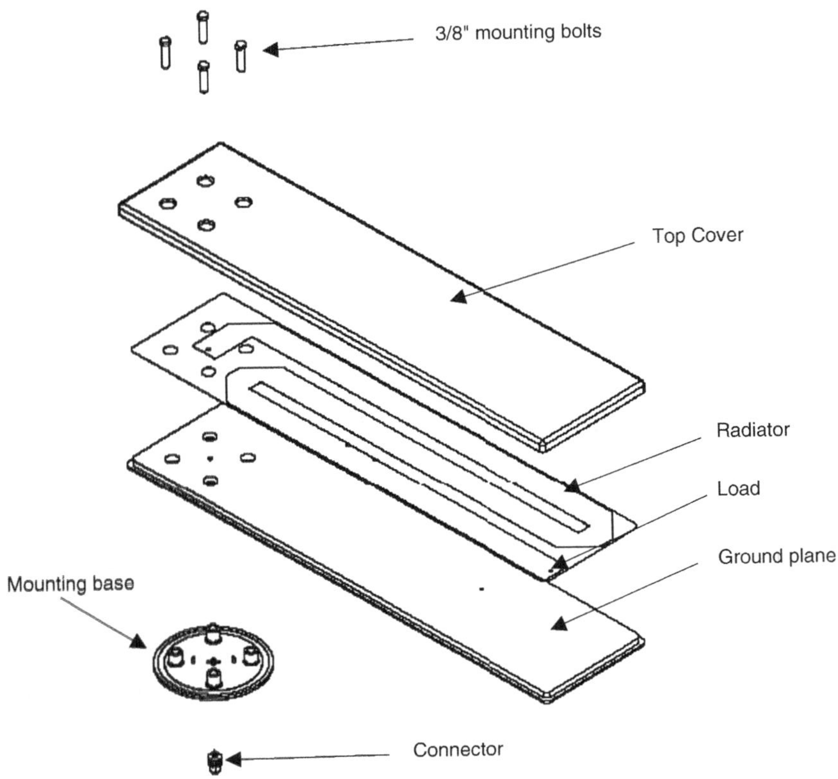

Figure 5.14 Schematic of the rectangular printed folded Beverage antenna.

Figure 5.15 Photograph of the printed folded Beverage antenna.

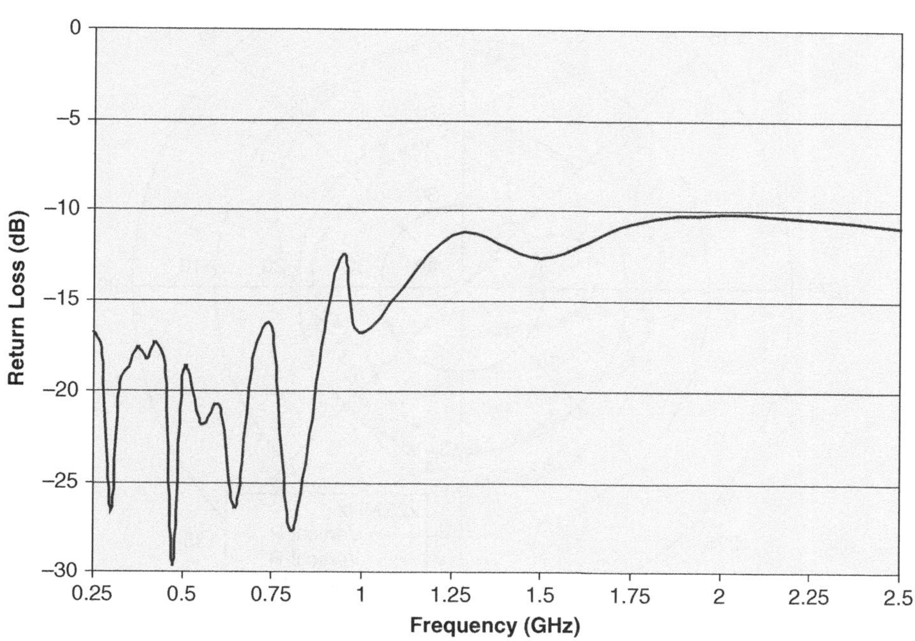

Figure 5.16 Measured return loss performance of a low-profile folded Beverage antenna.

Figure 5.17 Photograph of the low-profile Beverage antenna showing the measurement convention.

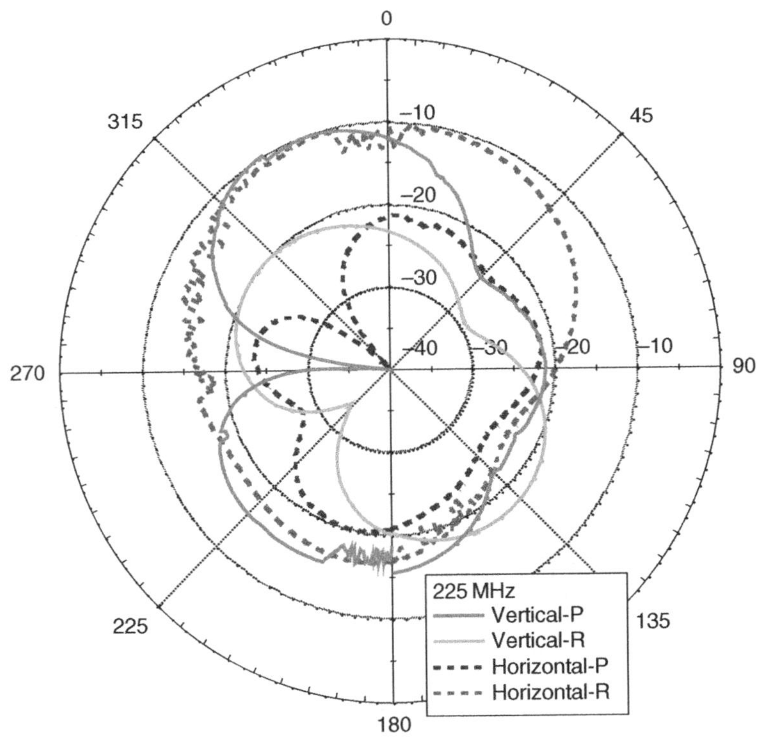

Figure 5.18 Radiation performance of the low-profile Beverage antenna at 225 MHz.

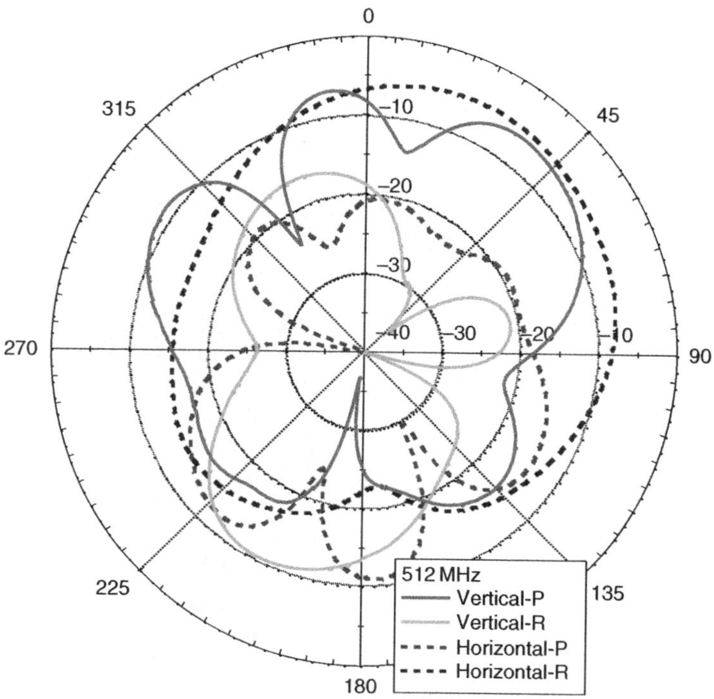

Figure 5.19 Radiation performance of the low-profile Beverage antenna at 512 MHz.

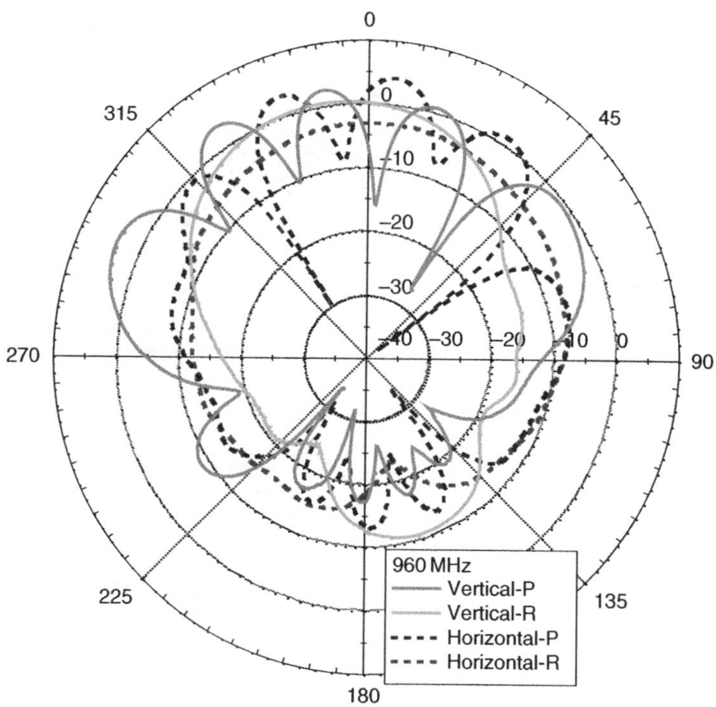

Figure 5.20 Radiation performance of the low-profile Beverage antenna at 960 MHz.

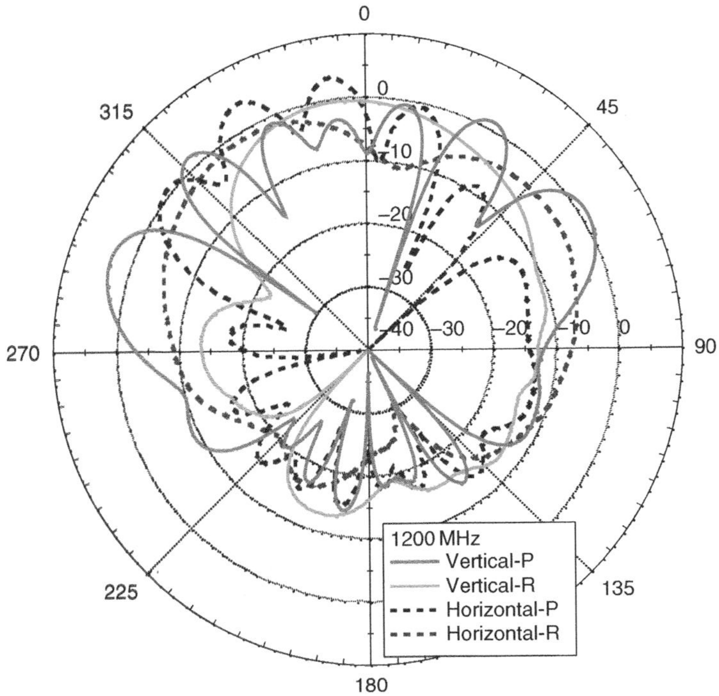

Figure 5.21 Radiation performance of the low-profile Beverage antenna at 1200 MHz.

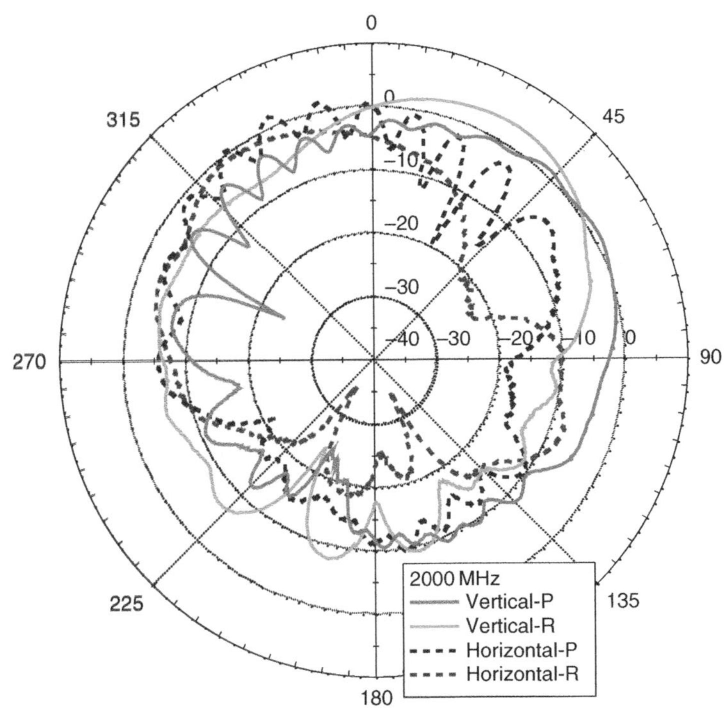

Figure 5.22 Radiation performance of the low-profile Beverage antenna at 2000 MHz.

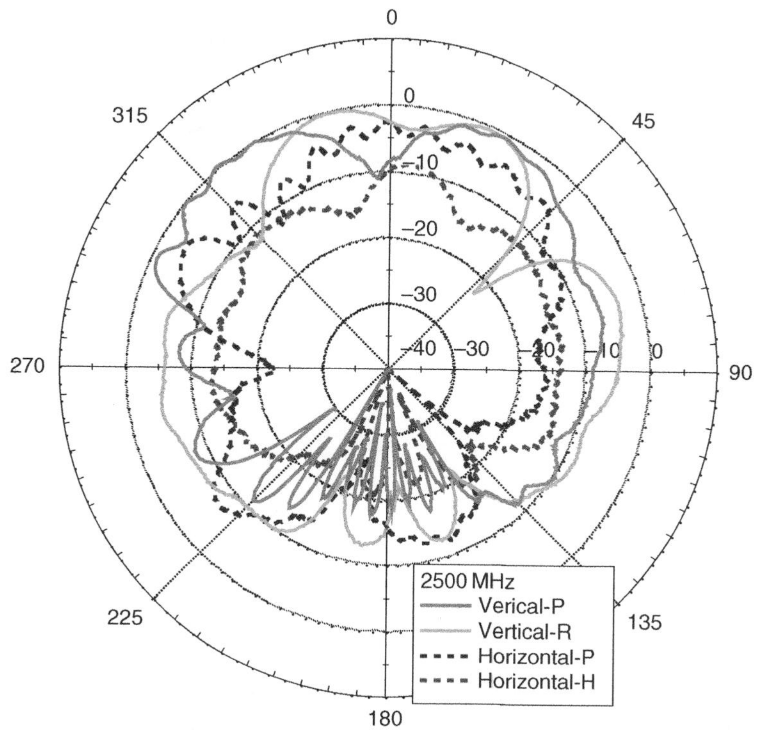

Figure 5.23 Radiation performance of the low-profile Beverage antenna at 2500 MHz.

5.5 THREE-DIMENSIONAL, CONSTRAINED AREA DESIGN

In the previous section the design of a folded Beverage antenna was investigated where the surface area for the antenna was confined. Now consider a scenario where the volume in which the antenna is to reside has been constrained, but the antenna must still operate over the 225–2500 MHz bandwidth. In this particular case, the area will be further restricted to 30 cm × 30 cm, but the allowable height of the antenna will be increased to 25 cm. A new form of Beverage antenna will then need to be created that can take advantage of the new volume available. To differentiate the new form of Beverage antenna from the previously developed radiator, the term three-dimensional folded Beverage antenna will be used to describe the new antenna.

Figure 5.24 shows a schematic of the three-dimensional folded Beverage antenna. Once again the current density on the radiator and the ground-plane have been plotted to highlight the physical features of the antenna. To analyse this structure the authors' FEM code has been used. The design principles for the three-dimensional Beverage antenna are similar to previously investigated solutions; the long 'transmission line' antenna is folded back on itself within the given volume to maximize the length of the radiator and also to help convert a typically directional radiation pattern into a more omnidirectional performance. As for the low-profile cases, the three-dimensional version is terminated with a matched load (in this case at the top of the structure shown in Figure 5.24). The ground-plane for this antenna is a three-dimensional box shown below the radiating conductor in Figure 5.24. As for the planar case, a low dielectric constant, thick material (Rohacell structural foam) is used as

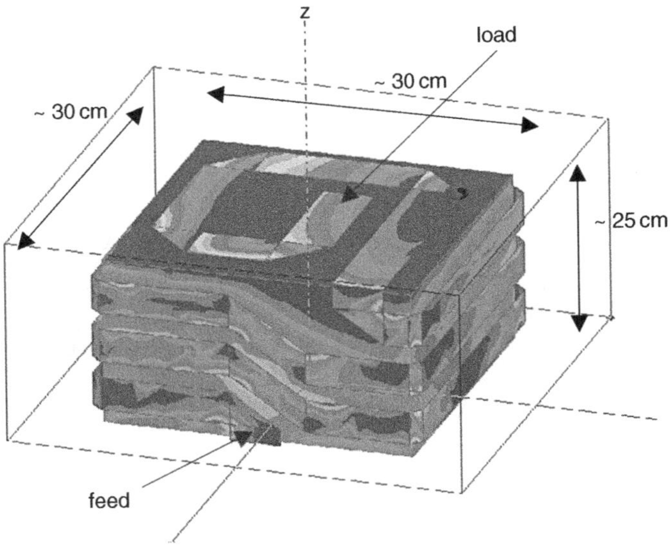

Figure 5.24 Current densities of the three-dimensional folded Beverage antenna highlighting the features of the radiator.

the substrate for the antenna, which helps maximize the antenna efficiency. The conductor width is chosen to be 50 Ω to ensure that a good impedance response can be achieved for the radiator and the gap between the conductors is approximately half the width of the conductor. In previous investigations the authors found that this helps to reduce ripple in the radiation patterns and also allows for a longer conductor to be used in the given volume.

The three-dimensional folded Beverage antenna was then prototyped; a solid drawing of the radiator is shown in Figure 5.25. There is one important difference between the radiator shown in Figure 5.25 and that modelled in Figure 5.24. For the prototype the locations of the feed and the load are swapped. This was done because of the ease in which the feed can be connected to the remainder of the mount when it is located at the top.

To realize the three-dimensional antenna is not a trivial task. Similar to the low-profile version, very thin (0.254 mm) FR4 was used to realize the conductors on the surface of the cube. The ground-plane was realized by a metallic box. Between the ground-plane and the FR4 material was a 1 cm layer of Rohacell foam. To join the conductors on each side of the cube together copper tape was used. The same procedure was used to realize the load as was used for the low-profile folded Beverage antenna. Finally, to create the feed an SMA connector was extended to the appropriate track on the three-dimensional antenna. Figure 5.26 shows a photograph of the developed antenna.

The measured return loss of the three-demenisonal folded Beverage antenna is shown in Figure 5.27. As can be seen from the response the performance is similar to a conventional folded Beverage radiator. The Beverage-based antennas have a similar performance and show typical responses for loaded antennas mounted on thick material, namely they are very well matched at low frequencies and then as the frequency increases the matching degrades. This is due to the feeding mechanism for the antenna, where the coaxial cable is effectively terminated at the ground-plane of the radiator and the centre conductor extends from this

(a)

(b)

Figure 5.25 Illustrations of the three-dimensional folded beverage radiator assembly without cover: (a) top view and (b) bottom view.

Figure 5.26 Photograph of the three-dimensional folded Beverage antenna without its radome.

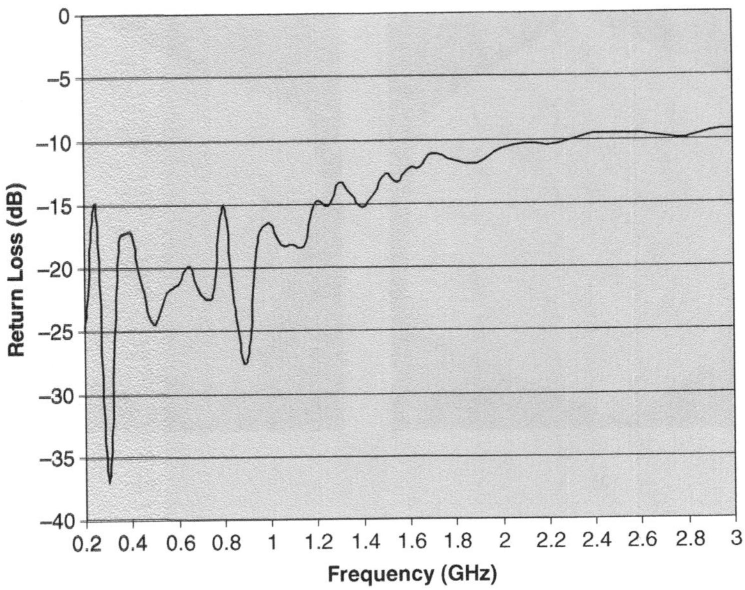

Figure 5.27 Measured return loss of the three-dimensional folded Beverage antenna.

junction to the conductor of the radiator. As the frequency increases this 'wire' becomes electrically longer and therefore has a greater impact on the input impedance response of the antenna, as is evident in the plots shown in Figure 5.27. This effect can be easily rectified by incorporating several alternatives: using a wideband balun at the feed, extending the coaxial cable closer to the junction with the antenna conductor (effectively reducing the electrical length of the 'wire'), or using a multilayered feed arrangement.

The radiation performance of the three-dimensional folded Beverage antenna was measured at NAVAIR, Pautxent River. Once again, due to the unusual shape of the antenna, for the measurements the following convention was used. The middle of the top of the antenna was aligned to zero degrees. The orthogonal axes (0 and 90°, or x and y) on the antenna were then marked (see Figure 5.28 for the marking convention) and the patterns in these two cuts were rotated and measured, for both vertical and horizontal excitations. These two planes are defined as pitch (P) and roll (R). All radiation patterns in this section use this terminology.

The radiation patterns (including gain) for the three-dimensional Beverage antenna are shown in Figures 5.29 to 5.34. The advantages associated with lifting the height constraint are evident in comparing these plots to the patterns of the low-profile version. The gain has been increased, in particular at the lower frequencies, while the polarization diversity has been maintained for this radiator. The gain at 225 MHz is −5 dBi. The maximum gain of 5 dBi occurs at 1200 MHz and then decays at the higher frequencies due to the reasons outlined for the low-profile Beverage. The decay is more rapid for the three-dimensional Beverage, which may be due to how the individual panels of the radiator were connected together to form the three-dimensional antenna; here copper tape was used to form the electrical connection between the panels. The three-dimensional Beverage shows less scalping than the low-profile version as the folding is more symmetrical here.

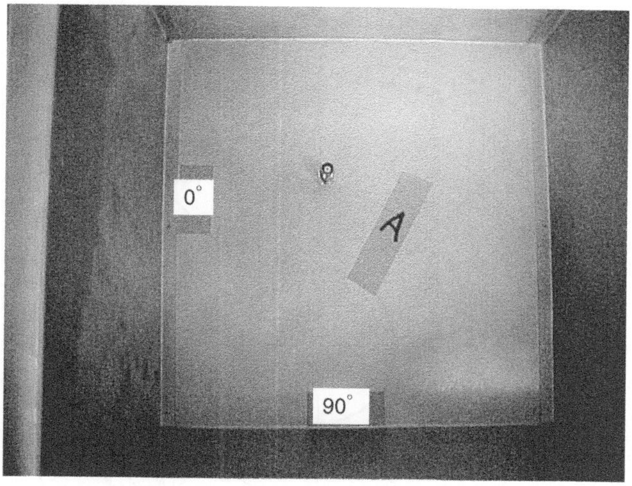

Figure 5.28 Photograph of the inside of the three-dimensional Beverage antenna showing the measurement convention.

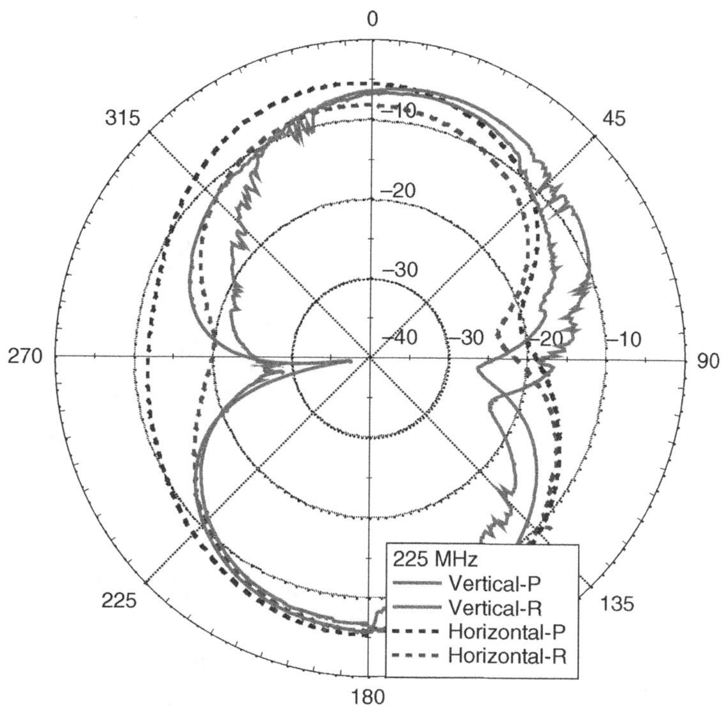

Figure 5.29 Radiation performance of the three-dimensional Beverage antenna at 225 MHz.

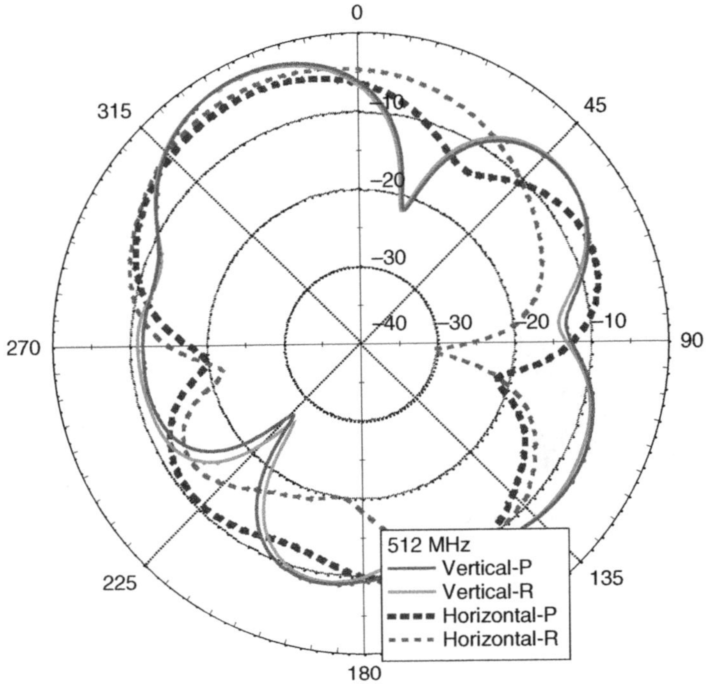

Figure 5.30 Radiation performance of the three-dimensional Beverage antenna at 512 MHz.

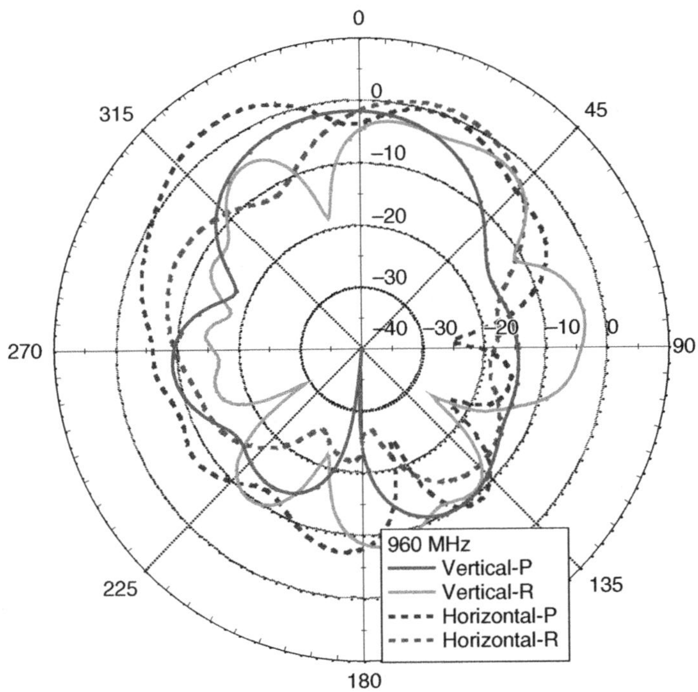

Figure 5.31 Radiation performance of the three-dimensional Beverage antenna at 960 MHz.

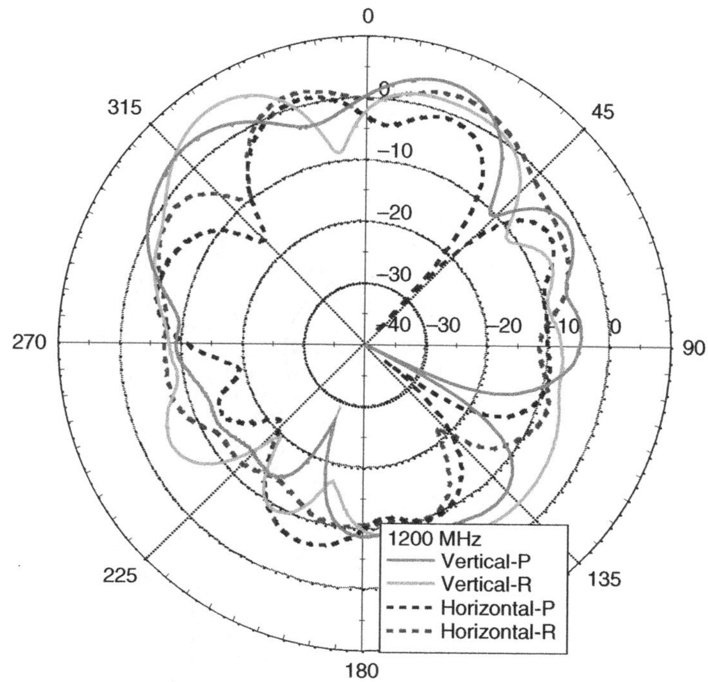

Figure 5.32 Radiation performance of the three-dimensional Beverage antenna at 1200 MHz.

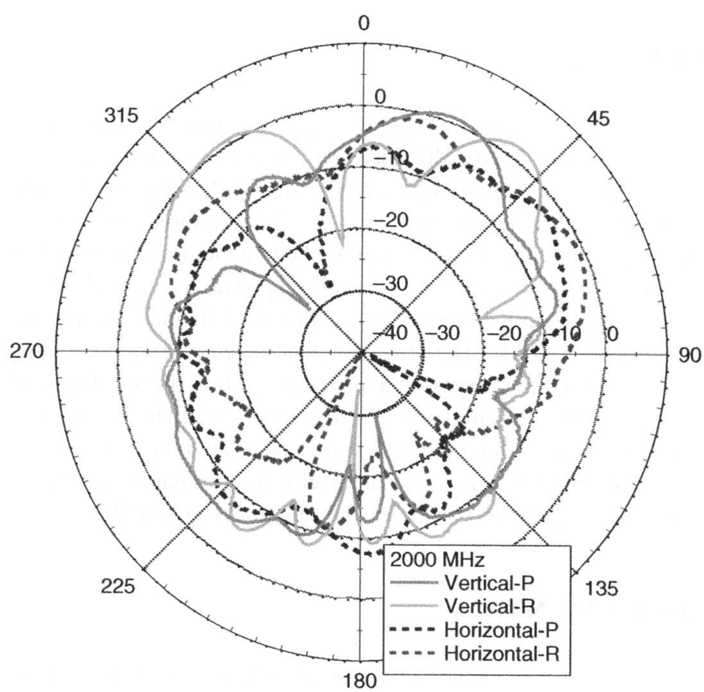

Figure 5.33 Radiation performance of the three-dimensional Beverage antenna at 2000 MHz.

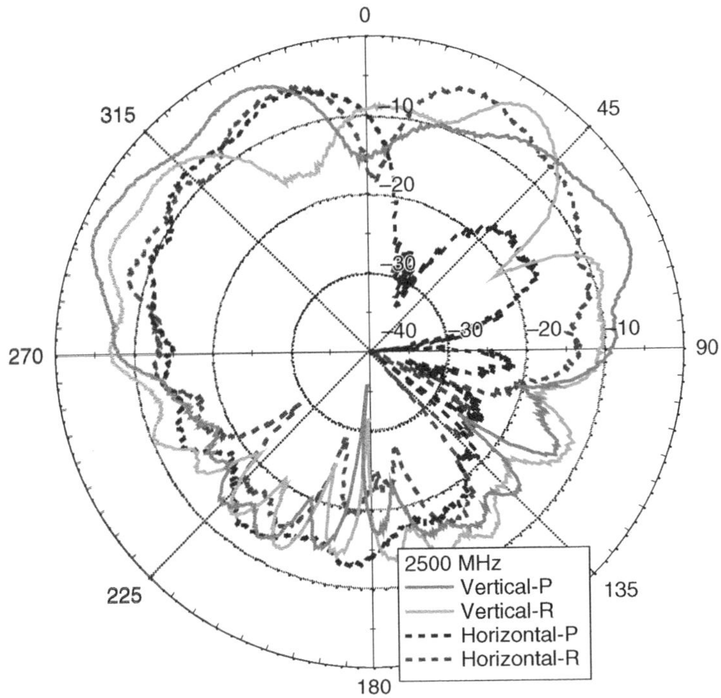

Figure 5.34 Radiation performance of the three-dimensional Beverage antenna at 2500 MHz.

5.6 SUMMARY

In this chapter a folded version of a printed Beverage antenna was presented. The antenna displays the wide impedance bandwidth nature of a conventional Beverage antenna, as well as a near-omnidirectional radiation pattern. In this work the general design parameters associated with this antenna were presented and also its measured return loss and radiation performance were reported. The authors then looked in detail at the techniques used to improve the return loss performance and radiation response, in particular focusing on how the shape of the conductor spiral impacts the overall performance. In the later sections of this chapter, variations of the folded Beverage antenna were presented: a rectangular-shaped, low-profile version and a three-dimensional prototype. The characteristics of these forms of wideband radiators were measured and the responses were found to be similar to the conventional folded Beverage antenna, highlighting the versatility of this form of wideband radiator. The folded Beverage antenna is very useful as a radiating element for wideband systems where the overall space and shape of the radiator is constrained.

ACKNOWLEDGEMENTS

The authors would like to thank Naval Air Systems Command at Patuxent River, Maryland, for the use of their anechoic chamber for the measurements of the antenna radiation patterns and gain presented here. They would also like to thank Steve Goodall and his team at

CERDEC, New Jersey. This project was sponsored in part by NavAir (through the SBIR program), CERDEC, and TEDCO, Maryland.

REFERENCES

1. http://jtrs.army.mil/
2. D. S. Filipovic and J. L. Volakis, 'Novel slot spiral antenna design for dual-band/multiband operation', *IEEE Transactions on Antennas and Propagation*, **51**, 430–440, March 2003.
3. J. J. H. Wang and V. K. Tripp, 'Design of multioctave spiral-mode microstrip antennas', *IEEE Transactions on Antennas and Propagation*, **39**, 332–335, March 1991.
4. C. H. Chio and D. H. Schaubert, 'Parameter study and design of wide-band widescan dual-polarized tapered slot antenna arrays', *IEEE Transactions on Antennas and Propagation*, **48**, 879–886, June 2000.
5. A. Serrano-Vaello and D. Sanchez-Hernandez, 'Printed antennas for dual-band GSM/DCS 1800 mobile handsets', *IEE Electronics Letters*, **34**, 140–141, January 1998.
6. S. D. Targonski, R. B. Waterhouse and D. M. Pozar, 'Design of wideband aperture-stacked patch microstrip antennas', *IEEE Transactions on Antennas and Propagation*, **46**, 1246–1251, September 1998.
7. N. Kaneda, W. R. Deal, Y. Qian, R. Waterhouse and T. Itoh, 'A broadband planar quasi-Yagi antenna', *IEEE Transactions on Antennas and Propagation*, **50**, 1158–1160, August 2002.
8. Y.-X. Guo, K.-M. Luk and K.-F. Lee, 'L-probe fed thick-substrate patch antenna mounted on a finite ground-plane', *IEEE Transactions on Antennas and Propagation*, **51**, 1955–1963, August 2003.
9. H. H. Beverage, C. W. Rice and E. W. Kellogg, 'The wave antenna, a new type of highly directional antenna', *Transactions of AIEE*, **42**, 215, 1923.
10. D. M. Pozar, *Microwave Engineering*, Addison Wesley, New York, 1993.
11. Personal communication, LCC, Virginia.
12. P. B. Kenington, 'Linearized transmitters: an enabling technology for software defined radio', *IEEE Communications Magazine*, **40**, 156–162, February 2002.
13. L. C. Godara (ed.), *Handbook of Antennas in Wireless Communications*, CRC Press, New York, 2002.
14. J. T. Aberle, D. M. Pozar and C. R. Birtcher, 'Evaluation of input impedance and radar cross section of probe-fed microstrip patch elements using an accurate feed model', *IEEE Transactions on Antennas and Propagation*, **39**, 1691–1696, December 1991.

6

Printed Tapered Slot Antennas

Amin M. Abbosh[1], **Marek E. Bialkowski**[1] **and Hing K. Kan**[2]
[1] *School of ITEE, The University of Queensland, Brisbane, Queensland, Australia*
[2] *School of Engineering and Physics, The University of the South Pacific, Suva, Fiji*

6.1 INTRODUCTION

Over the last two decades there has been an increasing interest in planar antennas for microwave and millimetre-wave systems. The various kinds of planar antennas presently in use may be classified into two categories: broadside radiating and endfire radiating ones. Examples of planar antennas radiating in the broadside direction include resonant-type antennas such as printed dipoles, slots, and microstrip patches. Broadside radiation for these antenna elements occurs at their fundamental mode of operation, which is related to the lowest frequency of their operation. These resonant-type antennas are formed by suitably configured conductors printed on a dielectric substrate. Because of the use of planar technology they are low-profile, easy to fabricate, and offer straightforward integration with active devices. Their disadvantage is a relatively narrow operational bandwidth. Because they feature a small effective aperture, their radiation pattern exhibits a large beamwidth and thus these antennas are of a low gain. This characteristic makes them unsuitable for a stand-alone use in applications requiring high directivity such as point-to-point wireless communications. The requirement for a large operational bandwidth and higher gain can be readily met by travelling-wave-type antennas which belong to the category of endfire antennas. One prominent example of this type of antennas is a tapered slot antenna, which is the subject of a thorough review in this article. This antenna can offer an exceptionally wide operational bandwidth, and because of this property it is frequently used in radar, remote sensing, and ultra wideband communications.

Tapered slot antennas (TSAs) were first introduced in the mid 1950s [1] and early 1960s [2] when Eberle *et al.* produced a waveguide-fed flared slot antenna for use in aircraft skins where conventional reflector or wire antennas could not be easily integrated. Early predecessors of this type of antennas used longitudinal tapered slots on a semirigid coaxial

Printed Antennas for Wireless Communications Edited by R. Waterhouse

cable, which was also employed as the feed structure. Although covering several octaves in bandwidth, this radiating structure was not simple to implement in practice since the taper (100:1) had to be cut into the shield of the coaxial cable. In 1974, Lewis *et al.* [3] designed several types of tapered slot antennas to get multioctave bandwidths and multiple polarizations. In 1979, Gibson [4] developed a strip-line-fed exponentially planar TSA for an 8–40 GHz frequency band, which he called the Vivaldi aerial. His design was the first recognized TSA that showed symmetric E- and H-plane beamwidths, low sidelobes, and moderate gain. In the same year, Prasad and Mahapatra [5] introduced the linearly tapered slot antenna (LTSA). Their antenna was short, about one wavelength, and etched on a 25 mil alumina substrate. Yngvesson *et al.* [6] presented experimental results on constant-width slot antennas (CWSAs). In all these works, it has been demonstrated experimentally that the tapered slot antenna can offer a very wide pattern bandwidth with a symmetric main beam despite its planar geometry. With regard to theoretical modelling of this new class of antennas, in Reference [7] Janaswamy applied the moment method to predict their radiation properties. This theoretical model for predicting the radiation characteristics has been of great assistance to obtain its successful design.

In this chapter a thorough examination is presented of tapered slot antennas. In Section 6.2 the various forms of TSAs are investigated. Then an overview is given of the advantages and disadvantages of TSAs compared to other forms of antennas in Section 6.3. In Section 6.4 a review is presented of the procedures to model tapered slot antennas and in Section 6.5 a parameter study associated with this radiator is given, where the impact on performance of varying different parameters associated with the TSA, including the slot profile, is discussed. Section 6.6 gives an overview on how to design TSAs and in Section 6.7 their feeding techniques are examined. In Section 6.8 a review is given of how TSAs can be incorporated into linear, planar, or circular arrays. Section 6.9 highlights some of the applications of TSAs and finally in Section 6.10 a summary is given of the overall findings presented in this chapter.

6.2 TYPES OF TAPERED SLOT ANTENNAS

In its general form, the planar tapered slot antenna consists of a tapered slot cut in a thin film of metal without or with an electrically thin dielectric substrate supporting it. The slot is wide at the outer or endfire end (typically this width is at least one-half of a free-space wavelength at the minimum operating frequency), and tapers inwardly such that the slot eventually becomes very narrow, forming a slot line. The slot line at the base of the tapered slot is usually coupled to a coaxial transmission line, microstrip line, or coplanar waveguide, which is used as a feeding element. When launched from the feeding point, the travelling wave propagating along the slot of a radiating antenna (at a phase velocity that is less than the speed of light) gradually radiates in the endfire or outward direction.

TSAs can be classified into different types depending on the shape of the taper and/or configuration of the antenna layers. The most important types are the linearly tapered slot line antenna (LTSA), the constant width slot line antenna (CWSA), the exponentially tapered slot antenna (ETSA or 'Vivaldi'), the antipodal tapered slot antenna (ATSA), the broken linearly tapered slot line antenna (BLTSA), the dual exponentially tapered slot antenna (DETSA), and the elliptically tapered slot antenna (ELTSA). In the following, configurations of these TSAs accompanied by their historical origins are described.

Figure 6.1 shows the configurations of the LTSA and the Vivaldi antenna, as introduced by Prasad and Mahapatra [5] and Gibson [4], respectively, in 1979. Due to the requirements

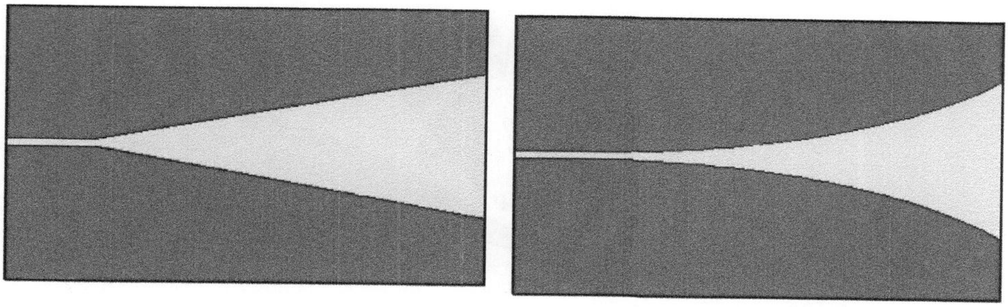

Figure 6.1 Schematic diagram of the LTSA (left) and ETSA (or Vivaldi) (right).

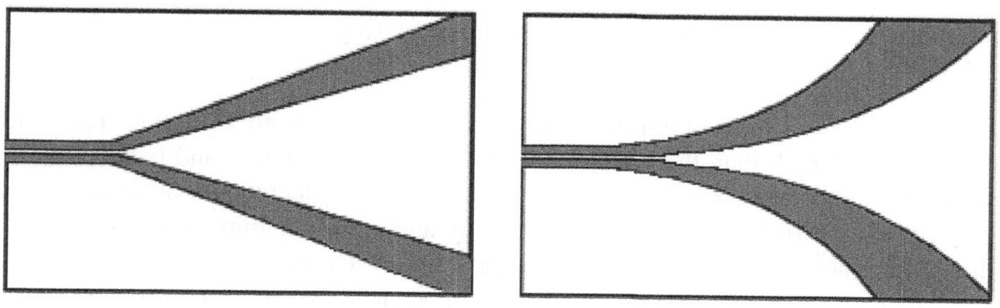

Figure 6.2 Schematic diagrams of the V-LTSA (left) and V-ETSA (or DETSA) (right).

for integrating the TSAs with the semiconductor devices, such as diode mixers, a new variant of the TSA, known as V-antennas, was introduced in 1979 [8]. They are shown in Figure 6.2. Here, the V-LTSA and a V-ETSA, also called DETSA, are excited by a coplanar strip line.

These antennas having a reduced ground-plane can enhance the antenna performance by suppressing the surface waves. It has been found that careful design of the V-antennas (including DETSA) can give improved directivity and half-power beamwidth (HPBW) performance over the ETSA or the LTSA. It has to be noted that V-antennas are compatible with uniplanar feed structures which allow the antenna to be excited using either a microstrip line or a waveguide.

Another variation of TSA, named CWSA, is shown in Figure 6.3. Introduced in 1985 [9], this antenna is composed of the main radiating element in the form of a constant width slot,

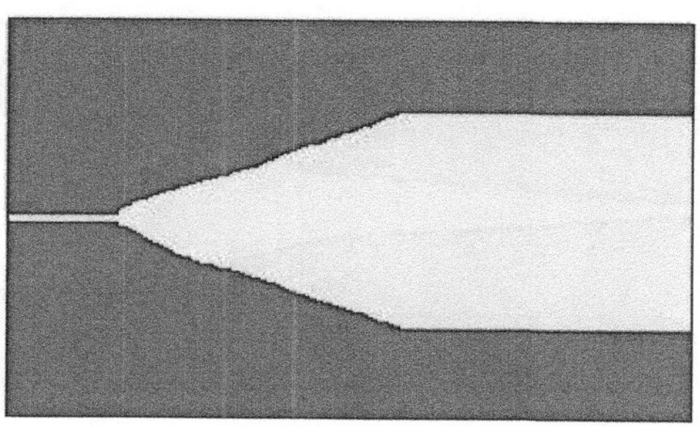

Figure 6.3 Configuration of the CWSA.

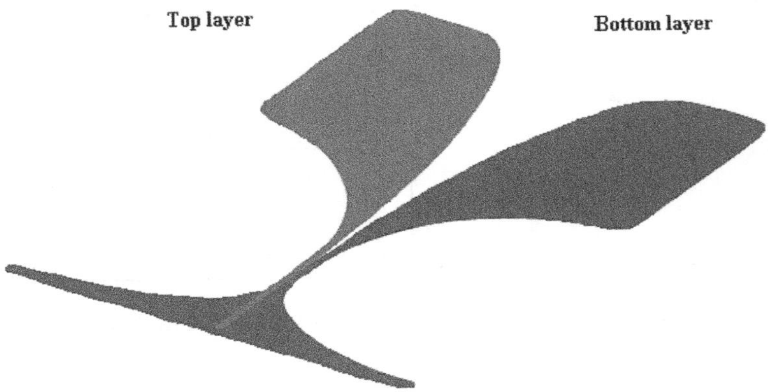

Figure 6.4 The antipodal TSA.

which is preceded by a linear or exponential tapered slot ending with a slot line. This antenna features a shorter length than the LTSA and exhibits a moderate gain and bandwidth.

Another variant of the conventional TSA is the antipodal tapered slot antenna, as shown in Figure 6.4. The differences between this new antenna and its uniplanar counterparts are as follows. In practice, the conventional planar TSA is fed by a balanced slot line. This creates a challenge in terms of fabrication and impedance matching of the slot line. The slot line fabricated on a low dielectric constant substrate has a relatively high impedance, which makes matching to a microstrip feed a difficult task. The antipodal TSA of Figure 6.4 overcomes this problem as the impedance match and transition from a slot line to a microstrip line is straightforward. Because of the antipodal arrangement, one of the radiating fins can easily be converted to a microstrip line while the other one can be shaped to create a ground-plane for the microstrip. This new layout allows for direct feeding of the TSA from a microstrip line [10]. One disadvantage of this type of TSA is increased cross-polarization radiation, which occurs mainly in the diagonal plane.

The BLTSA [11], shown in Figure 6.5, is another variation of a tapered slot antenna. It consists of three linear taper sections. In comparison with the LTSA and the CWSA the BLTSA exhibits a lower (around 2 dB) cross-polarization ratio in the diagonal plane. The reason for this is a shorter distance between the phase centres of the E- and H-planes [12]. Furthermore, because of smaller dimensions than the Vivaldi or the LTSA, the BLTSA is less prone to damage the fragile supporting membrane.

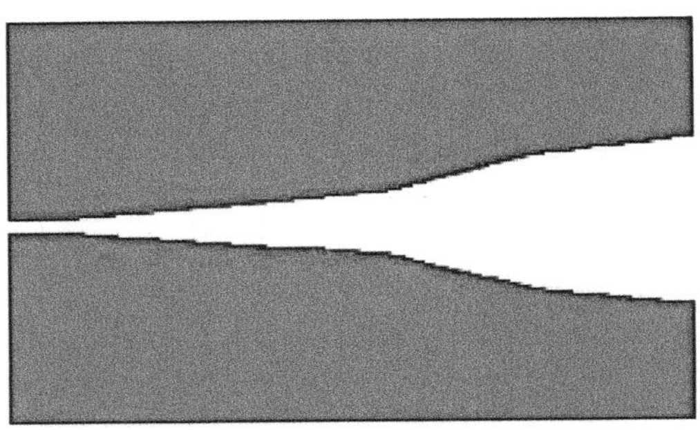

Figure 6.5 Configuration of the BLTSA.

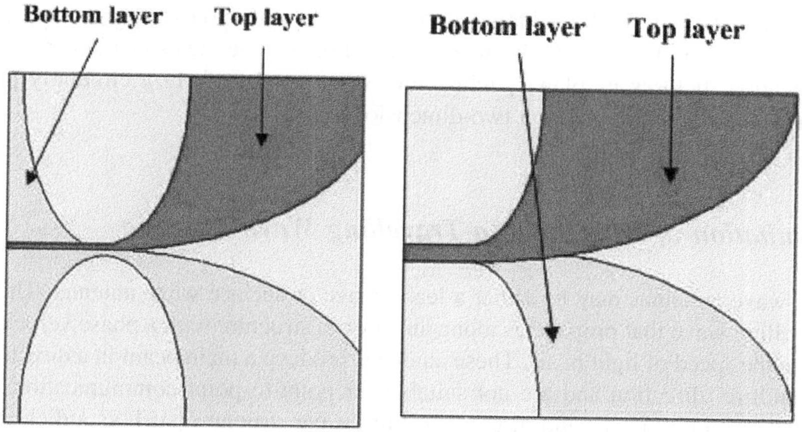

Figure 6.6 Two examples of the ELTSA: antipodal with a microstrip feeder (left) and antipodal with a parallel strip feeder (right).

Figure 6.6 shows two varieties of antipodal elliptical tapered slot antenna, described in References [13] and [14], which similar to the BLTSA fulfil the compact size requirement. Because of the compact size, these antennas are considered good candidates for use in an ultra wideband microwave imaging system [13, 14]. The two antenna varieties, being around one effective wavelength height with length at the lowest frequency of operation, 3.1 GHz, cover an ultra wide frequency band of 3.1–10.6 GHz.

6.3 ADVANTAGES AND DISADVANTAGES OF TAPERED SLOT ANTENNAS

6.3.1 Introduction

One of the principal advantages of tapered slot antennas is that they can be easily fabricated using standard integrated circuit lithography techniques. This approach is convenient from the point of view of integration with hybrid or monolithic microwave circuits to form a complete transceiver. Because of the use of planar substrates, similar to microstrip antennas, the TSA can be low in profile and light in weight. In terms of size, a general constraint applicable to all tapered slot antennas is that the width of the slot at the radiating end should reach at least one-half of the free-space wavelength at the lowest frequency of operation. As the aperture size becomes electrically larger at higher frequencies, the TSA antenna offers a higher directivity or gain at these frequencies. Another significant advantage of tapered slot antennas is that they are capable, despite their planar geometry, of producing a symmetric beam in the electric (the E-plane) and magnetic (the H-plane) field planes when appropriate dimensions and parameters for the slot shape, slot length, dielectric thickness, and dielectric constant are chosen. Other desired radiation characteristics including beamwidth variation can also be obtained by varying the slot length and the opening angle.

Some of disadvantages of this type of antenna are as follows. TSAs generally exhibit a high cross-polarization level in the diagonal plane. Although the TSA is a directional antenna, some types of it may have a broad main beam in both the azimuth and elevation planes, and contain several relatively high backlobes. In comparison with microstrip antennas, the TSA lacks the versatility in terms of multifunction operation such as dual- or multifrequency, or dual-polarization. It loses its planar architecture when used to form a circularly polarized antenna or when it is employed in a two-dimensional array.

6.3.2 Limitation of the TSA as a Travelling Wave Antenna

Travelling wave antennas may be either a leaky wave or surface wave antenna. The former uses a travelling wave that propagates along an antenna structure with a phase velocity that is greater than the speed of light in air. These antennas produce a main beam in a direction other than the 'endfire' direction and are not suitable for point-to-point communication. Surface wave antennas, on the other hand, use a travelling wave structure (such as a dielectric on a ground-plane or a periodic structure) for which the phase velocity of the travelling wave is less than or equal to the speed of light. Surface wave antennas therefore produce (or receive) endfire radiation, and so can be used in point-to-point communication systems.

As mentioned earlier, TSAs belong to the class of travelling wave antennas. Their operation is based on a travelling wave propagating along the surface of the antenna taper with a phase velocity less than the speed of light. Under this condition the radiation is endfire. Yngvesson *et al.* [9] empirically investigated three groups of TSAs: LTSA, CWSA, and ETSA. They found that for certain conditions (i.e. thin substrates with low dielectric constants), TSAs were well-behaved travelling wave antenna. They determined that in order to remain as well-behaved travelling wave antenna, the effective dielectric thickness t_e, which is given by; $\frac{t_e}{\lambda_0} = (\sqrt{\varepsilon_r} - 1)t/\lambda_0$ should be in the order of 0.005–0.03 wavelengths for $4 - 10\lambda_0$ antennas. Antennas outside these dimensions quickly diverge from travelling wave antenna characteristics.

6.4 THEORETICAL MODELS FOR TAPERED SLOT ANTENNAS

Since the introduction of the TSA in the 1950s, its properties have been thoroughly investigated, and this has been done mainly using experimental means. In particular, the trial-and-error and empirical approaches dominated the early years of design of the TSA. The theoretical treatment of this antenna has not been extensive due to the complexity of dealing with a relatively large electrically radiating structure, with the nonuniform distribution of the wave in the tapered slot and the variable propagation constant and impedance in the propagation direction. This trend has changed in recent years because of the development of many numerical full-wave electromagnetic field analyses that can be run on powerful computers. The techniques that are in use today can be classified as the moment method (MoM) (both spectral and spatial domains), the transmission line matrix, the finite difference time domain (FDTD), and the finite element method (FEM).

Janaswamy and Schaubert [15] presented a method for calculating the radiation pattern of endfire TSAs with or without a dielectric substrate. Their theory is valid for any smooth taper of the slot that can be approximated by piecewise linear segments. The approach is illustrated in Figure 6.7.

The method of analysis consists of two steps. In the first step, the tangential component of the electric field distribution in the tapered slot is obtained. In the second step, far-fields

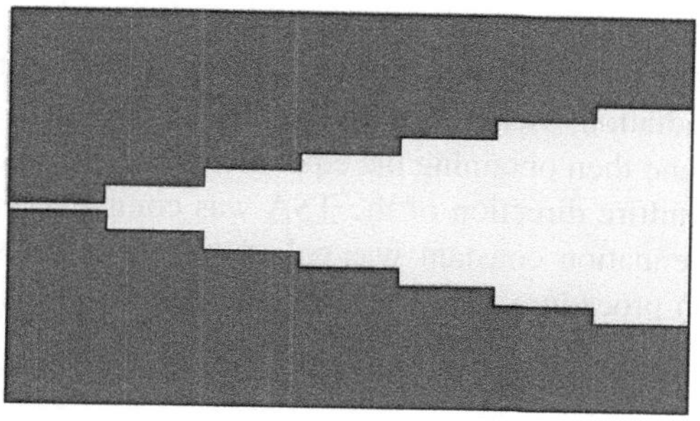

Figure 6.7 Piecewise linear approximation of the taper.

radiated by the equivalent magnetic current in the slot are obtained by using an appropriate Green function.

Simulations and experimental results show that this method is useful only for antenna with large dimensions (more than six free-space wavelengths in height). To overcome this difficulty Janaswamy proposed another method based on MoM [7]. In the new proposed method, the moment method solution of the electric field integral equation for the electric surface currents flowing on the conducting plates of the antenna is applied. In order to model the surface currents of the antenna properly, quadrilateral or triangular patches were used to segment the plates of the antenna. This method results in a large number of unknowns. To simplify the analysis, the geometry of the antenna is modified by assuming that the antenna structure has two rectangular-shaped conductors. This approach has a negative impact on the method's accuracy, especially when the height of the antenna is comparable to the wavelength. Koksal and Kauffman [16] used another version of the MoM to get more accurate results by avoiding the modification in antenna geometry.

The spectral domain method was used by Wang *et al.* [17] to analyse the performance of TSAs on dielectric substrates. The method is based on the exact Green function and the moment method procedure in the spectral domain to obtain the reflection coefficient and the unknown electric currents on the antenna. The currents are approximated by subsectional basis functions in both the transverse and propagation directions. When compared to measurements done on the LTSA this method shows a good accuracy, although it is still inferior to finite difference time domain (FDTD) methods, such as those presented in References [18] and [19].

Colburn and Rahmat-Samii [18] used the FDTD method to improve the radiation performance of the LTSA. In this method, the linear taper slot was stair-step approximated using five steps per wavelength. With this method they found that some perturbations in the ground-plane of the antenna could have a significant effect on the radiation characteristics of the antenna. In Reference [19] a three-dimensional FDTD method was employed to simulate the complete structure of a planar Ka-band antipodal LTSA. Good agreement was found between results of the presented method and measurements.

It has to be noted that the MoM or FDTD methods require considerable computer resources in terms of memory and execution time. In Reference [20], Oraizi and Jam presented an alternative method to obtain an optimum design of the TSA profile and geometry. The optimization was achieved through modelling by a step line terminated in free-space intrinsic impedance. The impedance matching of the TSA input impedance to the generator impedance was achieved by minimizing an error criterion constructed as the magnitude squared of the difference between the generator and input impedances over the desired bandwidth. The attenuation constant of the step line was computed by the broadside radiation of the TSA and the longitudinal power flow. The spectral domain immitance approach [21] was used to compute the broadside radiation, by first determining the electric field components in the slot by the Galerkin method and then obtaining the equivalent magnetic surface current densities. The power flow in the endfire direction of the TSA was computed using Poynting's vector in the substrate. The attenuation constant was calculated for various values of slot width. Finally, the minimization procedure gave the slot line widths and lengths.

6.5 PARAMETRIC STUDY OF TAPERED SLOT ANTENNAS

A good understanding of the effects of the various parameters (length, width, dielectric thickness, ground-plane size, taper profile, etc.) of a tapered slot antenna on its electrical characteristics can be of great advantage to the designer. Here, a thorough study devoted to this topic is provided. The main emphasis in the presented parameter analysis is on the various factors of radiation pattern. The issue of return loss performance is not included here. It is well known that the quality of return loss and bandwidth of the TSA are mainly governed by the electrical length of this antenna. For the complete structure, the return loss performance is further affected by the slot line-to-feed line transition, the finite width of the antenna, and the type of taper used. These issues will be covered in a separate section devoted to various feeding methods of the TSA.

6.5.1 Directivity and Beam Width of Tapered Slot Antennas

Despite the completely planar geometry, the TSA can produce symmetric radiation patterns in the planes parallel to the substrate (the E-plane) and perpendicular to the substrate (the H-plane). The beam width narrows and the directivity increases as the length (L) of the antenna is increased. For a travelling wave antenna with a constant phase velocity, there is an optimum phase velocity ratio, $c/v_{\mathrm{p}} = 1 + \lambda_0/(2L)$, which results in maximum directivity [22]. For a larger total phase shift than this, the beam moves into invisible space and a null develops on the axis, rendering the antenna useless.

The phase velocity along a TSA may be varied by changing the dielectric thickness, the dielectric constant, or the taper shape, so that beam widths are obtained that are in the required range. These considerations include the general constraint for the TSA that the slot width should reach at least one-half of the free-space wavelength for effective radiation to occur.

In the following, the finite element method analysis is used by the authors to study the effect of length of the TSA on its performance. The findings are accompanied by the views and comments presented in the antenna literature by other researchers. The antenna that is assumed in the investigation is the LTSA with Rogers RO4003C substrate having a dielectric constant ($\varepsilon_{\mathrm{r}} = 3.38$). Different substrate thicknesses are considered. Results of the analysis are shown in Figures 6.8 to 6.10 and the three-dimensional radiation patterns for different antenna lengths are shown in Figure 6.11. The gain of a TSA increases with the length, L, of the antenna, typically from 7 dB to around 14 dB as L increases from 1 to $4\lambda_0$. The maximum measured gain of 16–17 dB with a radiation efficiency of 80 % can be obtained for a long TSA with L greater than $6\lambda_0$.

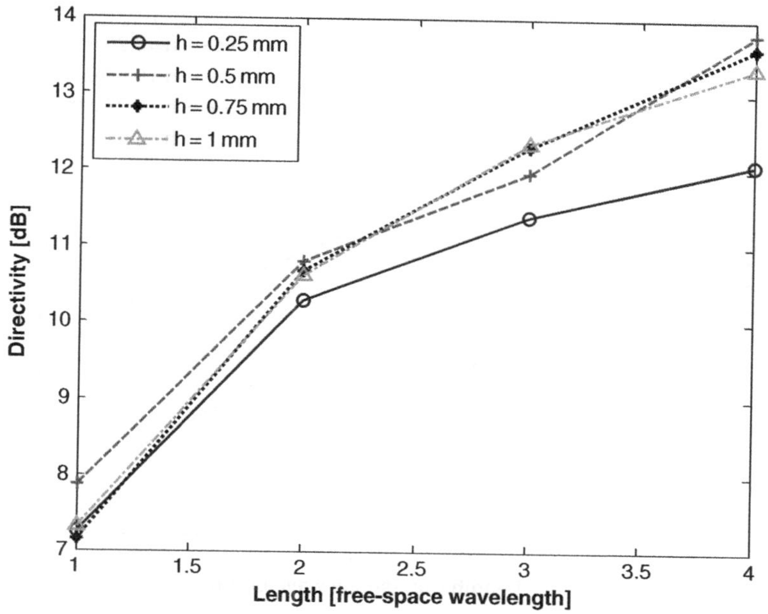

Figure 6.8 Variations of the directivity of the LTSA with antenna length.

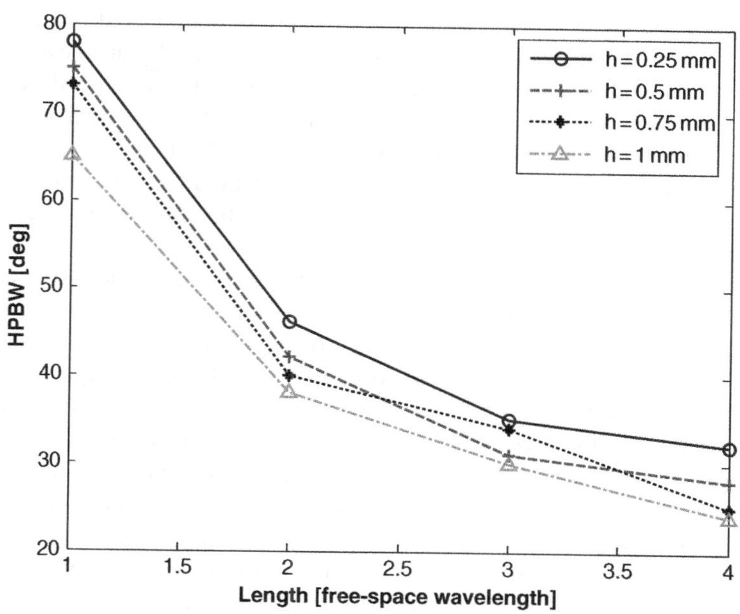

Figure 6.9 Variations of the E-plane HPBW of the LTSA with antenna length.

Results shown in Figures 6.8 and 6.9 indicate that the beamwidths, being inversely proportional to the gain, decrease rapidly as the length is increased; however, the H-plane beamwidth varies more slowly in comparison to the E-plane beamwidth, particularly for L less than $2\lambda_0$. In Reference [23] it was expected that the H-plane beamwidth follows $1/\sqrt{L}$

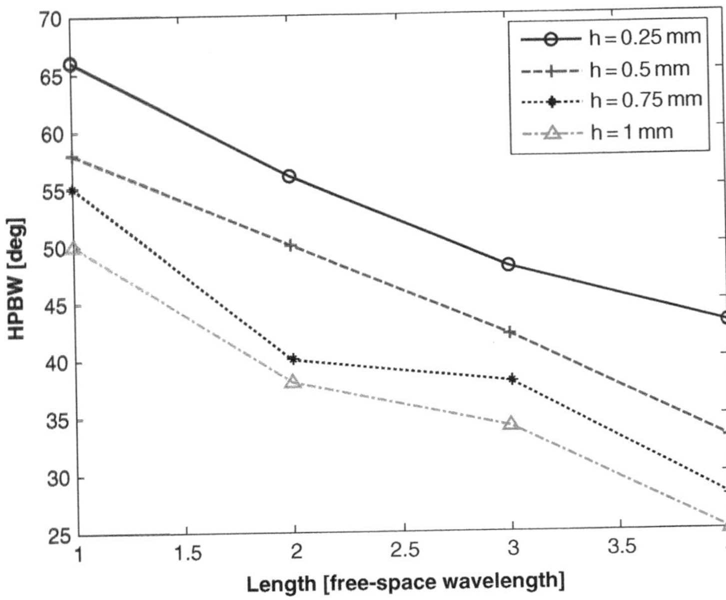

Figure 6.10 Variations of the H-plane HPBW of the LTSA with antenna length.

dependence, while the E-plane beamwidth depends more on the tapered angle. The results presented in Figure 6.10 agree well this estimation concerning the H-plane beamwidth, while for the E-plane beamwidth the results indicate that it follows $1/\sqrt{L}$ dependence when the antenna length is more than $2\lambda_0$ while it changes more rapidly for shorter lengths. It will be shown in another subsection that varying the tapered angle changes the phase velocity and hence the effective wavelength, which in turn changes the E- and H-plane beamwidths. Thus, a constant beamwidth in both the E- and H-planes can be achieved with proper choices of L and tapered angles.

The performance of a TSA is sensitive to the thickness and dielectric constant of the dielectric substrate. The presence of a dielectric substrate has the primary effect of narrowing the main beam of the antenna. The effect of dielectric substrates on the E- and H-plane beamwidths of the LTSA on Rogers RO4003C is investigated at 10 GHz, as shown in Figures 6.8, 6.9, and 6.12. By increasing the substrate thickness the E- and H-plane beamwidths become narrower. Concerning the effect of the dielectric thickness on the directivity it can be observed that, in general, increasing the dielectric thickness results in increased gain, but with higher sidelobes. Yngvesson *et al.* [9] identified an approximate optimum range of about 0.005–0.03 for the effective dielectric thickness of the substrate (t_e) normalized to the free-space wavelength. For thinner substrates, the beamwidth becomes wider, as shown in Figures 6.8 and 6.9, while for thicker substrates, the main beam breaks up. It is therefore especially important to observe the upper limit to the thickness of the substrate to avoid the main beam splitting.

For substrate thickness above the upper bound of $0.03\lambda_0$, unwanted substrate modes develop that degrade the antenna's performance, resulting in low efficiency and narrow bandwidth. This requires special attention with respect to TSAs operating at millimetre-wave frequencies, as mechanically fragile substrates need to be of only a few hundreds of micrometres thick. To avoid these problems, TSAs are proposed to be fabricated either on thick

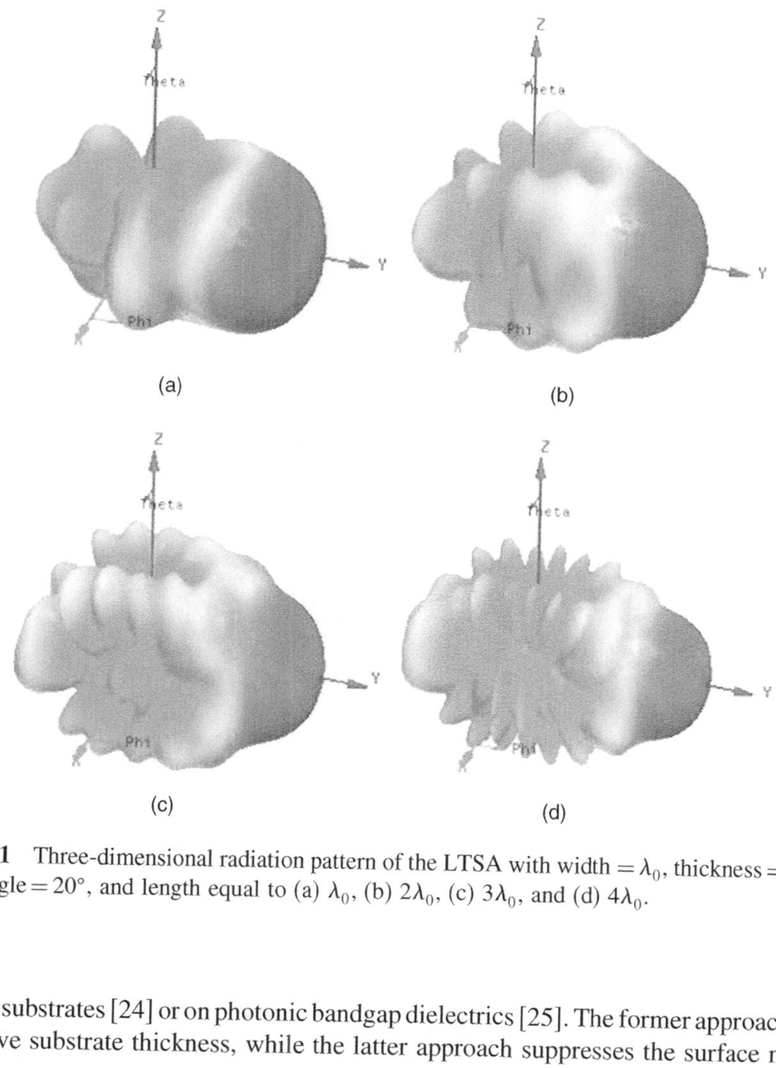

Figure 6.11 Three-dimensional radiation pattern of the LTSA with width $=\lambda_0$, thickness $=0.25$ mm, opening angle $=20°$, and length equal to (a) λ_0, (b) $2\lambda_0$, (c) $3\lambda_0$, and (d) $4\lambda_0$.

perforated substrates [24] or on photonic bandgap dielectrics [25]. The former approach reduces the effective substrate thickness, while the latter approach suppresses the surface modes.

6.5.2 Effect of Different Taper Shapes

In order to investigate the effect of different taper shapes on the TSA radiation pattern another set of FEM simulations was performed by the authors. In these simulations, Rogers RO4003C was assumed as the TSA substrate. The investigated antennas had the same dimensions (length $=2\lambda_0$ and width $=\lambda_0$) and opening angle (20°) but different taper shapes. In the performed FEM simulations, it was found that the taper profile had a strong effect on both the beamwidth and sidelobe level of the antenna. This is shown in the results in Figure 6.13. In this figure, antenna numbers 1, 2, 3, and 4 refer to the LTSA, ETSA, CWSA, and ELTSA, respectively. The results reveal that, in general, the E-plane beamwidth is narrower for the CWSA, followed by the LTSA, ETSA, and then ETSA for antennas with the same length, the same aperture size, and being developed on the same substrate. For the H-plane the ELTSA has the narrowest beamwidth followed by the CWSA, ETSA, and then LTSA.

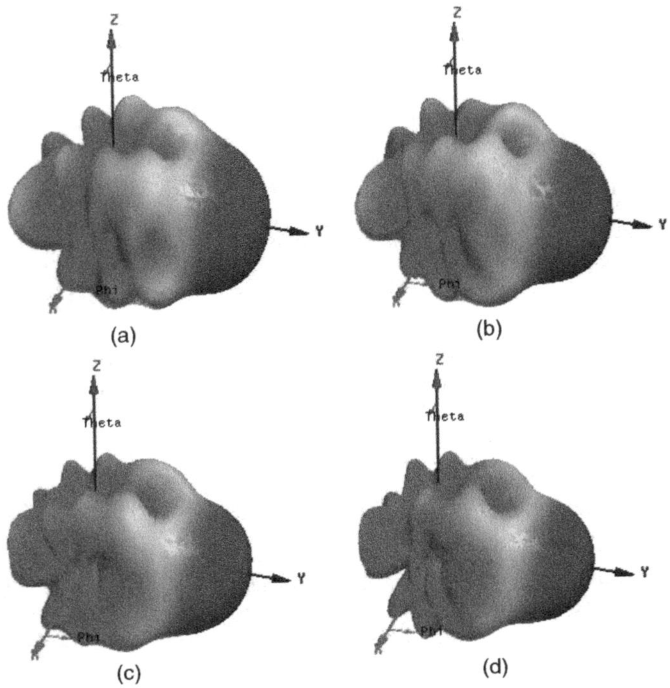

Figure 6.12 Three-dimensional radiation pattern of the LTSA with width $=\lambda_0$, opening angle $=20°$, length $=2\lambda_0$, and thickness equal to (a) 0.25 mm, (b) 0.5 mm, (c) 0.75 mm, and (d) 1 mm.

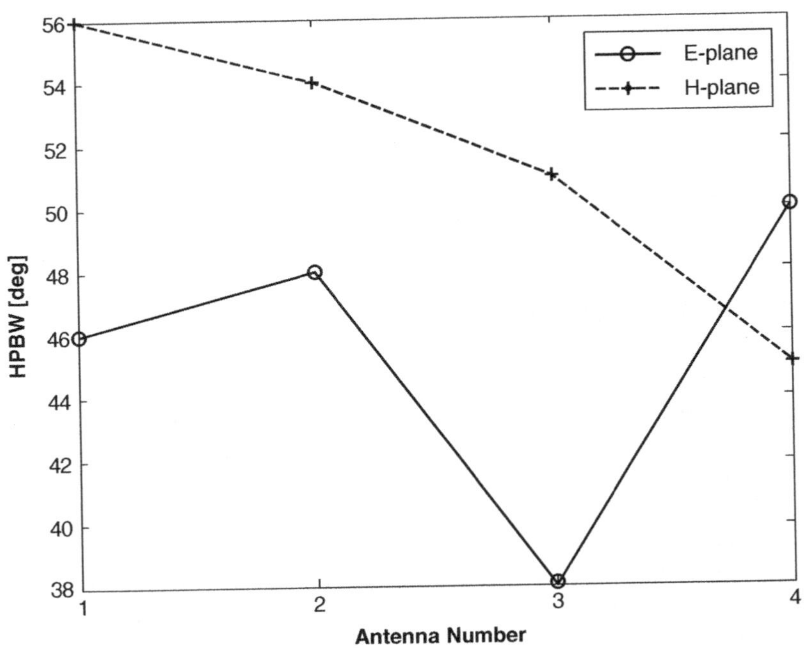

Figure 6.13 Effect of taper shape on the beamwidth of the TSA.

6.5.3 Effect of the dielectric constant

Figures 6.14 to 6.16 show the finite element analysis results for TSA antennas assuming a 0.25 mm thick substrate, length $= 2\lambda_0$, width $= \lambda_0$, and opening angle $= 20°$ with different values for the dielectric constant. The results presented in Figure 6.14 indicate that the E- and H-plane HPBWs of the TSA decrease when using a substrate compared to the TSA without substrate (dielectric constant $= 1$). For the E-plane HPBW there is a certain value for the dielectric constant where the TSA has the least beamwidth. The same conclusion is true for

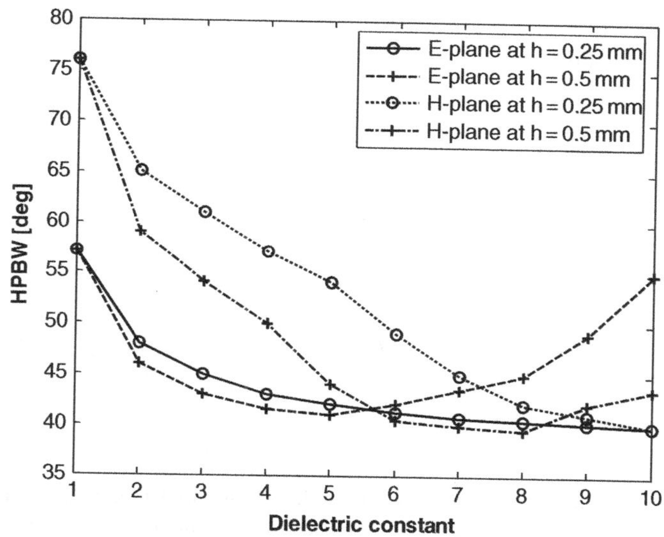

Figure 6.14 Variations of the HPBW of the LTSA with dielectric constant of the substrate.

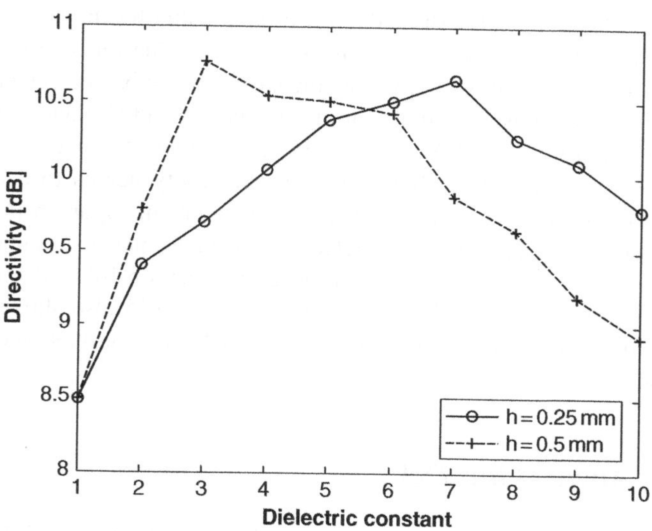

Figure 6.15 Variations of the directivity of the LTSA with dielectric constant of the substrate.

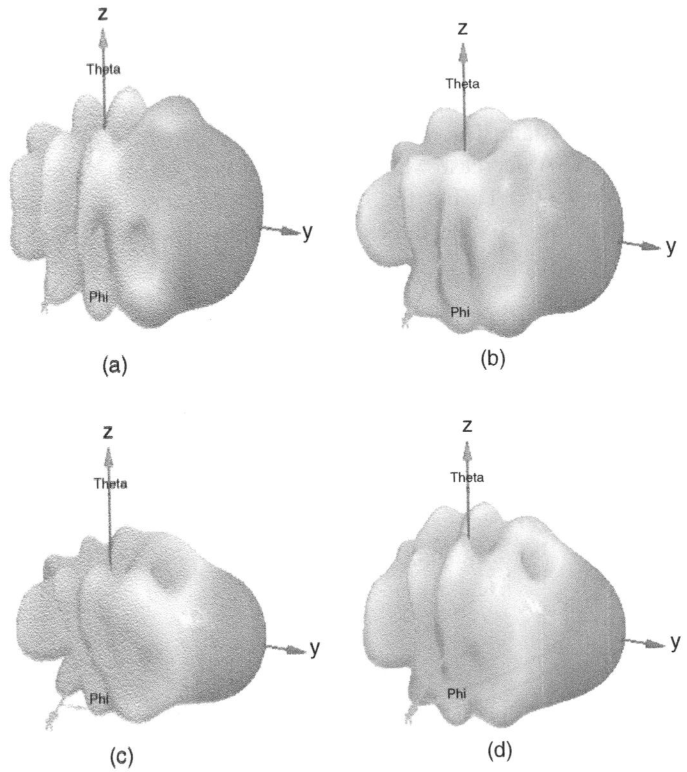

Figure 6.16 Three-dimensional radiation pattern for the LTSA with width $= \lambda_0$, opening angle $= 20°$, length $= 2\lambda_0$, and thickness $= 0.25$ mm, for dielectric constants equal to (a) 1, (b) 3, (c) 6, and (d) 9.

the H-plane HPBW although the minimum beamwidth could occur at a higher value for the dielectric constant. The results also show that for a fixed value for the antenna dimensions, length and width and substrate thickness, there is a certain value for the dielectric constant where the E- and H-plane HPBWs are equal. Increasing the substrate thickness decreases the value of the dielectric constant required to have equal E- and H-plane HPBWs.

Concerning directivity of the TSA, Figure 6.15 reveals that it increases with the dielectric constant up to a certain value, after which the directivity begins to decrease with an increasing dielectric constant. This phenomenon is apparent in Figure 6.16(c), where the three-dimensional radiation pattern for a dielectric constant equal to 6 appears to have a higher directivity in the endfire direction than those in Figures 6.16(a), (b), and (d), where the dielectric constants are 1, 3, and 9, respectively. Figure 6.15 also shows that the peak value for the directivity occurs at a lower value for the dielectric constant when the substrate thickness increases.

6.5.4 Opening Angle of the TSA

Figures 6.17 and 6.18 show the finite element analysis results for the LTSA on Rogers R4003C substrate when using different values for the opening, or tapering, angle. The

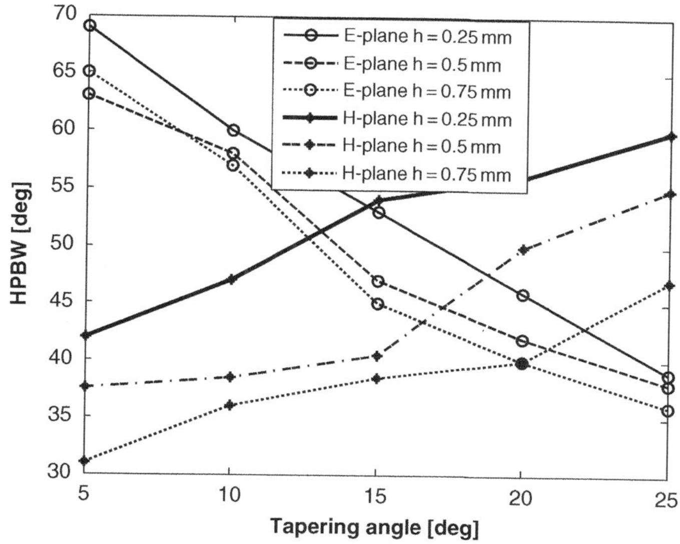

Figure 6.17 Effect of the tapering angle on the HPBW of the LTSA with width $= \lambda_0$ and length $= 2\lambda_0$.

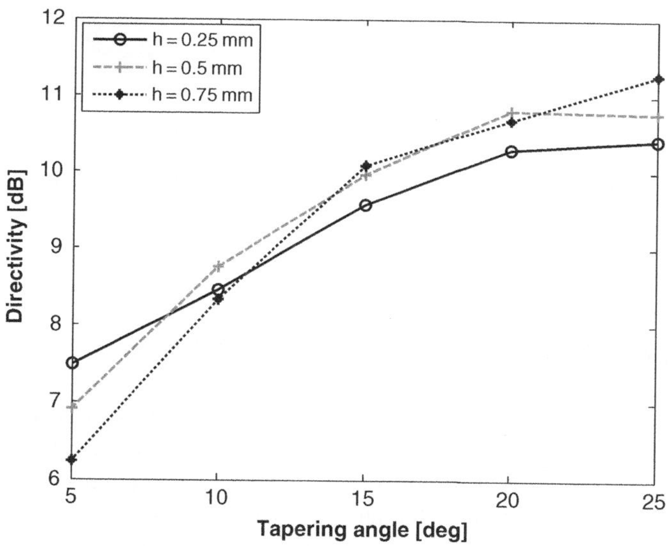

Figure 6.18 Effect of the tapering angle on the directivity of the LTSA with width $= \lambda_0$ and length $= 2\lambda_0$.

antenna width and length are λ_0 and $2\lambda_0$, respectively, while different values for the substrate thickness are considered. The E-plane beamwidth narrows somewhat as the opening angle is increased while the H-plane beamwidth increases slightly. Concerning the directivity, it seems that increasing the opening angle increases the directivity until a certain level, as can be seen in Figure 6.18. Those conclusions can be verified using the three-dimensional radiation pattern shown in Figure 6.19, where it is observed that the endfire directivity increases and the sidelobe level decreases with an increasing tapering angle.

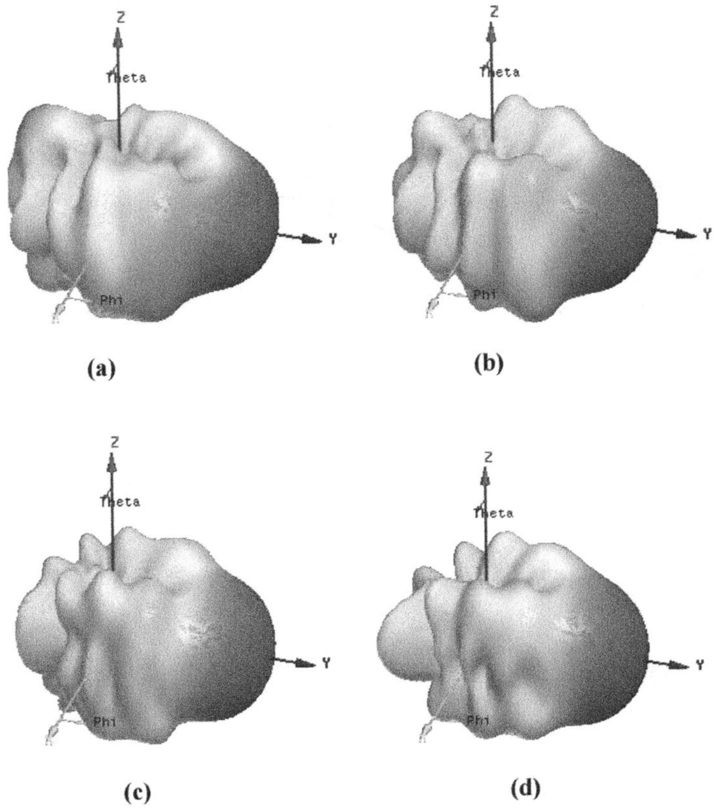

Figure 6.19 Three-dimensional radiation pattern for LTSA with width $= \lambda_0$, length $= 2\lambda_0$, and thickness $= 0.25$ mm. The opening angle is equal to (a)5°, (b) 10°, (c) 15°, and (d) 20°.

6.6 DESIGN OF TAPERED SLOT ANTENNAS

The design of the TSA is primarily based on a trial-and-error approach, which includes the following simple guidelines as an initial step. The antenna aperture width W (see Figure 6.20) is chosen to be at least equal to λ_0, where λ_0 is the free-space wavelength at the lowest frequency of operation of the antenna. The slot characteristics are chosen such that $1.05 \leq c/v_p \leq 1.2$, where c is the velocity of light and v_p is the phase velocity of the field along the slot. The effective thickness is chosen in the range $0.005\lambda_0 \leq t_e \leq 0.03\lambda_0$. The tapering angle or the opening angle is assumed to be typically $5 - 20°$. The length of the TSA, L, is chosen in the range from 1 to $10\lambda_0$ (according to the required gain).

Following these initial considerations, the detailed design of the TSA involves two major tasks: (1) the design of a broadband transition and feed structure offering a high return loss over a very wide frequency range and (2) determination of the dimensions and shape of the antenna in accordance with the required beamwidth, sidelobe, backlobe, etc., over the operating frequency range.

In the following, a specific TSA design is presented for an antipodal elliptically shaped TSA, whose layout is shown in Figure 6.20. The approach is similar to the one given in

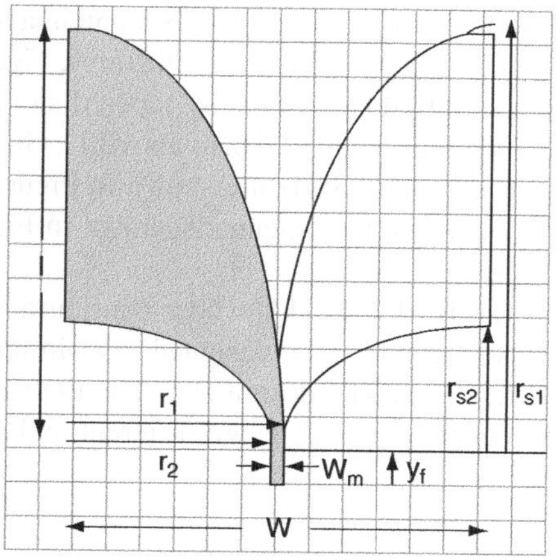

Figure 6.20 Configuration of the antipodal ELTSA showing the design parameters.

Reference [13]. The design objective is to obtain an antenna that is compact in size while maintaining the 10 dB return loss bandwidth requirement of 3.1–10.6 GHz.

The design steps of this TSA are summarized as follows:

Step 1. Given the lowest frequency of operation (f_l), thickness of the substrate (h), and its dielectric constant (ε_r), the width (w) and length (l) of the antenna structure, excluding the feeder, can be calculated using the following equation:

$$w = l = \frac{c}{f_l}\sqrt{\frac{2}{\varepsilon_r + 1}} \tag{6.1}$$

Step 2. The radiating structure of the antenna is formed from the intersection of quarters of two ellipses. The major (r_1 and r_2) and the secondary (r_{s1} and r_{s2}) radii of the two ellipses are chosen according to the following equations:

$$r_1 = w/2 + w_m/2 \tag{6.2}$$

$$r_2 = w/2 - w_m/2 \tag{6.3}$$

$$r_{s1} = k_1 l \tag{6.4}$$

$$r_{s2} = k_2 r_2 \tag{6.5}$$

Values of the parameters k_1 and k_2 are in the ranges 1.2–1.5 and 0.4–0.8 depending on the substrate characteristics.

Step 3. The width of the microstrip transmission feeder w_m needed to give the characteristic impedance, Z_0, equal to 50 Ω can be calculated using the following equation:

$$w_m = \frac{120\pi}{\sqrt{\varepsilon_r}} \frac{h}{Z_0} \tag{6.6}$$

The presented design method is verified for an ELTSA antenna assuming a low dielectric constant material, Rogers RO4003C ($\varepsilon_r = 3.38$, $h = 0.508$ mm). Figure 6.21 shows the simulated return loss of the designed antipodal elliptically tapered slot antenna. As can be seen from Figure 6.21, the antenna operates from 3 GHz to above 11 GHz. The three-dimensional radiation patterns calculated at 3, 6, and 9 GHz are shown in Figure 6.22. The front-to-back ratio is greater than 15 dB. The gain of the antenna is shown in Figure 6.23. The peak gain of the antenna is 11 dBi at 9 GHz.

Figure 6.24 shows another UWB ELTSA [14] whose configuration is an alternative to the one shown in Figure 6.20. This antenna can be designed in a similar manner as the previous ELTSA, but with some modifications. The design step 1 is the same as that used with the first structure of Figure 6.20, but the other steps are different and are given as follows:

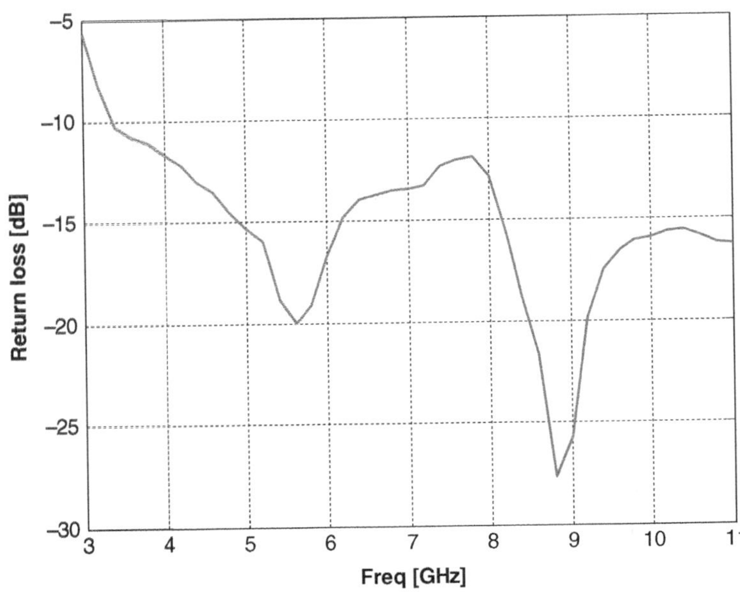

Figure 6.21 Variation of the return loss with frequency for the designed ELTSA.

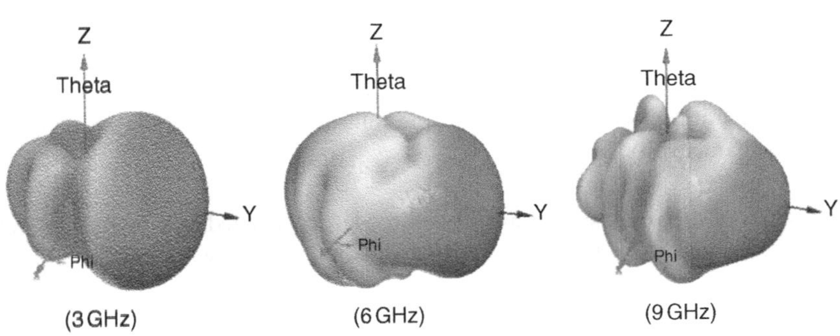

Figure 6.22 Three-dimensional radiation pattern for the designed ELTSA.

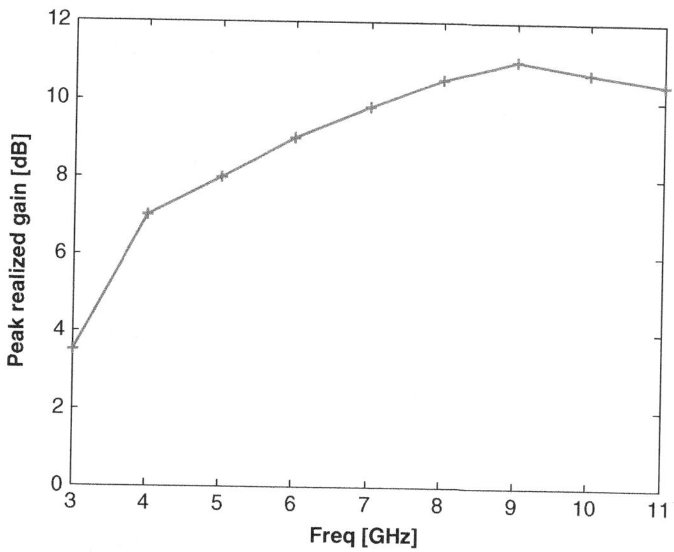

Figure 6.23 Variation of the gain with frequency for the designed ELTSA.

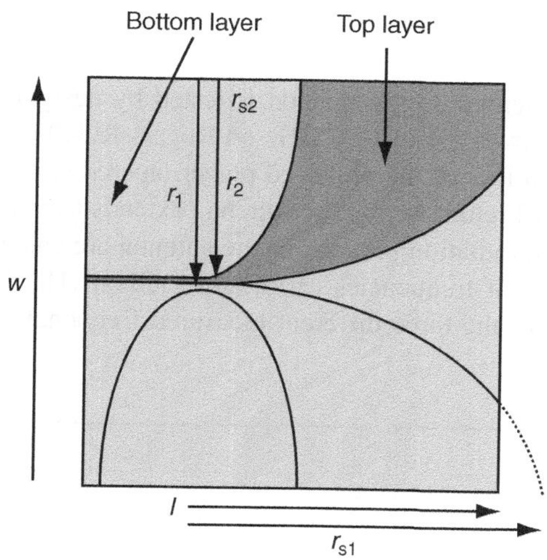

Figure 6.24 Configuration of the designed ELTSA.

Step 2. The radiating structure of the antenna is formed from the intersection of quarters of two ellipses. The major and the secondary radii of the two ellipses are chosen according to the following equations:

$$r_1 = w/2 \tag{6.7}$$

$$r_2 = w/2 - w_m \tag{6.8}$$

The secondary radii r_{s1} and r_{s2} can be calculated using Equations (6.4) and (6.5) as in the previous design example.

Step 3. The width of the microstrip transmission feeder w_m needed to give the characteristic impedance, Z_0, equal to 50 Ω can be calculated using the following equations:
For $w_m/h \leq 1$:

$$Z_0 = \frac{60}{\sqrt{\varepsilon_m}} \ln\left(\frac{8h}{w_m} + \frac{w_m}{4h}\right) \tag{6.9a}$$

For $w_m/h \geq 1$:

$$Z_0 = \frac{120\pi}{\sqrt{\varepsilon_m}[w_m/h + 1.39 + 0.67\ln(w_m/h + 1.44)]} \tag{6.9b}$$

where the effective dielectric constant for the transmission line, ε_m, is given by

$$\varepsilon_m = \frac{\varepsilon_r + 1}{2} + \frac{\varepsilon_r - 1}{2 \times \sqrt{1 + 12h/w_m}} \tag{6.10}$$

Step 4. The ground-plane of the antenna consists of two parts. The first part is a structure that is similar to that of the radiating element of the antenna. The second part is a tapered-shape structure formed from the intersection of a rectangular conductor with two antifaced quarter ellipses with dimensions r_1 and r_{s2} at one side, and r_2 and r_{s2} at the opposite side. To improve the impedance matching of the antenna, the ground-plane is extended by y_g.

The validity of the presented design method is tested by designing an antenna covering the UWB frequency band from 3.1 to 10.6 GHz on Rogers RO4003C substrate. Figure 6.25 shows the simulated return loss of the proposed planar tapered slot antenna. As can be seen from Figure 6.25, the 10 dB return loss of this antenna extends from 3.1 to over 11 GHz. The three-dimensional far-field radiation patterns of the antenna are calculated and are shown in Figure 6.26 at three different frequencies (3, 6, and 9 GHz). The front-to-back ratio is at least greater than 10 dB for the three different measured frequencies. The variation of gain

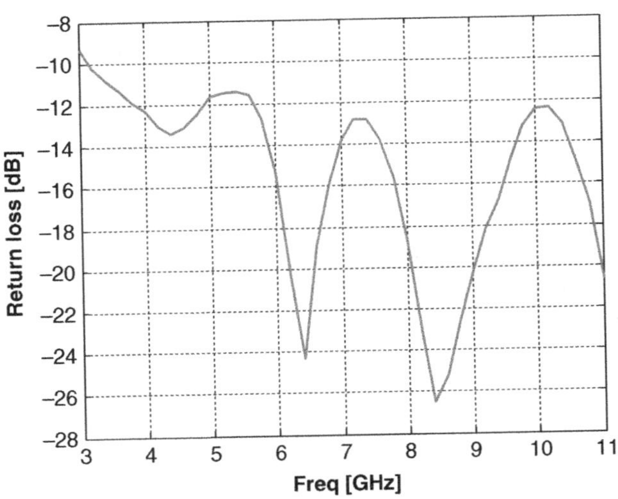

Figure 6.25 Variation of return loss with frequency for the designed ELTSA.

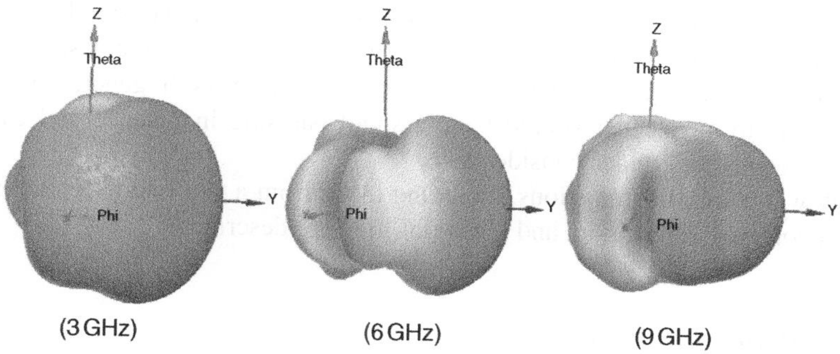

Figure 6.26 Three-dimensional radiation pattern for the designed ELTSA.

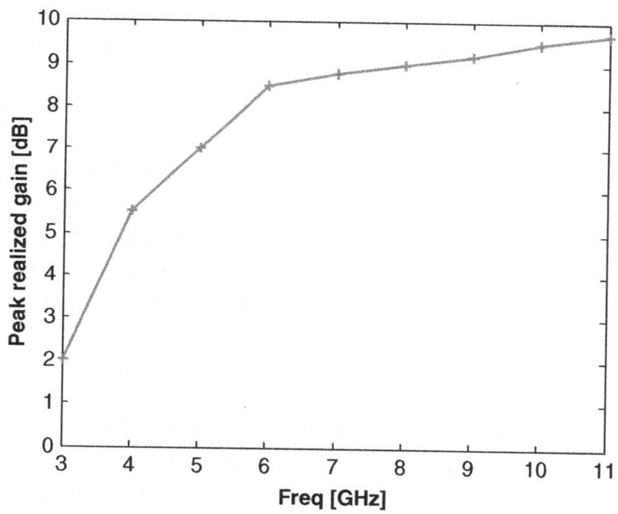

Figure 6.27 Variation of the peak gain with frequency for the designed ELTSA.

with frequency of the antenna is shown in Figure 6.27. The simulated gain for the antenna for frequencies between 3 and 11 GHz shows that the gain increases with frequency and is around 10 dBi at 11 GHz.

6.7 FEEDING PROCEDURES FOR TAPERED SLOT ANTENNAS

With respect to the input impedance/return loss frequency characteristics, TSAs resemble tapered transmission lines, which exhibit high-pass properties. For these structures, the return loss decays with frequency in an oscillatory manner. The rate of decay depends on the shape of the taper [26]. Because of this property, theoretically TSAs have an unlimited return loss bandwidth. In practice, their operational bandwidth is limited by the transition between the feed and the antenna slot. The main task when designing the feeding structure for TSAs is to obtain a feed of a shape and dimensions capable of maintaining a match to a specific input impedance (for example 50 Ω) over the entire (return loss, gain, or pattern) bandwidth of the antenna.

When the feed utilizes other types of transmission line than the slot line, a suitable transition between the two types of transmission lines is required. In the case of microstrip or coaxial lines, the transition is a balun, as used in the early TSAs. In general, the transition from the slot line to any feeder should have a small parasitic inductance and capacitance over the whole bandwidth under consideration.

In the following, various transitions, including those from a slot line to a microstrip line, a strip line, a coplanar waveguide, and a coaxial line, are described.

6.7.1 Microstrip Line Feed

A microstrip line is the most common form of printed transmission line used for feeding a tapered slot antenna element. A microstrip line, being formed by a conductive metal strip on one side of a dielectric substrate and a conductive ground-plate on other side of the substrate, is an unbalanced line [26]. This is the opposite of the slot line, which is a balanced transmission line. Because of this situation, feeding a TSA with a microstrip line requires a wideband balanced-to-unbalanced transition (balun) to avoid compromising the broadband performance.

A conventional microstrip line feeding arrangement for a TSA is shown in Figure 6.28. In this transition, a microstrip line on one side of a dielectric substrate crosses over the slot line of the planar slot antenna etched within the metallization layer on the other side of the substrate. The slot line extends beyond the microstrip line by a distance of $0.25\lambda_s$, where λ_s is the wavelength in the slot line at the centre operating frequency of the antenna. The microstrip line extends beyond the slot line by a distance of $0.25\lambda_m$, where λ_m is the wavelength in the microstrip line at the centre operating frequency of the antenna. The metallization layer of the TSA forms the ground of the microstrip line. As observed in Figure 6.28, at their ends the microstrip line is open-circuited and the slot line is shorted [27].

It is also possible for the slot line to terminate in an open-circuit, for example by adding a relatively large circular patch at the end of the slot line, and for the microstrip line to be shorted by means of a via that runs through the substrate to the ground metallization layer. In either case, a balun is created at the crossover that matches the unbalanced microstrip line to the balanced slot line of the antenna element. This electromagnetic coupling arrangement permits signal transmission from the microstrip transmission line to the slot line (for feeding the antenna). In general, the stronger the electromagnetic coupling, the better is the transition.

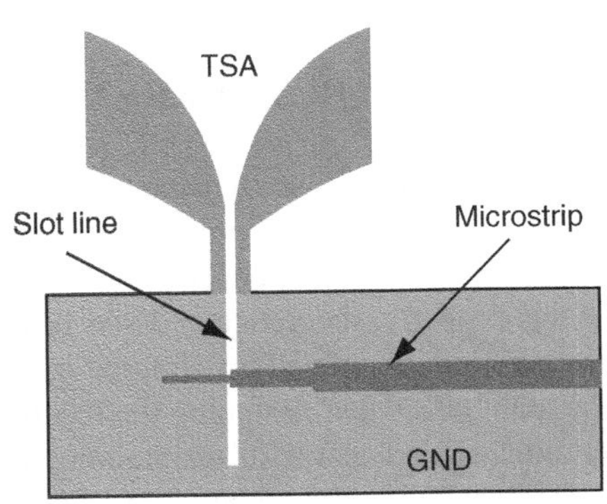

Figure 6.28 Slot line-to-microstrip transition.

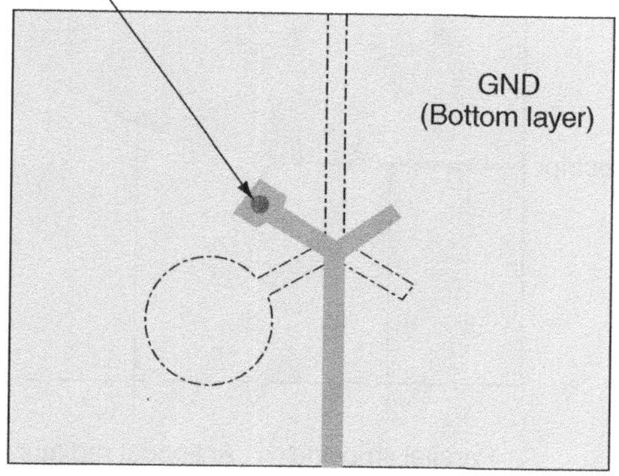

Figure 6.29 Slot line-to-microstrip transition using a wideband balun.

Figure 6.29 shows another type of balun that can be used for slot line-to-microstrip transition [28]. This configuration has no inherent bandwidth limitation other than parasitic inductances and capacitances. From configurations shown in Figures 6.28 and 6.29, it is apparent that the balun in the unilateral TSA complicates the antenna structure and limits the antenna bandwidth if it is not designed properly.

The use of balun is not necessary in the antipodal TSA. In this type of TSA, the feeding from the microstrip transmission line to the slot line is shown in Figure 6.30. As seen in Figure 6.30, in this feeding structure a microstrip line gradually tapers to a paired-strip transmission line. The transition region is responsible for connecting the highly capacitive feed structure to the inductive radiating section, and decouples the microstrip structure from the radiating portion of the antenna. In addition, a properly designed transition region converts the unbalanced feed into a balanced structure that can then be connected to the radiating region of the antenna. The paired-strip section is slowly tapered away on each side to develop into the radiating sections of the antenna. The taper into the radiating section of the antenna is the most critical aspect of the design. The taper should be gradual and as smooth as possible to avoid significant discontinuities at higher frequencies, which will cause reflections resulting in performance degradation.

The method of feeding shown in Figure 6.30 has wideband behaviour with respect to return loss but it causes a high cross-polar radiation pattern. In order to provide a clear explanation of this undesired phenomenon, the electric field lines at different cross-sections along the feed and the antenna are illustrated in Figure 6.31. The electric field lines, which are spread out in the conventional microstrip structure, concentrate between the metal strips of the balanced microstrip and finally rotate while travelling along the TSA. This field rotation is responsible for higher cross-polarization in the antipodal type of TSA.

6.7.2 Coplanar Waveguide (CPW) Feed

The configuration of a CPW structure for feeding a TSA is shown in Figure 6.32. The convenience of CPW feed is that it enables direct integration of active solid-state devices with

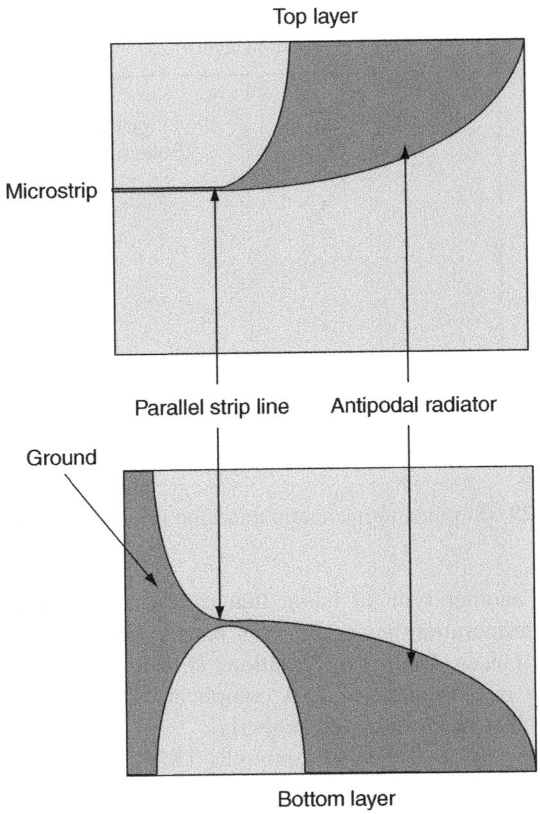

Figure 6.30 Microstrip feeder for the antipodal TSA.

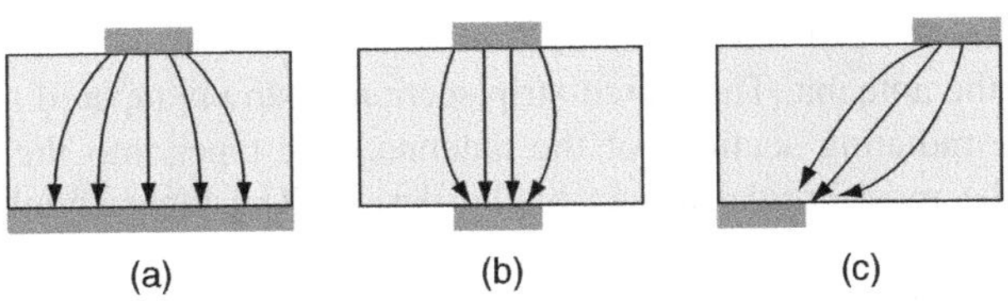

Figure 6.31 Distribution of electric field lines at (a) the input of microstrip feeder; (b) the balanced microstrip transition; and (c) the antenna slot.

the antenna. Other advantages of CPW include such useful characteristics as low radiation loss, less dispersion, and a uniplanar configuration. It can be etched on to the opposite side of the dielectric to the tapered slot antenna. The finite ground-plane of the coplanar waveguide can be connected to the antenna ground-plane metallization through via holes to provide impedance match and odd-mode operation. In the configuration shown in Figure 6.32, a plated through-hole interconnect from the metallized ground layer on the lower surface to a metallized layer on the top of the substrate is used. The signal is coupled to the antenna through a centre conductor of the coplanar waveguide which extends to form a crossover with the antenna slot line.

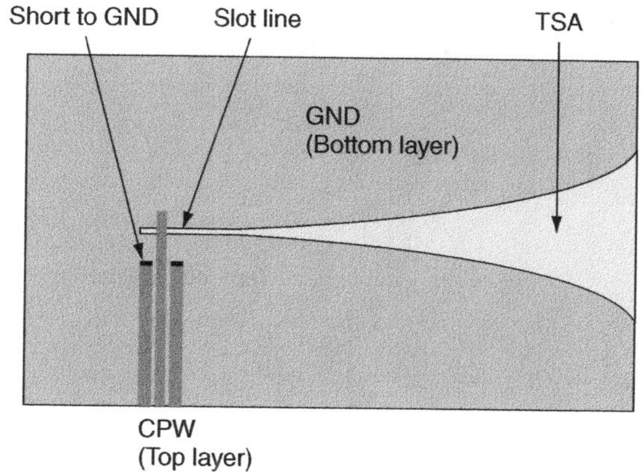

Figure 6.32 CPW to feed the TSA.

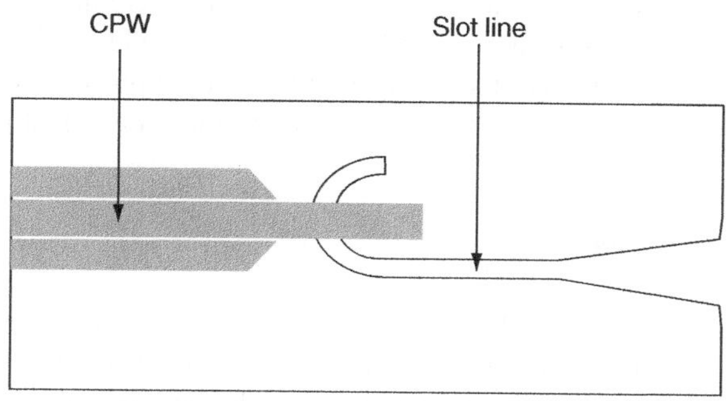

Figure 6.33 Strip-to-slot transition.

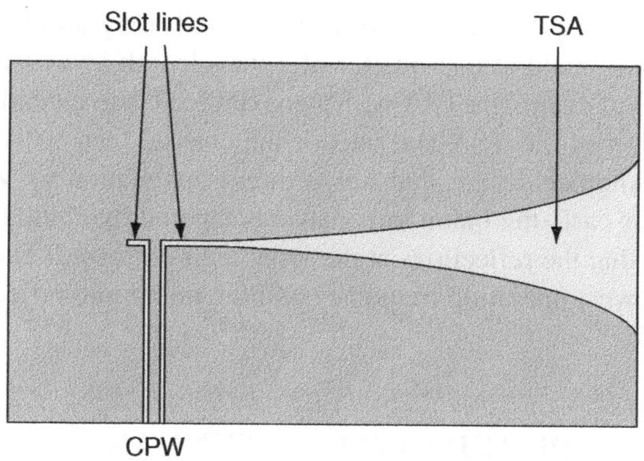

Figure 6.34 Uniplanar TSA and CPW.

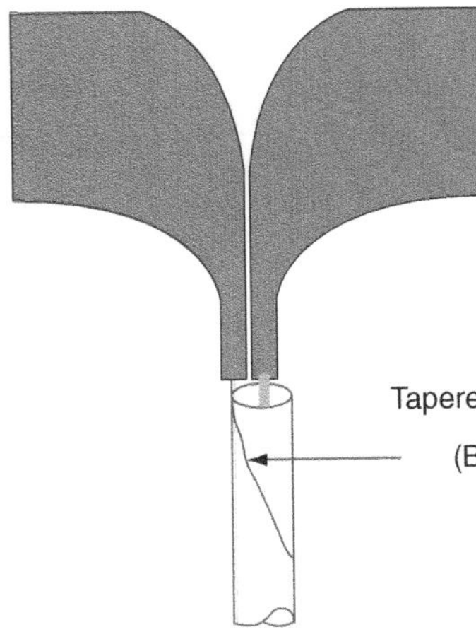

Figure 6.35 VTSA fed using a coaxial line with a tapered balun.

Other configurations in which the CPW can be used to feed the TSA are illustrated in Figures 6.33 and 6.34 [29]. Figure 6.33 shows a strip-to-slot coupling used with the LTSA. The TSA can also be fed by a direct slot-to-slot arrangement where the CPW and the antenna are on the same side of the substrate. In order to excite the odd mode where the electric fields in both slots of the CPW have the same amplitude but opposite direction, a bond wire is usually placed across the slots to keep the two ground-planes at equal potential.

6.7.3 Coaxial Line Feed

The coaxial line is compatible with the slot line, coplanar microstrip, or balanced microstrip. Therefore, it can be used to excite most of the TSAs. Its main disadvantages are that it is nonplanar and unbalanced. All of the currents flow inside the inner connector and the inside of the shield. As a result, feeding a balanced antenna, such as the TSA, with an unbalanced coaxial feed causes currents to flow on the outside of the shield. This results in spurious radiation leading to a significant power loss and serious distortion in the radiation pattern. One common approach to cut off the flow of that current is with an open-circuit, which is created by placing a 'skirt' one-quarter of a wavelength long around the outside coaxial shield and shorted to the shield at one-quarter of a wavelength away from the load.

The most direct way of exciting a TSA with a coaxial feed is to extend its centre conductor over the slot line section of the TSA and anchor the coaxial feed with a solder connection to the ground-plane. Another design that has a direct application to V-TSAs is shown in Figure 6.35 [30]. In this case, the balun impedance is tapered by cutting open the outer wall of the coaxial cable so that the reflections at the input are minimized. The length of the balun is determined by the lowest operating frequency and the maximum reflection coefficient that occurs in the passband.

6.8 ARRAYS OF TAPERED SLOT ANTENNAS

As opposed to microstrip patches for which all of the radiating elements in an array can be located on a single substrate (forming the so-called *tile* architecture), two-dimensional arrays of TSAs require a number of substrates equal to the linear dimension of the array. As a result, these arrays are the combination of *tile* and *tray* types of array architectures. In comparison with microstrip patch arrays, TSA arrays have some important advantages that make them potentially very well suited to a number of applications. These include a larger operational bandwidth and much narrower beamwidths because of higher antenna element gains. These attributes make TSA arrays attractive for ultra wideband radar applications. In addition they can be used as feeding elements illuminating reflector antennas and lenses directly with low spillover losses. In these and other applications, an extra advantage of the TSA elements is that their radiating portions can be well separated from the integrated circuit that follows it, as there is ample space for the latter. This is in contrast to a microstrip patch, which offers a limited space for inclusion of electronics mainly behind the radiating element. The transverse spacing between TSA array elements can be made very small, especially using the tray configuration of the array. This feature is advantageous in many applications including a high-resolution millimetre-wave imaging involving TSA/reflector systems and satellite communication antennas involving beam shaping and beam switching in which TSA arrays can provide lightweight alternatives for traditional focal-plane feeds.

6.8.1 Mutual Coupling

In typical broadside arrays, such as those formed by dipoles with close spacing, individual antenna elements show substantial changes in the input impedance compared to isolated elements. This occurs due to mutual coupling. TSA elements radiating in an endfire direction are more immune to this effect [22]. Shorting or open-circuiting other elements may have very little effect on the input impedance and radiation pattern of the remaining elements. This property is of high importance, especially in active arrays, because the failure of one or a few elements of the array does not lead to catastrophic failure of the array.

In general, mutual coupling produces profound impacts on the performance of an antenna array. These include impedance mismatch and scanning blindness, particularly for antennas on thick dielectric substrates. With respect to TSAs, mutual coupling can be investigated with respect to coplanar or stacked configurations involving two TSA elements, as shown in Figure 6.36. The mutual coupling between two LTSAs in the two mentioned configurations can be studied using a full-wave electromagnetic analysis. Here, the authors apply a finite element method to study the mutual coupling for the coplanar and stacked configurations of two TSAs shown in Figure 6.36. In the undertaken analysis, the antennas are assumed to use RO4003C substrate with 0.25 mm thickness and having a width and length equal to λ_0 and $2\lambda_0$, respectively.

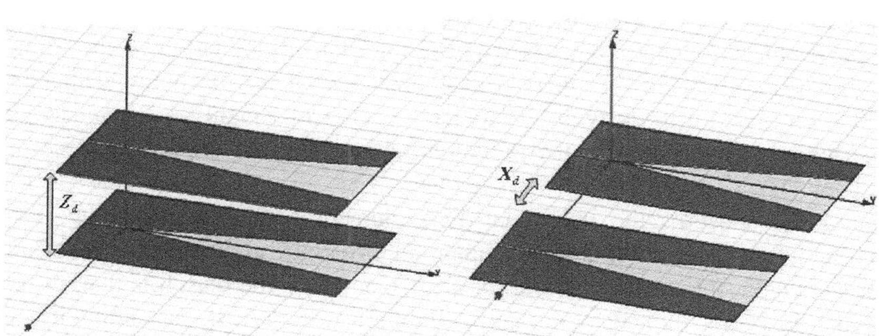

Figure 6.36 Two configurations for array of TSAs: stacked (left) and coplanar (right).

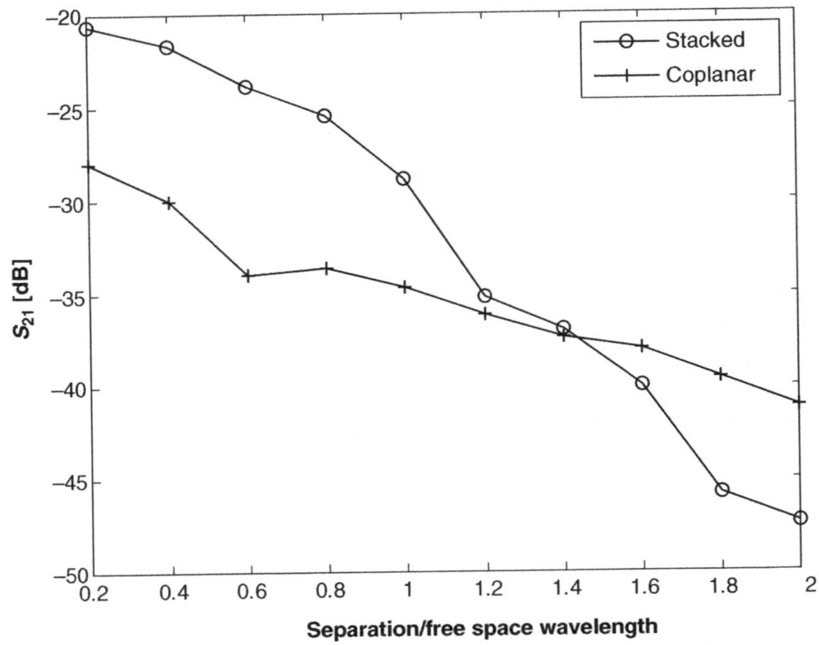

Figure 6.37 Mutual coupling between two LTSAs in two configurations.

The results are shown in Figure 6.37. It is clear that when the distance between the antennas is small ($\sim 0.2 - 1\lambda_0$) the mutual coupling between the stacked antennas is higher than when the antennas are coplanar. At a larger spacing ($> 1.4\lambda_0$) the stacked configuration shows mutual coupling lower than the coplanar variety. This result agrees well with the measured results presented in Reference [29].

6.8.2 Gain and Beamwidth of TSA Elements in Arrays

The presence of mutual coupling in the array affects the electrical characteristics of individual elements. In many applications, the TSA array is used as a focal plane array in conjunction with a reflector or lens focusing element. In this case, it is important to design a TSA array that results in equal E- and H-plane beamwidths at about the -10 dB level (with respect to the main beam peak value). This is required to obtain illumination of the reflector for maximum aperture efficiency. With respect to this task, it has been found that the most efficient TSA arrays are those with an element spacing of roughly 1–2 wavelengths. For element spacings larger than this, the radiation patterns become similar to those of single elements [22].

Another issue concerns the individual TSA element's gain or directivity. While the directivity and thus gain of a single element in principle may increase indefinitely by increasing its length, this is not possible for array elements. The reason is that, in the array, the element gain is constrained by the unit cell area it occupies. If, for simplicity, an array with square symmetry and with an element spacing of d is assumed, then the element area is d^2. Thus the effective aperture of the element and its gain is limited to the value of this area [22]. This area is related to the area of a unit cell of a periodic array.

6.9 APPLICATIONS OF TAPERED SLOT ANTENNAS

TSAs have found a wide range of applications and some examples, such as inclusion in feeds for reflectors or lenses, have already been mentioned in the previous section. Besides their wideband characteristics, the main drive for using a TSA in stand-alone and array configurations is that they are very inexpensive to fabricate. Also, because of the endfire operation they offer space for hosting electronic modules and therefore can form low-cost integrated antenna/receiver/transmitter units. The resulting broadband transceivers can be used in millimetre-wave and submillimetre-wave imaging systems in radio astronomy [31] or in medical applications [14]. When the cost or available space is precluded from use, single-element TSAs can also do a proper job if moderate gain for the radiating structure is sufficient. In this case, a single TSA can offer gains of up to 17 dBi.

An example of the use of a TSA in a millimetre-wave imaging system was described in Reference [32]. A seven-element array with a hexagonal arrangement of the elements in a cassegrain telescope with a 30 cm diameter main reflector at a frequency of 94 GHz was used. Test signals from a source in the far-field region were detected directly by diodes soldered across the narrowest portion of the tapered slot. A minimum beam spacing of 0.5° was measured when the elements were $1.6\lambda_0$ apart.

6.9.1 UWB Applications

Besides their traditional applications such as in radar and radio astronomy, TSAs may be found as very useful radiators in emerging ultra wideband (UWB) technology. Due to its inherent attributes, UWB is capable of bringing significant advances not only to wireless communications (delivery of high data rate transmission in the presence of existing communication systems) [33] but also to other areas such as microwave imaging [14].

UWB systems use narrow pulses to transmit data. Since no carrier frequencies are involved, the transmitter and receiver hardware can be made very simple. UWB antennas must cover multiple-octave bandwidths in order to transmit pulses that are of the order of a nanosecond in duration. Since data may be contained in the shape of the UWB pulse, antenna pulse distortion must be kept to a minimum.

The capabilities of the TSA to support very narrow pulses without distortions were investigated in Reference [13]. To this purpose two copolarized UWB ELTSA antennas developed on RT6010 substrate were used. Two ELTSA antennas, each with $5\,\text{cm} \times 5\,\text{cm}$ dimensions, were separated by the distance of 45 cm to investigate the pulse transmission. The results of the transmitted and received (normalized to the peak values) pulses are shown in Figure 6.38. The figure shows that the pulse duration of the antenna is 0.8 nanoseconds. The pulse distortion of 0.12 compared to the peak of the normalized received pulse, which is almost negligible, can be observed, indicating that the developed antenna supports a narrow pulse almost without distortion, making it an excellent radiator for UWB applications involving very short time duration pulses.

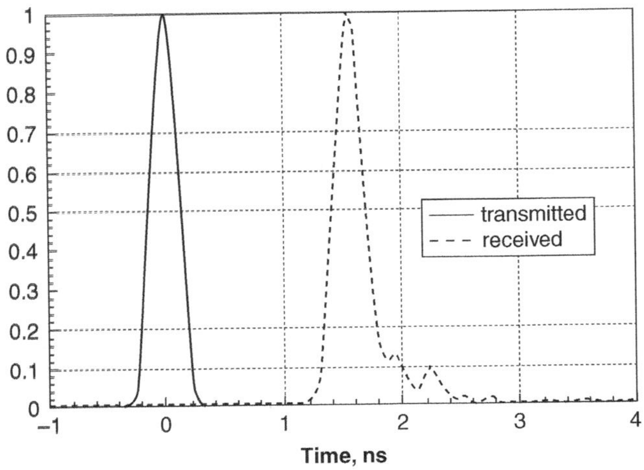

Figure 6.38 Impulse responses of the designed antenna.

The presented results indicate that this small size UWB TSA antenna can be used for portable UWB applications such as laptop computers or distributed sensor networks. Because of its relatively small size, it can be easily integrated with transmitter/receiver hardware.

6.9.2 Other Applications of TSAs

As shown in Reference [34], TSA arrays can also be used for spatial power combining to achieve power amplification and harmonic generation of solid-state devices. Combining the power of many sources offers a method for making higher output powers available when the amount of power from a single source is low, as is the case for most solid-state sources. The high aperture efficiency of the array and the small element spacing make TSA arrays attractive for this application.

An example of a 120 W X-band spatially combined solid-state amplifier employing trays of TSAs is shown in Figure 6.39. A linear four-element LTSA array has been used for combining interinjection-locked power [35]. As another important example of the applications of TSA arrays in mobile communication the K-band circular LTSA array, shown in Figure 6.40, was proposed [36]. The 16-element array is fed by a 1:16 microstrip line power splitter composed of T-junctions and right-angle bends. The output ports of the splitter are electromagnetically coupled to the slot line of the LTSA through a microstrip-to-slot line transition with the slot line and microstrip line characteristic impedances chosen to be 120 and 100 Ω respectively. The array was placed over a reflecting ground-plane to displace the beam above the horizon. This circular LTSA array produces an omnidirectional pattern in the azimuthal plane and a beam displacement of about 28° in the elevation plane.

The TSA array has found applications in intelligent transportation systems. A wide-scan spherical-lens antenna at millimetre-wave frequency (77 GHz) was introduced for automobile radars [37]. The multibeam antenna system is composed of planar TSAs which are positioned around a homogeneous spherical Teflon lens. Beam scanning is achieved by switching

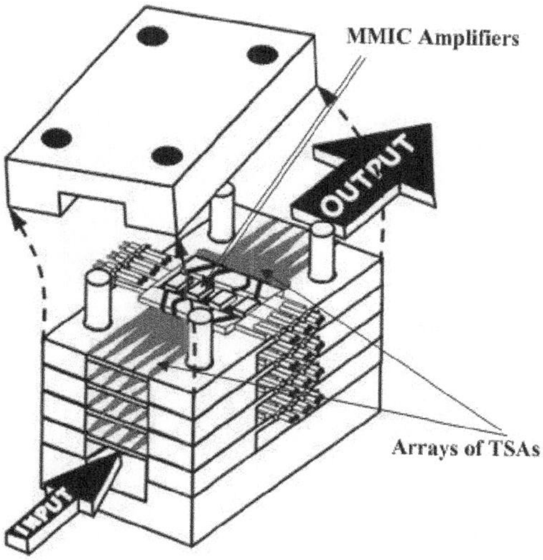

Figure 6.39 Configuration of an amplifier combiner using tapered slot antennas.

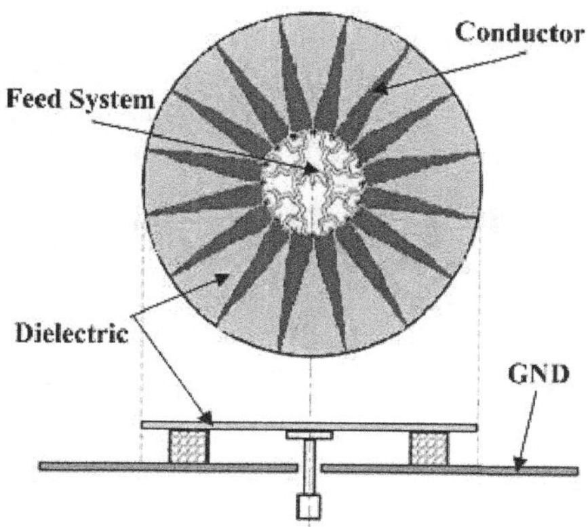

Figure 6.40 Circular LTSA array.

between elements. Figure 6.41 shows a 33-beam spherical Teflon lens system which has demonstrated an angular resolution of 5.5° and −3.5 dB crossover of adjacent beams.

In another development of a low-cost, multifrequency, and full-duplex phased array transceiver that transmits at 10 or 19 GHz and receives at 12 or 21 GHz, TSAs were used to achieve wideband performance. In addition, the system, illustrated in Figure 6.42, features a low-loss and low-cost multiline phase-shifter controlled by dual piezoelectric transducers and four channel microstrip multiplexers [38]. The array system has demonstrated scan angles of about 40° at 10–21 GHz.

In recent developments, arrays of TSAs have been envisaged for biomedical engineering applications. Figure 6.43 shows one of such applications in which a circular array of TSAs is used in a microwave imaging system for breast cancer detection [14]. In this system, one of the antennas is used to transmit a microwave signal while the rest of the antennas in the array receive the scattered signal. The measured data are collected and then the measurement procedure is repeated, with the second antenna transmitting the signal while the remainder

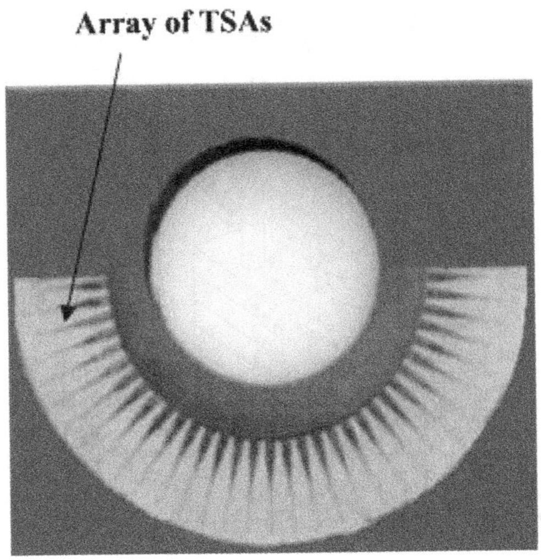

Figure 6.41 A 33-beam array with a spherical Teflon lens.

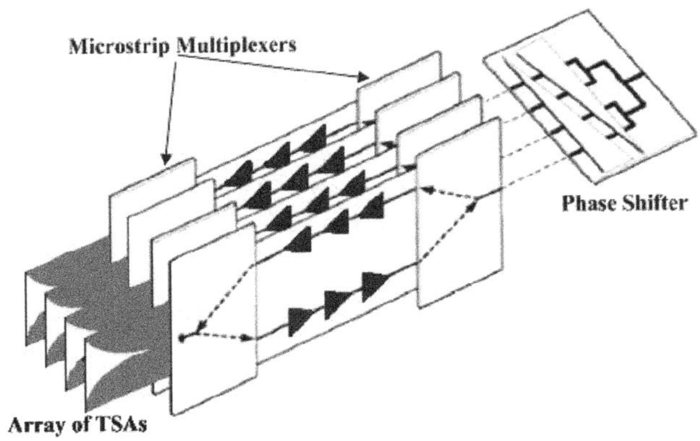

Figure 6.42 Multifrequency and full-duplex phased array of TSAs.

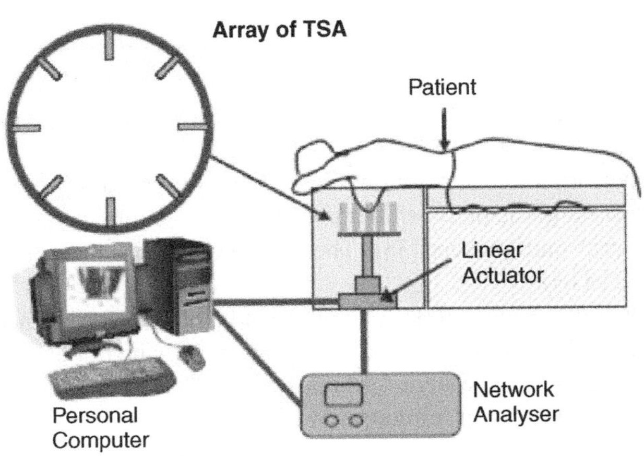

Figure 6.43 Microwave imaging system for breast cancer detection using an array of TSAs.

are used for receiving the scattered signal. The array can be moved up or down with a linear actuator. The process is used to create an image indicating the presence and location of a highly scattering target representing a malignant tissue.

6.10 SUMMARY

In this chapter a thorough examination of tapered slot antennas has been presented and the various forms of TSAs have been investigated. An overview of the advantages and disadvantages of using TSAs compared to other forms of antenna has been explained. A review of the procedures used to model tapered slot antennas has then been presented. A parametric study of the tapered slot antennas has been used to investigate the effect of each design parameter on the performance of the antenna. An overview has also been given on how to design TSAs. Many techniques have been examined on how to couple power to and from these radiators. How TSAs can be incorporated into arrays, linear, planar, and circular, has been explained and some TSA applications have been highlighted.

REFERENCES

1. B. Stephenson and C. Walter, 'Endfire slot antennas', *IRE Transactions on Antennas and Propagation*, **3**, 81–86, 1955.
2. J. Eberle, C. Levis and D. McCoy, 'The flared slot; a moderately directive flush-mounted broadband antenna', *IRE Transactions on Antennas and Propagation*, **8**, 461–468, 1960.
3. L. Lewis, M. Fassett and J. Hunt, 'A broadband stripline array elements', in *IEEE AP-S International Symposium*, USA, pp. 335–337, 1974.
4. P. Gibson, 'The Vivaldi aerial', in *9th European Microwave Conference*, UK, pp. 101–105, 1979.
5. S. Prasad and S. Mahapatra, 'A novel MIC slot line antenna', in *9th Eurpean Microwave Conference*, UK, pp. 120–124, 1979.
6. K. Yngvesson, J. Johansson and E. Kollberg, 'A new integrated feed array for multi-beam systems', *IEEE Transactions on Antennas and Propagation,* **AP-34**, 1372–1376, 1986.
7. R. Janaswamy, 'An accurate moment method for the tapered slot antenna', *IEEE Transactions on Antennas and Propagation*, **37**(12), 1523–1528, 1989.
8. T. Hwang, D. Rutledge and S. Schwarz, 'Planar sandwich antennas for submillimeter applications', *Applied Physics Letters*, **34**(1), 9–11, 1979.
9. K. Yngvesson, *et al.*, 'Endfire tapered slot antennas on dielectric substrates', IEEE Transactions on Antennas and Propagation, **33**(12), 1392–1400, 1985.
10. E. Gazit, 'Improved design of the Vivaldi antenna', *IEE Proceedings, Part H*, **135**(2), 89–92, 1988.
11. P. R. Acharya, J. F. Johansson and E. L. Kollberg, 'Slot line antennas for millimeter and submillimeter waves', in *Proceedings of the 20th European Microwave Conference*, Hungary, pp. 353–358, 1990.
12. P. Acharya *et al.*, 'Tapered slot antennas at 802GHz', *IEEE Transactions on Microwave Theory and Techniques*, **41**(10), 1715–1719, 1993.
13. A. Abbosh, H. Kan and M. Bialkowski, 'Design of compact directive ultra wideband antipodal antenna', *Microwave and Optical Technology Letters*, **48**(12), 2448–2450, December 2006.
14. A. Abbosh, H. Kan and M. Bialkowski, 'A compact UWB planar tapered slot antenna for use in a microwave imaging system', *Microwave and Optical Technology Letters*, **48**(11), 2212–2216, November 2006.
15. R. Janaswamy and D. Schaubert, 'Analysis of the tapered slot antenna', *IEEE Transactions on Antennas and Propagation*, **35**(9), 1058–1064, 1987.
16. A. Koksal and F. Kauffman, 'Moment method analysis of linearly tapered slot antennas', in *IEEE AP-S International Symposium*, USA, pp. 314–317, 1991.

17. H. Wang, D. Syahkal and I. Dilworth, 'A rigorous analysis of tapered slot antennas on dielectric substrates', in *10th International Conference on Antennas and Propagation*, Vol. 1, pp. 286–289, 1997.
18. J. Colburn and Y. Rahmat-Samii, 'Linear taper slot antenna directivity improvement via substrate perforation: an FDTD evaluation', in *IEEE AP-S International Symposium*, USA, Vol. 1, pp. 1176–1179, 1998.
19. L. Kuo, M. Tsai and H. Chuang, '3-D FDTD design simulation and experimental measurement of a Ka-band planar antipodal linearly-tapered slot antenna (ALTSA)', *IEEE Microwave and Wireless Components Letters*, **11**(9), pp. 382–384, 2001.
20. H. Oraizi and S. Jam, 'Optimum design of tapered slot antenna profile', *IEEE Transactions on Antennas and Propagation*, **51**(8), 1987–1995, 2003.
21. T. Itoh, 'Spectral domain immitance approach for dispersion characteristics of generalized printed transmission lines', *IEEE Transactions on Microwave Theory and Techniques*, **28**, 733–736, 1980.
22. K. Yngvesson *et al.*, 'The tapered slot antenna – a new integrated element for millimetre wave applications', *IEEE Transactions on Microwave Theory and Techniques*, **37**(2), 365–374, 1989.
23. T. Thungren, E. Kollberg and K. Yngvesson, 'Vivaldi antennas for single beam integrated receiver', in *Proceedings of the 12th European Microwave Conference*, pp. 475–480, 1982.
24. J. Colbum and Y. Rahmat-Samii, 'Printed antenna pattern improvement through substrate perforation of high dielectric constant material: an FDTD evaluation', *Microwave and Optical Technology Letters*, **18**(1), 27–32, 1998.
25. T. Ellis and G. Rebeiz, 'MM-wave tapered slot antennas on micromachined photonic bandgap dielectrics', in *IEEE MTT–S International Symposium*, pp. 1157–1160, 1996.
26. D. Pozar, *Microwave Engineering*, 3rd edn, John Wiley & Sons, Inc., New York, 2005.
27. A. Smolders and M. Arts, 'Wide band antenna element with integrated balun', in *IEEE AP-S International Symposium*, USA, 1998.
28. V. Trifunovic and B. Jokanovic, 'Review of printed Marchand and double Y baluns; characteristics and application', *IEEE Transactions on Microwave Theory and Techniques*, **42**(8), 1454–1462, 1994.
29. K. Lee and W. Chen, *Advances in Microstrip and Printed Antennas*, Wiley Interscience, New York, 1997.
30. P. Knott and A. Bell, 'Coaxially-fed tapered slot antenna', *Electronics Letters*, **37**(18), 1103–1104, 2001.
31. D. Winkler *et al.*, 'A new submillimeter wave SIS quasi-particle receiver for 750 GHz', in *Proceedings of the 11th International Conference on Infrared Millimeter Waves*, Italy, 1986.
32. T. Korzeniowski *et al.*, 'Imaging system at 94 GHz using tapered slot antenna elements', in *8th International Conference on Infrared Millimeter Waves*, USA, 1983.
33. T. Tsang and M. El-Gamal, 'Ultra-wideband (UWB) communications systems; an overview', in *3rd International IEEE-NEWCAS Conference*, pp. 381–386, 2005.
34. N. Cheng *et al.*, 'A 120W X-band spatially combined solid-state amplifier', *IEEE Transactions on Microwave Theory and Techniques*, **47**(12), 2557–2559, 1999.
35. W. Morgan and K. Stephan, 'Inter-injection locking – a novel phase-control technique for monolithic phased arrays', in *Proceedings of the 12th International Conference on Infrared Millimetre Waves*, USA, pp. 81–82, 1987.
36. R. Simons *et al.*, 'Radial microstrip slot line feed network for circular mobile communications array', in *IEEE AP–S International Symposium*, **2**, 1024–1027, 1994.
37. B. Schoenlinner *et al.*, 'Wide-scan spherical-lens antennas for automotive radars', *IEEE Transactions on Microwave Theory and Techniques*, **50**(9), 2166–2175, 2002.
38. T. Yun *et al.*, 'A 10–21 GHz low-cost multifrequency, and full-duplex phased-array antenna system', *IEEE Transactions on Antennas and Propagation*, **50**(5), 641–649, 2002.

Part II

Small Printed Antennas for Wireless Systems

7

Planar Inverted-F Antennas

Peter S. Hall[1], E. Lee[1] and C. T. P. Song[2]
Department of Electronic, Electrical and Computer Engineering
[1] University of Birmingham, UK
[2] Mobile Devices, Motorola, Birmingham, UK

7.1 INTRODUCTION

Planar inverted-F antennas (PIFAs) are compact antennas that are compatible with printed circuit technology and which are now widely used in a range of applications, including hand-held terminals, such as cell phones and laptops, and also in vehicles [1]. Their operation can be understood by considering their development from two well-known antennas, namely the quarter-wavelength monopole and the rectangular microstrip patch antenna. Figure 7.1 shows the development from the monopole. The conventional monopole (Figure 7.1(a)), is fed at the base and has an input impedance of 37.5 Ω for the thin-wire type at resonance [2]. As this is a resonant structure, the current distribution is sinusoidal, with a peak at the feed point and a zero at the monopole tip. This means that the impedance varies along the wire length and it is possible to find a position with an impedance of 50 Ω suitable for connection to standard connectors and cables. Figure 7.1(b) shows how such an antenna can be fed. While the authors are not aware of use of this configuration as a passive antenna, it has been used as an active antenna with a three-terminal device at the wire junction [3]. It is shown here merely to illustrate the PIFA development. It is also possible to reduce the height of the base-fed monopole by bending the wire, as shown in Figure 7.1(c), to form an L-shaped antenna [4]. To a first order the current distribution is the same as the straight monopole so that the horizontal section introduces some cross-polarized components into the radiation pattern. If these two concepts are combined, the inverted-F antenna of Figure 7.1(d) is obtained [5]. This form has been widely used in vehicles and hand-held terminals and is well suited for coplanar mounting on top of a printed circuit board. The planar inverted-F antenna can be obtained from the wire inverted-F, by replacing the top wire by a rectangular conducting plate and operating it over a ground-plane normal to the feed pin, as shown in Figure 7.1(d).

Printed Antennas for Wireless Communications Edited by R. Waterhouse

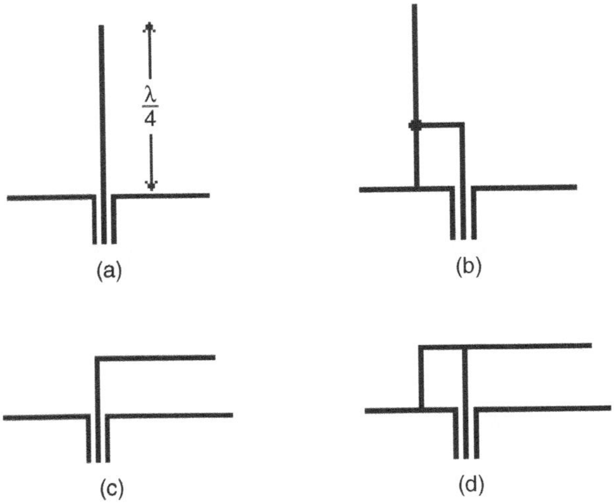

Figure 7.1 Development of the planar inverted-F antenna from the monopole: (a) quarter-wavelength monopole; (b) intermediate position-fed quarter-wavelength monopole; (c) inverted-L antenna; and (d) planar inverted-F antenna.

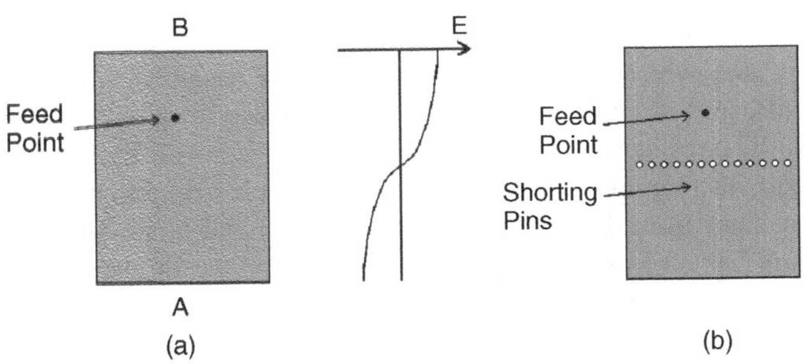

Figure 7.2 Derivation of the planar inverted-F antenna from the patch: (a) microstrip patch, with the field distribution between the patch and the ground-plane, and (b) patch with shorting pins; the top half forms the PIFA as the pin number is reduced.

An alternative development also helps understanding of the PIFA operation. Figure 7.2(a) shows a rectangular microstrip patch. The patch is approximately half a wavelength long, where the wavelength considered is that in the microstrip transmission line formed by the strip and the ground-plane, between the so-called radiating edges A and B. The feed point is located on the centre line of the transmission line about one-third of the distance between the ends, and is positioned to achieve an input impedance of 50 Ω on resonance. Within the resonator formed by the open-circuited transmission line is a half-sinusoid distribution of a vertical electric field, between the strip and the ground-plane, as indicated in the figure. At the centre of the patch is a voltage of zero. It is possible to insert a shorting plane at this point without perturbing the field distribution in the excited half of the patch. Such a shorting plane can be manufactured using shorting pins or vias (through hole plating). If the unexcited half of the patch is then discarded, then a shorted quarter-wavelength patch is obtained [6]. If the number of shorting pins is reduced then the frequency of operation is lowered, resulting in a very compact antenna, as shown in Figure 7.3. Most frequency

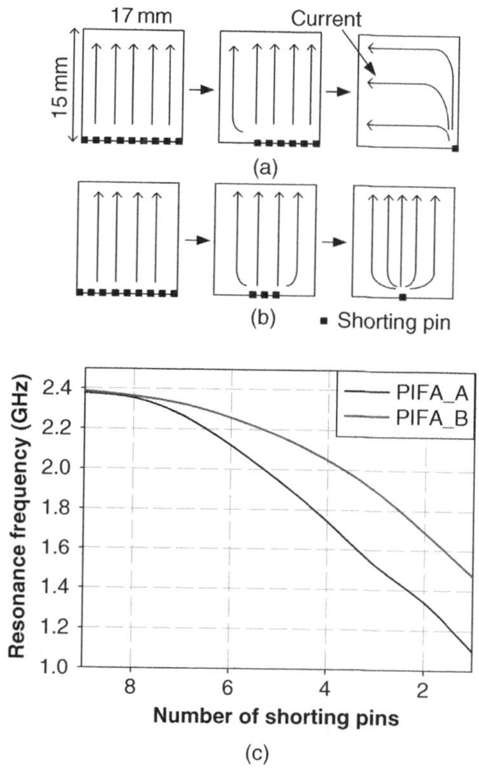

Figure 7.3 Effect of reducing the number of shorting pins in a quarter-wavelength patch to form a planar inverted-F antenna: (a) PIFA_A; (b) PIFA_B; and (c) resonant frequency.

reduction is obtained if the remaining shorting pin is at the corner of the patch as the current path is longer than in the case with the shorting pin at the centre of the edge. To obtain an impedance of 50 Ω on resonance, the feed pin position is different from that of the quarter-wavelength patch. Details of the design and operation of the planar inverted-F antenna are given in the following sections.

7.2 SINGLE-FREQUENCY PLANAR INVERTED-F ANTENNAS

The development of the multimodal PIFA [7, 8] can be traced to its origin back to the simple, single-frequency PIFA antenna design as illustrated in Figure 7.4. In this design, the PIFA antenna consists of a simple rectangular patch with dimensions given by $W \times L$. The antenna is usually suspended in free space. A shorting pin is placed at one corner of the patch with the coaxial feed placed relatively close to the shorting pin. The operating frequency for this simple planar inverted-F antenna can then be determined approximately by

$$f_0 = \frac{c}{4(W+L)} \tag{7.1}$$

where c is the speed of light, W and L are the width and length of the radiating element, and f_0 is the operating frequency. For the design given in Figure 7.4, the above equation gives a

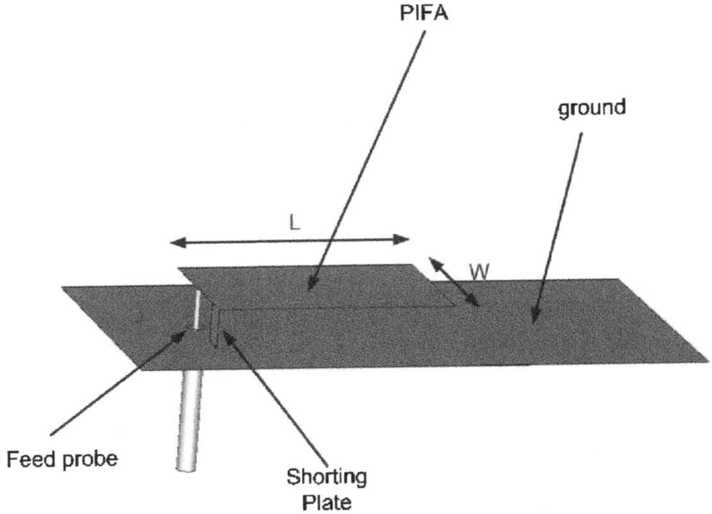

Figure 7.4 Geometry of a simple, single-frequency PIFA.

good first-order approximation of the dimensions required for a PIFA to operate at a desired resonant frequency.

Recently, transmission line modelling has also been used to model the resonant performance of the planar inverted-F antenna, PIFA [9], or shorted patch antenna [10]. Vaughan [10] has showed that an accurate transmission model for the shorted patch can be derived from the cavity model. If only the effect of the radiating slot of the shorted antenna is taken into account, the transmission line model will not be accurate. This is because the side slots also have a significant impact on the impedance performance. Hence, the effects of the radiating slot and side slots need to be taken into account for an accurate transmission line model. This transmission line model gives a simplified circuit perspective into the inner operation of the printed inverted-F antenna, PIFA, and shorted patch antenna.

One of the limitations of the traditional PIFA is its narrow bandwidth. This is usually not a problem for current PCS and mobile antenna applications because the input impedance bandwidth is usually very narrow, typically less than 7–10 %. In order to meet the needs of future wireless communication systems, such as the broadband or ultra wideband (UWB) communication system, a bandwidth in excess of 20 % or greater than 500 MHz with a carrier frequency in the range of 3.1–10.6 GHz would be needed [11]. In order to meet this future demand, there has been an intense research focus on improving the bandwidth performance of the traditional PIFA.

There are numerous methods that can be applied to improve the bandwidth of the PIFA. One of the simplest methods is to raise the height of the antenna [12]. This has the effect of increasing the radiation loss resistance and reducing the amount of stored energy under the PIFA structure. Thus the bandwidth of the antenna is improved. Another method involved the use of two antennas placed in close proximity to one another to generate a dual resonance effect [13, 14]. If the resonances are close to one another, a much broader bandwidth can also be achieved. Similarly, the dual resonance modes can also be generated by perturbing the patch antenna's geometry, as shown in Reference [15]. However, if the PIFA geometry is tapered, as given in Reference [16], the bandwidth of the antenna can also be significantly increased.

In most of the bandwidth enhancement methods highlighted above, it is usually the top patch element that is being modified to improve the antenna bandwidth performance. It should be noted that the size of the ground-plane on which the antenna is mounted can also affect its antenna characteristic. Recently, Wang et al. [17] have shown that the bandwidth of the PIFA can also be increased by modifying the ground-plane. In this design, a PIFA mounted on a T-shaped ground-plane has been proposed as shown in Figure 7.5(a) and its

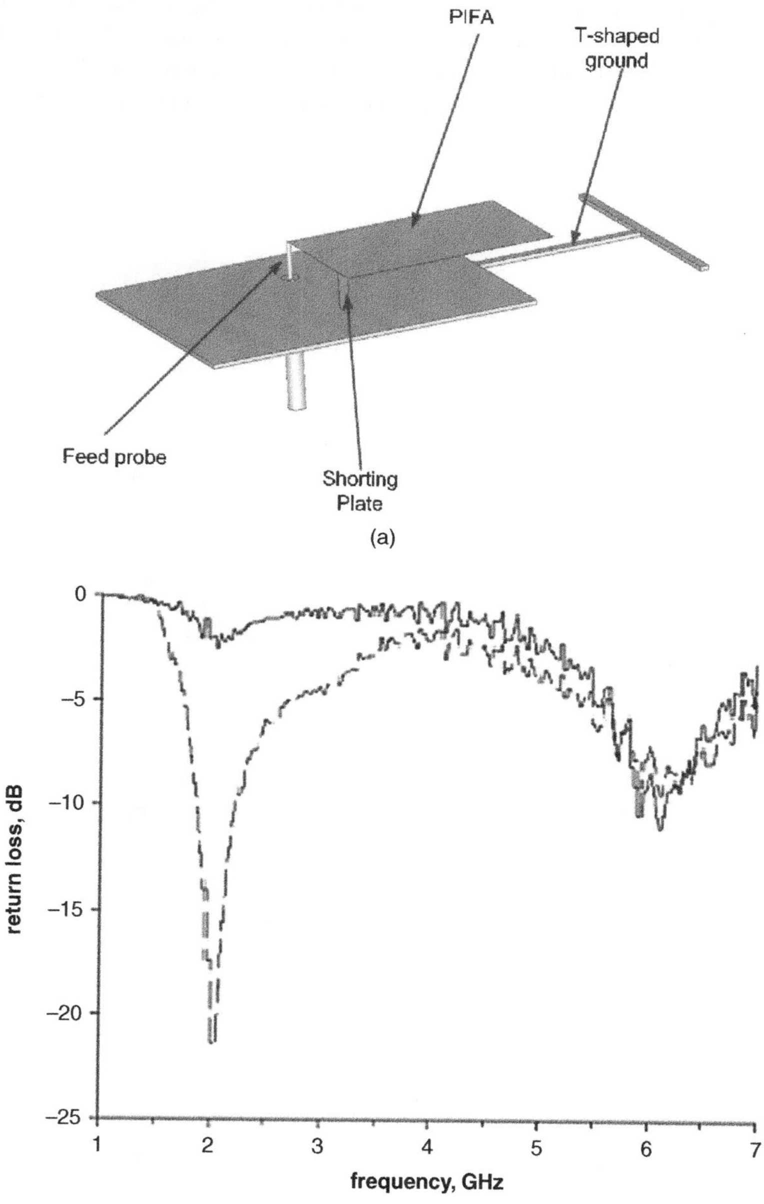

Figure 7.5 (a) Geometry of the PIFA with a T-shaped ground plate and (b) measured return loss performance (- - -, PIFA with a T-shaped ground-plate; ——, traditional PIFA with an unmodified ground-plane). Extracted from Reference [17]. © 2004 IET. Reproduced with permission.

measured results are showed in Figure 7.5(b). When compared to a similar printed inverted-F antenna, a PIFA mounted on an unmodified ground-plane. The PIFA with the T-shaped ground-plate has a better return loss performance and the measured results showed that it has a bandwidth greater than 16 %.

Feick *et al.* [18] showed that the bandwidth of a PIFA can also be improved simply by changing the feed plate silhouette. A number of designs shown in Figure 7.6 using different feed plate silhouettes have been studied. Input impedance matching bandwidths for the various feed plate silhouettes are also given in Table 7.1. The results suggested that the bandwidth of the ds-PIFA-uf, PIFA-rf, and PIFA-btf surpass the bandwidth of the PIFA-wf by factors of up to 2.3. The bandwidth improvement is close to that obtainable by a relatively thick quarter-wave monopole without the associated increase in antenna size.

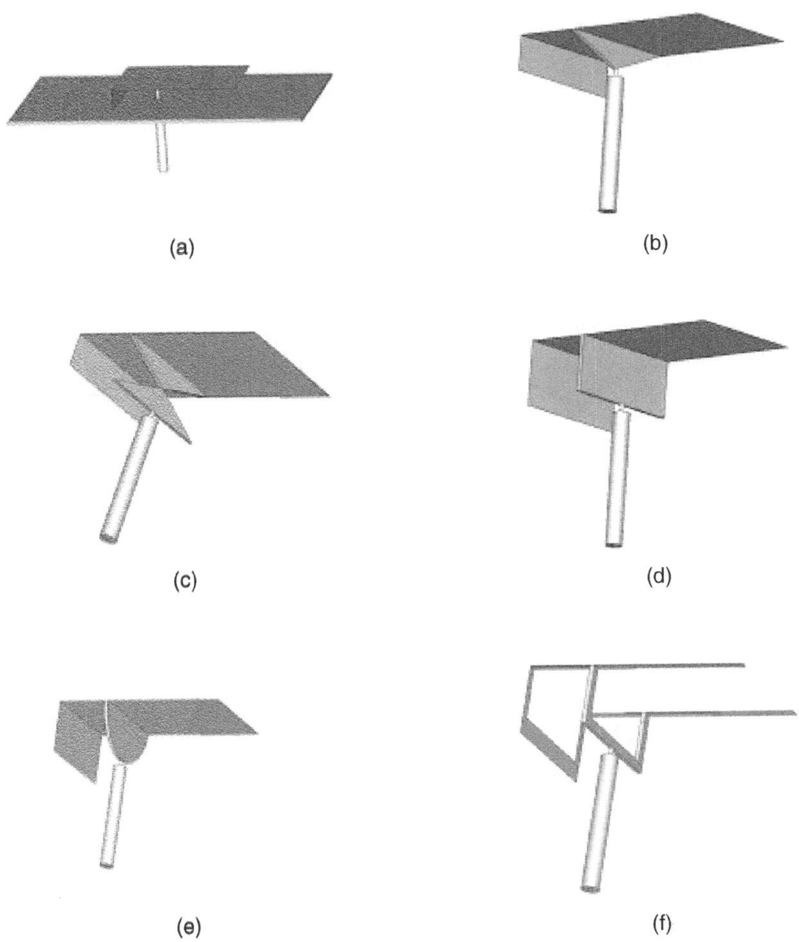

Figure 7.6 PIFA designs with different feed silhouettes: (a) wire feed (PIFA-wf); (b) triangular silhouette (PIFA-tf); (c) bitriangular silhouette (PIFA-btf); (d) rectangular silhouette (PIFA-rf); (e) rounded rectangular silhouette (PIFA-rrf); and (f) U-shaped strip element (ds-PIFA-uf). Extracted from Reference [18]. Reproduced with permission.

Table 7.1 Input bandwidth of PIFA designs with different feed silhouettes.

PIFA designs	Input bandwidth (%)	
	VSWR = 1.5 Numerical/measured	VSWR = 2 Numerical/measured
PIFA-wf	6.2/5.6	11.2/10.7
PIFA-tf	7.8/—	13.5/—
PIFA-btf	13.8/12.0	24.1/25.2
PIFA-rf	9.9/9.2	19.1/19.5
PIFA-rrf	9.3/—	16.6/—
PIFA-uf	9.8/—	18.2/—
PIFA quarter-wave monopole	14.3/14.6	27.2/27.2

7.3 MULTIPLE-FREQUENCY PLANAR INVERTED-F ANTENNAS

7.3.1 Introduction

The PIFA is not only a compact but also a versatile antenna, realizing multiple bands. Numerous researches have been reported and are largely driven by the rapid global growth of mobile telephony. One of the key requirements is a multifunction terminal, allowing subscribers to roam between different networks and systems. There are many techniques to-date in designing a dual- and multiple-band PIFA. These techniques can be largely grouped into five methodologies. Firstly, there are multiple resonator elements [7, 19], where each element is fed individually. Secondly, spur lines, slots, and shorting pins can be used to achieve multiple resonances using a single feed antenna [19–24]. Thirdly, other forms of capacitive loading on the PIFA [25–27] and also its ground [28], including LC resonators, can be used to achieve multiple bands [29]. Fourthly, parasitic elements can be used to enhance the bandwidth of the antenna [30, 31] or to generate additional resonance [32]. Finally, there is a need to identify innovative multiple-band resonators, such as the fractal Sierpinski gasket [33, 34]. With the ever-increasing demands of requiring more operating bands and bandwidths from mobile communication systems, combinations of these techniques are often required to achieve this. Figure 7.7 shows a summary of the techniques used.

7.3.2 Multiple-Band Resonators using Multiple Elements, Spur Lines, Slot, and Strips

One of the first reported dual-band PIFA was designed to operate at the GSM and DCS bands [7]. The author uses two resonators with separate feeds, while also ensuring that the DCS antenna coexisted within the dimension of the GSM element (see Figure 7.7(a)). A dual-band single-feed PIFA design was later realized by inserting an L-shaped spur line separating out the lower frequency resonator to the higher frequency one [7,19]. This provided an equivalent

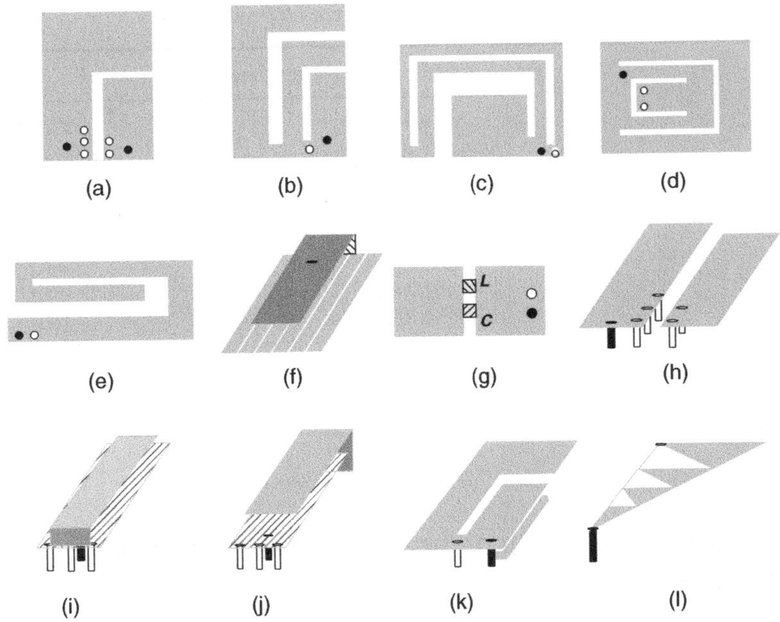

Figure 7.7 Examples of various multiple-frequency PIFA: (a) [7], (b) [19], (c) [20], (d) [22], (e) [23], (f) [28], (g) [29], (h) [30], (i) [31], (j) [31], (k) [32], (l) [34].

of two resonators where each resonance can be individually tuned. Various frequency ratios between the upper and lower resonance can be achieved by optimizing the element sizes as shown in Figure 7.8(a). By using additional spur lines, a triple-band single-feed PIFA can be obtained [19,20]. Figure 7.8(b) shows the return loss of a triple-band PIFA using dual spur lines similar to that shown in Figure 7.7(b) [19].

Often a single-mode resonator is unable to provide a large enough bandwidth. Multiple spur lines not only provide a lower band resonator but also tune the resonance of the two upper resonators closed to each other such that the two modes are fused together, providing a larger bandwidth [20]. With this concept of generating a larger bandwidth from two modes, a $\lambda/2$ resonant slot at the higher band was introduced within the lower frequency resonator of a conventional dual-band PIFA (see Figure 7.7(c)) [24]. This effectively broadens the upper resonance band while still maintaining a lower resonance. Another approach using slots within the antenna element to generate triple bands was reported using dual C-slots and two shorting posts, as shown in Figure 7.7(d) [22]. The slot width and pin position optimizes the antenna for dual resonance with bandwidths (VSWR$<$3) of 11.5 % (880–990 MHz) and 25 % (1710–2200 MHz). Finally, a combination of meandered strip forming a slot-like configuration and a single-mode resonator patch achieving five modes was demonstrated [21]. The two strips generating dual resonant modes are tuned to approximately a quarter-wavelength for the GSM (global system mobile) bands, while the middle patch and higher order modes of the strip accounts for the three resonant modes of the PCS (personal communications services) bands. These various techniques of generating multiple modes are widely used to broaden bandwidths of the required bands.

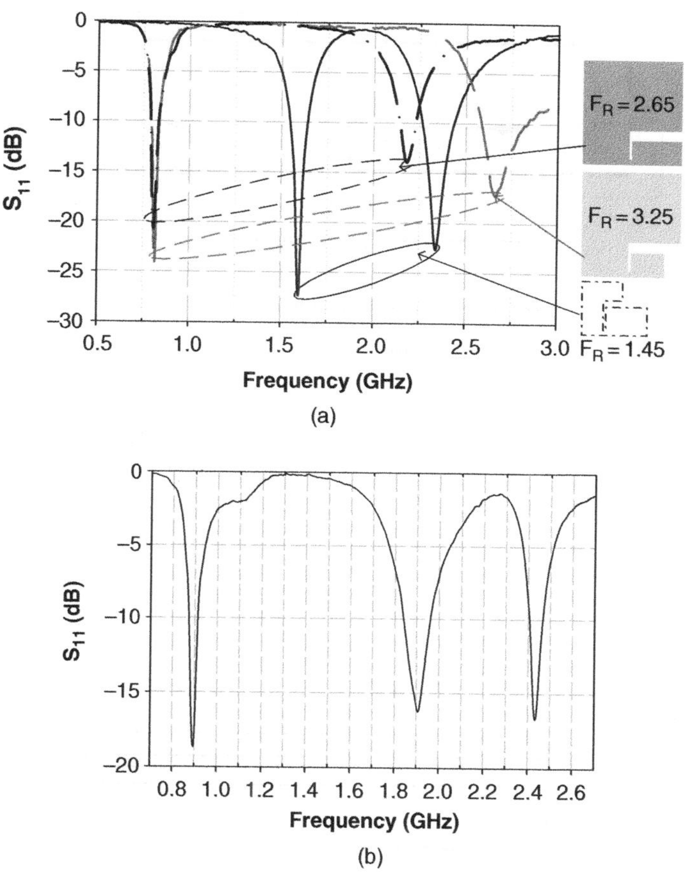

Figure 7.8 Spur line PIFAs: (a) frequency ratio tuning and (b) a triple-band single-feed dual spur line PIFA.

7.3.3 Capacitive and Inductive Loading

In most PIFA design, resonators for each band can always be easily identified from the current paths formed by the spur line or slot cuts. However, designing a multiband PIFA from a single element folding upon itself is not always straightforward. A capacitively loaded magnetic dipole antenna described in Reference [25] has arms that are configured to produce a circulating current flow. The electrical length of the antenna generates the lower band while the upper resonance is realized by an induced mode on the inner element, as shown in Figure 7.9(a). A dual-band PIFA can also be realized from a single strip folding on itself, as shown in Figure 7.9(b) [26]. The height of the antenna, folding dimensions optimized by the length, and the gap between the top and bottom strips form the critical parameter to its dual-band property. Additional capacitive loading was introduced to tune the upper resonance of the antenna while keeping the lower band relatively unaffected. The same technique can also be applied on a planar antenna configuration similar to that in Reference [25]. This single strip with a single folding loop, shown in Figure 7.9(c), also provides a dual resonance design [27].

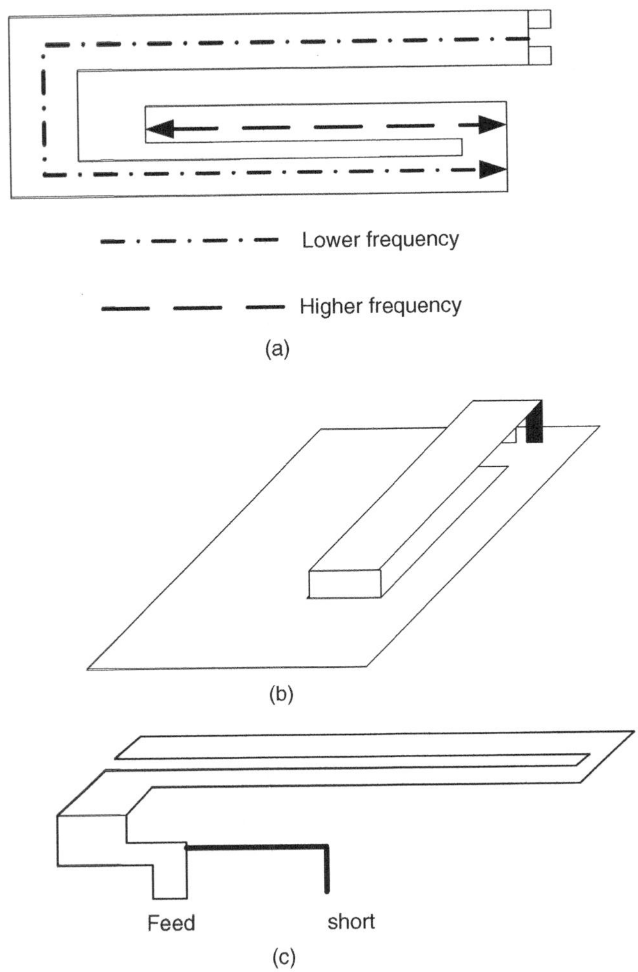

Figure 7.9 Capacitive loaded PIFA antennas: (a) a multiband capacitively loaded magnetic dipole antenna [25]; (b) a curled PIFA with capacitive loading [26]; and (c) a bent radiating arm PIFA [27].

The PIFA being a small antenna is highly sensitive to its ground. A single-band PIFA was reported to resonate with dual-band characteristics by designing parallel slot cuts on the ground (see Figure 7.7(f)) [28]. The higher frequency is generated by the fundamental TM_{01} mode of the PIFA while the lower frequency is contributed by the resonance of the slotted ground-plane. The proposed design resonates at 1.9 and 2.4 GHz, with a bandwidth of 9 and 8.4 % and a gain of 1.7 and 4.3 dBi, respectively.

Another method of realizing a dual-band design was to introduce RF traps with LC resonators on the PIFA [29], similar to that implemented in the multiple-band monopole or dipole (see Figure 7.7(g)). To achieve dual bands at 900 and 1800 MHz, the PIFA is split into two halves where each has a specific transmission length. Using the transmission line method, suitable LC components are then selected to provide the prescribed open/short conditions of the operating frequencies. The authors have also implemented a number of solutions to realize these LC components, using a meandered line and metal on the insulator capacitor.

7.3.4 Parasitic Elements

The use of parasitic elements has long been a popular technique to designing broadband antennas by generating additional resonance. The Yagi–Uda antenna is a classic example. The technique has since been employed in the design of electromagnetically coupled patch antennas. The PIFA with its limited bandwidth from a single resonator also uses this technique of parasitic elements to increase its bandwidth significantly or to migrate the degenerated dual resonance into a dual-band antenna.

The method usually consists of the driven antenna with the parasitic element in close proximity (see Figure 7.7(h)). The electrical lengths of the two resonators are often a small offset from each other, and this dictates the overall required broadband, or dual-band resonance. At close to the frequency of interest, the driven antenna is design to be overcoupled at the feed. The parasitic antenna is then placed near the driven antenna. This will result in a looped behaviour on the Smith chart that forms the basis of the additional resonance property [30]. The size and location of this coupling loop is optimized by the feed position, element separation, and the length offset between them.

Parasitic elements can be realized in a coplanar configuration (see Figure 7.7(h)) or stacked configuration (Figures 7.7(i) and (j)). In the stacked PIFA configuration design, the bandwidth of the antenna could be doubled [31]. A different configuration can also be realized by orienting the parasitic antenna in the opposite phase (see Figure 7.7(j)). While the stacked parasitic solution results in a thicker product, a coplanar solution as shown in Figure 7.10(a) can also provide the required coupling loop on the Smith chart [30]. The parasitic element in this case is tuned close to the higher resonant band and aligned beside the high-band resonator of the dual-band PIFA. This results in a dual resonance effect at the upper band while leaving the lower band undisturbed. Multiple parasitic resonators as shown in Figure 7.10(b) can also be strategically placed where strong mutual coupling excites these elements for multiple-band operation [23]. The overall antenna size measures about 38.5 mm × 28.5 mm × 8.5 mm with a lower bandwidth of 90 MHz (870–960 MHz) at a VSWR better than 2.5, while the upper bandwidth of 460 MHz (1710–2170 MHz) has a VSWR less than 2.

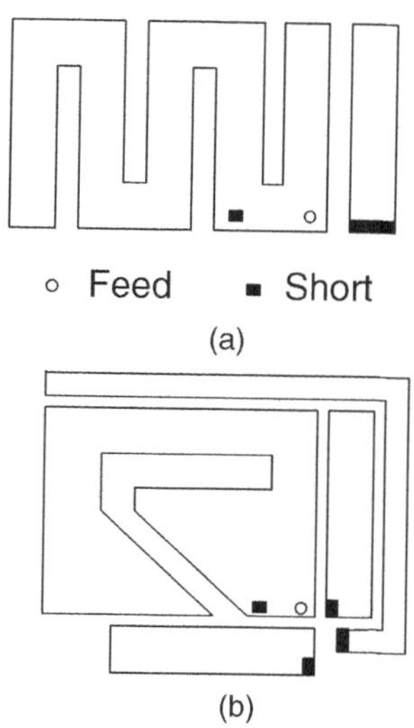

Figure 7.10 PIFA with parasitic elements: (a) single parasite [30] and (b) multiple parasites [23].

Parasitic resonators are usually implemented separately to the driven element. However, parasitic resonators can also be directly implemented on the feed of the antenna (see Figure 7.7(k)). A compact six-band PIFA was achieved with a combination of a parasitic stub loaded along the feed and a capacitive parasitic coupling plate under the feed element [32]. The parasitic plate measures approximately the same dimension as the feed element and shares the same shorting location as the fed element. The antenna was realized in a space of 36 mm × 17 mm × 8 mm, and resonates with $S_{11} < -6$ dB at 930–1000 MHz, 1518–2314 MHz, and 2422–2587 MHz. Additional parasitic elements can be loaded on to the feed to generate additional resonance. Parasitic elements can also be directly loaded in a similar way on to the shorting pins of the PIFA to generate multiple resonances.

7.3.5 *Fractal PIFAs*

Multiple resonances from a single-element PIFA could result in complex structures employing techniques such as additional parasitic elements and slot and spur lines. Fractal-shaped antennas, in particular the Sierpinski gasket, is inherently a single-element multiple-band antenna [33]. The monopole configuration of the Sierpinski gasket has been widely reported. Its evolution to a PIFA design, employing only half the size of the gasket, was reported by Song *et al.* [34]. The antenna is folded over and arranged in a planar configuration

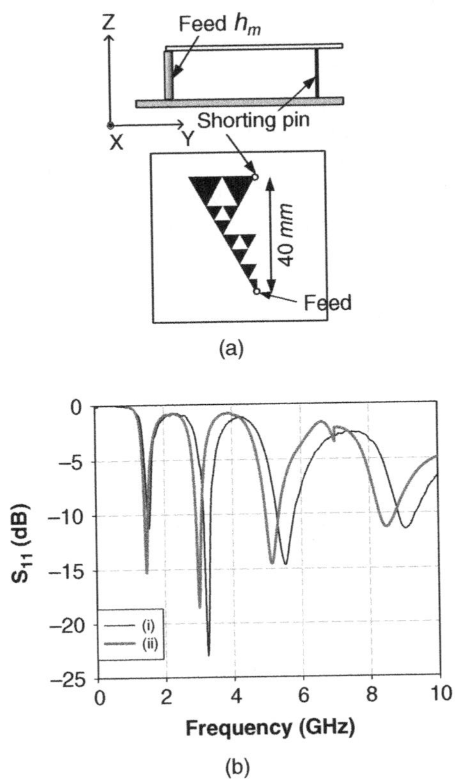

Figure 7.11 Planar shorted Sierpinski gasket antenna [34]: (a) configuration and (b) return loss (i) and measured (ii) simulated antennas. Extracted from Reference [34]. © 2004 IEEE. Reproduced by permission of the IEEE.

where a shorting pin is then introduced to match the lower operating band. Figure 7.11(a) shows the configuration of the design along with the measured performance, shown in Figure 7.11(b). Tuning of the resonance band can also be achieved by optimizing the scaling ratio of the gasket.

7.4 REDUCED-SIZE PLANAR INVERTED-F ANTENNAS

In recent years, the demand for compact wireless handsets has grown tremendously. This has led to the increased demand for smaller wireless mobile devices. However, the size of the wireless handset is usually limited by the size of the battery and its antenna. In addition, the need to employ antenna diversity or adaptive antennas on the handset to improve receiver performance are currently driving the need for a more compact antenna design. In principle, the miniaturization of the PIFA can be achieved by using different approaches. These methods are described in more detail in the following section.

One of the simplest methods for reducing the size of the PIFA is to load the antenna with a high dielectric constant material [35]. The higher the electric permittivity, ε_r, the smaller is the PIFA antenna. However, the drawbacks of using a high dielectric constant material, ε_r, would be to degrade the antenna gain and its bandwidth performances. This would be unacceptable. The combination of the substrate and superstrate with a high dielectric constant material can be used to overcome this gain degradation problem [36, 37]. However, in this case, the bandwidth of the antenna still remains very narrow. In order to overcome the bandwidth issue, Lo and Hwang [35] have opted to increase the height of their miniaturized PIFA by doubling it. An example of the proposed antenna is shown in Figure 7.12. Through the FDTD simulation, they have shown that a size reduction of 75 % compared to a conventional PIFA using a high dielectric constant material ($\varepsilon_r = 38$) is feasible. At the same time, the gain and bandwidth of the PIFA have been improved to 7 dBi and 5.7 %, respectively. Similarly, the magnetic material could also be used to load the PIFA to reduce its physical size [38].

It is also possible to reduce the size of the PIFA through reactive loading. One possible approach is to use capacitive loading [9, 39]. The antenna structure with the capacitive

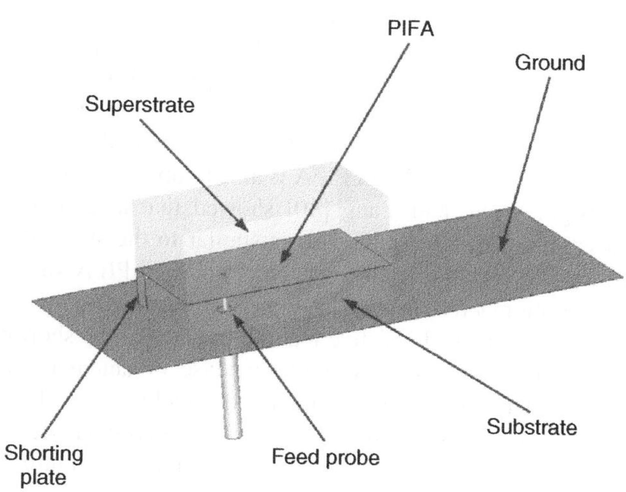

Figure 7.12 Configuration of a PIFA loaded with a high dielectric material. Extracted from Reference [35]. © 1998 IEEE.

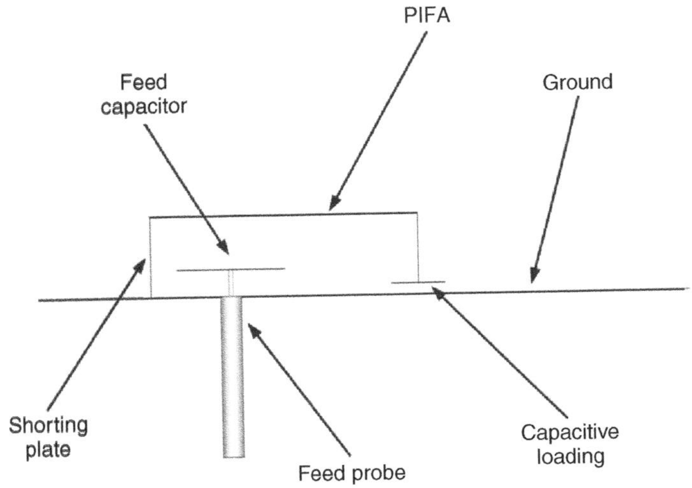

Figure 7.13 Capacitively loaded PIFA structure. Extracted from Reference [39]. © 1996 IEEE.

load is illustrated in Figure 7.13. With this capacitor loading technique, Rowell and Murch have managed to reduced the resonant length of the PIFA from $\lambda/4$ to less than $\lambda/8$. The capacitive load is formed by folding the open end of the PIFA towards the ground-plane, to produce a parallel plate capacitor for the load. When the dimensions of the capacitor load are varied, the resonant frequency of the antenna can be changed. With increasing capacitive load, the resistive and reactance peaks in the antenna impedance increase while the widths of the impedance peak curves decrease. Hence a much narrower bandwidth is achieved with increasing capacitive load. It should be noted that although the capacitor loading allows the resonant frequency of the PIFA to be reduced, this is achieved at the expense of a poorer impedance match and narrower bandwidth. However, the poor matching as a result of the capacitive load can be easily overcome by the use of a capacitive feed to manipulate the resistance and reactance curves.

Another possible method of achieving a significant size reduction is by perturbing the PIFA antenna. This can be achieved by the introduction of slots appropriately placed in the PIFA structure. Figure 7.14 illustrates two different compact PIFA designs that utilize slots as a loading element. In Figure 7.14(a), the introduction of slots into the PIFA structure has the equivalent effect of increasing the electrical length of the antenna. Hence for a similarly sized PIFA structure, the modified PIFA is able to operate at a much lower resonant frequency. In this design, Wong and Yang [40] showed that a reduction in the antenna resonant length to less than $\lambda/8$ is feasible. This is similar to the size reduction achievable using the capacitive loading technique. However, the modified PIFA suffered from narrow bandwidth due to the strong reactive near-field. In order to overcome the narrow bandwidth problem, Wong and Yang introduced resistive loading in place of the shorting pin. Although the bandwidth has been improved, this is at the expense of antenna gain due to ohmic losses in the resistor. Another different compact PIFA design [41] with slots /or corrugations added to its structure is shown in Figure 7.14(b). In this particular design, the slots are introduced into a tapered PIFA. Similar to the design illustrated in Figure 7.14(a), the main design concept for this antenna is based on lengthening the path of the current. This is achieved through the introduction of the slots/corrugations. Due to the tapered PIFA design,

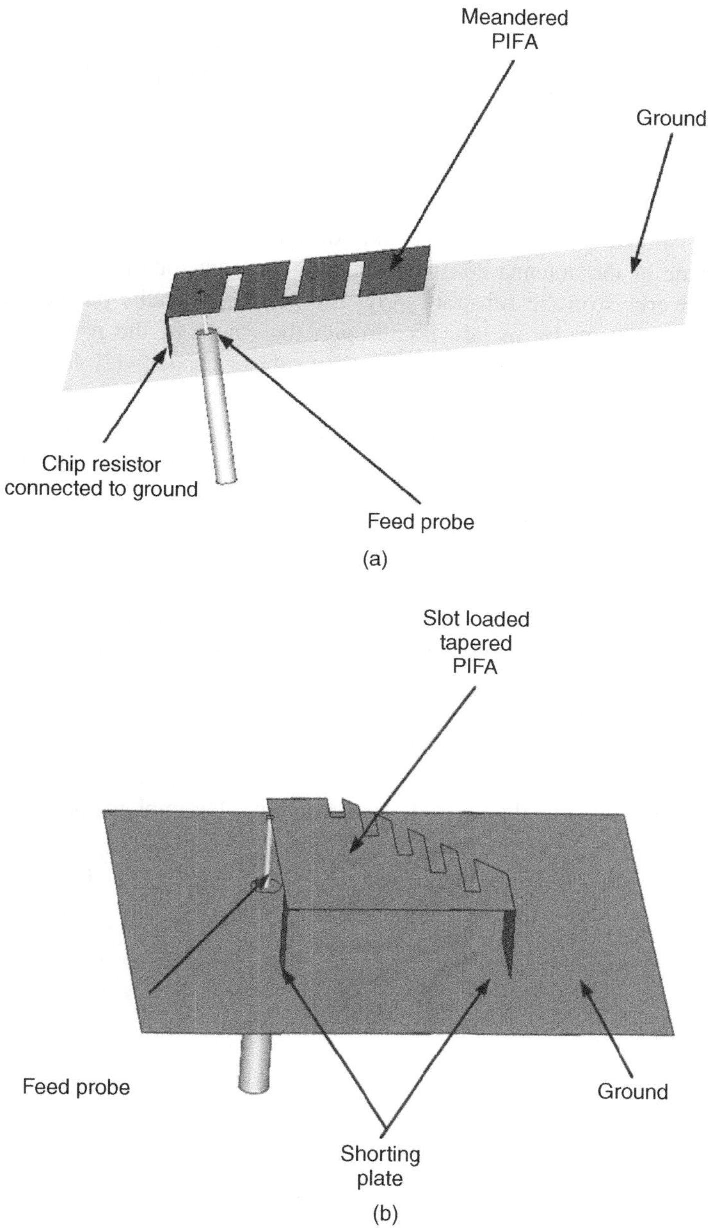

Figure 7.14 (a) Geometry of a modified printed inverted F-antenna [40] and (b) geometry of a tapered PIFA loaded with slot [40]. © 1998 IET. Reproduced with permission.

this antenna is capable of a much broader impedance bandwidth than the conventional PIFA. In this case, the simulated (using CST Microwave Studio) and measured bandwidths are 16.64 and 20.3 %, respectively. Also, the overall volume of this corrugated tapered PIFA is 50 % smaller than the conventional $\lambda/4$ PIFA structure.

Wire antennas do not function effectively when positioned parallel or close to the perfect electric conductor ground-plane. This is due to problem with out-of-phase image current. With the recent development of researches on the electromagnetic bandgap (EBG), a high-impedance surface with very interesting properties has been proposed [42]. One of the more interesting properties of the electromagnetic bandgap (EBG) is its in-phase image current property. This allowed antennas to operate very close to the high-impedance surface without suffering from poor radiation performance. The proposed compact PIFA on the EBG-type ground-plane is shown in Figure 7.15. In this design, a simple PIFA is used. The EBG-type ground-plane of the antenna consists of a dielectric substrate on a metallic plate with periodic patch overlays on the substrate [43]. The periodic metallic patches are connected to the bottom metal layer by metal vias through the centre of the patches. The antenna with a high-impedance surface is a type of inductively or capacitively loaded antenna. By incorporating the EBG-type ground-plane the size of the PIFA can be reduced. In this case, the operating frequency of the PIFA can be adjusted by modifying the thickness of the EBG-type ground-plane. Note that the bandwidth and resonant frequency increases as the dielectric substrate thickness of the ground-plane decreases.

Recent development has shown that fractal geometry can be successfully applied to antenna design. One of the main advantages of using fractal geometry is the possibility of a new fractal antenna with a much reduced size. This is due to the space-filling property of the fractal geometry. When this is applied to an antenna element, it effectively increases the overall electrical length of the antenna. Therefore, a much reduced antenna structure for the same resonant frequency can be achieved. In a recent publication, Guterman *et al.* [44], successfully applied the Koch curves in a PIFA configuration, as shown in Figure 7.16(a). By doing this, a reduction in size of approximately 62 % in comparison with the simple rectangular patch has been achieved. In an alternative fractal implementation, a Hilbert curve could also be used to achieve the desired size reduction. One implementation of the Hilbert

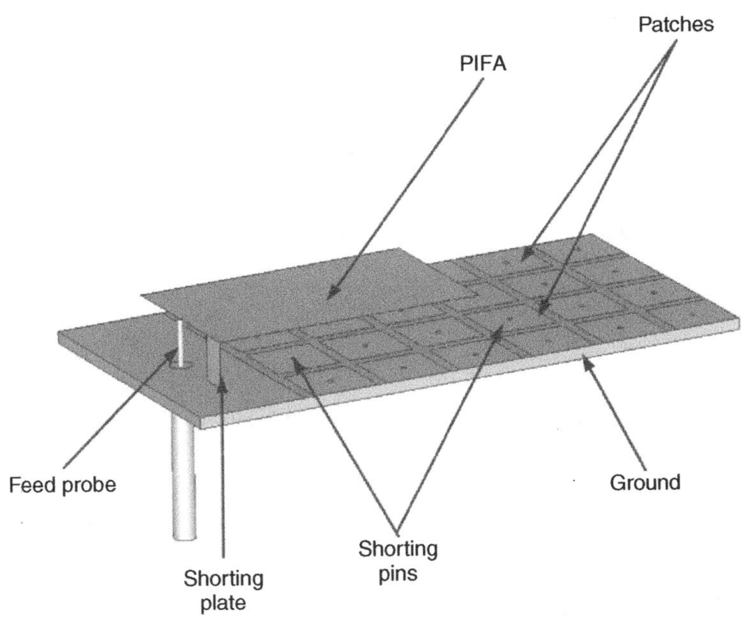

Figure 7.15 PIFA with a PBG-type ground-plane configuration. Extracted from Reference [43]. © 2003 IEEE.

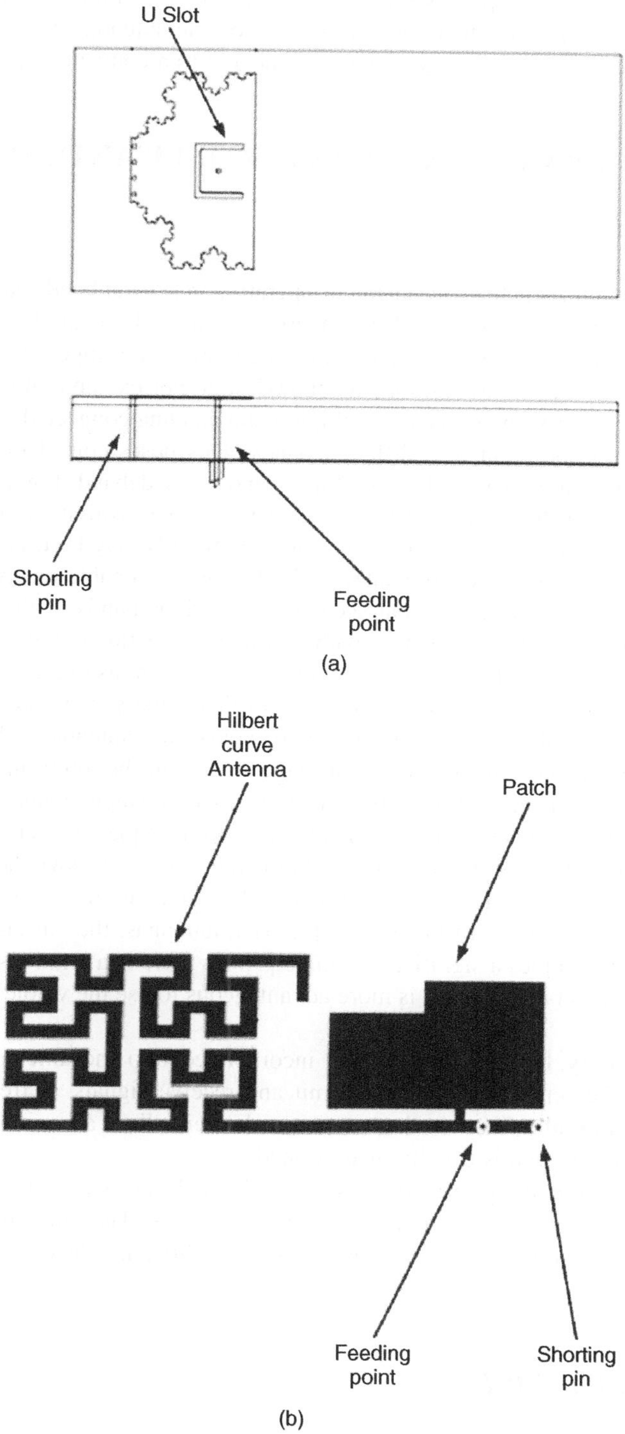

Figure 7.16 (a) Structure of a fractal PIFA with a U-slot [44] and (b) a dual-band PIFA consisting of a Hilbert-type curve. Extracted from Reference [44]. © 2004 IEEE.

fractal PIFA by Azad and Ali [45] is shown in Figure 7.16(b). Similar to the Koch curves, the Hilbert curve is used to lengthen the current path on the antenna structure. In this case, an overall size reduction of 50 % was achieved compared to a conventional PIFA.

7.5 SWITCHED-BAND PLANAR INVERTED-F ANTENNAS

7.5.1 Introduction

The increasing demand for antennas capable of operating in several bands has given rise to a lot of work in the design of multiband planar inverted antennas. In parallel to this much work is also being done on multiband RF front-ends. There are some important choices for the circuit engineer that have a bearing on the antenna design. For example, it may be simplest, from an antenna point of view, to have a multiple-band antenna connected to a superheterodyne receiver that can be retuned to different bands. This poses considerable demands on the receiver design such as the need for multiple-band, or wideband, low-noise amplifiers, filters, and mixers, and this is also likely to result in lower sensitivity when compared to a single-band receiver. On the other hand, multiple front-ends may be used to simplify the design, albeit at the expense of equipment size. In the case of tunable receivers, an antenna is required with a single connection that accesses the various bands of the antenna [7]. In the case of multiple transceivers, antennas with multiple connection points [46] can be used.

Alternately an antenna with switches or tuning elements can also be used and connected to a tunable front-end. There are two advantages to this. Firstly, the antenna presents the minimum bandwidth to the incoming signal, as of course the antenna is the first filtering element in the receiver chain, and this will reduce the noise contribution from external sources in the environment. In the case of a multiband antenna with a single connector, the external noise will be greater. However, a multiband antenna with multiple connection points will be equivalent to a switchable one in this respect. On the other hand, the switches will introduce additional loss into the system. Secondly, in a switched antenna, the whole of the antenna volume is used in every band. In many multiple-band antennas, the various resonators are separated, and each occupies a significant volume [7]. As there is a relationship between the antenna volume and its bandwidth, it is more advantageous to use the whole antenna volume at each band.

A further necessary function that can be incorporated into the antenna is that of the duplexer, namely the separation of the transmit and receive signals. If frequency division duplex is used then usually filters in the transmit and receive lines are used. If time division duplex is used then a switch is usually incorporated.

Many diodes, such as the positive-intrinsic-negative (PIN) diode, can be used both as a switch and as a device giving a continuous capacitance change. The following sections have, however, been separated into those antennas that use switching and those that use continuous tuning.

7.5.2 Switched-Band PIFAs

Frequency tuning can be obtained by inserting a switch into the surface of the planar inverted-F antenna. A typical arrangement is shown in Figure 7.17 [47] and another related switching arrangement is seen in Reference [48]. An extra length of PIFA is added, which is connected

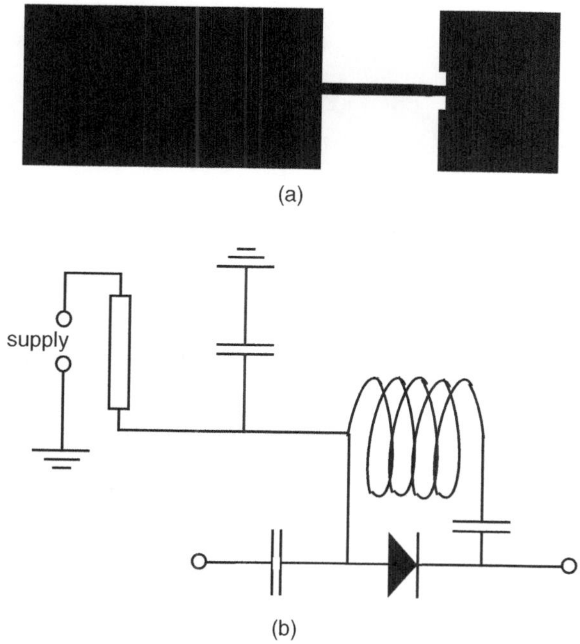

Figure 7.17 PIFA a with band switch [47]: (a) antenna and (b) diode circuit.

by a section of high-impedance line, containing the Philips BA682 UHF band PIN diode. The high-frequency PIFA is 20×14 mm^2, the connecting line length is 5 mm, and the extra PIFA section is 10×14 mm^2. This is etched on FR4 substrate with $\varepsilon_r = 4.6$; the substrate thickness and whether it is air spaced are not given. The switch has an LC compensation, which not only helps tune-out the parasitic capacitance at the diode's reverse biased state but also prevents a DC short-circuit introduced by planar inductance across the diode. Under a reverse bias of -5 V, two resonance frequencies occur, at 900 MHz and 1 .74 GHz. The higher frequency resonance is due to the patch, while the lower resonance is due to the LC network and other reactive components. Under a forward bias of $+5$ V a single resonance occurs at 1.12 GHz with an input return loss of 20 dB. The radiation pattern at both diode bias conditions is well behaved. Continuous variation of the diode bias also gives continuous frequency tuning. For a bias of 0 V to -15 V DC, the frequency varies from 1.43 to 1.63 GHz, which is 200 MHz or 14 %.

Figure 7.18 [49] shows a switching arrangement using multiple diodes to achieve wider frequency shifting, which are located on the antenna ground-plane to simplify bias connections. Two PIFA top plates are also used to form a dual-band antenna; the lower plate operates in the upper band and vice versa. In each part the normal single shorting pin is extended into a shorting strap and many are used, each of which is attached to a BA682 PIN diode. The inset shows the switch bias connections. A current limiting resistor of 680 Ω is used to limit the DC bias current to less than 10 mA and surface-mounted capacitors of 10 pF are connected between the diode switch and the shorting strap to isolate the DC bias current to flow to adjacent shorting straps, the patch, and the feed pin. Switch losses were measured to be 0.3 dB under forward bias. By successively switching on the upper patch the frequency changed from 730 to 810 MHz and for the lower patch from 1050 to 1250 MHz. Limited information on radiation patterns is given, but were in general well behaved.

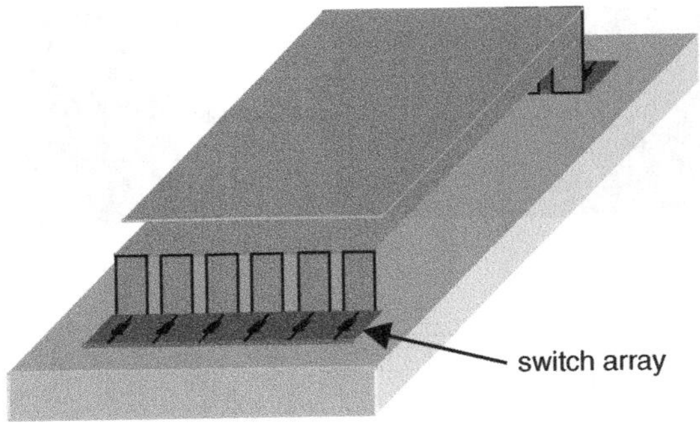

Figure 7.18 Dual-band PIFA tunable by switched shorting straps [49].

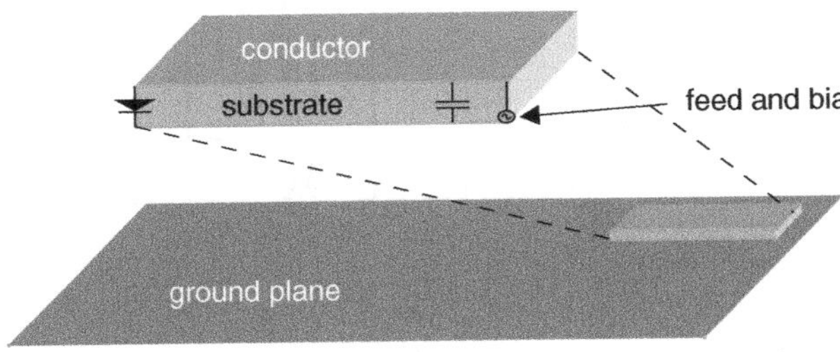

Figure 7.19 Dual-mode switched PIFA [50].

In Reference [50], a single HSMP 386C PIN diode switch, shown in Figure 7.19, has been added to the PIFA at the opposite end to that shown in Figure 7.18. Few results are given but a reduction of the radiation efficiency is noted, due to high forward bias resistance. Diode switches have also been added close to the connection point in a self-diplexing PIFA [46]. Four diodes are used, two for each band, which allows switching between the transmit and receive functions.

7.5.3 Continuously Tuned PIFAs

In an arrangement similar to that of Figure 7.17, a pair of varactor diodes has been inserted between the two sections of a split PIFA [51]. The diodes are placed at the edges of the conductor as the current is strongest at those points. The PIFA sections were $32.25 \times 25.81\ \text{mm}^2$ and $25.82 \times 25.81\ \text{mm}^2$, separated by a 6.46 mm gap in which the diodes were located. For a -20 V to -1.25 V bias change, in which the capacitance of each diode varied from 5 to 35 pF, a 14.7 % change in resonant frequency was noted around 900 MHz. No information on radiation patterns or efficiency is given.

Figure 7.20 shows a PIFA in which both position of a second shorting pin, pin 2 in the figure, and the effect of a varactor between the plate and ground-plane have been studied [52]. This shorting pin was positioned at different points along the perimeter of the PIFA. The resonant frequency increases with increasing distance from the feed point, due to the higher inductance presented to the path of currents on the patch, but it reaches a plateau as the pin approaches the corners. An effective relative bandwidth of 50 % is thus achievable. If the second shorting pin is replaced by a variable capacitor of fixed position at the antenna

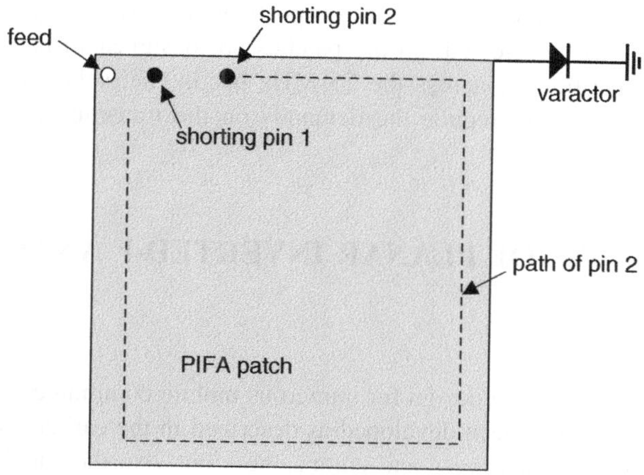

Figure 7.20 PIFA tuned by a single varactor [52].

perimeter then tuning ranges of the order of 15–20 % can be achieved. The gain was found to vary between 0.5 and −2 dBi across the tuning range.

Figure 7.21 shows a duplexing antenna [53], consisting of two narrowband PIFAs, whose frequencies are located in the transmit and receive bands, respectively, and whose individual frequencies can be tuned across the band using a number of varactor diodes located on the sides of the PIFA plates. The PIFAs are mounted on either side of a pedestal ground-plane and have a shape similar to the one shown in Figure 7.21, with wide sections $20 \times 10\ mm^2$ and a narrow connecting section $20 \times 4\ mm^2$, located 7 mm above the ground-plane, designed for operation in the US 824–894 MHz band. In this arrangement 10 dB isolation between transmit and receive is achieved. If the PIFA patches are replaced by rectangular ones, the shorting pins are removed and four switched capacitors are introduced between the patch edges and the pedestal ground, then much narrower band antennas are obtained, which can be tuned across the bands. A capacitively coupled short to the ground acts in a similar way

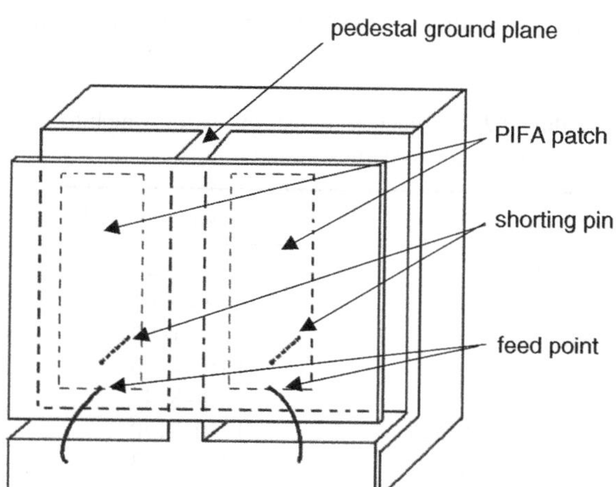

Figure 7.21 Duplexing PIFA [53].

to the conventional PIFA short and also allows postmanufacture tuning. The four switches allow 16 frequencies to be selected, so the patches are designed to have an approximate bandwidth of 1.6 MHz tunable across the 25 MHz band. The isolation between bands is 20 dB, which can be used to reduce the demands on the transmit–receive duplexer that follows the antenna.

7.6 APPLICATIONS OF PLANAR INVERTED-F ANTENNAS

7.6.1 Introduction

The PIFA has been an attractive design for numerous mobile communication devices. Many design techniques have since been developed as described in the earlier sections. Its configuration provides not only robustness and compactness but also the capability of receiving both horizontal and vertical electromagnetic fields. This is highly favourable in a multipath rich urban environment. One of the key successes is its ability to generate multiple bands in a small space. Despite challenges to obtain a large enough operating bandwidth at each band, designing multiple bands within the same space of the lowest operating band remains a challenge. In the following sections, some of the most popular applications of PIFA will be described.

7.6.2 Mobile Handsets

Since the astronomical growth in cellular communication, consumers have fuelled the demands of a multiple-mode handset, enabling users to roam between networks and countries. With existing 2.5G/3G systems around the world and various other services such as Bluetooth, 802.11a/b/g, GPS, FM radio, and even UWB in the future offered to mobile subscribers, antenna engineers are challenged to provide multiple resonances in an ever-reducing handset size and antenna volume. Table 7.2 lists the frequency bands used by today's cellular communication systems.

Pioneering handset designs in the early 1990s have widely adopted the monopole antenna configuration. They are mostly realized with an external stubby and later evolved into small

Table 7.2 Frequency range of various mobile communication systems.

System	Frequency range (MHz)
GSM	824–894
EGSM	880–960
DCS	1710–1880
PCS	1850–1990
UMTS	824–894
Bluetooth, Wireless	1710–2170
LAN 802.11b/g	2400–2483
Wireless LAN 802.11a	5150–5250
GPS	1559–1610

helical antenna. In recent years, a mobile handset has increasingly become a fashion icon. The requirement for an internal antenna thus becomes a tremendously attractive solution. The PIFA, due to its small size and multiband capability, has since been widely adopted in many handset designs where volume for the antenna is critically limited. In order to ensure product consistency throughout all manufacturers, the measurement procedure for the radiated RF power and receiver performance such as that from the Cellular Telecommunications and Internet Association (CTIA) [54] is becoming a standard requirement for most networks. Figure 7.22 shows an example of a triple-band PIFA mounted on a rear housing of a mobile phone. The size of the antenna measures approximately 20 mm $\times$38 mm $\times$6 mm ($L \times W \times H$). To date, the design of a popular handset configuration can be classified into three generic categories. They are the candy bar phone, the flip phone, also known as the clamshell, and, finally, the slider phone. Recently, more dynamic designs have emerged using a rotary mechanical chassis.

As the size of a mobile telephone varies significantly, so does its radiated performances. The PIFA being an electrically small antenna couples strongly to its ground-plane. Due to this induced current, the resonance of the ground-plane should be taken into consideration in the overall design. This will have a significant effect on the bandwidth and efficiency of the PIFA and is also dependent on the operating frequency [14, 55, 56]. Figure 7.23 shows the bandwidth variations with respect to the changing PCB (printed circuit board) ground length. In both cases, the PIFA is placed at the top end of the PCB ground while the lower end of the PCB is varied. Both results concluded that the bandwidth becomes very small for the lower band (900 MHz) operations when the PCB length is small. From the radiation patterns perspective, a dipole-like pattern is obtained from a typical handset PIFA at 900 MHz. It could be hypothesized that the PIFA is a physically small feed end of a dipole while the PCB ground forms the current's return path. Hence at this lower frequency band the entire handset is resonating.

The PIFA positioned at different locations of the ground will also excite different ground currents and hence variations in overall antenna performance. In applications such as the clamshell and the slider phone, the different ground conditions during the standby and talk position will be a design challenge. Despite bandwidth and efficiency variations, frequency shifts are also expected. With handset size, in particular its length slowly converging to < 85 mm, designing a wider band and efficient antenna at the lower frequency band remains a challenge.

Another design requirement for handset antennas is the health and safety mandate ensuring that all manufacturers conform to prescribed RF exposure limits. The specific absorption

Figure 7.22 A triple-band antenna (courtesy of Sendo).

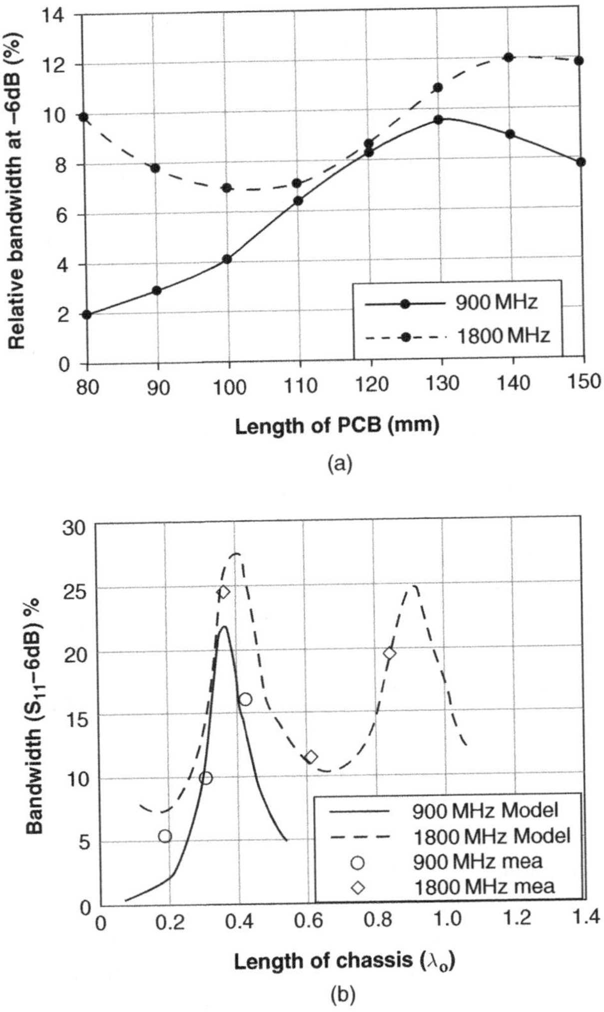

Figure 7.23 (a) Bandwidth variations with respect to different PCB lengths [14]. (b) Simulated and measured impedance bandwidth ($S_{11} < -6$dB) of 900 and 1800 MHz antennas in free space as a function of chassis length [56].

rate (SAR) is a measure of radiation on to human tissue and is defined by the International Commission on Non-Ionizing Radiation Protection (ICNIRP) [57] and the ANSI/IEEE [58]. A specific anthropomorphic mannequin, or 'SAM', is used as the head model for use in SAR testing. Other systems such as the Dosimetric Assessment System (DASY) are also widely used in various test laboratories for SAR compliance certification. The CTIA also depicts radiated performance in the presence of the simulated head mode [54]. The current SAR limit set by IEEE C95.1:1999 is 1.6 W/g in 1 g averaging mass while the ICNIRP sets 2 W/kg in 10 g averaging mass. Several methods of measurements have been defined by the FCC [59] and European standard EN50361:2001 [60]. A lower SAR can be achieved from a PIFA in comparison to the external helical antenna, since its orientation is usually towards the opposite direction of the user's head and is shielded by the PCB. This is certainly so when the operating frequency is higher at the 1800 MHz range where the PCB effect becomes more prominent. However, at the 900 MHz range, the PCB ground becomes resonant and no significant benefits can be expected from the PIFA compared to the external stubby.

7.6.3 Mobile Connectivity and Computing Device

Consumer desire for connectivity and a high data rate wireless access has been increasing tremendously. Systems operating at the licence-free Industrial, Scientific and Medical (ISM) bands such as Bluetooth have been widely adopted in recent years for wireless connectivity. For example, the mobile handset industry offers a wireless hands-free kit. At the operating frequency of 2.45 GHz, the PIFA measures about 24 mm, and can be easily fitted on to the PCB of a small wireless device. Figure 7.24 shows a typical design integrated within the PCB. In such designs, a suitable ground clearance >2 mm is required in order not to short-out the antenna performance [61]. A peak gain of the antenna measures typically >1 dBi. Other techniques such as meandering can also be applied to reduce the length of the element. The simplicity of the design makes the integration of such antennas very attractive and involves no additional material cost.

The IEEE 802.11 wireless local area network (WLAN) system offering a higher data rate of >54 Mbps has also become very popular to mobile devices offering wireless internet access. Coverage of these systems is often termed 'hotspots' and is rapidly increasing worldwide. The standards are broken down into 'a', 'b', and 'g', each offering different data rates. The two ISM band frequency bands allocated are 2400–2483 MHz for the 802.11b,g system and 5150–5250 MHz for the 802.11a system. Figure 7.25(a) shows a simple design of a dual-band PIFA for the 802.11 systems [62]. It consists of a shorter element exciting the upper band while a longer element piggybacks on the shorter element. The longer element is also folded to fit into a smaller overall space. Apart from realizing embedded antennas on to laptops and other mobile devices, designs have also been implemented on a PCMCIA network identity card (NIC). A typical implementation is shown in Figure 7.25(b) and the measured S-parameters of this design are shown in Figure 7.25(c). In most WLAN products, a pair of antennas is implemented to combat the dynamic fading environment. These coexisting diversity antenna solutions must satisfy the envelope signal correlation $\rho_e < 0.5$. This can be easily evaluated with

$$\rho_e < 25\,|S_{21}|^2 \tag{7.2}$$

where the magnitude of the S-parameters must satisfy the following condition [63]:

$$|S_{11}| + |S_{22}| + |S_{11}|\,|S_{21}| + |S_{12}|\,|S_{21}| < 0.25 \tag{7.3}$$

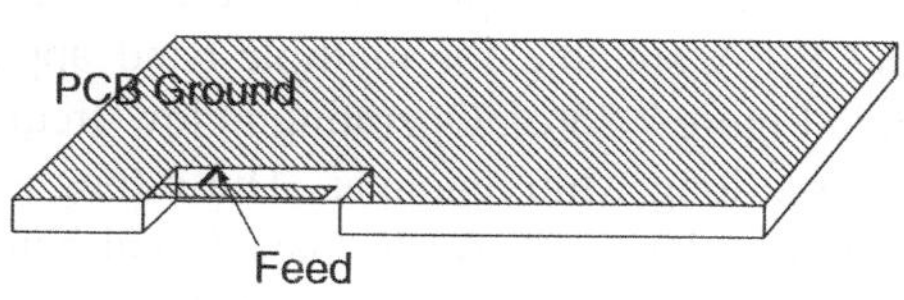

Figure 7.24 An integrated PIFA on a PCB.

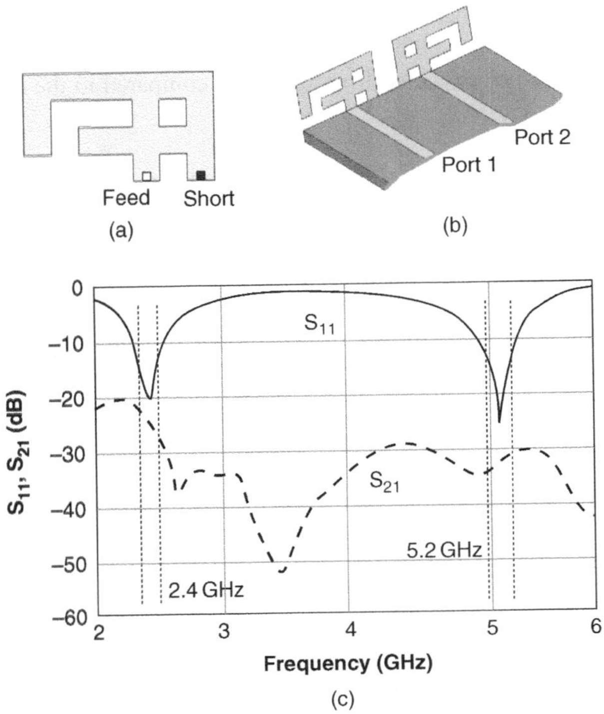

Figure 7.25 Dual-band 802.11a/b/g PIFA [62]: (a) antenna element; (b) NIC implementation; and (c) measured results.

7.6.4 Multiband PIFA for vehicular telematics

There has also been significant development of PIFA solutions for vehicle telematics. Hidden antennas that are flushed within the contours of the vehicle are very popular among manufacturers, giving better styling and vehicle aerodynamics. These antennas are usually implemented on the roof of the vehicle, windscreen, wing mirrors, or even bumpers. New safety regulations for vehicle and pedestrian safety also mandate a requirement of flexible construction if these antennas are >30 mm [64]. This makes the PIFA a suitable candidate for such applications. Leelaratne and hangley [64] have proposed a capacitive gap feed PIFA for cellular communication integrated with other antennas for GPS and FM/AM, as shown in Figure 7.26(a). The PIFA measures a height of <30 mm and the roof mount configuration gives good azimuth coverage in all bands (925, 1795, and 200 MHz). The authors later implemented the PIFA flush mounted on a windscreen, as shown in Figure 7.26(b). Since the antenna is within the body frame of the car, ripples along the radiation patterns are expected. However, a very broadband PIFA suitable for multiple-band application is realized. The measured results are shown in Figure 7.26(c). Operating in low-frequency systems especially UHF bands could result in large long whip antennas. These long exposed antennas are not only unsightly but could be easily damaged. An R-shaped dual-band low-profile PIFA [65] has also developed for military applications operating at 225 and 450 MHz. The height of the antenna measures only 3.8 cm and is mounted on the roof of the vehicle.

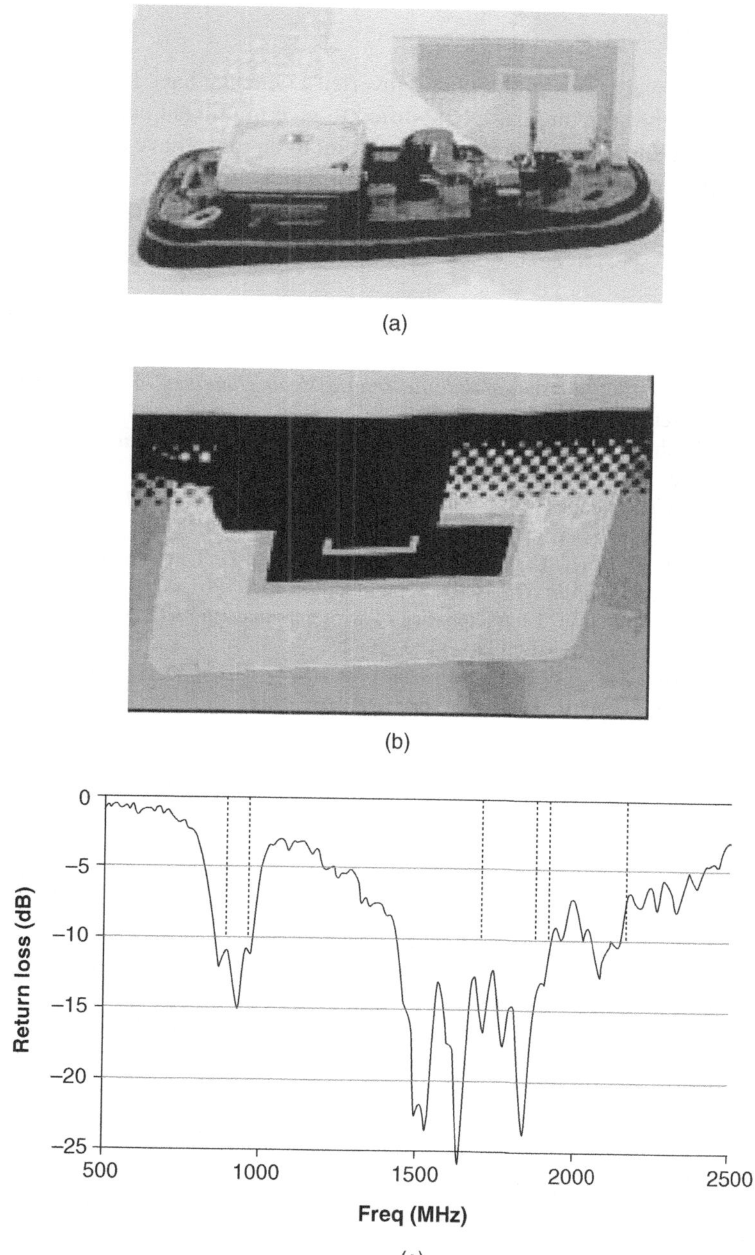

Figure 7.26 Vehicle PIFAs [64]: (a) capacitive-fed PIFA on a multiple antenna module; (b) capacitive-fed PIFA flush mounted on a windscreen; and (c) measured S_{11} results of a capacitive-fed PIFA flush mounted on a windscreen.

7.7 SUMMARY

In this chapter several variations of the planar inverted-F antenna have been presented and it has been shown how the structure can be modified to give multiband performance. Actively integrated PIFAs that allow the frequency of operation to be tuned have also been presented. Finally several applications have been given where planar inverted-F antennas have been utilized.

REFERENCES

1. K. Fujimoto and J. R. James (eds), *Mobile Antenna Systems Handbook*, 2nd edn, Artech House, Boston, Massachusetts, 2001.
2. S. Ramo, J. R. Whinnery and T. Van Duzer, *Fields and Waves in Communications Electronics*, 3rd edn, John Wiley & Sons, Inc., New York, p. 612, 1994.
3. P. A. Ramsdale, 'Wire antennas', in *Advances in Electronics and Electron Physics* (ed. L. Marton) Vol. 47, Academic, New York, pp. 123–196, 1978.
4. S. Prasad and R. W. P. King, 'Experimental study of inverted L, T and related transmission line antennas', *Journal of Research, NBS*, **65D**, September/October 1961.
5. K. Hirasawa and K. Fujimoto, 'Characteristics of wire antennas on rectangular conducting body', *IECE Transactions, Japan*, **J65-B**, 1113–1139, 1982.
6. J. R. James, P. S. Hall and C. Wood, *Microstrip Antenna Theory and Design*, IEE Electromagnetic Wave Series 12, Peter Perigrinus, London, pp. 102–107, 1981.
7. Z. D. Liu, P. S. Hall and D. Wake, 'Dual frequency planar inverted F antenna', *IEEE Transactions on Antennas and Propagation*, **45**(10), 1451–1458, October 1997.
8. P. Salonen and M. Keskilammi, 'Novel dual-band planar inverted-F antenna', in *Asia-Pacific Microwave Conference*, pp. 706–710, 3–6 December 2000.
9. C. R. Rowell and R. D. Murch, 'A capacitively loaded PIFA for compact mobile telephone handsets', *IEEE Transactions on Antennas and Propagation*, **45**(5), 837–842, May 1997.
10. R. Vaughan, 'Model and results for single mode PIFA antenna', in *IEEE Antennas and Propagation Society Symposium*, Vol. 4, pp. 4028–4031, 20–25 June 2004.
11. D. Porcino and W. Hirt, 'Ultra wideband technology: potential and challenges ahead', *IEEE Communication Magazine*, 66–74, July 2003.
12. K. L. Virga and Y. Rahmat-Samii, 'Low-profile enhanced-bandwidth PIFA antennas for wireless communications packaging', *IEEE Transactions on Microwave Theory and Techniques*, **45**(10), 1879–1888, 1997.
13. A. F. Muscat and C. G. Parini, 'Novel compact handset antenna', in *International Conference on Antennas and Propagation Proceedings*, Manchester, Vol. 2, pp. 336–339, 2001.
14. L. L. Rauth, J. S. McLean, K. R. Dorner, J. R. Casey and G. E. Crook, 'Broadband, low-profile antenna for portable data terminal', in *IEEE Antennas and Propagation Society International Symposium, Digest*, Vol. 1, pp. 438–441, 13–18 July 1997.
15. F. Yang, X. X. Zhang, X. Ye and Y. Rahmat-Samii, 'Wide-band E-shaped patch antennas for wireless communications', *IEEE Transactions on Antennas and Propagation*, **49**(7), 1094–1100, 2001.
16. C. K. Byung, D. P. Ju and D. C. Hyumg, 'Tapered type PIFA design for mobile phones at 1800 MHz', in *Vehicular Technology Conference*, Orlando, Florida, Vol. 2, pp. 1012–1014, 2003.
17. F. Wang, Z. Du, Q. Wang and K. Gong, 'Enhanced-bandwidth PIFA with T-shaped ground plane', *Electronics Letters*, **40**(23), 1504–1505, 11 November 2004.
18. R. Feick, H. Carrasco, M. Olmos and H. D. Hristov, 'PIFA input bandwidth enhancement by changing feed plate silhouette', *Electronics Letters*, **40**(15), 921–922, 22 July 2004.

19. C. T. P. Song, P. S. Hall, H. Ghafouri-Shiraz and D. Wake, 'Triple band planar inverted F antennas for handheld devices', *Electronics Letters*, **36**(2), pp. 112–114, 20 January 2000.
20. F. L. Lin and C. W. Chiu, 'Compact dual-band PIFA with multi-resonators', *IEE Electronics letters*, **38**(12), 538–539, June 2002.
21. S. H. Yeh, K. L. Wong, T. W. Chiou and S. T. Fang, 'Dual-band planar inverted F antenna for GSM/DCS mobile phones', *IEEE Transactions on Antennas and Propagation*, **51**(5), 1124–1126, May 2003.
22. Y. S. Shin, B. N. Kim, W. I. Kwak and S. O. Park, 'GSM/DCS/IMT-2000 triple band built in antenna for wireless terminals', *IEEE Antennas and Wireless Propagation Letters*, **3**, 104–107, 2004.
23. P. Ciais, R. Staraj, G. Kossiavas and C. Luxey, 'Design of an internal quad band antenna for mobile phones', *IEEE Microwave and Wireless Components Letters*, **14**(4), 148–150, April 2004.
24. D. Manteuffel, 'Design of multiband antennas for the integration of mobile phones with optimized SAR', in *IEEE Antennas and Propagation Society International Symposium*, Vol. 3, pp. 66–69, 22–27 June 2003.
25. L. Desclos, J. Shamblin, G. Poilasne, V. Pathak and S. Rowson, 'Low-profile, built-frequency multi-band, magnetic dipole antenna', US Patent 6,717,551 B1, 6 April 2004.
26. S. Villeger, P. Le Thuc, R. Staraj and G. Kossiavas, 'Dual planar inverted F antenna', *Microwave and Optical Technology Letters*, **38**(1), 40–42, 5 July 2003.
27. P. L. Teng, T. W. Chiou and K. L. Wong, 'Planar inverted F antenna with a bent meandered radiating arm for GSM/DCS operation', *Microwave and Optical Technology Letters*, **38**(1), 5 July 2003.
28. J.-S. Row, 'Planar inverted-F antenna with slotted ground plane', *Microwave and Optical Technology Letters*, **39**(1), 72–75, 5 October 2003.
29. G. K. H. Lie and R. D. Murch, 'Compact dual-frequency PIFA designs using LC resonators', *IEEE Transactions on Antennas and Propagation*, **49**(7), 1016–1019, 2001.
30. J. Ollikainen, O. Kivekäs, A. Toropainen and P. Vainikainen, 'Internal dual-band patch antenna for mobile phones', in *Proceedings of the AP2000 Millennium Conference on Antennas and Propagation*, Davos, Switzerland, 9–14 April 2000.
31. J. Ollikainen, M. Fischer and P. Vainikainen, 'Thin dual-resonant stacked shorted patch antenna for mobile communications', *Electronics Letters*, **35**(6), 437–438, 1999.
32. Y. X. Gou and H. S. Tan, 'New compact six-band internal antenna', *IEEE Transactions on Antennas and Wireless Propagation Letters*, **3**, 295–297, 2004.
33. C. Puente, J. Romeu, R. Pous and A. Cardama, 'On the behavior of the Sierpinski multiband fractal antenna', *IEEE Transaction on Antennas and Propagation*, **46**(4), 517–524, 1998.
34. C. T. P. Song, P. S. Hall and H. Ghafouri-Shiraz, 'Shorted fractal Sierpinski monopole antenna', *IEEE Transactions on Antennas and Propagation*, **52**(10), 2564–2570, October 2004.
35. T. K. C. Lo and Y. M. Hwang, 'Bandwidth enhancement of PIFA loaded with very high permittivity material using FDTD', in *IEEE Antennas and Propagation Society International Symposium*, Vol. 2, pp. 798–801, 21–26 June 1998.
36. Y. Hwang, Y. P. Zhang, K. M. Luk and E. K. N. Yung, 'Gain-enhanced miniaturized rectangular dielectric resonator antenna', *Electronics Letters*, **33**(5), 350–352, 27 February 1997.
37. Y. Hwang, Y. P. Zhang, G. X. Zheng and T. K. Lo, 'Planar inverted F antenna loaded with high permittivity material', *Electronics Letters*, **31**(20), 1710–1712, 28 September 1995.
38. T. Tanaka, S. Hayashida, K. Imamura, H. Morishita and Y. Koyanagi, 'A study on miniaturization of a handset antenna utilizing magnetic material', in *Communications 2004 and the 5th International Symposium on Multi-Dimensional Mobile Communications Proceedings; 2004 Joint Conference of the 10th Asia-Pacific Conference*, Vol. 2, pp. 665–669, 29 August–1 September 2004.
39. C. R. Rowell and R. D. Murch, 'A capacitively loaded PIFA for compact PCS handsets', in *Antennas and Propagation Society International Symposium, AP-S Digest*, Vol. 1, pp. 742–745, 21–26 July 1996.
40. K. L. Wong and K. P. Yang, 'Modified planar inverted F antenna', *Electronics Letters*, **34**(1), 7–8, 8 January 1998.

41. B. C. Kim, J. H. Yun and H. D. Choi, 'Small wideband PIFA for mobile phones at 1800 MHz', in *59th IEEE Vehicular Technology Conference*, Vol. 1, pp. 27–29, 17–19 May 2004.
42. D. Sievenpiper, L. Zhang, R. F. J. Broas, N. G. Alexópolous and E. Yablonovitch, 'High-impedance electromagnetic surfaces with a forbidden frequency band', *IEEE Transactions on Microwave Theory and Techniques*, **47**, 2059–2074, November 1999.
43. Z. W. Du, K. Gong, J. S. Fu, B. X. Gao and Z. H. Feng, 'A compact planar inverted-F antenna with a PBG ground plane for mobile communication', *IEEE Transactions on Vehicular Technology*, **52**(3), pp. 483–489, May 2003.
44. J. Guterman, A. A. Moreira and C. Peixeiro, 'Microstrip fractal antenna for multi-standard terminals', *IEEE Antennas and Wireless Propagation Letters*, **3**, 351–354, 2004.
45. M. Z. Azad and M. Ali, 'A miniaturized Hilbert PIFA for dual-band mobile wireless applications', *IEEE Antennas and Wireless Propagation Letters*, **4**, 59–62, 2005.
46. K. R. Boyle, M. Udink, A. de Graauw and L. P. Ligthart, 'A novel dual-fed, self-diplexing PIFA and RF front-end (PIN-DF/sup 2/-PIFA)', *IEEE Antennas and Propagation Society Symposium*, 1935–1938, 20–25 June 2004.
47. N. C. Karmakar, L. S. Firmansyah and P. Hendro, 'Tunable PIFA using low cost band switch diodes', in *IEEE Antennas and Propagation Society International Symposium*, **4**, 516–519, 16–21 June 2002.
48. O. Kivekäs, J. Ollikainen and P. Vainikainen, 'Frequency tunable internal antenna for mobile phones', in *JINA 2002*, Nice, France, Vol. 2, pp. 53–56, 12–14 November 2002.
49. N. C. Karmakar, 'Shorting strap tunable stacked patch PIFA', *IEEE Transactions on Antennas and Propagation*, **52**(11), 2877–2884, November 2004.
50. Q. Chen, M. Kurahashi and K. Sawaya, 'Dual-mode patch antenna switched by PIN diode', in *IEEE Topical Conference on Wireless Communication Technology*, pp. 148–149, 15–17 October 2003.
51. K. L. Virga and Y. Rahmat-Samii, 'Low-profile enhanced-bandwidth PIFA antennas for wireless communications packaging', *IEEE Transactions on Microwave Theory and Techniques*, **45**(10), 1879–1888, October 1997.
52. P. K. Panayi, M. O. Al-Nuaimi and I. P. Ivrissimtzis, 'Tuning techniques for planar inverted-F antenna', *Electronics Letters*, **37**(16), 1003–1004, 2 August 2001.
53. A. E. McGirr, P. L. Camwell and J. McRory, 'Duplexing antenna for portable radio transceiver', US Patent 6,061,024, 9 May 2000.
54. 'Test plan for mobile station over the air performance', Cellular Telecommunications and Internet Association (CTIA) Certification Program, Rev 2.1, April 2005.
55. D. Manteuffel, A. Bahr, D. Heberling and I. Wolff, 'Design considerations for integrated mobile phone antennas', in *IEE 11th International Conference on Antennas and Propagation*, pp. 252–256, 2001.
56. O. Kivekäs, J. Ollikainen, T. Lehtiniemi and P. Vainikainen, 'Bandwidth, SAR, and efficiency of internal mobile phone antennas', *IEEE Transactions on Electromagnetic Comparability*, **46**(1), 71–86, February 2004.
57. ICNIRP's Safety Guideline, 'Guidelines for limiting exposure to timevarying electric, magnetic, and electromagnetic fields (up to 300 GHz)', *Health Physics*, **74**, 494–552, 1998.
58. IEEE C95.1:1999, 'Safety levels with respect to human exposure to radio frequency electromagnetic fields, 3 kHz to 300 GHz', IEEE Standard, 1999.
59. FCC, 'Evaluating compliance with FCC guidelines for human exposure to radio frequency electromagnetic fields', OET Bulletin 65 (edn 9701), Supplement C, Washington, DC, 1997.
60. EN50361:2001, 'Basic standard for the measurement of specific absorption rate related to human exposure to electromagnetic fields from mobile phones (300 MHz–3 GHz)', European Standard, 2001.
61. M. Ali and G. J. Hayes, 'Small printed integrated inverted F antenna for Bluetooth application', *Microwave and Optical Technology Letters*, **33**(5), 347–349, 5 June 2002.
62. K. L. Wong, Y. Y. Chen, S. W. Su and Y. L. Kuo, 'Diversity dual band planar inverted F antenna for WLAN operation', *Microwave and Optical Technology Letters*, **38**(3), 223–225, 5 August 2003.

63. C. K. Ko and R. D. Murch, 'Compact integrated diversity antenna', *IEEE Transactions on Antennas and Propagation*, **49**, 954–960, June 2001.
64. R. Leelaratne and R. J. Langley, 'Multiband PIFA vehicle telematics antennas', *IEEE Transactions on Vehicle Technology*, **54**(2), 477–485, March 2005.
65. M. Ali, G. Yang, H. S. Hwang and T. Sittironnarit, 'Design and analysis of an R-shaped dual-band planar inverted-F antenna for vehicular applications', *IEEE Transactions on Vehicular Technology*, January 2004.

8

Small Patch-Based Antennas

Hing K. Kan[1], Rod Waterhouse[2], Dean Pavlickovski[3] and Andrew Y. J. Lee[4]
[1] *School of Engineering and Physics, The University of the South Pacific, Suva, Fiji*
[2] *Pharad, Maryland, USA*
[3] *RMIT University, Melbourne, Victoria, Australia*
[4] *Nanyang Technological University, Singapore*

8.1 INTRODUCTION

The advent of broadband wireless communication has increased the need for portability and mobility. Access to wireless communications such as 3G and internet (Wi-Fi) has amplified the consumer need for new portable devices. The expectation is that new portable devices will be able to cater for the increase in data connectivity [1–3]. Even for developing countries such as Fiji, consumers can access broadband wireless communication with a 1 Mbps connection. With such a high-speed data rate, it is essential that those antennas embedded in portable devices are able to maximize the offered bandwidth. Hence, there is a need for aesthetically pleasing antennas that can provide the necessary size within the constraints of space, bandwidth, and radiation requirements for applications of tomorrow. Patch-based antennas can provide the solutions desired due to several key desirable features such as light weight, low profile, conformal to planar and nonplanar surfaces, inexpensive to manufacture, and when the particular shape and mode are selected they are very versatile in terms of resonant frequency, polarization, pattern, and impedance [4–5]. However, there are limitations at low microwave frequencies, principally in characteristics such as its size and narrow bandwidth. Therefore, techniques to achieve the necessary requirement for patch-based antenna are discussed.

In this chapter small patch antennas and techniques for improving the performance of these antennas to meet specific requirements will be examined. In Section 8.2 an overview is given of shorted patch antennas, which have been shown to have the most significant reduction. It is shown that even further reductions can be achieved by maximizing the direction in which the current travels. Due to the space constraint in a portable device, it is advantageous to have dual frequency in a single patch, thus enabling two applications and reducing the number of antenna on the device. The technique for creating a dual-frequency

Printed Antennas for Wireless Communications Edited by R. Waterhouse

patch antenna is examined. This is followed by a discussion on bandwidth enhancement, which can be applied to shorted patch antennas. Circular polarization is an important feature that allows the antenna to receive a signal irrespective of the direction of the transmitter and is usually applicable to satellite communication. The methods of generating circular polarization on patch antennas will be discussed. Lastly, a means of reducing the high cross-polarization level commonly associated with the shorted patch antenna is offered.

New small patch antennas are discussed in Section 8.3. A dual C-slot printed antenna and perforated shorted patch antenna are presented. Both antennas, either embedding slots or perforating the antenna, employ the concept of maximizing the direction of the current path to reduce the shorted patch antenna. In Section 8.4 new dual-frequency small printed antennas and their characteristics are presented. In particular, a spiral-like shorted patch loaded with an additional shorting pin to generate the second resonance is presented as well as a dual-frequency stacked shorted patch. Finally, techniques to reduce the level of cross-polarization are presented in Section 8.5. A single feed and a balanced feed shorted patch antenna are discussed.

8.2 MICROSTRIP PATCH ANTENNAS

8.2.1 Introduction

Microstrip patch antennas or patch antennas are gaining popularity due to the many advantages mentioned. Different applications will have different specifications; therefore, to cater for every need, some common ways to achieve the required specification is discussed. Shorted patch antennas will be discussed in detail in Section 8.2.2. Various versions of the shorted patch antennas are reviewed as well as the methods used to reduce the size of the antenna. Techniques to generate dual frequency in planar antennas are then presented in Section 8.2.3. One of the disadvantages of small patch antennas is their inherent narrow bandwidth. This problem can be alleviated by applying certain bandwidth enhancement techniques and are discussed in Section 8.2.4. In Section 8.2.5, methods of generating circular polarization and its application to shorted patches are reviewed. Finally, ways to alleviate the high level of cross-polarization due to the use of a shorting pin is discussed in Section 8.2.6.

8.2.2 Reducing the Conductor Size

A microstrip-shorted patch antenna consists of a patch incorporating a single shorting post in close proximity to the feed point. A probe-fed circular microstrip antenna incorporating a single shorting post was investigated [5]. The antenna was mounted on a substrate with a dielectric constant $\varepsilon_r = 4.81$. The antenna achieved a reduction in the printed conductor in the order of 3 compared to a conventional microstrip patch mounted on the same dielectric material. The reduction in size has been achieved at the expense of bandwidth, in this case 1.2 %. Since then, various conductor shapes incorporating a shorting post have been reported [6–8]. The conductor shapes investigated include the triangle, semicircle, and annular ring shorted patches. The yielding solutions demonstrate a real estate reduction similar to that in Reference [5]. The impedance bandwidths reported have been less than 5 %.

Although there has been a significant reduction in antenna size with the utilization of a single shorting pin, further reductions can be achieved. One such example is the use of strategically positioned notches on a shorted patch antenna [9]. The resulting antenna was 75 % smaller in surface area than a conventional microstrip patch. Various shapes of slots and slits have been embedded on shorted patch antennas, which has been shown to reduce the size of the antenna. However, the reported bandwidth has been relatively low in the order of a few percents. It is worth noting that when perturbations such as slot or slits are embedded on a shorted patch antenna, the current path of the antenna is increased. The antenna utilizes the meander line technique traditionally used to decrease the size of the wire dipoles and monopoles operating in the low MHz frequency range. Here the wire (of set length) is folded back on itself to reduce the physical length of the antenna in a given direction. This in turn leads to a very important postulation: to decrease the resonant frequency of an antenna for a given surface area, the current path must be maximized within that area. This postulation has been applied to the shorted patch concept. A reduction in the resonant frequency by a factor of 10 compared to a standard microstrip has been achieved using spiral-shaped conductors [10]. This frequency reduction corresponds to a decrease in area by a factor of 261 albeit at the cost of reduced bandwidth. This is the largest reduction recorded in the literature.

8.2.3 Dual-Frequency Configurations

There are three basic categories of creating patch-based antennas with dual frequency [11]. The methods are orthogonal-mode dual-frequency patch antennas, multipatch dual-frequency antennas, and reactively loaded dual-frequency patch antennas.

The orthogonal mode method can be achieved by any kind of patch shape that offers two cross-polarized resonant modes. The antenna can be fed using a single-point or dual-point feed. The simplest dual-frequency patch antenna can be accomplished by carefully selecting the location of the feed point resulting in matched input impedance at the two frequencies. This can be achieved with a probe-fed configuration as shown in Reference [12]. The antenna utilizing the dual-point feed can be excited by a separated microstrip feed. This solution can provide more flexibility in designing the frequency ratio when compared to the single feed.

The multipatch configuration is obtained by means of multiple radiating elements, either vertically or horizontally, each of them supporting strong currents and radiation at the resonance. This category includes multilayer stacked patches (vertical configuration) with various radiator shapes [11]. The same multilayer structure can be used to broaden the bandwidth of the single-frequency antenna when two frequencies are forced to be closely spaced. In the latter case, the lower patch can be fed by a conventional arrangement while the upper patch can be fed by proximity coupling with the lower patch. In order to avoid the disappearance of the upper resonance, the size of the two patches should be close, so that a frequency ratio close to unity can be obtained. A horizontal or coplanar multipatch configuration can be obtained by printing more resonators on the same substrate. Radiation patterns have been shown to be consistent at all operating frequencies and the antenna is attractive for its simplicity.

Finally, the reactively loaded patch antenna method is the most popular technique for obtaining dual-frequency behaviour. The simplest way is to connect a stub to one radiating edge in such a way as to introduce a further resonant length that is responsible for the second operating frequency. An adjustable coaxial was utilized in Reference [13] to provide both tuning and design of the frequency ratio. Other types of reactive loading are based on the same principle of creating the same effect, as the stub includes introducing notches, pins, capacitors, and slots with the advantage of reduced size.

8.2.4 Bandwidth Enhancement Procedures

One of the major disadvantages with using microstrip technology is the narrow bandwidth. One way of increasing the bandwidth of the patch antennas is by placing driven antenna elements in close proximity to the antenna [14, 15]. The driven elements can be placed either horizontally or vertically and are gap-coupled, producing double resonance and thereby increasing the bandwidth. A bandwidth of greater than 20 % has been reported in the literature. Utilizing this technique increases the overall size of the antenna as the number of driven elements is increased.

Another technique to enhance the bandwidth of printed antenna is by stacking the antennas. A broadband-stacked shorted patch was reported [16]. The antenna is probe-fed with two shorting pins. A 10 dB return loss bandwidth greater than 30 % was reported. The wideband characteristic of this reported publication is due to the mutual resonance occurring in the two patches. The shorted patches are overcoupled, therefore producing somewhat large coupling loops similar to a conventional stacked patch when the patches are located too close together. This phenomenon yields a loop in the impedance locus when plotted on the Smith chart. It should be noted that an increase in substrate thickness was also reported utilizing this technique.

The third method of increasing the bandwidth is by interleaving two shorted patch antennas [17]. The impedance bandwidth is enhanced over a conventional shorted patch due to mutual coupling between the two radiators, in doing so reducing the overall impedance variation at the input of the antenna. A large conductor size reduction for the antenna is achieved by maximizing the current path of each printed radiator.

8.2.5 Circular Polarization Techniques

Circular polarization (CP) occurs when the magnitudes of the two components (horizontal and vertical) are the same and the time phase difference between them is 90°. This is an important feature for applications where the transmitter and receiver are not stationary. An antenna with circular polarization is able to receive power in all direction irrespective of the orientation of the antenna. This is an important trait for satellite applications due to the movement of the satellites relative to the earth.

A single-feed antenna can produce CP with some perturbations on the patch [18–20]. The perturbations can be slots, resistor loading, or slits, which introduce the 90° phase shifts required to generate circular polarization. The single feed is the simplest form of antenna to generate CP. However, the major disadvantage of using the single-feed configuration is due to its narrow impedance and axial ratio bandwidth, typically less than 5 %.

The second technique to generate CP is feeding the antenna with two feeds at adjacent edges [21, 22]. A hybrid coupler or offset feeding line can be used to produce the required power and 90° phase distribution to the antenna. Both the impedance and axial ratio characteristics of dual-fed antennas are broader that those of single-fed antennas because the hybrid coupler is typically broadband in nature. The drawback with using this method is that the improvements in the antenna performance have been achieved at the expense of increased volume due to the feeding mechanism required to produce circular polarization.

Sequential rotation is the third method that can be utilized to generate circular polarization. Single circularly polarized [23] or linearly polarized [24–26] antenna elements can be used to rotate each of the antennas sequentially. The advantage with using linearly polarized patches is that the feed complexity is reduced and also the bandwidth performance is improved. The reason for reduced feed complexity is that this technique only requires a single feed for each antenna element while four feeds might be required for each element in a conventional array [23], with a wide axial ratio bandwidth requirement. Another feature of this uniquely arranged array is a reduction in the occurrence of mutual coupling. The theories of sequential rotation of linearly polarized elements to generate CP have been discussed in Reference [24], with the design principles presented in References [25] and [26].

The principles of sequential rotation have been applied to a linearly polarized shorted patch antenna element to generate circular polarization. Four shorted patch antennas were sequentially rotated and fed with a 90° phase difference feed network located beneath the ground-plane [27]. The feed network for the synchronous subarray had additional lengths of microstrip line to provide the required 90, 180, and 270° phase shifts at three of the other excitation ports. The antenna size can be further reduced by utilizing three shorted equilateral triangle patch antennas [28]. In this instance, the antenna elements are sequentially rotated at an angle of 120° and were fed in a 0, 120, and 240° phasing arrangement. Thus the entire antenna demonstrated symmetry about the boresight. Finally, a compact synchronous subarray using two concentric shorted ring printed antennas provides the smallest antenna in terms of size [29]. It is important to note that the three configurations, namely utilizing four-element, three-element, and two-element configurations, are able to achieve an impedance and axial ratio bandwidth in excess of 10 %.

8.2.6 Cross-Polarization Reduction Procedures

There are several wireless communication systems, both terrestrial and satellite, that require good polarization to give enhanced system performance. Some require pure linear polarization to enhance the system capacity. It is well known that shorted patch antennas have high levels of cross-polarization. A common technique to reduce cross-polarized fields and hence improve the purity of linearly polarized fields radiated from a patch antenna is to use a balanced feed. This method was applied to a shorted patch antenna [30]. A cross-polarization level of more than 20 dB below the copolarized fields in both planes was obtained. The reduction of cross-polarized fields was achieved at the expense of an increase in the size of the patch conductor. The patch antenna real estate had increased from an eighth of a wavelength for a conventional shorted patch using a single shorting to approximately a quarter-wavelength. The increase in antenna real estate was attributed to each excited probe 'seeing' the shorting post used for the other feed. This 'inductive' loading somewhat cancels the 'capacitive' effect of the closeness of the excited probe and its associated shorting post [31]. A compact dual concentric ring printed antenna utilizing the balanced feed configuration achieved improved bandwidth and size [29].

8.3 VARIATIONS OF THE SHORTED PATCH ANTENNA

8.3.1 Introduction

In this section, two new versions are presented of small-sized shorted patch antennas. This work is an extension of the shorted patch concept summarized in Reference [32]. In Section 8.3.2, a dual C-slot printed antenna is presented. The C-slots are embedded on a shorted patch antenna. A perforated shorted patch is presented in Section 8.3.3. The concepts of using slots or perforating the shorted patch antenna are based on the concept of maximizing the direction of the current path. This reduces the resonant frequency of the antenna while maintaining a comparable bandwidth of the antenna.

8.3.2 Dual C-Slot Printed Antenna

8.3.2.1 Introduction

In this section, a dual C-slot printed antenna is presented. The antenna consists of a shorting post with two C-shaped slots surrounding the coaxial feed and the shorting post. The dual C-slot is strategically placed around the feed within the printed conductor to maximize the current path of the antenna. The concept utilized to reduce the size of the printed conductor is to maximize the current path of the shorted patch. In doing so, the resonant frequency of the antenna is reduced.

8.3.2.2 Configuration

Figure 8.1 shows the proposed antenna configuration. Here, the antenna has an overall length L and width W and is printed on a substrate of thickness d, with a relative dielectric permittivity ε_r. The antenna is fed by a coaxial probe of radius r_0 located at (x_p, y_p) from the centre of the antenna and consists of a shorting pin located in close proximity to the feed positioned at (x_{ps}, y_{ps}) from the centre of the patch to reduce the overall size of the antenna. Two C-shaped slots are embedded into the patch as shown in Figure 8.1 (labelled slots 1 and 2). The width of the slots is 0.5 mm. The length of slots 1 and 2 are 13 mm and 11.5 mm, respectively. The vertical length of slot 1 is 8 mm and 5 mm for slot 2. The gap spacing between the slots is 1 mm and there is a 1 mm gap between slot 1 and the edges of the patch. As mentioned in Reference [10], the key to reducing the size of printed antennas is to increase the current path of the antennas. Therefore, C-slots are introduced to maximize the current path of the shorted patch antenna as shown in Figure 8.1. It is interesting to note that having only one C-slot, namely slot 1, decreases the resonant frequency of the shorted patch antenna, but a further reduction in resonant frequency can be observed with slot 2 inserted.

As seen from Figure 8.1, the current path of the shorted patch is increased from the feed point by inserting the two C-slots, thereby reducing the resonant frequency of the antenna. It should be further noted that different configurations of slots can be utilized, providing many degrees of freedom. However, the present configuration provides a compromise between size reduction and attainable bandwidth. The proposed antenna configuration achieved a reduction of 15 % in resonant frequency compared to a standard shorted patch antenna mounted on the same substrate.

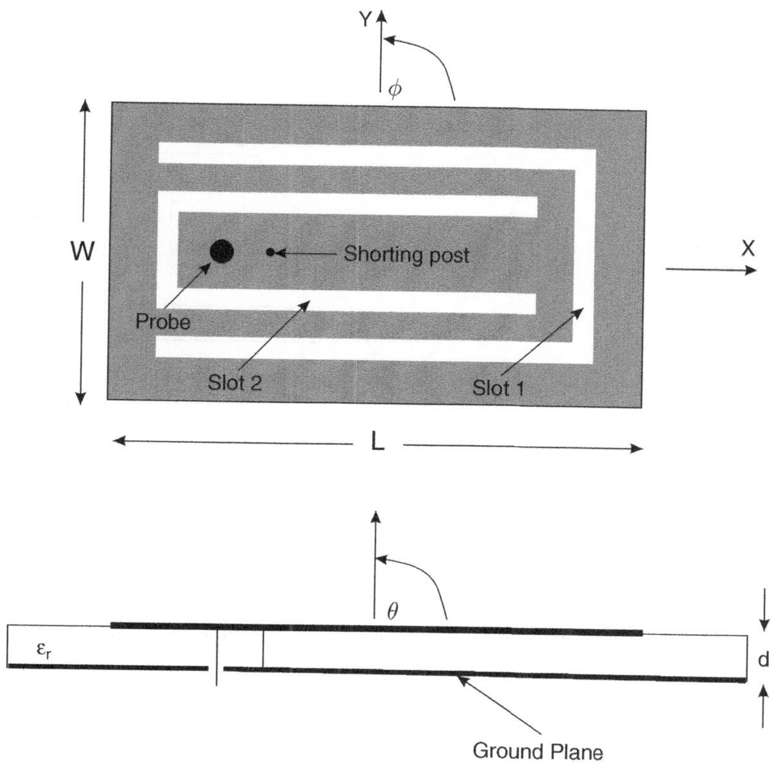

Figure 8.1 Schematic diagram of a shorted dual C-slot printed antenna.

8.3.2.3 Results and Discussion

Figure 8.2 shows the measured 10 dB return loss bandwidth of the proposed dual C-slot printed antenna. The 10 dB return loss bandwidth of the antenna was 5.4 %. This value is significantly higher than the shorted single spiral antennas [10]. On the other hand, the impedance bandwidth of the proposed antenna is lower than the dual spiral printed antenna [17], but the present configuration as shown in Figure 8.1 does not require an external feed network. Hence, the dual C-slot antenna is compact and does not require any complicated manufacturing process compared to a conventional patch antenna. As mentioned previously, the proposed antenna achieved a reduction of resonant frequency in the order of 15 % compared to a conventional shorted patch antenna (in this case from 2.6 to 2.22 GHz) mounted on the same material, making it suitable for applications where space is a premium.

The far-field radiation patterns of the dual C-slot printed antenna were measured at 2.2 GHz and are shown in Figure 8.3 (note that the broadside is 0° in this plot). The radiation patterns are similar to that of the shorted patch antennas [5]. However, the cross-polarization level for this antenna is relatively high, particularly in the E-plane due to the small-sized ground-plane. For this measurement, the ground-plane was very small, extending 15 mm on each side of the printed conductor. The gain of the antenna was 1.3 dBi.

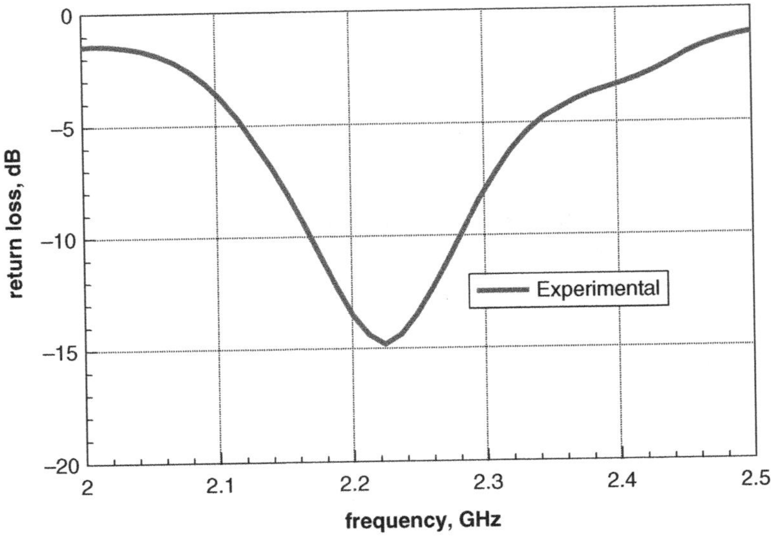

Figure 8.2 Measured return loss of a shorted dual C-slot printed antenna (parameters: $\varepsilon_r = 1.07$, $d = 10$ mm, $L = 15$ mm, $W = 10$ mm, $r_0 = 0.6$ mm, $r_{0s} = 0.4$ mm, $x_p = -2.5$ mm, $y_p = 0$ mm, $x_{ps} = 1.5$ mm, $y_{ps} = 0$).

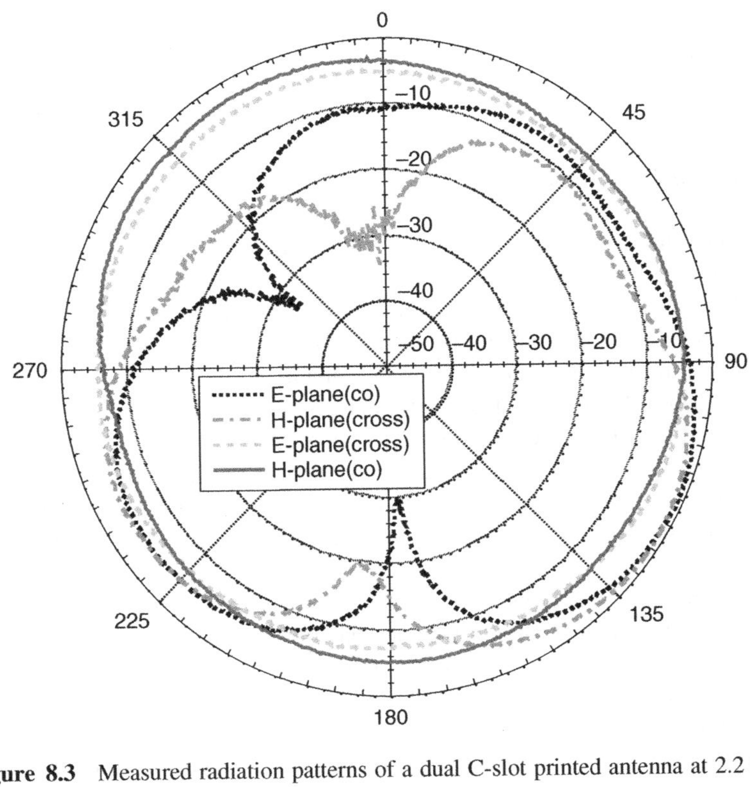

Figure 8.3 Measured radiation patterns of a dual C-slot printed antenna at 2.2 GHz.

8.3.3 Shorted Perforated Antenna

8.3.3.1 Introduction

Perforating or meshing the microstrip patch conductor has been shown to give reasonable size reductions for a conventional patch antenna without compromising the radiation and impedance performance of the printed antenna. In this section, a perforated shorted printed antenna is presented. The printed antenna utilizes a shorting post as well as perforated slots that are embedded within the single-layer printed antenna to reduce the surface area of the antenna.

8.3.3.2 Configuration

Figure 8.4 shows the proposed antenna configuration. As can be seen from the figure, the antenna has an overall length L and width W and is printed on a substrate of thickness d with a relative dielectric permittivity ε_r. The perforated slots are embedded within the printed antenna. Each of the perforated slots has a length of L_s and width W_s and is located 3 mm (W_1) from the edge of the antenna. The separation between each of the perforated slots is 2 mm apart. The antenna is probe fed by a via of radius r_0 positioned at (x_p, y_p) relative to the centre of the patch. A shorting post of radius r_{0s} located in close proximity to the feed probe is positioned at (x_{ps}, y_{ps}) relative to the centre of the antenna and is used to reduce the overall size of the patch. A thin layer of *RT/Duroid™ 5880* ($\varepsilon_r = 2.2$, thickness of 0.254 mm) was used to etch the printed conductor. The size of the perforated slots has been selected from running many simulations of the structure (using *Ensemble 6.0™*). As mentioned in Reference [10], the key to reducing the conductor size of a shorted patch is to create extra capacitance to counter the inductive nature of a patch. In addition, by creating perforated slots within a shorted patch, the current patch of an antenna is increased, thereby reducing the size of the antenna [33]. One of the advantages with the proposed antenna apart from the reduction in antenna size is that components may be placed on to the perforated slots for applications where limiting the size of the overall antenna platform is a premium.

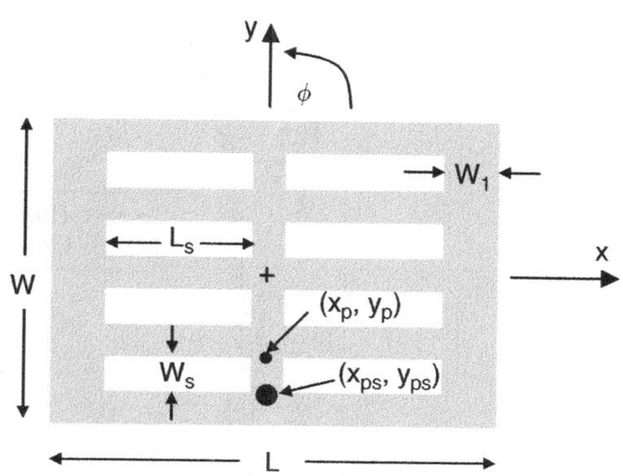

Figure 8.4 Schematic diagram of a perforated shorted patch (parameters: $\varepsilon_r = 1.07, d = 10$ mm, $L = 20$ mm, $W = 18$ mm, $L_s = 7$ mm, $W_s = 2$ mm, $W_1 = 3$ mm, $r_0 = 0.25$ mm, $r_{0s} = 0.5$ mm, $x_p = 0, y_p = -5, x_{ps} = 0, y_{ps} = -7$).

As can be seen from Figure 8.4, the fabrication of the proposed antenna does not require any complicated manufacturing process other than the conventional method for fabricating a printed antenna.

The dimensions of the proposed antenna are given in the caption of Figure 8.4. The size reduction of the new perforated shorted patch is considerable; the patch conductor area is reduced by 30 % compared to a typical shorted patch [5] and by 93 % compared to a conventional probe-fed microstrip patch antenna, both designed at the same frequency and using the same material. Another advantage of the perforated (or meshed) shorted patch over a conventional shorted patch is the increase in spacing between the shorting pin and the feed for the new antenna. This distance was increased from 0.7 mm (for the conventional shorted patch) to 2 mm.

8.3.3.3 Results

Figure 8.5 shows the predicted (using *Ensemble 6.0*) and measured return loss of the proposed perforated shorted printed antenna. The predicted and measured resonant frequencies are well matched. The measured 10 dB return loss bandwidth of the antenna is 8.1 % compared to a simulated bandwidth of 5.1 %. The surprising increase may be due to the small ground-plane used here (only extending 1 mm beyond each side of the printed conductor). Using Ensemble, the small ground-plane cannot be taken into account, but it was thought that this scenario would better approximate the environment in which small antennas are typically mounted. An investigation of this effect is currently underway. The bandwidth of a conventional shorted patch mounted on the same material is 5.9 % and for a traditional probe-fed patch, 7.7 %.

Figure 8.6 shows the measured radiation patterns of the proposed shorted perforated patch antenna at 1.73 GHz. As can be seen from this plot, the antenna is omnidirectional, consistent with other forms of the shorted patch antenna mounted on a small ground-plane. The gain of the antenna across its 10 dB return loss bandwidth varied from 0.5 to 1 dBi.

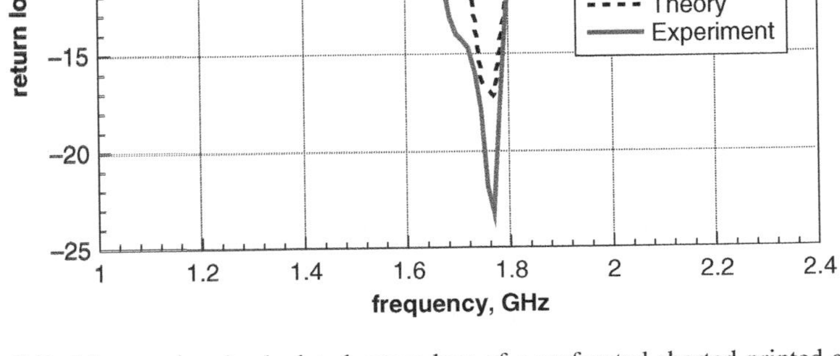

Figure 8.5 Measured and calculated return loss of a perforated shorted printed antenna.

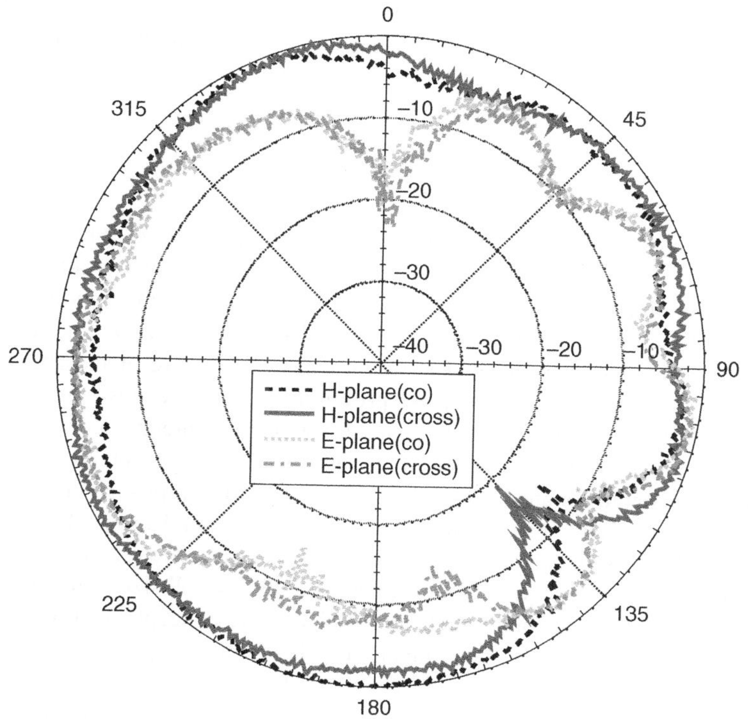

Figure 8.6 Measured radiation patterns of a perforated shorted printed antenna at 1.73 GHz.

8.4 DUAL-FREQUENCY SHORTED PATCH ANTENNAS

8.4.1 Introduction

Dual-frequency antennas have the advantage of having a number of applications on a single device. In this section, two techniques providing means of achieving this while maintaining the size reduction and reasonable bandwidth are given. In Section 8.4.2, dual frequency is achieved by perturbation of an existing shorted patch to generate the second resonance. Stacking two shorted patch antennas results in a dual-frequency patch antenna and this is discussed in Section 8.4.3.

8.4.2 Dual-Resonance Spiral-Like Shorted Patch Antenna

8.4.2.1 Introduction

A simple structure that is small in size and has a dual-resonance characteristic is discussed in this section. The antenna consists of a spiral-like printed conductor incorporating a shorting post in close proximity to the probe feed. Spiral-like conductors were selected due to the current travelling a longer distance, thereby reducing its resonant frequency. A further reduction of the conductor can be achieved with the shorting post in close proximity to the feed location. An additional shorting pin is introduced to generate the second resonance of the antenna.

8.4.2.2 Configuration

Figure 8.7 shows a schematic of the proposed spiral-like printed antenna. As can be seen from the figure, the antenna has an overall length ($L = 25$ mm) and width ($W = 17$ mm) with the conductor track folded back on to itself. The conductor track width and length is 1.2 mm. The separation between the conductor tracks (gap) is 1 mm. The antenna is fed by a coaxial probe of radius r_0 located at (x_p, y_p) from the centre of the antenna and has a shorting pin of radius r_{0s} located in close proximity to the feed positioned at (x_{ps1}, y_{ps1}) from the centre of the patch to reduce the overall size of the antenna. Common techniques such as introducing perturbations or shorting pins to an existing antenna can lead to an introduction of a second resonance [11]. It has been shown that spiral-like printed conductors achieve a significant reduction in antenna size due to the maximization of the current path of the antenna [10]. This antenna configuration was selected due to its small size. A second shorting pin positioned at (x_{ps2}, y_{ps2}) from the centre of the overall size of the antenna allows for the introduction of the second resonance, utilizing the concept mentioned previously. The antenna consists of three layers: a layer to etch the ground-plane, a foam layer, and a layer to etch the radiator. *Rohacell*™ foam ($\varepsilon_r = 1.07, d = 10$ mm) was used to provide a reasonable bandwidth and maximize the overall efficiency. To etch the ground-plane and the radiator, *RT/Duroid 5880* ($\varepsilon_r = 2.2, d = 0.508$ mm) was used. The antenna was fabricated and tested experimentally. The photograph of the developed antenna is shown in Figure 8.8.

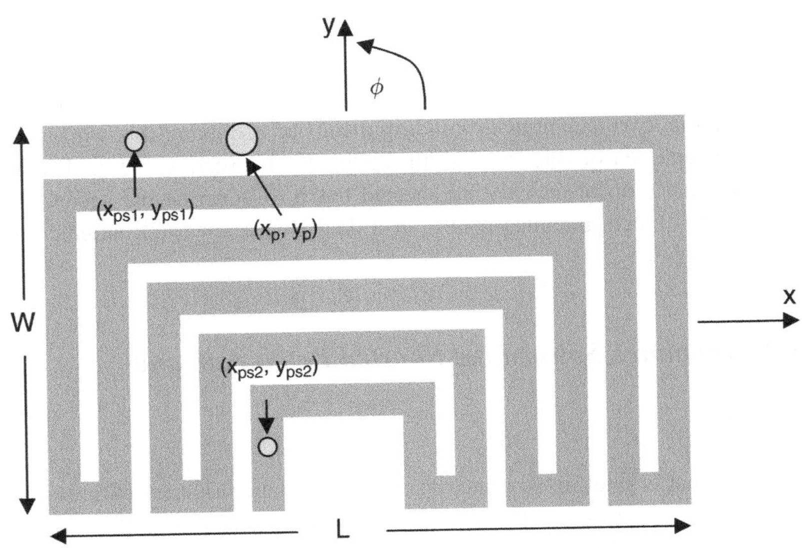

Figure 8.7 Schematic of proposed dual-resonance spiral-like printed antenna.

Figure 8.8 Photograph of a proposed dual-resonance spiral-like printed antenna.

8.4.2.3 Experimental Results

Figure 8.9 shows the measured return loss of the proposed antenna. As can be seen from the plot, there are two resonant frequencies, namely at 1.55 and 2.5 GHz. The lower resonant frequency ($f = 1.55$ GHz) is due to the shorting pin in close proximity to the probe feed. It has been shown that an antenna with the same dimension has a resonant frequency at 715 MHz although the location of the probe and the shorting pin is dissimilar [10]. This is consistent with previous findings that multiple shorting pins increase the size of the antenna. The present locations of the probe feed and the shorting pin is selected to better match the antenna albeit at the cost of a higher resonant frequency. The second resonance (upper frequency) at 2.5 GHz is due to the introduction of the second shorting pin. The bandwidth of the antenna is 23 MHz (1.5 %) and 55 MHz (1.4 %) for the lower and upper resonance frequencies, respectively. The bandwidth is comparable to single-layer shorted patch antennas [5]. The overall size of the antenna at the lower resonance is $0.13\lambda_0 \times 0.09\lambda_0 \times 0.06\lambda_0$ and $0.21\lambda_0 \times 0.14\lambda_0 \times 0.09\lambda_0$ for the higher resonance at 2.5 GHz, making it compatible for wireless applications where space is a premium (λ_0 is the free-space wavelength).

The far-field radiation patterns for the proposed dual-resonance spiral-like antenna were measured at the Nanyang Technological University anechoic chamber. The results are shown in Figures 8.10(a) and (b) for frequencies measured at 1.55 and 2.5 GHz, respectively. The patterns are omnidirectional, which is consistent with small antennas due to their small ground-plane (note that the broadside is 0° on the plots). The cross-polarization level is relatively high, particularly for the H-plane, which is a common trait of small antennas. The ground-plane for this antenna extends for 1 mm on each side of the printed conductor. The gain of the antenna is −0.6 dBi at 1.55 GHz and 0.6 dBi for 2.5 GHz.

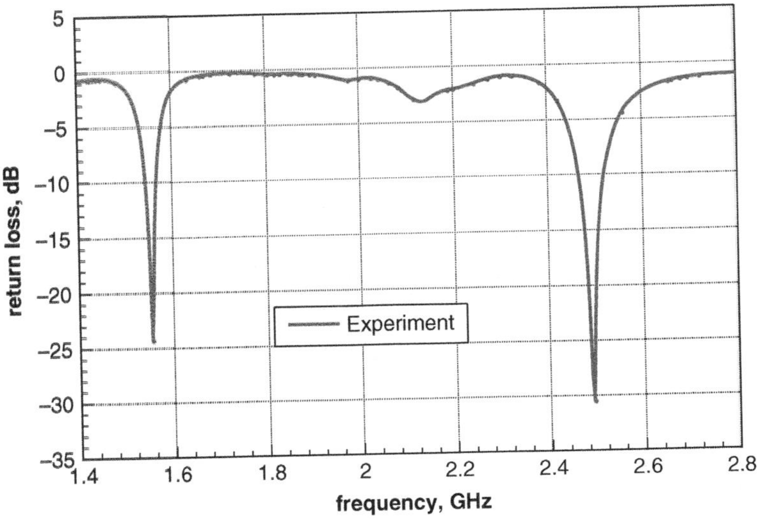

Figure 8.9 Return loss performance of printed antenna (parameters: $r_{0s} = 0.325$ mm, $x_{ps1} = -10$ mm, $y_{ps1} = 8.6$ mm, $x_{ps2} = -3.6$ mm, $y_{ps2} = -5$ mm, $r_0 = 0.6$ mm, $x_p = 6$ mm, $y_p = 8.6$ mm).

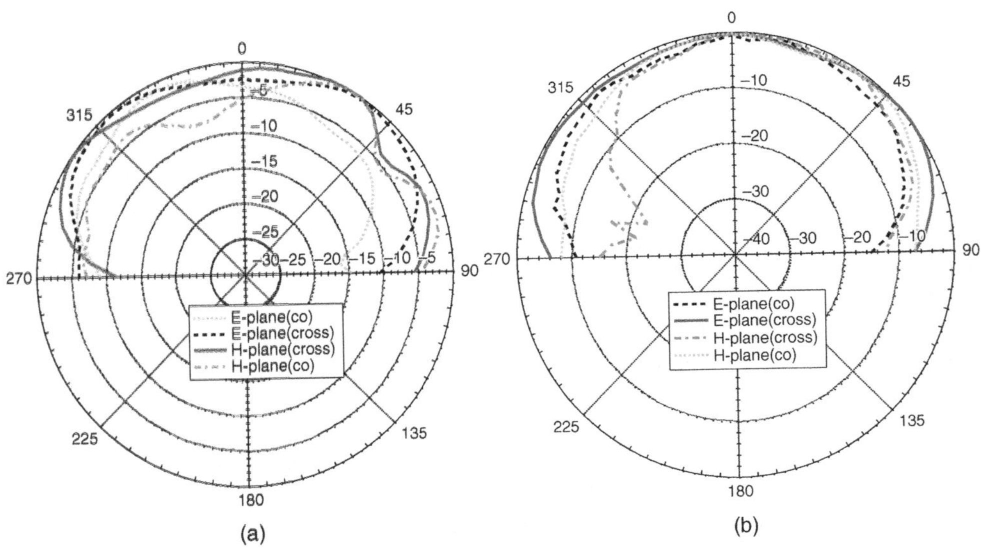

Figure 8.10 Radiation patterns of the proposed printed antenna at (a) 1.55 GHz and (b) 2.5 GHz.

8.4.3 Dual-Frequency Stacked Shorted Patch Antenna

8.4.3.1 Introduction

In this section a single-feed, small, dual-frequency printed antenna is examined. The printed antenna consists of a stacked shorted patch geometry, where the lower frequency element is fed by a probe and the upper frequency radiator is electromagnetically coupled to the feed. The lower frequency radiator is a perforated (or meshed) shorted patch. The upper frequency radiator is located below the other element and is in the shape of a shorted ring. This combination of shaped conductors is chosen in an attempt to minimize the conductor sizes and also to try to minimize the overlap of the current paths of each antenna, which should in turn help provide relatively independent radiators.

8.4.3.2 Configuration

As mentioned before, the antenna structure is based on stacked shorted patch geometry [16], although the overall height of the radiator is restricted to less than $0.04\lambda_0$ (where λ_0 is the free-space wavelength at the lowest frequency of operation). Restricting the height greatly impacts the achievable bandwidth of the radiator [16, 34] and so a multiple-resonator solution would appear to be the best option. The patch shapes chosen for the small dual-frequency radiator were a modified perforated patch and a rectangular annular ring. These geometries can give very small surface radiators, although the concept is not limited to these geometries. One reason for choosing these conductor shapes was to attempt to minimize the overlap of the current distributions on the conductor of each radiator. The authors believe this can help provide two relatively independent radiators and therefore the capability of independently controlling the resonant frequency of each radiator. Of course, the two radiators cannot be completely independent due to the shared feed and also the close proximity of the radiators due to the restricted volume to house the radiator.

Figure 8.11 shows the overall geometry of the dual-frequency shorted patch antenna. The antenna consists of five layers: a layer to etch the ground-plane, a foam layer, a layer to etch the higher frequency radiator, another foam layer, and a final layer to etch the lower frequency radiator. *Rohacell* foam ($\varepsilon_r = 1.07$; $d = 2.5$ mm) was used throughout the radiator to maximize the overall efficiency. For the conductor layers, *RT/Duroid 5880* ($\varepsilon_r = 2.2$, $d = 0.508$ mm) was used. The high-frequency radiator is located below the low-frequency patch in order to minimize the compromise in bandwidth performance for this antenna, as the bandwidth is directly proportional to the electrical thickness of the radiator. The feeding arrangement for the dual-frequency small radiator is relatively straightforward. A coaxial probe (radius 0.3 mm) extends through the ground-plane of the antenna and is connected to the conductor of the perforated patch. A shorting pin (radius 0.6 mm) located in close proximity to the feed [16] is connected to the perforated patch conductor and dramatically reduces the resonant frequency of the radiator. The annular ring also has a shorting pin (radius 0.6 mm) connecting its conductor to the ground-plane. This pin must be in close proximity to either the shorting pin or the feed pin of the perforated patch to ensure that the annular ring can couple efficiently to the feed port. The photograph of the developed antenna is shown in Figure 8.12.

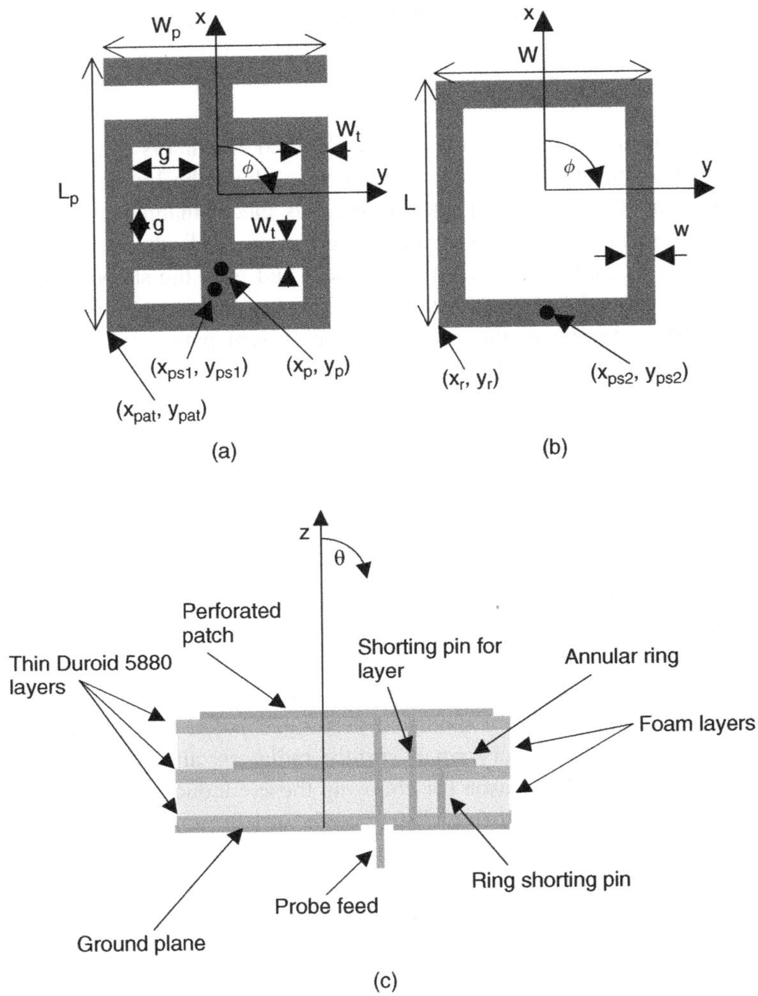

Figure 8.11 Schematic of the proposed printed antenna: (a) perforated patch layer; (b) annular ring layer; and (c) side view showing the appropriate layers.

8.4.3.3 Experimental Results and Discussion

The dual-frequency shorted patch using the geometry highlighted in Figure 8.11 was investigated using *Ensemble 6.0*. Although this analysis tool does not include the effects of a finite-size ground-plane it can still give valuable design trends even for small radiators. To prove the concept of the new radiator, the antenna was designed for operation at 1.8 and 2.6 GHz (refer to the caption of Figure 8.13 for the dimensions). Figure 8.13 shows the theoretical return loss response of the radiator. The resonant frequency of the shorted perforated patch without the annular ring is 1.78 GHz, which is similar to that for the stacked configuration, indicating that the annular ring has very little impact on the return loss performance of the perforated patch.

Figure 8.12 Photograph of the stacked shorted patch.

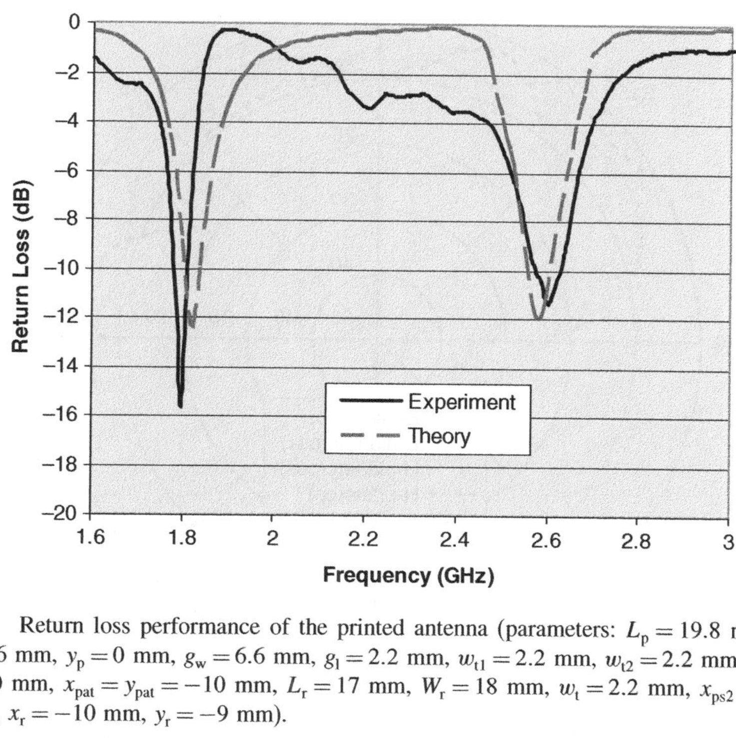

Figure 8.13 Return loss performance of the printed antenna (parameters: $L_p = 19.8$ mm, $W_p = 20$ mm, $x_p = -6$ mm, $y_p = 0$ mm, $g_w = 6.6$ mm, $g_l = 2.2$ mm, $w_{t1} = 2.2$ mm, $w_{t2} = 2.2$ mm, $x_{ps1} = -7.2$ mm, $y_{ps1} = 0$ mm, $x_{pat} = y_{pat} = -10$ mm, $L_r = 17$ mm, $W_r = 18$ mm, $w_t = 2.2$ mm, $x_{ps2} = -8.6$ mm, $y_{ps2} = 0$ mm, $x_r = -10$ mm, $y_r = -9$ mm).

The proposed stacked shorted patch antenna and the measured return loss results are also shown in Figure 8.13. As can be seen from this graph, the measured results are in good agreement with the theory. The discrepancies in the responses can be attributed to the very small ground-plane used (20 mm × 20 mm), but its effect does not appear to be very dramatic. There is some evidence of coupling between the ground-plane and the radiators within the band of 1.9–2.2 GHz with a reduction in the return loss response over this band, which is not present in the modelled results. Nevertheless, the results in Figure 8.13 validate the proposed structure. The bandwidth at 1.8 and 2.6 GHz was 1.6 and 2.0 %, respectively. It was found that using the configuration in Figure 8.11 the upper frequency radiator could be tuned to within 16 % of the lower frequency and still get well-defined resonances. Beyond that point the resonances merged. Using *Ensemble 6.0* it was easy to extend the resonant frequency of the upper frequency radiator beyond 4 GHz.

The radiation patterns of the proposed dual-frequency antenna were measured, and the results at 1.8 GHz are shown in Figure 8.14. The radiation patterns at 2.4 GHz were similar to those in Figure 8.14, with slightly more pattern scalping (up to 3 dB more), possibly due to the field interaction with the perforated patch and the ground-plane. For the sake of brevity these results are not presented here. The gain of the antenna was measured at each frequency to be approximately 0.5 dBi at 1.8 GHz and −0.1 dBi at 2.6 GHz. These levels are consistent with very small radiators.

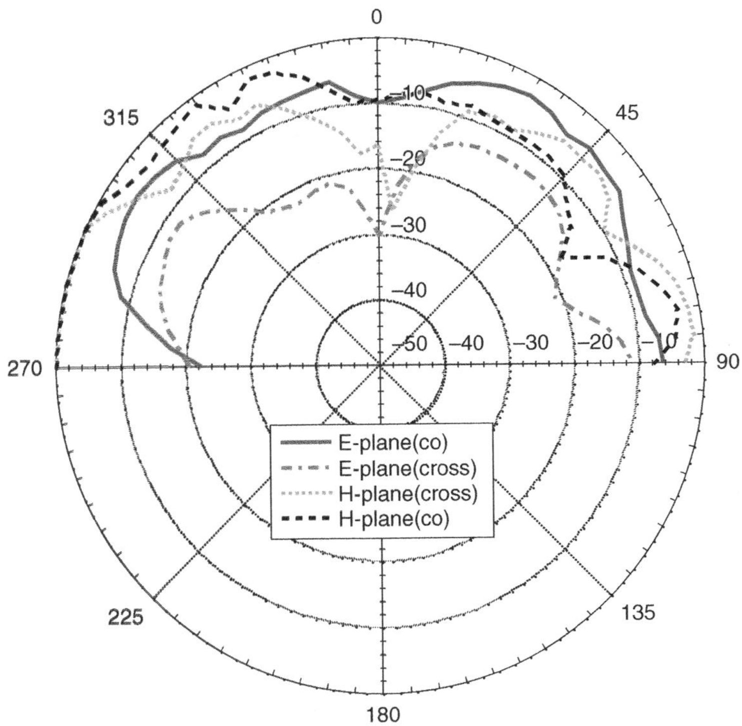

Figure 8.14 Radiation patterns of the proposed printed antenna at 1.8 GHz.

8.5 LOW CROSS-POLARIZED ANTENNAS

8.5.1 Introduction

One of the problems associated with shorted patch antennas is that the radiation patterns have high cross-polarization levels. This problem can be alleviated if the antenna is balanced. In Section 8.5.2, a single-feed shorted patch antenna with parasitic patches is presented. This is a novel method and has been shown to have a significant reduction in the cross-polarization levels. A balanced dual L-shaped patch antenna is presented in Section 8.5.3. Here, a feed network is used and power is coupled to two shorted patches with a phase difference of 180° between the patches.

8.5.2 A Single-Feed Shorted Patch Antenna with Low Cross-Polarized Fields

8.5.2.1 Introduction

In this section, a small printed radiator with low cross-polarized radiation levels is presented. The antenna consists of a conventional shorted patch with the addition of two small shorted parasitic elements. These elements significantly reduce the cross-polarization levels by an order of 10 dB. One very important advantage of the new antenna is the simple feed arrangement, as the parasitic radiators ensure that a single feed only is required.

8.5.2.2 Configuration

Figure 8.15 shows a schematic of the top view of the proposed radiator. Here a conventional probe-fed rectangular shorted patch antenna (with two shorting pins to ease the manufacturing tolerances of the radiator [31]) is etched on a grounded substrate of thickness d and dielectric constant ε_r. For the design trends related to the shorted patch, refer to Reference [31]. As can be seen from Figure 8.15, there are two parasitic rectangular shorted elements separated from the driven element by a distance g_i (where $i = 1, 2$).

The origin of this new radiator came from an unusual source [35] where the authors were investigating a shorted patch with an integrated artificial magnetic conductor (AMC) and the impact of the size of the AMC on the overall performance of the antenna. All of the trends observed were readily explainable with respect to the properties of an AMC with the exception of the low cross-polarized levels. As the number of unit cells of the AMC was reduced from five to one, the cross-polarization fields were still very low. Even taking into consideration the simplistic nature of an AMC, which if appropriately designed converts energy directed towards the endfire (in the plane of the paper in Figure 8.15) and converts it to radiated energy, the explanation for the low cross-polarization levels for such a small AMC structure was not intuitive. It was discovered through simulation and examining the current densities on the shorted patch and the unit cells of the AMC that there were strong densities on the unit cells of the AMC adjacent to the shorted patch in the E-plane. The radiator with only these adjacent unit cells were then investigated; the geometry is shown in Figure 8.15. It was found that the theoretical cross-polarization levels were also 30 dB, below

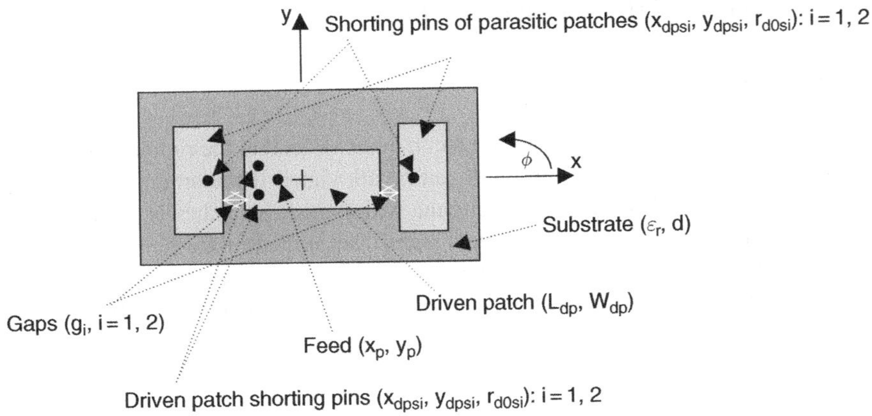

Figure 8.15 Top view schematic of a single-feed shorted patch antenna (parameters: material: $\varepsilon_r = 2.2$, $d = 9.51$ mm; dimensions of patches: $L_{pd} = 12$ mm, $W_{pd} = 6$ mm, $L_{pp1} = L_{pp2} = 5$ mm, $W_{pp1} = W_{pp2} = 9$ mm, $g_1 = g_2 = 3.5$ mm; shorting pins and feed pin: $x_p = 3.6$ mm, $y_p = 0$ mm, $x_{pds1} = -x_{pds2} = 5$ mm, $y_{pds1} = -y_{pds2} = 1.5$ mm, $x_{pps1} = -x_{pps2} = 10$ mm, $y_{pps1} = y_{pps2} = 0$, $r_0 = r_{0ds1} = r_{0ds2} = 0.6$ mm, $r_{0ps1} = r_{0ps2} = 0.5$ mm).

that for the stand-alone shorted patch. The authors believe that the presence of the shorted parasitic element provides electromagnetic symmetry to the radiator, thereby reducing the cross-polarization radiated levels.

8.5.2.3 Results

The printed antenna shown in Figure 8.15 (refer to the figure caption for the dimensions) was fabricated using standard PCB etching techniques. An SMA connector was used to feed the radiator from below. Figure 8.16 shows the predicted (using a full-wave simulator tool based on the integral equation technique) and the measured return loss of the antenna (using an Agilent 8720B vector network analyser). As can be seen from Figure 8.16, there is good agreement between the results and the antenna is well matched at the design frequency of near 2.6 GHz.

The predicted and measured radiation patterns for both copolar and cross-polarization far-fields in the principal planes (E($\phi = 0°$)- and H($\phi = 90°$)-planes) of the proposed antenna at 2.57 GHz are shown in Figures 8.17(a) and (b), respectively. The measurements were undertaken at the Nanyang Technological University anechoic chamber. The copolar patterns are in good agreement, with the measured results showing slightly more scalping in the copolar patterns than predicted. The cross-polarization levels are also higher for the measured results. Both of these discrepancies can be attributed to the very small ground-plane of the radiator, which extend only 5 mm beyond the edge of the conductors of the antenna. In the simulation, infinite ground-planes are assumed. To put the results into perspective, the measured H-plane cross-polarization levels for a conventional shorted patch developed on the same material are 10 dB higher than that presented in Figure 8.17(b). The measured gain of the new radiator was 2.25 dBi.

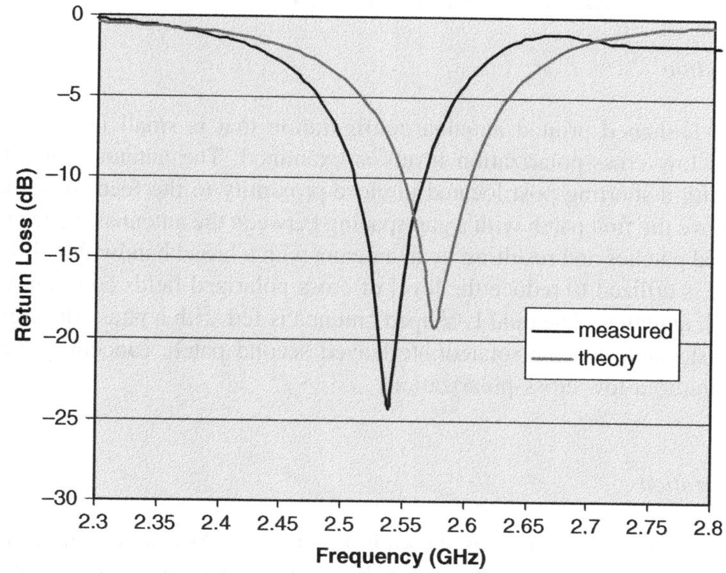

Figure 8.16 Measured and calculated return loss of a single-feed shorted patch antenna.

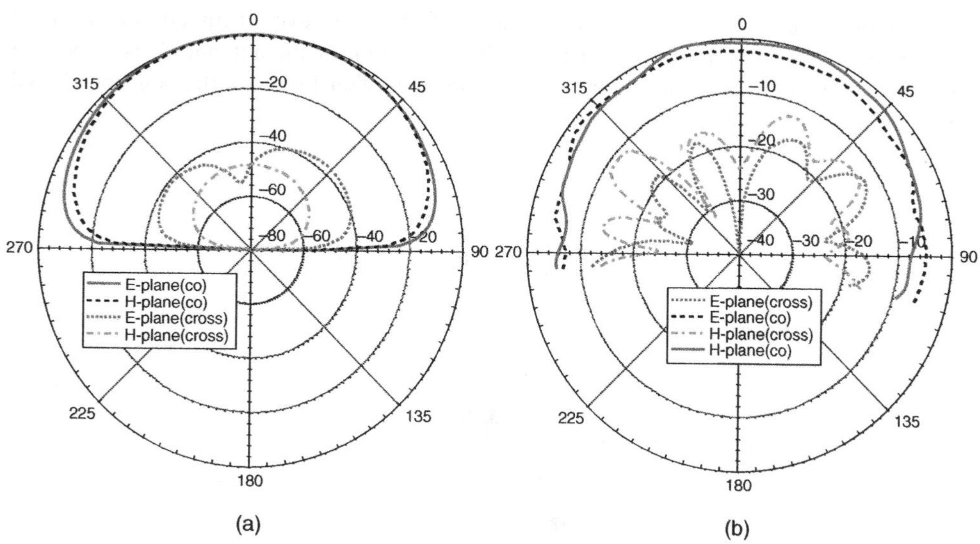

Figure 8.17 Radiation patterns of a single-feed low cross-polarization printed antenna at 2.57 GHz: (a) theoretical and (b) measured.

8.5.3 Small Dual L-Shaped Printed Antenna

8.5.3.1 Introduction

A shorted dual L-shaped printed antenna configuration that is small in size, has a broad bandwidth, and low cross-polarization levels is examined. The antenna consists of an L-shaped patch with a shorting post located in close proximity to the feed. A second patch is rotated 180° above the first patch with a gap spacing between the antennas, hence interleaving the two L-shaped patches and resulting in an antenna with a broad bandwidth. The concept of a balanced feed is utilized to reduce the level of cross-polarized fields commonly associated with small patch antennas. The dual L-shaped antenna is fed with a phase difference of 180° between the first patch and the rotated/interleaved second patch, cancelling the unwanted TM_{11} and leading to a low cross-polarization.

8.5.3.2 Configuration

The proposed antenna configuration is shown in Figure 8.18. As can be seen from the figure, the antenna has an overall length L and width W and is printed on a substrate of thickness d with a relative dielectric permittivity ε_r. The two L-shaped printed antennas are interleaved with a gap spacing of 1 mm existing between the patches. Each of the patches has a length L_2 and width W_2 and contains a shorting post of radius r_{0s} positioned at (x_{psi}, y_{psi}) relative to the centre of the overall area $(L \times W)$ of the patch. The shorting post located in close proximity to the feed probe is to reduce the overall size of the patch [5]. As mentioned previously, a dual antenna configuration provides a significant improvement in the impedance bandwidth when compared to a single patch. This concept is based upon the principal of any dual resonance antenna, namely that if a mutual resonance can be formed the overall impedance variation of the antenna will be minimized, enhancing the bandwidth of the antenna. It is important to note that the patches must not be in contact with each other to utilize the concept of dual resonance [17].

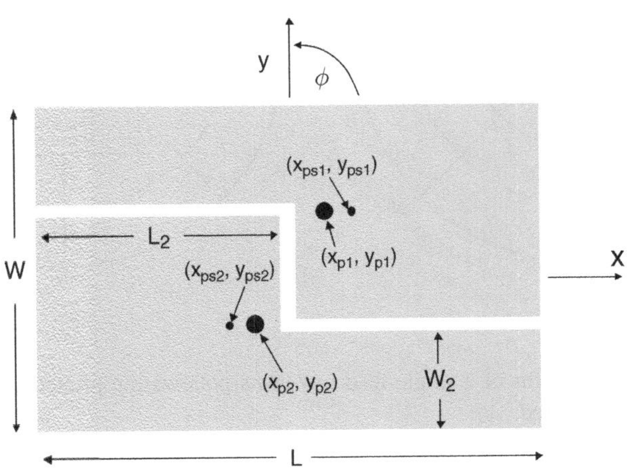

Figure 8.18 Schematic diagram of the dual L-shaped printed antenna.

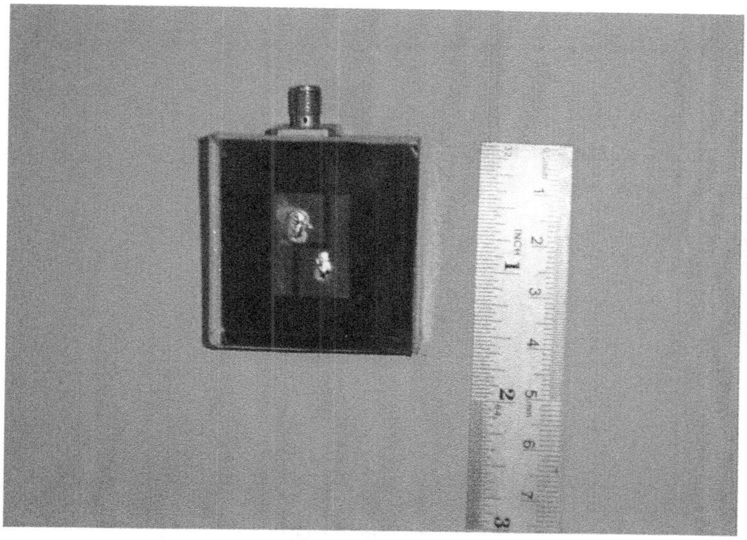

Figure 8.19 Photograph of the dual L-shaped antenna.

To couple power into the antenna, the printed L-shaped patches were connected by probes of radius r_0 and positioned at (x_{pi}, y_{pi}) relative to the centre of the overall size (L × W) of the antenna, on to the feed network located beneath the ground-plane. It is worth noting that port 1 is excited 180° out of phase with respect to port 2 to ensure a balanced configuration. As already mentioned, this feed configuration provides a balanced feed with the 180° phase difference cancelling the unwanted high cross-polarized fields commonly associated with shorted patch antennas [30]. To minimize the size of the feed network, a high dielectric constant substrate was used (*RT/Duroid 6010*, $\varepsilon_r = 10.2$). In addition, a relatively thick substrate was chosen ($d = 2.5$ mm) to ensure that the widths of the feed lines required were not too small, thereby improving the robustness of the proposed antenna. The L-shaped shorted patches were designed for 50 Ω so transformers were required on the feed network to transform the input port into the radiators. It is important to note that an additional length of microstrip line was used for port 1 to provide the required 180° phase shift. Figure 8.19 shows the developed prototype of the proposed antenna.

8.5.3.3 Results and Discussion

Figure 8.20 shows the measured return loss bandwidth of the proposed shorted dual L-shaped printed antenna. The 10 dB return loss bandwidth of the antenna is 16.6 %. This value is considerably higher than a single shorted printed antenna [5] and is comparable with dual antenna configurations [17]. The somewhat larger bandwidth is achieved due to the mutual resonance phenomenon between the two patches, which is evident in the plot of Figure 8.20. It is important to note that the dual-resonance nature of the antenna is achieved without sacrificing the size of the antenna. The overall size of the dual L-shaped printed antenna is $0.2\lambda_0 \times 0.15\lambda_0 \times 0.10\lambda_0$.

The far-field radiation patterns of the shorted dual L-shaped printed antenna were measured at 2.95 GHz and are shown in Figure 8.21 (note that the broadside is 0° in this

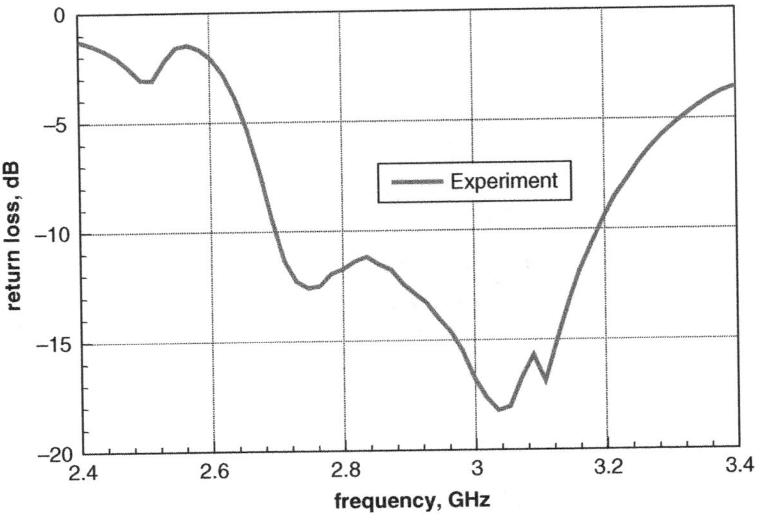

Figure 8.20 Measured return loss of the shorted dual L-shaped printed antenna (parameters: $\varepsilon_r = 1.07$, $d = 10$ mm, $L = 20$ mm, $W = 15$ mm, $L_2 = 9.5$ mm, $W_2 = 4$ mm, $r_0 = 0.6$ mm, $r_{0s} = 0.325$ mm, $x_{p1} = 3$ mm, $y_{p1} = 2.5$ mm, $x_{ps1} = 5$ mm, $y_{ps1} = 2.5$, $x_{p2} = -3$ mm, $y_{p2} = -2.5$ mm, $x_{ps2} = -5$ mm, $y_{ps2} = -2.5$).

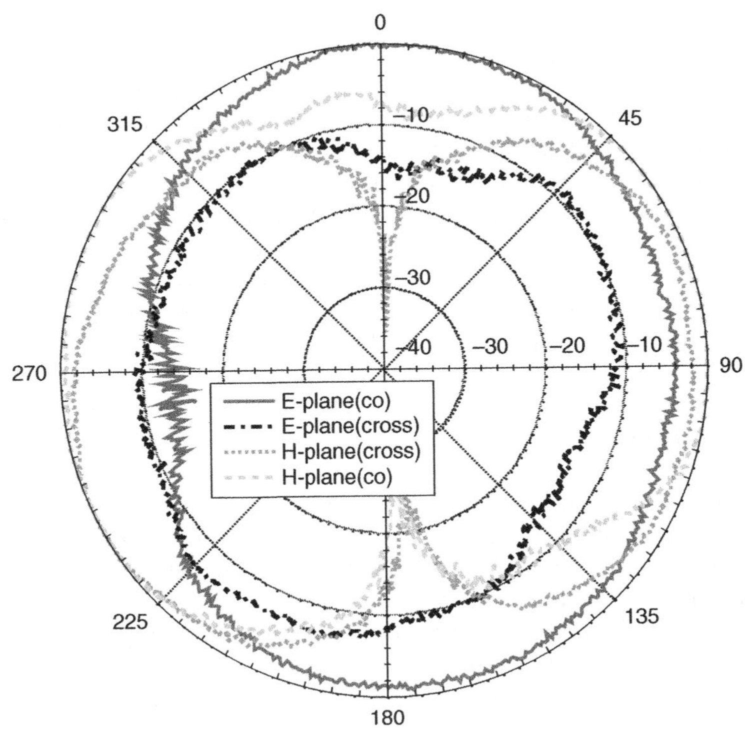

Figure 8.21 Measured radiation patterns of the shorted dual L-shaped printed antenna at 2.95 GHz.

plot). It is evident from the plot that the cross-polarization level has been reduced. It can be seen from Figure 8.21 that the cross-polarized fields are 15 dB (for both the E-plane and H-plane) below the copolarized field. A further reduction in cross-polarized fields can be achieved if a larger ground-plane is utilized. However, for this measurement, the ground-plane was very small, extending 10 mm on each side of the printed conductor. The gain of the antenna is 2.5 dBi.

8.6 SUMMARY

This chapter introduced small patch-based antennas. A shorted patch antenna was discussed describing the methods of size reduction. This was then followed by an explanation of techniques for developing dual-frequency patch antennas. Due to the inherent narrow bandwidth nature of patch antennas, ways to alleviate this problem were provided. Circular polarization methods applicable to shorted patch and ways to reduce the high cross-polarization level associated with shorted patch were presented.

Small-sized shorted patch antennas were presented in Section 8.3. Firstly, a small dual C-slot printed antenna was discussed. The antenna incorporates strategically placed C-slots within the shorted patch to maximize the current path of the printed conductor. The resulting antenna achieves a reduction in resonant frequency of 15 %. The 10 dB return loss bandwidth was 5.4 %. Next, a perforated shorted patch was presented. The concept of perforating the patch conductor has been applied to a shorted patch, yielding an even smaller radiator (by 30 %) without appearing to compromise the structural complexity or the impedance and radiation properties of the antenna. The measured 10 dB return loss of the antenna was 8.1 % and the gain across this bandwidth varied from 0.5 to 1.0 dBi.

Spiral-like printed antenna with a dual-resonance characteristic was examined in Section 8.4. The antenna utilizes a spiral-like shorted patch and an additional shorting pin to generate the extra resonance. The bandwidth for the antenna is at least 1.4 % for each of the resonant frequencies and the radiation patterns are omnidirectional. The structure is simple to manufacture. Another technique for the dual-frequency printed antenna was discussed. The radiator consisted of a stacked shorted patch configuration, which always had relatively independent control of the resonant frequencies of the antenna. A proof of the concept of the proposed radiator was designed and developed and the measured response agreed well with the predicted performance.

A single-feed shorted patch antenna that uses two parasitic elements to reduce the cross-polarization levels of the radiator was discussed in Section 8.5. The performance of the new radiator was experimentally verified, showing that the cross-polarization levels of a shorted patch can be dramatically reduced, even for an antenna developed on a truncated substrate. Finally, a small shorted dual L-shaped printed antenna was presented. The proposed antenna utilizes the dual-resonance concept and the balanced feeding technique for increased bandwidth with low cross-polarized fields. The 10 dB return loss bandwidth of the antenna is 16.6 % and the cross-polarized fields are 15 dB below the copolarized levels.

REFERENCES

1. J. L. Finol and J. G. Mielke, 'Past and future directions in cellar telephony', in *IEEE AP-S*, pp. 7–13, 1998 (invited paper).

2. T. Ojanpera and R. Prasad, 'An overview of third-generation wireless personal communications: a European perspective', *IEEE Personal Communications*, 59–65, December 1998.
3. D. M. Pozar, 'An overview of wireless systems and antennas', in *IEEE Proceedings AP-S*, pp. 566–569, 2000.
4. L. C. Godara, (ed.), *Handbook of Antennas in Wireless Communications*, CRC Press, New York, 2002.
5. R. B. Waterhouse, 'Small microstrip patch antenna', *Electronics Letters*, **31**, 604–605, April 1995.
6. K.-L. Wong and S.-C. Pan, 'Compact triangular microstrip antenna', *Electronics Letters*, **33**, 433–434, March 1997.
7. S. K. Satpathy, K. P. Ray and G. Kumar, 'Compact shorted variations of circular microstrip antennas', *Electronics Letters*, **34**, 137–138, January 1998.
8. D. M. Kokotoff, J. T. Aberle and R. B. Waterhouse, 'Rigorous analysis of probe-fed printed annular rings', *IEEE Transactions on Antennas and Propagation*, **47**, 384–388, February 1999.
9. H. K. Kan and R. B. Waterhouse, 'Size reduction technique for shorted patches', *Electronics Letters*, **35**, 948–949, June 1999.
10. H. K. Kan and R. B. Waterhouse, 'Shorted spiral-like printed antennas', *IEEE Transactions on Antennas and Propagation*, **50**, 396–397, March 2002.
11. S. Maci and G. Biffi Gentili, 'Dual-frequency patch antennas', *IEEE Antennas and Propagation Magazine*, **39**, 13–20, December 1997.
12. J.-S. Chen and K.-L. Wong, 'A single-layer dual-frequency rectangular microstrip patch antenna using a single probe feed', *Microwave and Optical Technology Letters*, **11**, 83–84, February 1996.
13. W. F. Richards, S. E. Davidson and S. A. Long, 'Dual-band reactively loaded microstrip antennas', *IEEE Transactions on Antennas and Propagation*, **33**, 556–560, May 1985.
14. G. Kumar and K. C. Gupta, 'Nonradiating edges and four gap-coupled multiple resonator broadband microstrip antennas', *IEEE Transactions on Antennas and Propagation*, **33**, 173–178, June 1985.
15. C. K. Aanandan, P. Mohanan and K. G. Nair, 'Broad-band gap coupled microstrip antenna', *IEEE Transactions on Antennas and Propagation*, **38**, 1581–1585, October 1990.
16. R. B. Waterhouse, 'A broadband stacked shorted patch', *Electronics Letters*, **35**, 98–100, January 1999.
17. H. K. Kan and R. B. Waterhouse, 'Small square dual spiral printed antennas', *Electronics Letters*, **37**, 478–479, April 2001.
18. J.-H. Lu, C.-L. Tang and K.-L. Wong, 'Single-feed slotted equilateral-triangular microstrip antenna for circular polarization', *IEEE Transactions on Antennas and Propagation*, **47**, 1174–1178, July 1999.
19. C.-Y. Huang, J.-Y. Wu and K.-L. Wong, 'Broadband circularly polarised square microstrip antenna using chip-resistor loading', *IEE Proceedings on Microwave, Antennas and Propagation*, **146**, 94–96, February 1999.
20. K.-P. Yang and K.-L. Wong, 'Dual-band circularly-polarised square microstrip antenna', *IEEE Transactions on Antennas and Propagation*, **49**, 377–382, March 2001.
21. X. M. Qing and Y. W. M. Chia, 'Circularly polarised circular ring slot antenna fed by stripline hybrid coupler', *Electronics Letters*, **35**(25), 2154–2155, December 1999.
22. H.-M. Chen and K.-L. Wong, 'On the circular polarisation operation of annular-ring microstrip antennas', *IEEE Transactions on Antennas and Propagation*, **47**, 1289–1292, August 1999.
23. T. Teshirogi, M. Tanaka and W. Chujo, 'Wideband circularly polarised array antenna with sequential rotations and phase shift of elements', in *International Symposium on Antennas and Propagation, ISAP*, Tokyo, 117–120, 1985.
24. J. Huang, 'A technique for an array to generate circular polarization with linearly polarized elements', *IEEE Transactions on Antennas and Propagation*, **34**, 1113–1124, September 1986.
25. P. S. Hall, J. S. Dahele and J. R. James, 'Design principles of sequentially fed, wide bandwidth, circularly polarised microstrip antennas', *IEE Proceedings, Part H*, **136**, 381–389, October 1989.

26. P. S. Hall, ‘Application of sequential feeding to wide bandwidth, circularly polarised microstrip patch arrays’, *IEE Proceedings, Part H*, **136**, 390–398, October 1989.
27. H. Kan and R. B. Waterhouse, ‘Small circularly polarised printed antenna’, *Electronics Letters*, **36**, 393–394, March 2000.
28. H. K. Kan and R. B. Waterhouse, ‘A small CP printed antenna using 120° sequential rotation’, *IEEE Transactions on Antennas and Propagation*, **50**, 398–399, March 2002.
29. H. K. Kan, R. B. Waterhouse and D. Pavlickovski, ‘Compact dual concentric rings printed antennas’, *IEE Proceedings of Microwave, Antennas and Propagation*, **141**, 37–42, February 2004.
30. R. B. Waterhouse, ‘Small printed antennas with low cross-polarised fields’, *Electronics Letters*, **33**, 1280–1281, July 1997.
31. R. B. Waterhouse, S. D. Targonski and D. M. Kokotoff, ‘Design and performance of small printed antennas’, *IEEE Transactions on antennas and Propagation*, **AP-46**, 1629–1633, November 1998.
32. R. B. Waterhouse, *Microstrip Patch Antennas: A Designer’s Guide*, Chapter 5, Kluwer, Boston, Massachusetts, 2003.
33. G. Clasen and R. Langley, ‘Meshed patch antennas’, *IEEE Transactions on Antennas and Propagation*, **52**, 1412–1416, June 2004.
34. H. A. Wheeler, ‘Small antennas’, Chapter 6 in *Antenna Engineering Handbook*, 2nd edn (eds R. C. Johnson and H. Jasik), McGraw-Hill, New York, 1984.
35. D. Pavlickovski and R. B. Waterhouse, ‘Shorted patches mounted on high impedance ground planes’, *IEEE International Antennas and Propagation Symposium*, Monterey, California, Vol. 4, pp. 3601–3604, June 2004.

9

Small Multiband Printed Monopole Antennas

Karu Esselle and Yuehe Ge
Electronics Department, Macquarie University, Sydney, NSW, Australia

9.1 INTRODUCTION

Wireless communications is a rapidly expanding area of technology. Cellular telephones, portable computers, hand-held computers, and numerous other wireless communication devices are used by ever-increasing numbers of people and have become ubiquitous. A corresponding trend is that such equipment is becoming more compact and often lighter. With reduced equipment sizes has grown a need for more compact antennas for use in cellular communications and other wireless technologies, such as Personal Communications Services (PCS), Bluetooth, and the IEEE 802.11 family of wireless local area networks (WLANs, also called 'wireless Ethernet') (e.g. 802.11a, 802.11b, 802.11g, and 802.11j).

Antennas with the features of low cost, small size, and good performance are needed for application in these wireless communication systems. Monopole antennas have been well investigated and designed for the mobile communication systems such as GSM900, DCS1800 or 1900, UMTS, etc. [1–4]. Among them, the planar monopole antenna is low profile, etched on a substrate, and can provide the feature of broadband or multiband [4], which makes it attractive to mobile handset antenna applications.

In this chapter, a variety of advanced planar antennas suitable for WLAN operations [5–7] and mobile communications [8] are presented. The design techniques of using multiple-branch strips and spiral monopole strips to achieve multiband operations are applied. In Section 9.2, the designs of binding a dual-arm monopole strip into a rectangular shape to achieve a compact structure and quad-band operations (IEEE 802.11a, 802.11b, 802.11g, and 802.11j) are described. The techniques to obtain good impedance matching and bandwidth in quad-band operations are discussed. As an example, a microstrip-fed dual-arm monopole antenna was implemented and presented. Another similar, but CPW-fed (coplanar waveguide), dual-arm monopole antenna providing the same quad-band operations is given

Printed Antennas for Wireless Communications Edited by R. Waterhouse

in Section 9.3. Coplanar waveguides (CPWs) are increasingly used in microwave circuits due to their various advantages, including the ease to integrate active components with CPW lines, low dispersion, good characteristic impedance control, etc. [9]. By wrapping one more strip around a dual-arm antenna element, one more operational band can be achieved. In Section 9.4, a triple-arm monopole antenna design operating at 2.4/4.9/5.2/5.8 GHz and 1.8 or 1.7 GHz (WLAN and DCS1800 or PCS1900) with a compact configuration is presented. In Section 9.5, a novel planar monopole antenna structure is presented to provide multiband operation for mobile communication systems. By properly spiralling a monopole strip, the antenna can provide multiple-resonant modes and the first two modes can resonate with good impedance matching and better bandwidth. It is well known that the multiarmed spiral antennas can offer a broad bandwidth and they are normally excited at the inside terminal. In this proposed antenna, the spiral antenna element is fed by a 50 Ω microstrip line at its outside terminal. The antenna is printed on an FR4/Epoxy substrate with an arbitrary thickness. It is easy to be integrated with the associated microstrip circuits. This has the benefit of cost reduction and improving system reliability. As examples, a quad-band antenna for covering GSM900, DCS1800, PCS1900, and UMTS2000 and a triple-band antenna for WLAN operations at 2.4, 4.9, and 5.2 GHz were designed and presented in this section.

Section 9.6 presents a novel monopole antenna to provide multiband operation for wireless communication systems. By properly folding a metallic strip around a cuboid base that is mounted on an FR4/Epoxy substrate and fed by a microstrip line, the monopole antenna can provide two resonant modes with good impedance matching, better bandwidth, a relatively low profile, and small size. The theory to achieve two resonant modes is the same as that of spiral antennas. An example is designed and fabricated for mobile communication systems. The simulated and measured results show that the antenna can operate at GSM900, DCS1800, and PCS1900 bands.

9.2 DUAL-ARM PRINTED MONOPOLE ANTENNAS

One of the approaches to obtain multiband operations is to use the multiple-arm or branch antenna element. In these antennas, the different branch lines resonate at different frequency bands. Sometimes, one branch can resonate at two or multiple bands close to one another. With the appropriate tuning techniques, good impedance matching can be obtained at those resonant frequencies. Normally, printed multiple-arm monopole antennas have the characteristics of a compact configuration, low profile, and multiband operation.

Figure 9.1 shows a dual-arm printed monopole antenna model for dual- or multiband operations. As can be seen, the radiating element is fed by a 50 Ω microstrip line, which has a width of W_f; both are printed on an FR4 Epoxy substrate, which has a thickness (H) of 0.8 mm and a dielectric constant (ε_r) of 4.4, making the structure very low profile. The ground is on the other side of the substrate and truncated at or near a point where the feed line is coupled to the first metal strip or the second metal strip. The dimension of the ground-plane is $L_g \times W_s$. Since the antenna is printed on a substrate, it can be easy to integrate it with the associated microstrip circuits, which are printed on the same substrate. The radiating element is composed of two metal strips, which will operate at two different frequency bands. The two metal strips have the same and uniform width (W) and are folded to make the element's shape rectangular. The longer strip is wrapped around the shorter one, leaving a small separation (d) between the metal strips for a substantial portion of the length of the first metal strip. There are several advantages to this arrangement. Firstly, it makes the configuration compact. Secondly, by selecting the value of parameter d, the distance

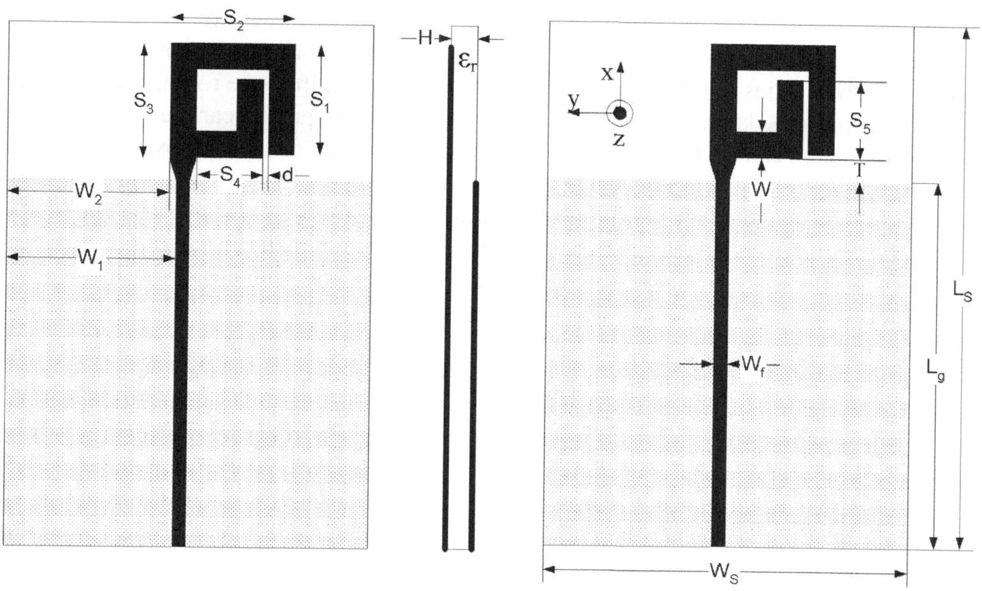

Figure 9.1 A two-arm antenna model. (From Reference [5]. © 2005 IEEE.)

between the two end-stubs of the two metal strips, the first resonant frequency, determined by the length of the longer strip, can be shifted in a small range. Thirdly, the selection of d can affect the impedance matching at the resonant frequencies. Although the multiple-arm antenna can operate at multiple-frequency bands, the good impedance matching (normally below -10 dB) is not always obtained at these resonant frequencies.

As an example, the dual-arm antenna element in Figure 9.1 is simulated using the commercial software Ansoft HFSS. The antenna has two operation bands, determined by the two branch arms, at 2.4 and 4.9–6.0 GHz. The parameters of the two branch arms are: $S_1 = 7.8$ mm, $S_2 = 9.5$ mm, $S_3 = 9.5$ mm, $S_4 = 5.0$ mm, and $S_5 = 6.0$ mm. Figure 9.2 shows the simulated S_{11} with respect to the variation of the parameter d. It can be seen that the

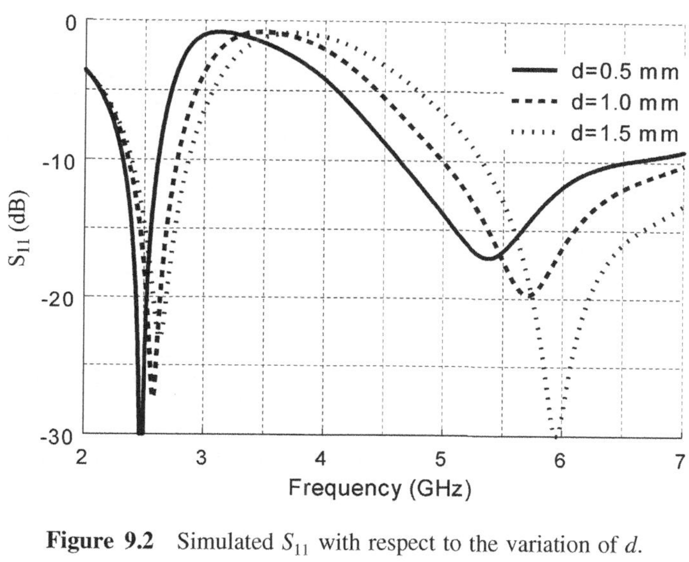

Figure 9.2 Simulated S_{11} with respect to the variation of d.

first resonant frequency shifts to lower when the parameter d varies from 1.5 to 0.5 mm. The resonant frequency at about 2.4 GHz is mainly determined by the length of the longer branch. However, adjusting the value of d can make a minor shift in the resonant frequency. In addition, the impedance matching at the first resonant frequency can be improved by changing the parameter d. Due to the coupling effects between the two strips, the resonant length of the longer strip is larger than that without the coupling. This will lower the resonant frequency and change the impedance matching condition. A similar phenomena can be found again in the design of the spiral antennas in Section 9.4. In addition, the separation T between the end of the longer strip and the edge of the ground-plane will affect the impedance matching at the resonant frequency determined by the longer strip. The small T will lead to an adverse effect on the impedance matching at the resonant frequency.

Based on the above investigation, a microstrip-fed dual-arm design for 2.4 and 4.9–6.0 GHz WLAN bands is developed. The motivation of the antenna design is to operate at the WLAN bands in IEEE 802.11a, 802.11b, 802.11g, and 802.11j. The parameters of the antenna element are: $S_1 = 7.8$ mm, $S_2 = 10.0$ mm, $S_3 = 9.0$ mm, $S_4 = 5.5$ mm, and $S_5 = 3.5$ mm. The total size of the element is around $10 \times 10\,\text{mm}^2$. The shorter metal strip is designed to operate at the upper frequency band, which is 4.9 ~ 6 GHz. Suppose the centre frequency is 5.5 GHz and the start point of the strip is at the feed point, which is located at the end of the transmission line and right above the end edge of the ground-plane, the resonant length of the shorter strip is around half a guide wavelength at 5.5 GHz. For this kind of monopole antenna, a frequency band of above 20 % can be obtained.

The longer metal strip operates at the lower band of 2.4 GHz. The length is also around half a guide wavelength at 2.4 GHz. The size of the substrate ($L_S \times W_S \times H$) is $55 \times 35 \times 0.8\,\text{mm}^3$ and the ground-plane has an area ($L_g \times W_S$) of $40 \times 35\,\text{mm}^2$. The other parameters are: $W = 2.0$ mm, $W_f = 1.5$ mm, $W_1 = 14.0$ mm, $W_2 = 14.25$ mm, $T = 1.0$ mm, and $d = 0.5$ mm.

The antenna design was developed using commercial software Ansoft HFSS 8.5. A prototype of the design was fabricated and tested. Figure 9.3 shows the simulated and measured return loss of the designed antenna. It can be seen from the measured one that two operating bandwidths are obtained with good impedance matching over them. The lower bandwidth,

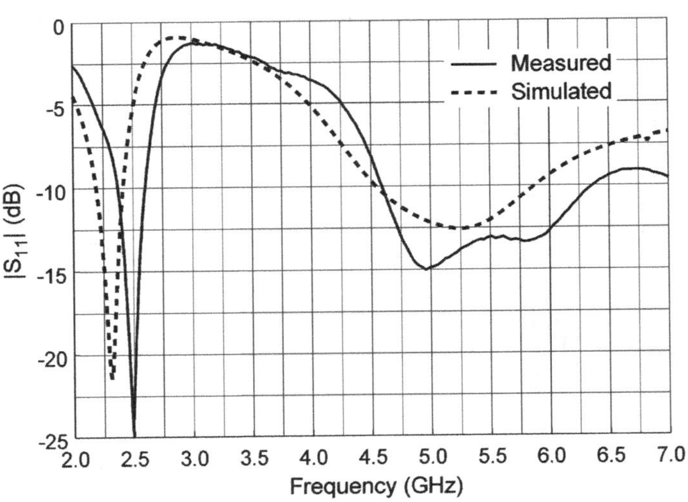

Figure 9.3 Measured and simulated return loss for the antenna shown in Figure 9.1. (From Reference [5]. © 2005 IEEE.)

which is determined by S_{11} of < -10 dB, reaches 218 MHz (9.1 %) and covers the IEEE 802.11b or 802.11g band (2.40–2.485 GHz). Comparing the two results in Figure 9.3, it is noted that the measured resonant frequency shifts about 100 MHz upwards. On the other hand, the upper band has a bandwidth as large as 1.755 GHz (32 %) and covers the bands of IEEE 802.11a and 802.11j in Japan.

The radiation patterns for this antenna were also measured and are shown in Figure 9.4. These patterns are measured in the xz plane (E-plane) and the yz plane (H-plane) at 2.45,

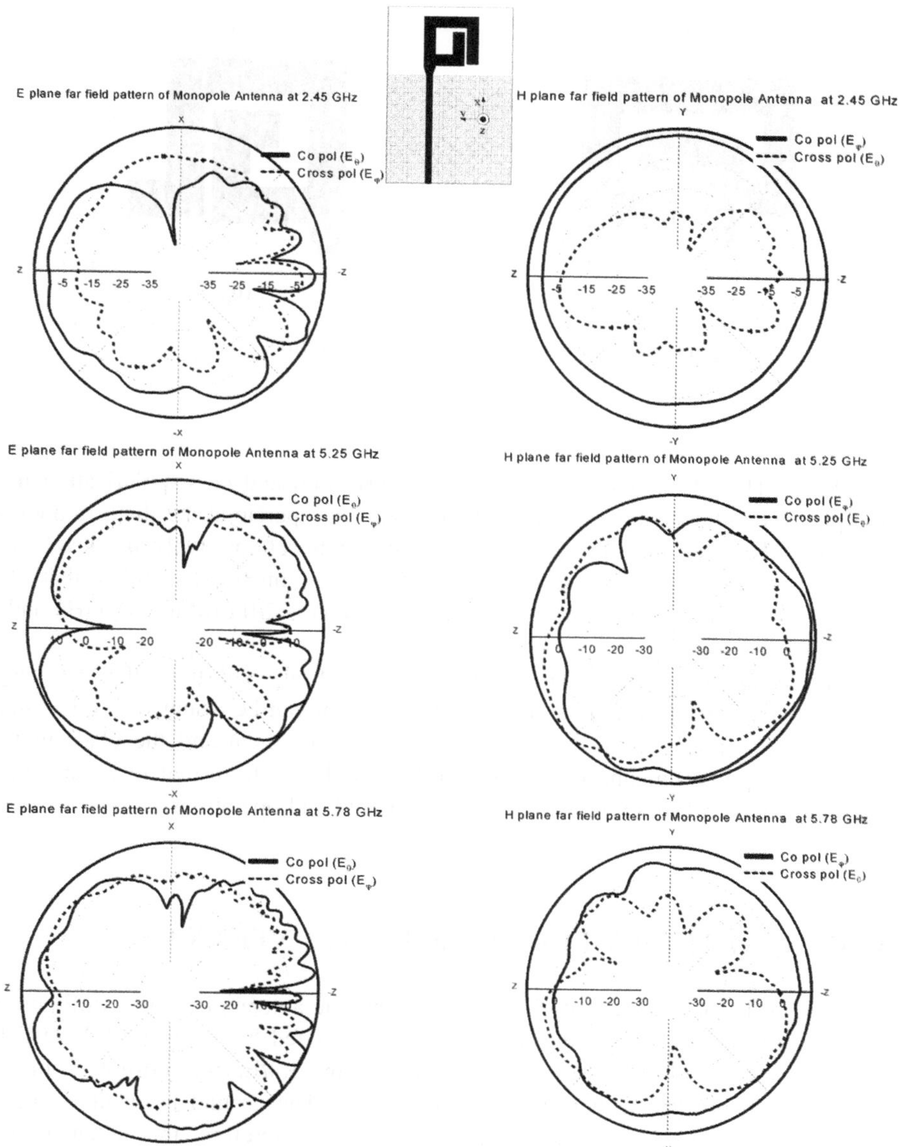

Figure 9.4 Measured radiation patterns at 2.45, 5.25, and 5.78 GHz for the antenna shown in Figure 9.1. (From Reference [6]. © 2005 ANTEM. Reproduced with permission.)

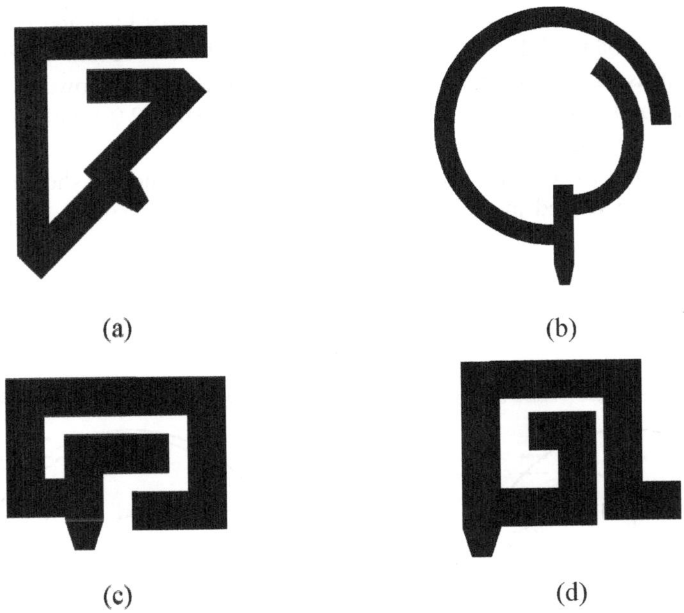

Figure 9.5 Two-arm antenna models of various shapes.

5.25, and 5.78 GHz. Most of the patterns of copolarization and cross-polarization in the measured planes exhibit omnidirectional radiation and can operate at the three frequencies, which is acceptable and an advantage in mobile communications. Only the level of the cross-polarization pattern at 5.25 GHz in the *YOZ* plane is much smaller than that of the copolarization one. The measured peak gain of the antenna is 2.0 dBi at 2.45 GHz and 2.1 dBi at 5.25 GHz.

Since the antenna element and the feedline are printed on a substrate, they are easy to integrate with the associated circuits. The antenna element can be changed to other shapes according to practical requirements, such as the lengths of them and the place they are located. Some proposed antenna models with different shapes from that in Figure 9.1 are shown in Figure 9.5. All these models can achieve multiband operations.

9.3 A DUAL-ARM CPW-FED MONOPOLE ANTENNA

The geometry of the proposed CPW-fed printed monopole antenna is shown in Figure 9.6. As can be seen from this figure, the radiating element is fed by a 50 Ω CPW feed line, and both the antenna and the feed line are printed on an FR4/Epoxy substrate, which has a thickness of 0.8 mm and a dielectric constant of 4.4. The ground is on the same side of the substrate and only surrounds the CPW feed line. Therefore, this antenna can be easily integrated with the CPW-based microwave circuits, which are printed on the same substrate.

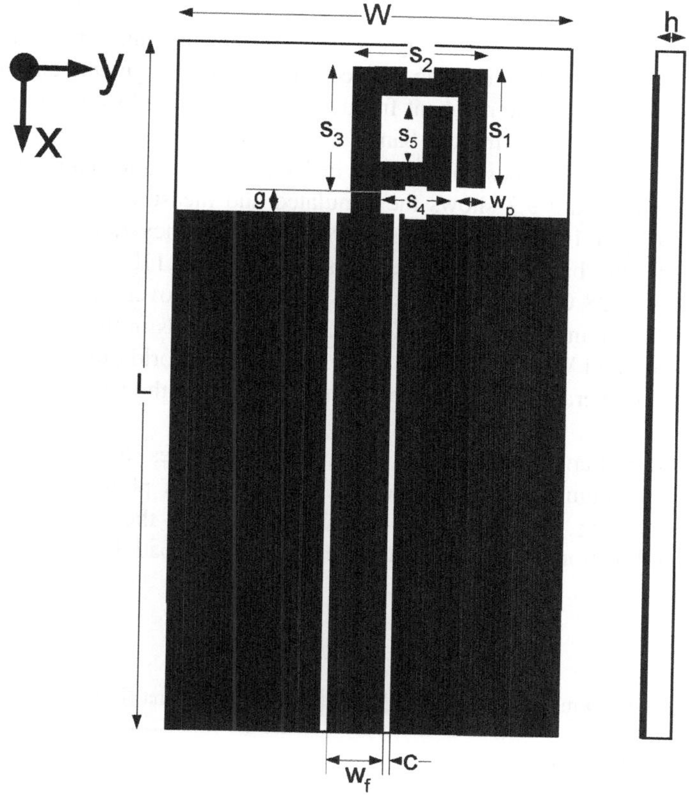

Figure 9.6 The configuration of the CPW-fed quad-band WLAN antenna.

The radiating element is composed of two metal strips, which efficiently radiate at two frequency bands. The two metal strips are bent to save space, and the mutual coupling between the two strips is accounted for when designing the radiating element. The overall size of the radiating element is around $1 \times 1\,\text{cm}^2$. The shorter metal strip is designed to operate at the upper frequency band, which is $\sim$ 4.9–6 GHz. Assuming that the centre frequency is 5.5 GHz and the start point of the strip is at the feed point, which is located at the end of the transmission line and right above the edge of the ground-plane, the resonant length of the shorter strip is around half a guided wavelength at the upper operating frequency band. For this type of monopole antenna, a frequency band of above 20 % can be obtained.

The longer strip operates in the lower band around 2.45 GHz. The longer strip is folded to partly encircle the shorter one and to bring the two ends closer. This is important for the antenna to operate at two bands above, because this arrangement improves matching and moves the resonant frequency to the correct positions. The length of the longer strip is also around half a guided wavelength at the lower operating band.

The antenna was designed using Ansoft HFSS commercial software. It was initially found from the simulations that with the shape of the two metal strips as shown in Figure 9.6, good matching in the higher band, say 4.9–6.0 GHz, is easy to obtain. However, in the lower band, i.e. 2.4–2.5 GHz, it is difficult to obtain a good (>10 dB) return loss. However, when the two ends of the two strips are brought closer, good matching can be achieved, even in the lower band. By adjusting the gap between the two end segments and the lengths of the two strips appropriately, the resonant frequency can be moved to the desired frequency while achieving good matching in both bands.

The parameters of the final design are listed in Table 9.1. The antenna has been fabricated in-house and tested. Figure 9.7 shows the simulated and measured return loss. The lower band, determined by a 10 dB return loss, covers the frequency range from 2.35 to 2.52 GHz, which fulfils the bandwidth requirements of IEEE 802.11b and 802.11j. The upper bandwidth is about 2.28 GHz, from 4.6 to 6.88 GHz, which covers the frequency bands of IEEE 802.11j in Japan and all three IEEE 802.11a U-NII bands in the USA. Such antennas are known as quad-band WLAN antennas in the commercial world (although it is essentially an antenna with two operating bands). The antenna also covers the HiperLAN2 WLAN band in Europe.

The simulated radiation patterns at operating frequencies of 2.45, 5.25, and 5.78 GHz are plotted in Figures 9.8, 9.9, and 9.10. The patterns plotted in three planes (*xy* plane, *xz* plane, and *yz* plane) are very consistent between the two operating frequencies. Good omnidirectional radiation characteristics and broad beams are noted at the three frequencies.

Table 9.1 Dimensions of the CPW-fed multiband antenna (in mm).

L	W	H	W_f	w_p	c	g	S_1	S_2	S_3	S_4	S_5
56.0	35.0	0.8	3.0	2.0	0.5	1.5	9.0	10.0	9.5	5.5	5.0

(From Reference [6]. © 2005 ANTEM. Reproduced with permission.)

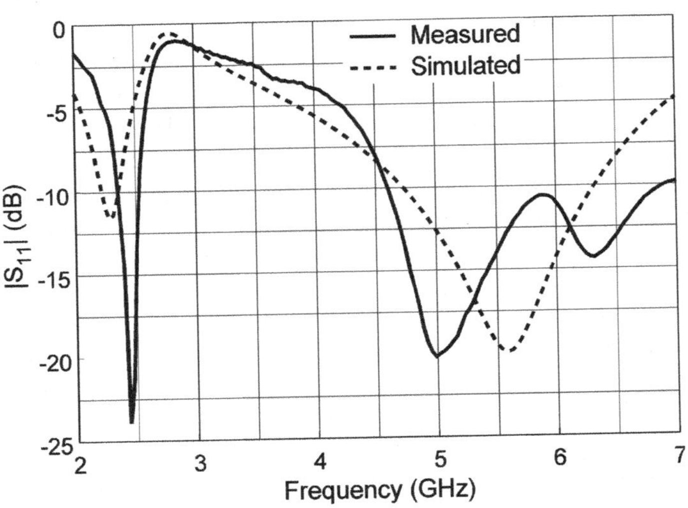

Figure 9.7 Simulated and measured return losses of the multiband antenna. (From Reference [6]. © 2005 ANTEM. Reproduced with permission.)

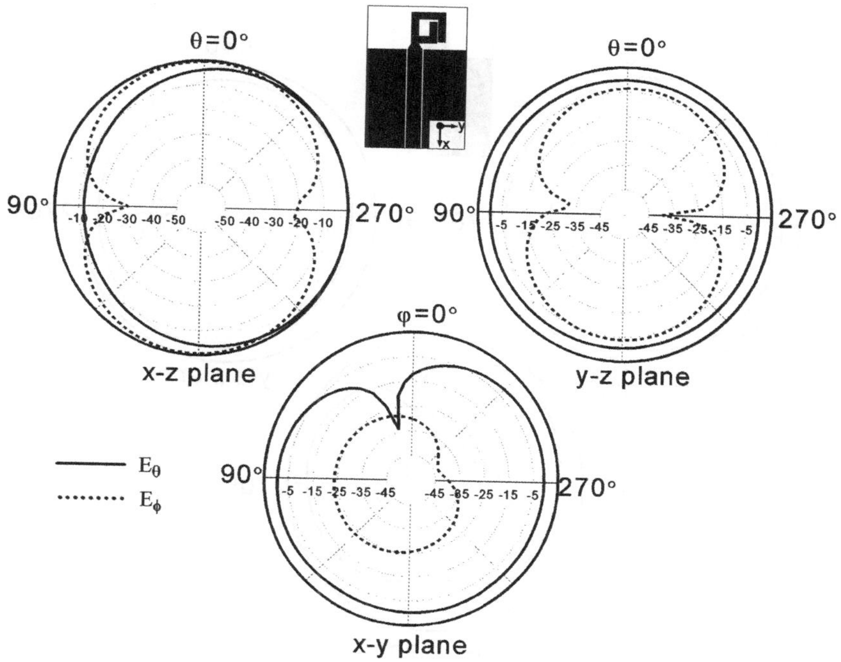

Figure 9.8 The radiation pattern of the quad-band antenna at 2.45 GHz. (From Reference [6]. © 2005 ANTEM. Reproduced with permission.)

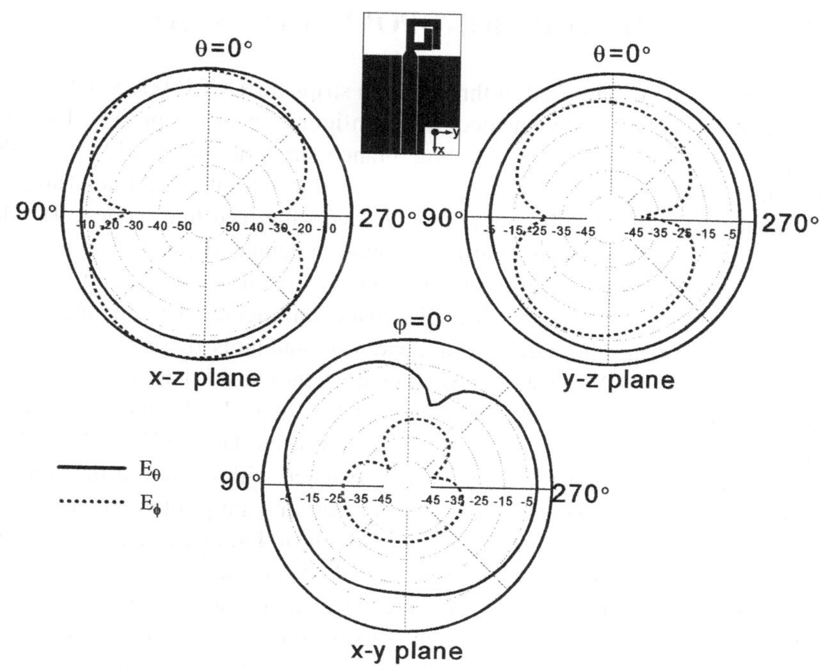

Figure 9.9 The radiation pattern of the quad-band antenna at 5.25 GHz.

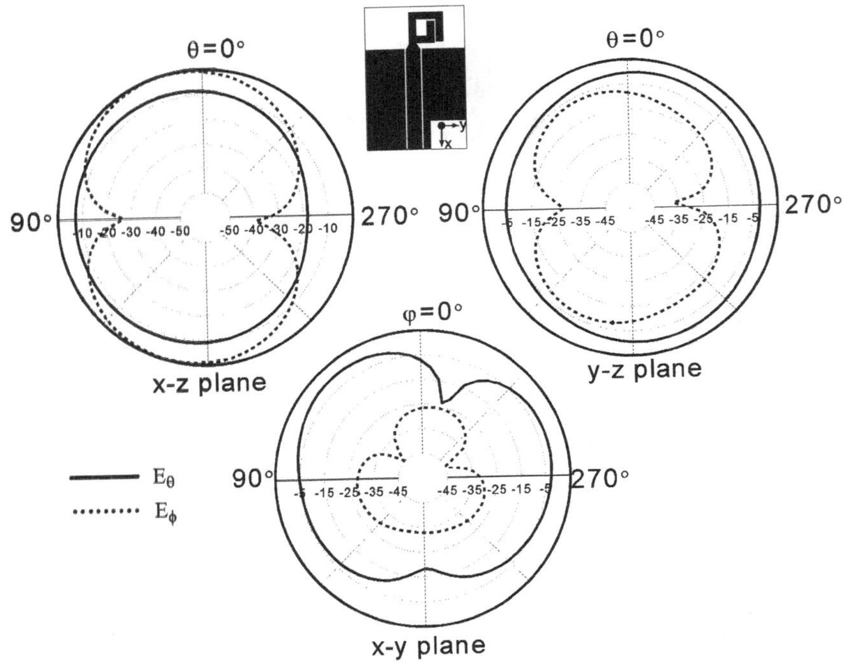

Figure 9.10 The radiation pattern of the quad-band antenna at 5.78 GHz. (From Reference [6]. © 2005 ANTEM. Reproduced with permission.)

9.4 THREE-ARM PRINTED MONOPOLE ANTENNAS

A planar monopole antenna composed of three branch strips for DCS1800/PCS1900/WLAN multiband operations has been implemented. The configuration of the proposed antenna is shown in Figure 9.11. As can be seen, the three branch strips are wrapped into a compact rectangular structure and printed in a thin substrate together with its feed line. The ground-plane is on the other side of the substrate and only covers the feed line section. The design procedures are similar to those of dual-arm monopole antennas described above. In fact, the triple-arm antenna is composed of a dual-arm antenna element and a third branch strip wrapping around the dual-arm element. The dual-arm antenna element still operates at 2.4 and 4.9–6.0 GHz bands, covering the required bandwidth of standards IEEE 802.11a, 802.11b, 802.11g, and 802.11j for WLAN systems. Another frequency band determined by the third branch strip can be tuned to appear within the range of 1.7–2.0 GHz, depending on the length of the strip. This band can cover the required bandwidth of the DCS1800 or PCS1900.

The proposed triple-arm monopole antenna was designed with the aid of the commercial software Ansoft HFSS. The antenna element and the feed line are printed on an FR4/Epoxy substrate with a thickness of 0.8 mm, a dielectric constant of 4.4, and a size of $50 \times 35\ \text{mm}^2$ ($L \times W$). The ground plane is only below the feed line and has a size of $30 \times 35\ \text{mm}^2$ ($L_1 \times W$). The dimension parameters of the antenna element, shown in Figure 9.11, are listed in Table 9.2. The three strips are designed to operate at 1.8, 2.4, and 4.9–6.0 GHz bands. The strip lengths, starting from the edge of the ground-plane, are about 39.0 mm (about 0.234 wavelength at 1.8 GHz), 28.5 mm (about 0.233 wavelength at 2.45 GHz), and 14.5 mm (about 0.251 wavelength at 5.2 GHz), respectively. The 1.8 GHz band is mostly

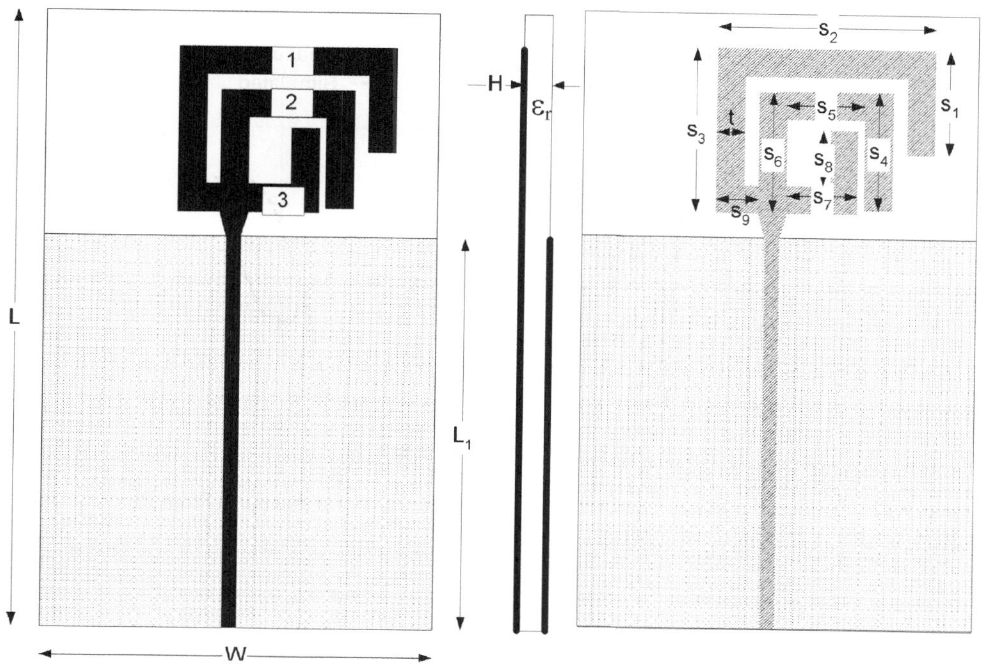

Figure 9.11 The configuration of the triple-arm WLAN antenna. (From Reference [10]. © 2006 IEEE.)

Table 9.2 The parameters of the designed triple-arm monopole antenna (in mm).

S_1	S_2	S_3	S_4	S_5	S_6	S_7	S_8	S_9	t	H	L	L_1	W
8.0	17.0	14.0	9.5	7.0	11.0	6.0	5.5	3.0	2.0	0.8	50.0	30.0	35.0

(From Reference [10]. © 2006 IEEE.)

determined by the length of strip 1 shown in Figure 9.11. By tuning its length, the 1.8 GHz band can be shifted to the 1.7, 1.9, or 2.0 GHz band. Similarly, the 4.9–6.0 GHz band is determined by the length of strip 3. The 2.4 GHz band is determined by the length of strip 2 and the coupling between strips 2 and 3. As mentioned in Section 9.2, by reducing the distance between the end stubs of strips 2 and 3, the resonant frequency will shift to the lower.

The designed antenna has been fabricated and tested. The simulated result of the return loss, obtained from the commercial software Ansoft HFSS, and the measured result are shown in Figure 9.12. It can be seen that the two results are reasonably conformed. The three operational bands have bandwidths (about −10 dB return loss) of 170 MHz (1.8–1.97 GHz), 170 MHz (2.37–2.54 GHz), and 2.06 GHz (4.64–6.7 GHz), covering the required bandwidth of PCS1900 and all the WLAN standards.

The radiation patterns of the proposed triple-arm monopole antenna were also simulated. The simulated radiation patterns at 1.8, 2.45, and 5.78 GHz are presented in Figures 9.13, 9.14, and 9.15, respectively. These patterns are simulated in the *xz* plane, *yz* plane, and *xy* plane, respectively. The simulations show that there is not much variation among the patterns in the *xz* plane and *yz* plane at the three frequencies. In the *xy* plane,

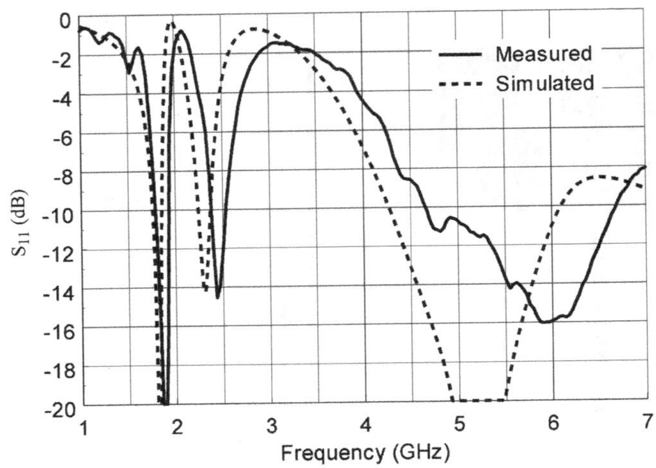

Figure 9.12 Simulated and measured return losses of the triple-arm antenna. (From Reference [11]. © 2006 WARS. Reproduced with permission.)

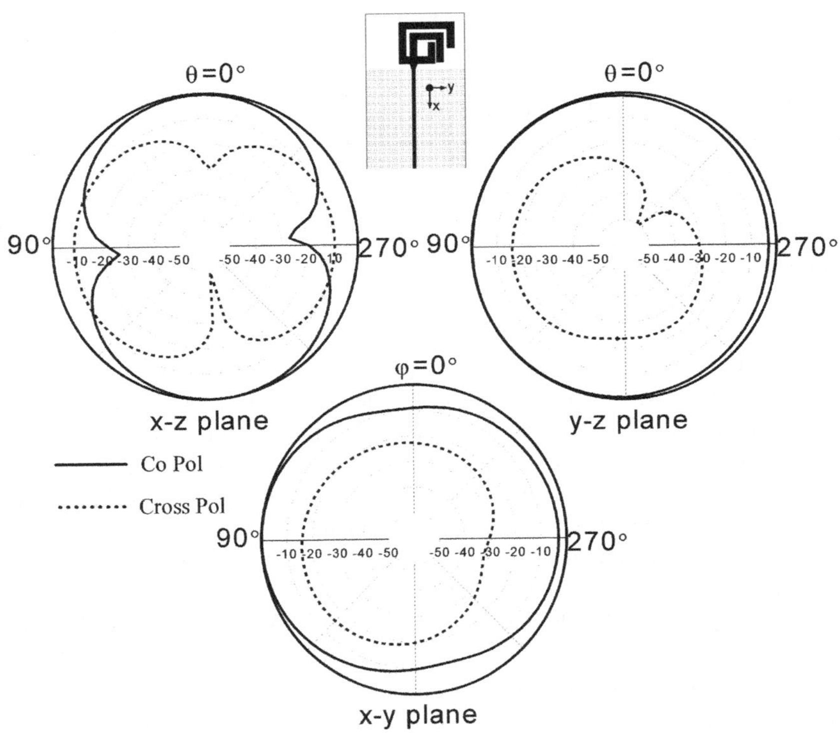

Figure 9.13 The radiation pattern of the triple-arm antenna at 1.8 GHz. (From Reference [10]. © 2006 IEEE.)

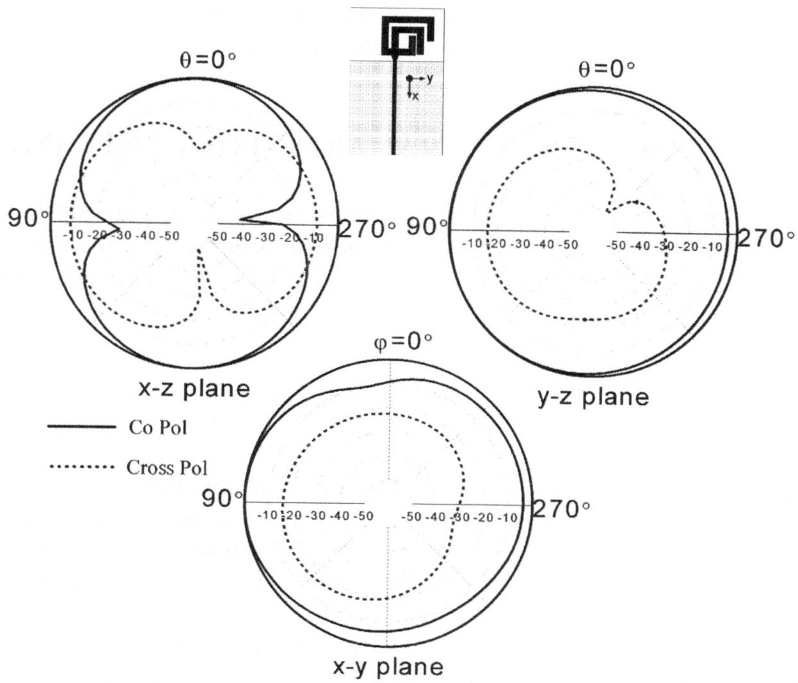

Figure 9.14 The radiation pattern of the triple-arm antenna at 2.45 GHz. (From Reference [10]. © 2006 IEEE.)

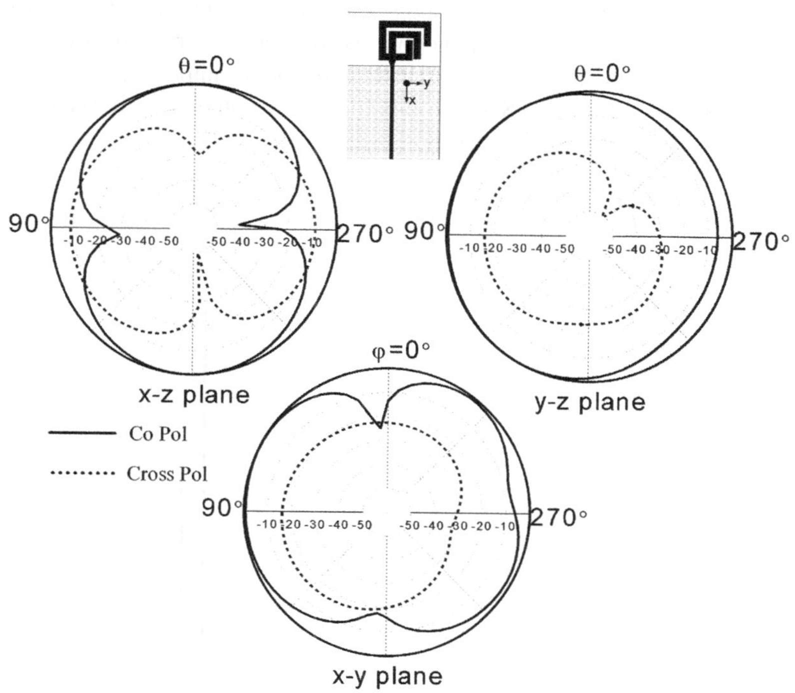

Figure 9.15 The radiation pattern of the triple-arm antenna at 5.78 GHz. (From Reference [10]. © 2006 IEEE.)

with the frequency increasing, the radiation along the direction of the x axis will reduce. The patterns in the yz and xy planes exhibit omnidirectional characteristics.

9.5 MULTIBAND PRINTED SPIRAL MONOPOLE ANTENNAS

An alternative method to achieve multiband operation is by binding a monopole strip into a spiral shape. It is well known that a straight monopole can have several resonant modes. The potential resonant frequencies of a printed monopole strip can be estimated by the following formula:

$$f_n = a\frac{(2n-1)c}{2\sqrt{\varepsilon_r}L}, \quad n = 1, 2, 3, \ldots \tag{9.1}$$

where c is the speed of light, L is the length of the monopole strip, ε_r is the dielectric constant of the substrate, and a is a correction factor approximately equal to 0.88. However, good impedance matching will only be achieved at the first resonant frequency. When the antenna shape is changed into spiral shapes, shown in Figure 9.16, the resonant frequencies will not always follow the formula (9.1). Because of the existence of the mutual coupling along the slot, the resonant frequencies will normally shift to lower. Some of them are well matched with their impedance, and some are mismatched or even disappear. Spiral antennas were previously implemented in References [12] and [13]. These antennas operate at 0.9/2.4 bands and 2.4/5.5 bands, respectively, and their antenna elements are shorted at their feed points through a matching bridge.

In this section, two printed spiral monopole antennas without any additional matching branch are presented. The printed monopole strips have spiral shapes so that the first two resonant modes are well matched with their impedance and cover the GSM900/DCS1800/PCS1900/ UMTS2000 bands for mobile communications and 2.4/4.9/5.2 GHz bands for WLAN systems, respectively.

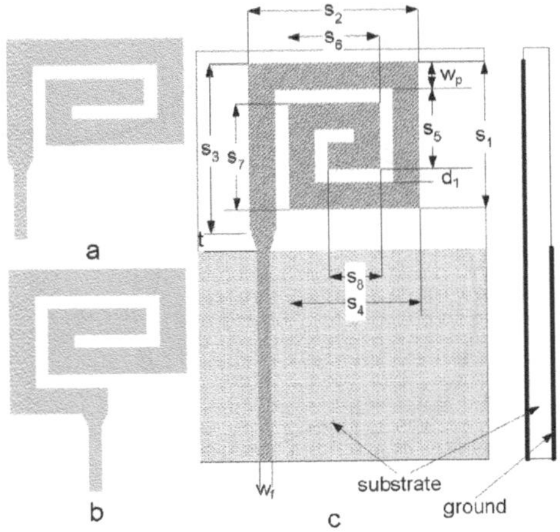

Figure 9.16 Configurations of the planar spiral monopole antennas. (From Reference [14]. © 2006 IEEE.)

9.5.1 A Spiral Antenna for GSM900/DCS1800/PCS1900/UMTS2000 Operations

Figure 9.16 shows the structure of the proposed planar monopole antenna, which is composed of an antenna element, a microstrip feed line, a substrate, and a ground-plane. The antenna element is a metal strip that is bent like a rectangularly spiral shape and etched on an FR4/Epoxy substrate with a thickness of 0.8 mm and a relative permittivity of 4.4. A 50 Ω microstrip feedline is printed on the same plane of the substrate. The ground is on the other side of the substrate and only below the microstrip line, as can be seen in Figure 9.16. It is well known that the multiarmed spiral antennas can offer a broad bandwidth and are normally excited at the inside terminal. In this antenna, the antenna element is similar to a single-armed rectangularly spiral antenna with a 50 Ω microstrip feed line at its outside terminal. This structure can excite several modes and works at multifrequency bands.

In this design, the ground-plane was selected to be $40 \times 39\,\text{mm}^2$. The antenna element is located in the extension of the substrate above the ground-plane. There is a transition part between the feedline and the antenna element. As mentioned above, these antenna structures can work at multifrequency bands. Here an effort will be made to design them to work at the 900 MHz band and 1.7–2.1 GHz frequency range, which are related to GSM900, DCS1800, PCS1900, and UMTS2000 applications. The simulations were performed by using the commercial software HFSS.

Figure 9.17 shows the simulated results for the three antenna structures shown in Figure 9.16. From these simulations, it was found that the lengths of the inside edges of these structures are all within the range of 73–74 mm, which is slightly less than the quarter free-space wavelength and half a guide wavelength of the operating frequency at 900 MHz. The first resonant band is between 0.9 and 1.0 GHz. The second band moves from 1.6 to 2.5 GHz. From the result of the structure in Figure 9.16(a), where the width (w_p) of the element is 3 mm and the slot d_1 is 0.5 mm, it can be seen that the two resonant bands

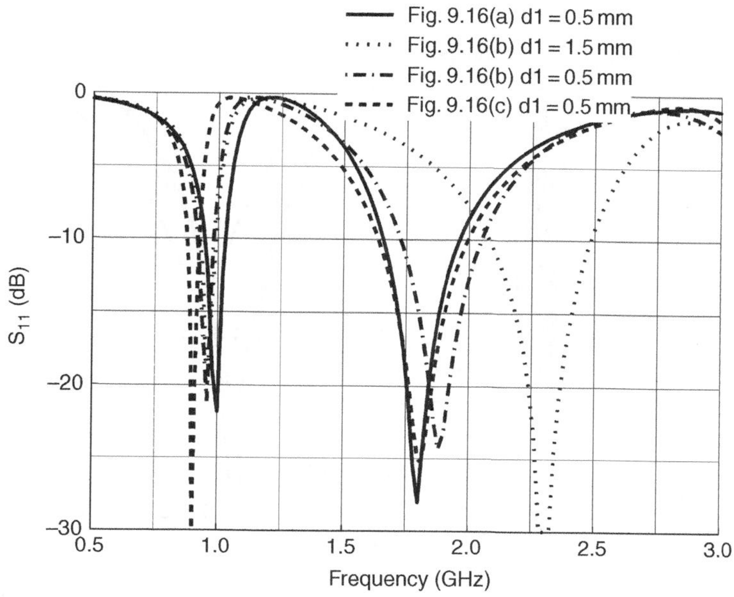

Figure 9.17 Simulated return loss for the antenna structures in Figure 9.16. (From Reference [14]. © 2006 IEEE.)

are closest. In the two results of the structure in Figure 9.16(b), where the width (w_p) is 3 mm and the slot d_1 is 0.5 and 1.5 mm, respectively, the second resonant bands move to the higher resonant band, compared with the result in Figure 9.16(a); that is to say, the distance between the two resonant frequencies in the structure in Figure 9.16(b) is more than that in Figure 9.16(a), and the larger the slot d_1, the larger is the distance between the two resonant bands. The result of the antenna in Figure 9.16(c) is between the above two results, and the second band, determined by −10 dB, covers the frequency range from 1.7 to 2.0 GHz. In summary, the location of the two resonant bands of such a kind of antenna structure can be controlled by proper selection of the strip length, the structure among the three in Figure 9.16, and the dimension of the slot.

The substrate thickness (h) is 0.8 mm in previous simulations. Since the value of h is not always 0.8 mm in practice, it is necessary to study the effects of the substrate thickness on the impedance characteristics of the proposed antenna. The values of the basic antenna parameters can be found in Table 9.3. The substrate thickness (h) will vary from 0.4 to 2.0 mm. The substrate and the ground-plane have the area of $46 \times 25.5\,\text{mm}^2$ and $20 \times 25.5\,\text{mm}^2$, respectively. The width of the feed line (W_f) corresponding to the substrate thicknesss 0.4, 0.8, 1.2, 1.6, and 2.0 mm are 0.75, 1.5, 2.3, 3.0, and 3.8 mm, respectively. The effects of the variation of h are shown in Figure 9.18. It is observed that the bandwidths and the impedance matching of the two resonant frequencies are not changed much with the variation of h. Only a minor shift of both resonant frequencies can be seen when changing the value of h.

Table 9.3 Parameters of the prototype monopole antenna (in mm).

S_1	S_2	S_3	S_4	S_5	S_6	S_7	S_8	W_f	w_p	d_1
18.0	22.75	24.0	18.25	9.0	13.75	13.5	9.25	1.5	4.0	0.5

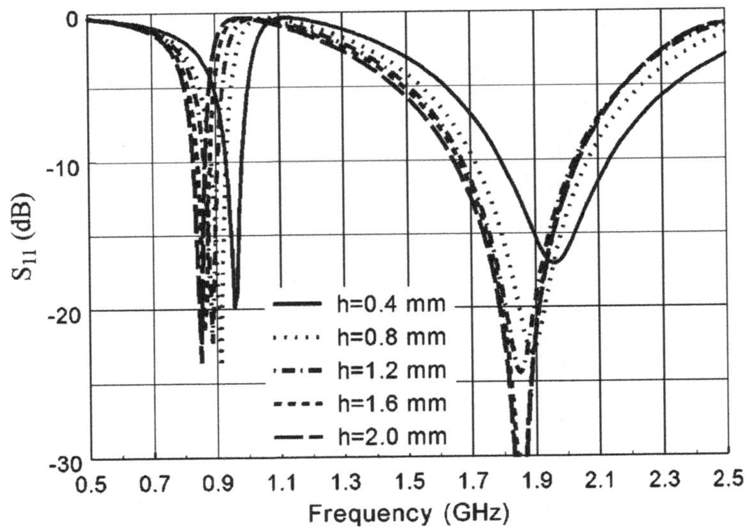

Figure 9.18 Simulated return losses with variations of the substrate thickness (h) for the antenna in Figure 9.16(c). (From Reference [14]. © 2006 IEEE.)

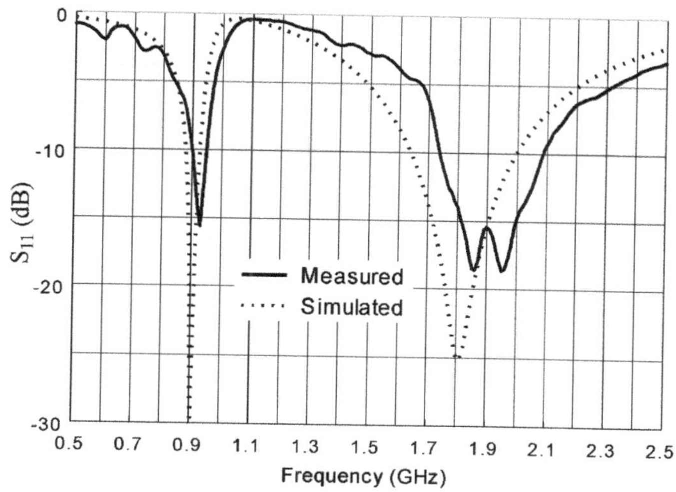

Figure 9.19 Measured and simulated return losses for the spiral monopole shown in Figure 9.16.

A prototype of the proposed antenna, shown in Figure 9.16(c), was constructed. The dimensions of the antenna are listed in Table 9.3. The substrate is FR4/Epoxy with a thickness of 0.8 mm, a dielectric constant of 4.4, and an area of 66×25.5 mm^2. The ground-plane has an area of 40×25.5 mm^2. Figure 9.19 shows the measured and simulated return loss (S_{11}). As shown in Figure 9.19, the measured data in general agree with the simulated results. The measured lower bandwidth, determined by 1:2.5 VSWR, reaches about 90 MHz and covers the GSM900 band (890–960 MHz). On the other hand, the upper band has a bandwidth as large as 460 MHz (1.71–2.17 GHz) and covers the DCS1800 (1710–1880 MHz), PCS1900 (1850–1990 MHz), and UMTS2000 (1920–2170 MHz) bands.

Figures 9.20 and 9.21 plot the simulated radiation patterns in the *xz* plane (elevation direction), *yz* plane (azimuthal direction), and *xy* plane (substrate surface) for the prototype

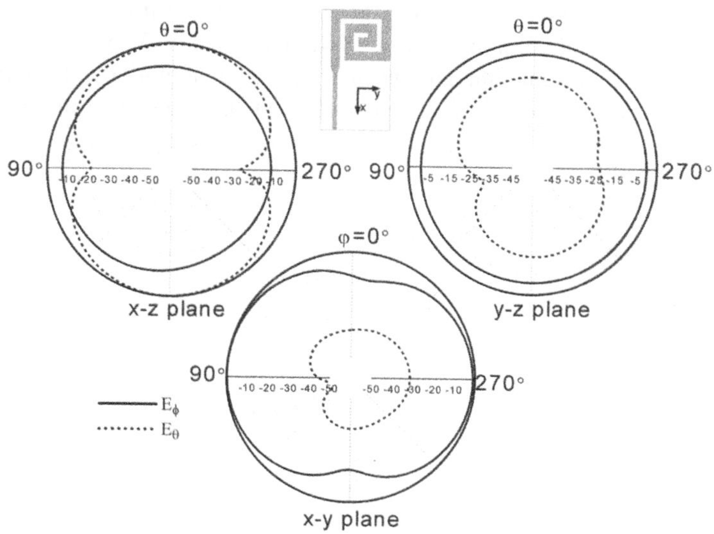

Figure 9.20 Computed radiation patterns of the spiral antenna at 0.9 GHz. (From Reference [14]. © 2006 IEEE.)

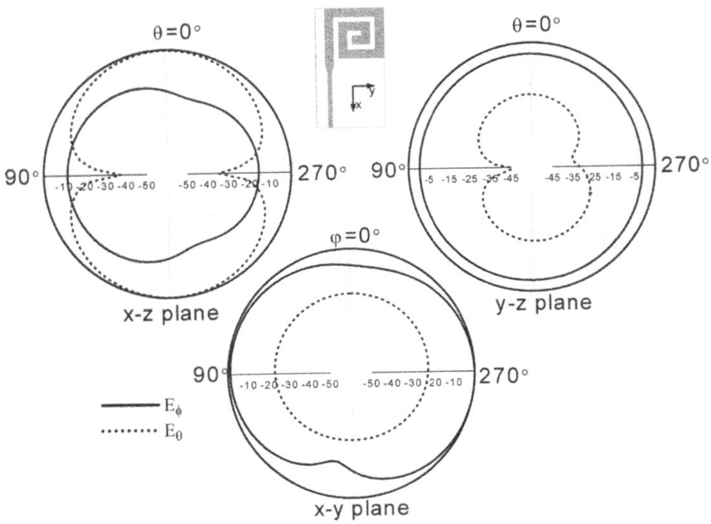

Figure 9.21 Computed radiation patterns of the spiral antenna at 1.85 GHz. (From Reference [14]. © 2006 IEEE.)

antenna at 900 and 1850 MHz. It can be seen that the antenna has similar far-field patterns at the two frequencies and good omnidirectional radiation patterns at the two operating frequencies in the *yz* plane.

9.5.2 A Spiral Antenna for WLAN Operations

As the second example, a spiral antenna working at 2.4/4.9/5.2 GHz was designed and presented here. Figure 9.22 shows the geometry of the spiral antenna, which is fed by a transmission line and printed on top of a grounded FR4 substrate. The substrate has a thickness of 0.8 mm, a relative permittivity of 4.4, and a size of 51.5×14.0 mm^2 ($l \times w$). On the other side of the substrate, the ground-plane has an area of 40.0×14.0 mm^2 ($l_1 \times w$) and does not cover the portion of the antenna element located. Since the antenna works

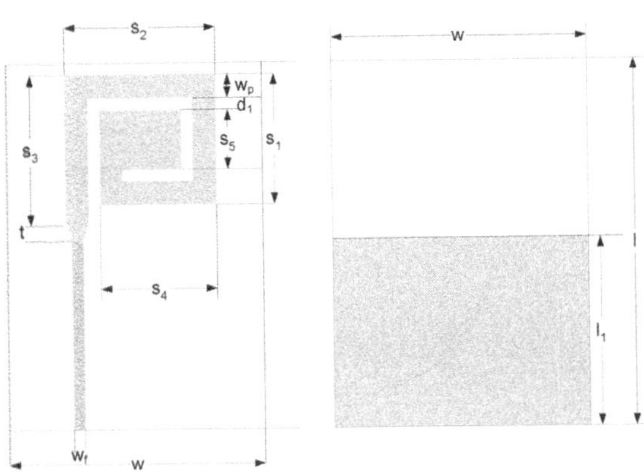

Figure 9.22 Geometry of the spiral antenna for the WLAN.

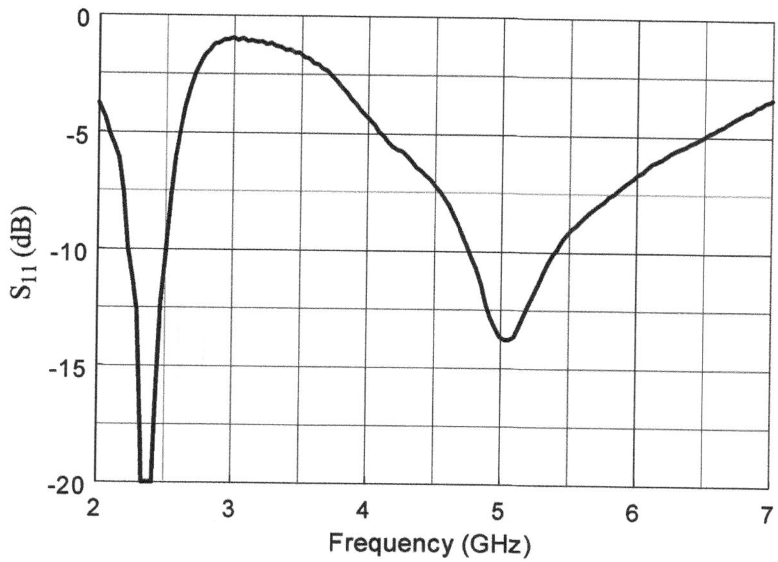

Figure 9.23 Measured return loss for the spiral monopole shown in Figure 9.22.

at 2.4 and 5.2 GHz, the length of the inner edge of the strip should be about one-quarter wavelength at 2.4 GHz. The strip starts from the feed point at the opposite side of the top edge of the ground-plane. In order to obtain the second operational band at 4.9 and 5.2 GHz, the strip is fine spiralled. The width of the end section of the strip can be increased to narrow the width of the slot along the spiral strip. This can fine-tune the locations of the two resonant frequencies. The dimension parameters of the antenna element shown in Figure 9.22 are: $S_1 = 6.5$ mm, $S_2 = 9.0$ mm, $S_3 = 11.5$ mm, $S_4 = 7.0$ mm, $S_5 = 1.5$ mm, $w_p = w_f = 1.5$ mm, $d_1 = 1.0$ mm, and $t = 0.0$ mm. A prototype based on the design was fabricated and tested. Figure 9.23 shows the measured return loss. The first band, determined by -10 dB return loss, has a bandwidth of 295 MHz (2.225–2.52 GHz), which covers the required bandwidth of WLAN standards IEEE 802.11b and 802.11g. The bandwidth of the second band is 640 MHz (4.74–5.38 GHz) and covers the bandwidth of WLAN standards IEEE 802.11j (4.9–5.1 GHz) and IEEE 802.11a (5.15–5.35 GHz).

9.6 A SURFACE-MOUNT MONOPOLE ANTENNA

In this section, a novel design of a triple-band small internal antenna is presented. This antenna covers GSM900, DCS1800, and PCS1900 mobile communication bands. The proposed antenna is made out of a thin copper strip, which is folded appropriately to achieve multiband operation while reducing the size. The antenna is mounted on a dielectric substrate and fed by a microstrip line.

The geometry of the surface-mount monopole antenna is shown in Figure 9.24(a). A 50 Ω microstrip feed line is etched on an FR4/Epoxy substrate, which has a thickness of 0.8 mm and a dielectric constant of 4.4. The truncated ground-plane (40×39 mm^2), which is on the back surface of the substrate, is limited to the feed line section. The antenna element is mounted on the same side as the feed line, in the top portion of the substrate where there is no ground on the opposite side.

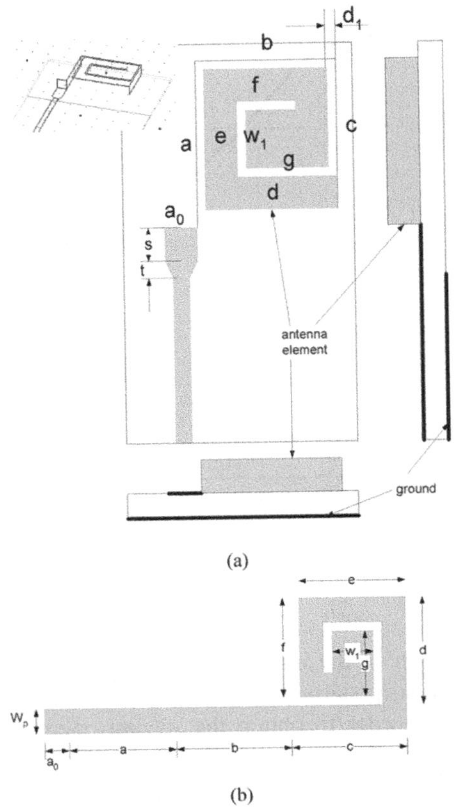

Figure 9.24 Configuration of the surface-mount monopole antenna. (From Reference [6]. © 2005 IEEE.)

Between the microstrip feed line and the antenna element there is a connection strip with a trapezoidal shape and a length of t (see Figure 9.24(a)). A small portion of the antenna is printed on the substrate, which has a length of s and a width of w_p. Most of the antenna element is composed of a folded metallic strip. It is noted that this folded element is easily constructed by folding a metallic strip, which has the dimensions and shape shown in Figure 9.24(b), on to a rectangular foam base with dimensions of about $b \times c \times w_p$. The foam base has a dielectric constant of 1.007 and is used to support the folded metallic strip. The folded portion a_0 can be soldered at the end of the portion s, as shown in Figure 9.24(a).

The antenna in Figure 9.24(a) supports two resonant modes, giving good impedance matching in two bands. It is noted that a small portion of the antenna element is printed on the substrate and most of the element is mounted on it. This makes the total length of the antenna element to be between a quarter and half free-space wavelengths in the lower band. By folding the strip into the shape shown in Figure 9.24(a), the area of the antenna has been reduced. The cost is the increased antenna height, which is the same as the width of the strip (w_p).

An example antenna was designed using Ansoft HFSS and its dimensions are given in Table 9.4. Figure 9.25 shows the measured return loss. The lower band, determined by −7 dB return loss, covers the frequency range from 0.88 to 0.975 GHz. The higher bandwidth

Table 9.4 Parameters of the surface-mount monopole antenna design (in mm).

w_p	W_1	t	s	a_0	A	b	C	d	e	f	G
3.0	6.5	1.0	5.0	3.0	14.5	19.0	14.0	10.0	13.5	17.5	14.0

(From Reference [6]. © IEEE 2005.)

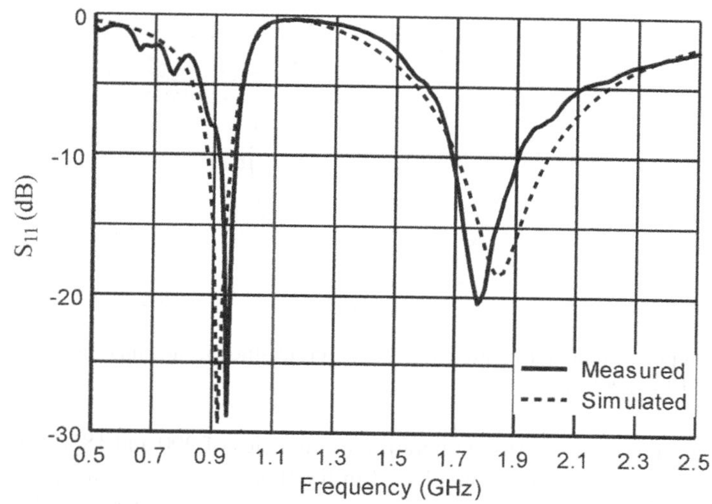

Figure 9.25 Measured return loss of the surface-mount antenna.

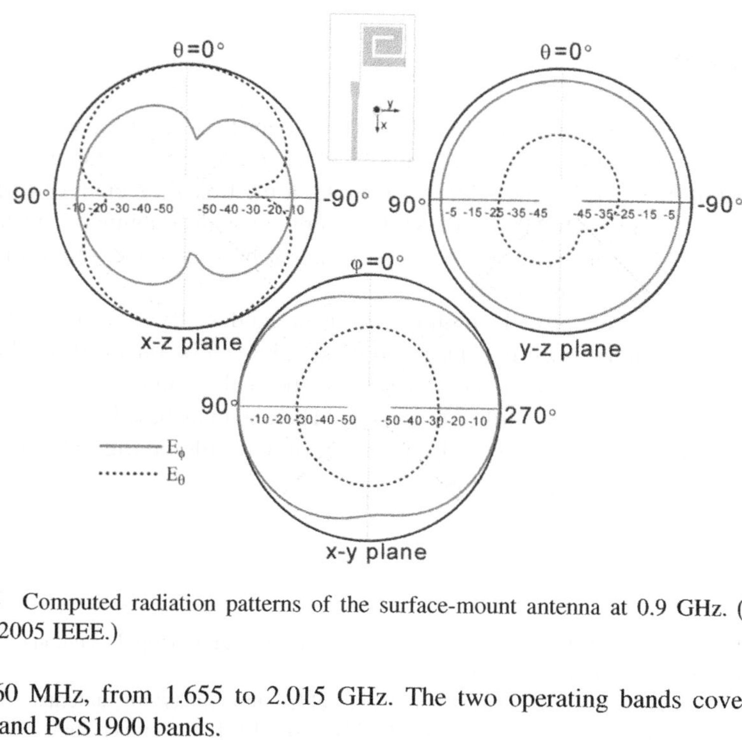

Figure 9.26 Computed radiation patterns of the surface-mount antenna at 0.9 GHz. (From Reference [6]. © 2005 IEEE.)

is about 360 MHz, from 1.655 to 2.015 GHz. The two operating bands cover GSM900, DCS1800, and PCS1900 bands.

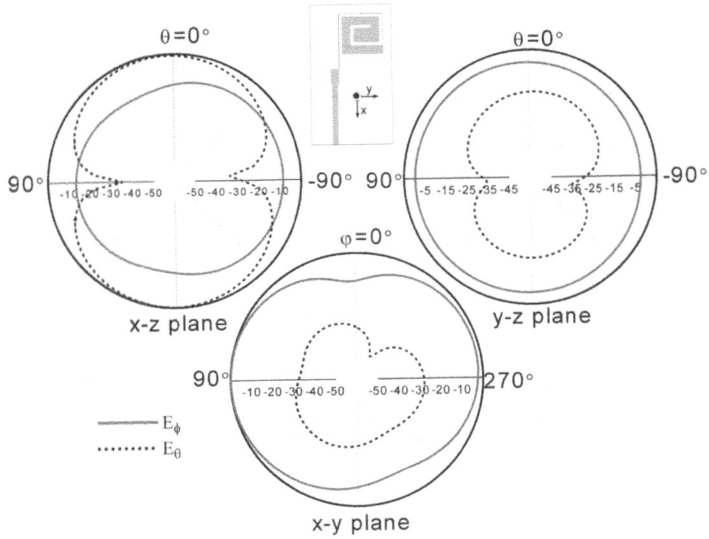

Figure 9.27 Computed radiation patterns of the surface-mount antenna at 1.85 GHz. (From Reference [6]. © IEEE 2005.)

The simulated radiation patterns at operating frequencies of 900 and 1850 MHz are plotted in Figures 9.26 and 9.27. The patterns plotted in the three planes, *xy* plane, *xz* plane, and *yz* plane, are very similar in the two operating frequencies. Good omnidirectional characteristics are noted at both frequencies.

9.7 SUMMARY

The multiple-branch strips have been found to offer the antenna engineer with a significant amount of design flexibility to achieve multiband operations. By binding two or three strips and printing them on an FR4/Epoxy substrate, a multiband antenna can be obtained for WLAN systems and DCS1800 or PCS1900 operation with a compact and low-profile configuration. The resonant frequencies can be changed by tuning the length of the strips and the mutual coupling among them.

A spiral monopole can provide multiple-resonant modes. By properly spiralling the monopole strip, good impedance matching and bandwidth can be obtained for the first two modes. Considering the effects of the slots arising from the spiral strip, the location of the resonant frequencies can be further fine-tuned. The antennas based on these techniques have been successfully implemented with compact, low-profile configurations for mobile communications and WLAN applications.

REFERENCES

1. E. Lee, P. S. Hall and P. Gardner, ‘Compact wideband planar monopole antenna’, *Electronics Letters*, **35**, 2157–2158, December 1999.
2. E. Lee, P. S. Hall and P. Gardner, ‘Dual band folded monopole/loop antenna for terrestrial communication system’, *Electronics Letters*, **36**(24), 1990–1991, November 2000.

3. V. Stoiljkovic, S. Suganthan and M. Benhaddou, 'A novel dual-band centre-fed printed dipole antenna', in *IEEE Antennas and Propagation Society (AP-S) International Symposium*, Ohio, Vol. 2, pp. 938–941, 22–27, June 2003.
4. K. Wong, G. Lee and T. Chiou, 'A low-profile planar monopole antenna for multiband operation of mobile handsets', *IEEE Transactions on Antennas Propagation*, **51**(1), 121–125, January 2003.
5. Y. Ge, K. P. Esselle and T. S. Bird, 'Small quad-band WLAN antenna', in *IEEE Antennas and Propagation Society (AP-S) International Symposium*, Washington, DC, 3–8 July 2005.
6. Y. Ge and K. P. Esselle, 'A small quad-band CPW-fed antenna for WLAN applications', in *11th International Symposium on Antenna Technology and Applied Electromagnetics* (*Antem 2005*), Saint Malo, France, 15–17 June 2005.
7. K. P. Esselle and Y. Ge, 'Multi-band, broadband, fully-planar antennas', International Patent, Ref. 686312C, pending on 23 December 2004.
8. Y. Ge, K. P. Esselle and T. S. Bird, 'A dual-band monopole antenna for mobile communications', in *2005 IEEE/ACES International Conference on Wireless Communications and Applied Computational Electromagnetics*, Honolulu, Hawaii, 3–7 April 2005.
9. E. A. Soliman, S. Brebels, G. A. E. Vandenbosch and E. Beyne, 'Antenna arrays in MCM-D technology fed by coplanar CPW networks', *IEEE Transactions on Microwave Theory and Techniques*, **48**, 1065–1068, June 2000.
10. Y. Ge, K. P. Esselle and T. S. Bird, 'Compact triple-arm multi-band monopole antenna', IEEE International Workshop on Antenna Technology: Smart Antennas and Novel Metamaterials, White Plains, New York, USA, pp. 17–175, 6–8 March 2006.
11. Y. Ge and K. P. Esselle, 'A class of multi-arm monopole antennas for multi-band wireless/mobile applications', *Workshop on Applications of Radio Science 2006 (WARS'06)*, Blue Mountain, Sydney, Australia, 15–17 February 2006.
12. Z. Ying, 'Miniature printed spiral antenna for mobile terminals', World Intellectual Property Organization, International Publication WO0003453A1, 20 January 2000.
13. S. Chung, M. Lee and J. Hsiao, 'Dual frequency band inverted-F antenna', US Patent 20040189530A1, 30 September 2004.
14. Y. Ge, K. P. Esselle and T. S. Bird, 'A compact, low-profile spiral antenna for mobile communications', *IEEE Antennas and Propagation Society (AP-S) International Symposium*, Albuquerque, New Mexico, USA, 9–14 July 2006.

10

Printed Antennas in Packages

Jennifer T. Bernhard
Electromagnetics Laboratory, ECE Department, University of Illinois at Urbana-Champaign, Urbana, Illinois, USA

10.1 INTRODUCTION

The attributes of printed antennas – low profile, light weight, inexpensive to manufacture – make them perfect candidates for integration into wireless communication devices. Not only are they more aesthetically pleasing than simple linear antennas but they are also more rugged and may be able to direct more energy away from the user.

Embedding an antenna into portable wireless packages, however, poses some interesting problems. Most of these problems arise from the fact that the device of interest, be it a cellular telephone, laptop computer, or other hand-held portable device, is usually comparable in size to an operating wavelength. Since the antenna and the rest of the device both act as radiating elements of the system, good antenna designs must incorporate not only fundamental operating characteristics but also specific internal environments and potential operating scenarios unique to a particular device.

These internal environments and operating scenarios often have more impact on final antenna designs than fundamental electromagnetic performance. For instance, the location of antennas in portable laptop computers not only depends on the antenna's performance but also on the cost and reliability of the coaxial cable that connects the antenna to the radio. When placed in the laptop's screen, for instance, this cable is lossy, expensive, and may be subjected to repeated bending at the hinge between the screen and the laptop base, making it a potential point of failure.

This chapter explores the effects of packaging on printed antennas as well as techniques for developing antenna–package combinations that deliver desired performance. Section 10.2 describes the fundamental changes in printed antenna performance in both frequency and radiation characteristics as a result of nonideal operating scenarios created by practical packaging strategies. Section 10.3 presents several examples of antenna–package interaction studies in this context. In Section 10.4, a number of design methods and approaches are

Printed Antennas for Wireless Communications Edited by R. Waterhouse

highlighted that provide guidance for those wishing to integrate antennas into wireless devices. Finally, Section 10.5 briefly discusses the directions for future work in this area to meet the challenges presented by multipurpose/multifrequency wireless devices.

10.2 EFFECTS OF THE EMBEDDED ENVIRONMENT ON PRINTED ANTENNAS

Three main aspects of the portable wireless device environment can impact antenna performance: (1) covers or other dielectric packaging, (2) finite and asymmetric ground-planes, and (3) the presence of the user. Each one of these factors is discussed in detail with respect to both frequency and radiation characteristics.

10.2.1 Dielectric Covers

Antenna integration with a package usually means that the antenna is placed beneath or beside a plastic cover that forms the chassis of the device. Depending on the design and the materials used, however, the effects of this cover on antenna performance can vary dramatically.

In the case of frequency characteristics, a number of investigators have developed design methodologies to anticipate and account for operating frequency shifts that occur with the addition of a dielectric cover to a printed antenna. In Reference [1], conformal mapping and the concept of equivalent capacitance are used to determine the effective permittivity of a covered rectangular microstrip antenna, which is then in turn used to predict the antenna's operating frequency. Generally, the model is shown to provide results that are as accurate as possible within the tolerance range of the structure's electrical and physical parameters [1]. Guha and Siddiqui [2] performed a similar study using circular microstrip antennas.

According to Reference [3], a dielectric superstrate flush with a simple printed patch antenna can also dramatically change the patch's radiation pattern. In particular, substrates up to about 0.14λ thick will change the frequency characteristics of the patch and may reduce the gain of the antenna. As the thickness increases above this value, the radiation pattern is divided into two and then three nearly identical lobes that broaden the overall pattern but result in a reduction in the absolute gain of the antenna [3]. A special case of this effect involves the use of superstrates with high permittivities [4–6] or multiple superstrate layers [7] that can be used to deliver high directivity and sometimes larger frequency bandwidth. While the insertion of high-permittivity materials may also allow for antenna miniaturization, the superstrate–substrate combination as well as superstrate spacing must be coordinated to achieve designs that deliver comparable directivities and efficiencies to their counterparts made with low-permittivity substrates alone [5, 7]. It is reasonable to assume that dielectric covers will have similar effects on other types of printed antennas (such as planar inverted-F antennas, etc.). At this time, inclusion of high-permittivity superstrate materials may not be economical for portable wireless devices, but new metamaterials, which are discussed in Section 10.5, hold promise for future designs.

One strategy adopted by some portable device manufacturers to reduce emissions from devices (perhaps in the direction of a user [8]) while still keeping them light weight is the use of conductive plastic as an outer chassis material. During fabrication, the liquid plastic is mixed with carbon and other materials, creating an electromagnetically lossy superstrate

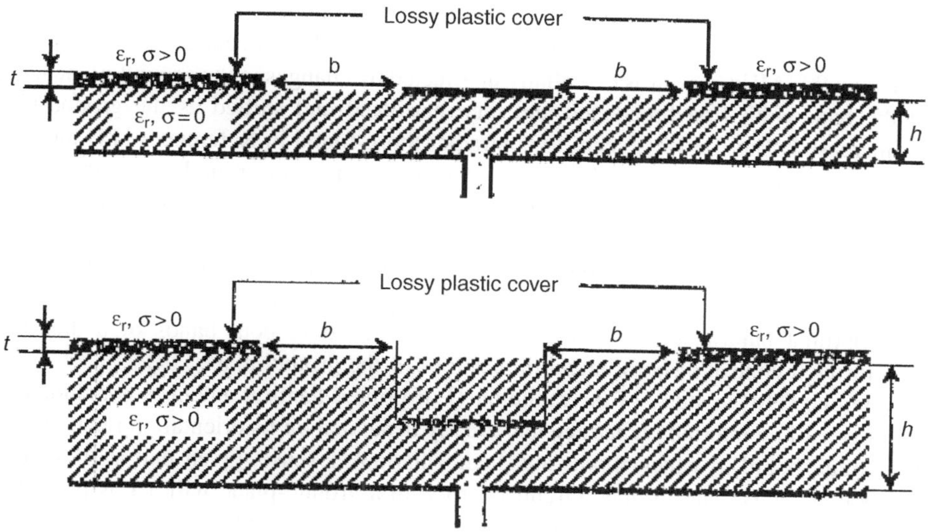

Figure 10.1 Microstrip antenna in the vicinity of lossy dielectric packaging materials: exposed to open air (top) and embedded in a lossless dielectric material (bottom) (from Reference [9]. © 2000 IEEE. Courtesy of the IEEE).

for any internally embedded printed antenna. Without proper consideration during the design phase, placement of an antenna beneath this kind of cover has the potential to render the antenna practically useless. In Reference [9], two possible internal antenna configurations that include a lossy plastic chassis were investigated. The design geometries are shown in Figure 10.1 (from Reference [9], Figure 1). In each of these designs, the lossy plastic cover has an opening to allow the internal antenna to radiate properly.

In the design at the top of Figure 10.1, the microstrip antenna is exposed to the open air, while the design at the bottom of Figure 10.1 is embedded in a lossless dielectric material [9]. Results of this study indicate that the lossy dielectric cover, even at a distance from the microstrip antenna, still affects the antenna gain, bandwidth, and impedance match. As the thickness or the conductivity of the lossy material decreases, the optimal feed point of the antenna shifts towards the edge of the element and a broader bandwidth is observed. Also, as the spacing between the antenna and the lossy material decreases, the gain of the antenna goes down dramatically while the bandwidth increases, since the lossy material will act as a resistive load for the antenna. One interesting exception to this effect occurs when the lateral spacing between the antenna and the cover is small ($\sim$ 1 mm) and the conductivity of the lossy plastic is high. In this case, the energy coupled to the chassis from the antenna may be reradiated to contribute to the antenna pattern [9].

10.2.2 Ground-Planes

Integrated printed antennas require metallic ground-planes in order to operate properly. Since theoretical designs assume infinite ground-planes, the small ground-planes found in most portable devices create deviations from ideal behaviour.

In Reference [10], a study of the effects of ground-plane size on an internal multiband PIFA antenna was conducted. The length of the ground-plane was varied from 60 mm

($\sim 0.18\lambda$ at 900 MHz and $\sim 0.375\lambda$ at 1.9 GHz) to 100 mm ($\sim 0.30\lambda$ at 900 MHz and $\sim 0.625\lambda$ at 1.9 GHz). For the antenna, which operated in bands centred at 900 MHz and 1.9 GHz, the 900 MHz band was essentially unaffected by changes in the ground-plane length while the bandwidth first broadened and then narrowed and shifted as the ground-plane length increased. These effects seem to contradict those found in References [11] and [12], where performance of a dual-band printed antenna on a handset ground-plane operating at 900 MHz was optimized through selection of the ground-plane dimensions, while operation at 1800 MHz seemed largely determined by the antenna design alone. However, since the two antenna designs and their placements on the ground-planes were not identical, there are obviously more variables involved in the overall design. Another effort in this area indicates that the optimal bandwidth performance can be achieved in both bands of a dual-band antenna through control of both lateral dimensions of the ground-plane [13].

In Reference [14], a study of the radiation Q of printed handset antennas concluded that the efficiency of an internal planar antenna greatly depended on its orientation relative to the printed circuit board to which it was attached. This makes sense, given that at frequencies above about 1.5 GHz, any printed circuit board with a ground-plane will act as part of the radiating system. Plots of simulated power density indicate that antenna configurations that best exploit the ground-plane as a radiator have higher efficiencies and bandwidths [14].

While it may be desirable in some situations to use the ground-plane as an intentional radiator, in other cases it may be useful to make the ground-plane look electrically larger, thereby making it less important to the radiating system. One way to do this is to serrate the edges of the ground-plane [15]. This technique may allow designers to achieve acceptable microstrip antenna behaviour in small packages despite a nonideal environment.

Asymmetric ground-planes presented by larger devices, such as laptop computers, can also have a significant effect on the radiation pattern of packaged antennas. As a simple example, the patterns from a monopole antenna positioned in the same plane as the laptop keyboard are directed by the asymmetric ground-plane presented by the shielding beneath the keyboard towards the user [16]. For antennas that are integrated into laptop packages, edge diffraction and asymmetries in the internal shielding may also create telltale scallops in radiation patterns [17].

10.2.3 Presence of a User

Over the past 15 years, numerous researchers have investigated antenna–user interaction. Most of this work has concentrated on the calculation of the specific absorption rate (SAR) of energy in the user. Many of these studies have used simple monopole models for basic linear antennas on handsets. Additionally, a number of excellent simulation and measurement studies (e.g. References [18] and [19]) have determined that, for simple configurations of printed antennas on handsets (mainly patches or variants of planar inverted-F antennas), approximately 50 % of the energy radiated by an antenna is absorbed by the user. Additionally, the presence of a hand in proximity to a shorted microstrip patch antenna on a handset at 1800 MHz has been shown to reduce the antenna's efficiency by at least 10 % [19].

Certainly, this situation is not one that can be taken lightly and is likewise not easy to improve through antenna design. A number of researchers, however, have proposed strategies to address this very challenging problem. For instance, Hirose and Miyaki considered the use of parasitic elements to increase the gain of an embedded PIFA antenna while directing energy away from a handset user [20]. In another example, Lazzi *et al.* [21] used stacked

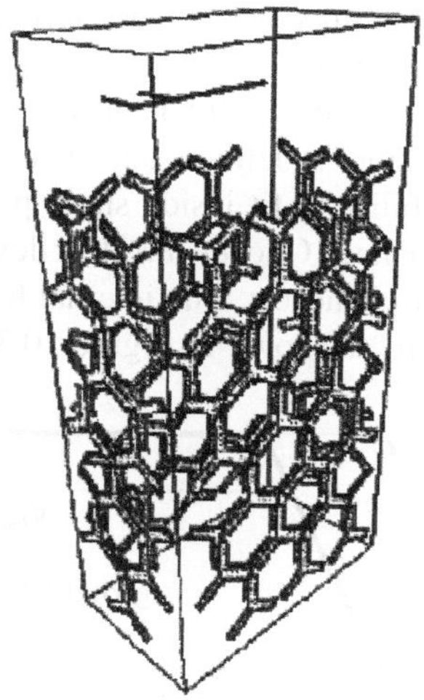

Figure 10.2 Flexible miniaturized metallodielectric electromagnetic bandgap material embedded in a handset case to mitigate antenna detuning due to a user's hand (from Reference [24]. © 2005 IEEE. Courtesy of the IEEE).

microstrip patches in handsets to increase antenna gain in a direction away from the intended user. Fuhl *et al.* [22, 23] have also developed integrated antennas with small shields to reduce user exposure to radiated fields.

A unique solution to coupling between internal handset antennas and the user's hand has been proposed in Reference [24]. A flexible miniaturized metallodielectric electromagnetic bandgap structure as depicted in Figure 10.2 [24] were embedded in the handset casing as a means to improve a PIFA's performance in the presence of the user's hand. The embedded structure was found to reduce the amount of frequency detuning caused by the user's hand and improved the measured radiating efficiency by 9% [24].

10.3 EXAMPLES OF OTHER WORK ON PRINTED ANTENNAS IN PACKAGES

Spurred by the emergence of commercial wireless markets, an extensive body of literature has developed over the past decade on small printed antenna designs suitable for portable communication devices. Much of this work, however, has concentrated solely on the antenna itself and may give little or no insight into how particular antenna designs would function in an actual package. This section contains a brief and certainly not exhaustive overview of some recent work in this area, which uniquely considers the packaging (and sometimes the user) along with the antenna that can provide insight into the unique electromagnetic environment posed by portable wireless devices.

Handset antenna design has received, by far, the most attention in the literature, and rightly so. Handsets pose a particularly difficult problem for antenna designers, since the internal space available for an antenna is limited. Additionally, the challenges to integrate an antenna into situations with such demanding operating requirements and often nonideal environments are substantial. Without prior consideration of accurate package and environmental characteristics during antenna development, system performance can certainly suffer. With a great deal of effort, industrial and academic designers have developed antennas for these packages that can deliver reasonable impedance characteristics, radiation patterns, and efficiencies over one or more operating bands.

One of the most popular planar antennas for handsets is the planar inverted-F antenna (PIFA) [25]. In its low-profile version, it delivers adequate bandwidth and radiation pattern performance while retaining its ability to fit inside small packages. PIFAs also lend themselves well to multiband performance. Guo *et al.* [10] developed two multiband designs based on PIFAs for integration into handsets. The internal handset circuitry in this case was simulated by a printed circuit board as shown in Figure 10.3 [10].

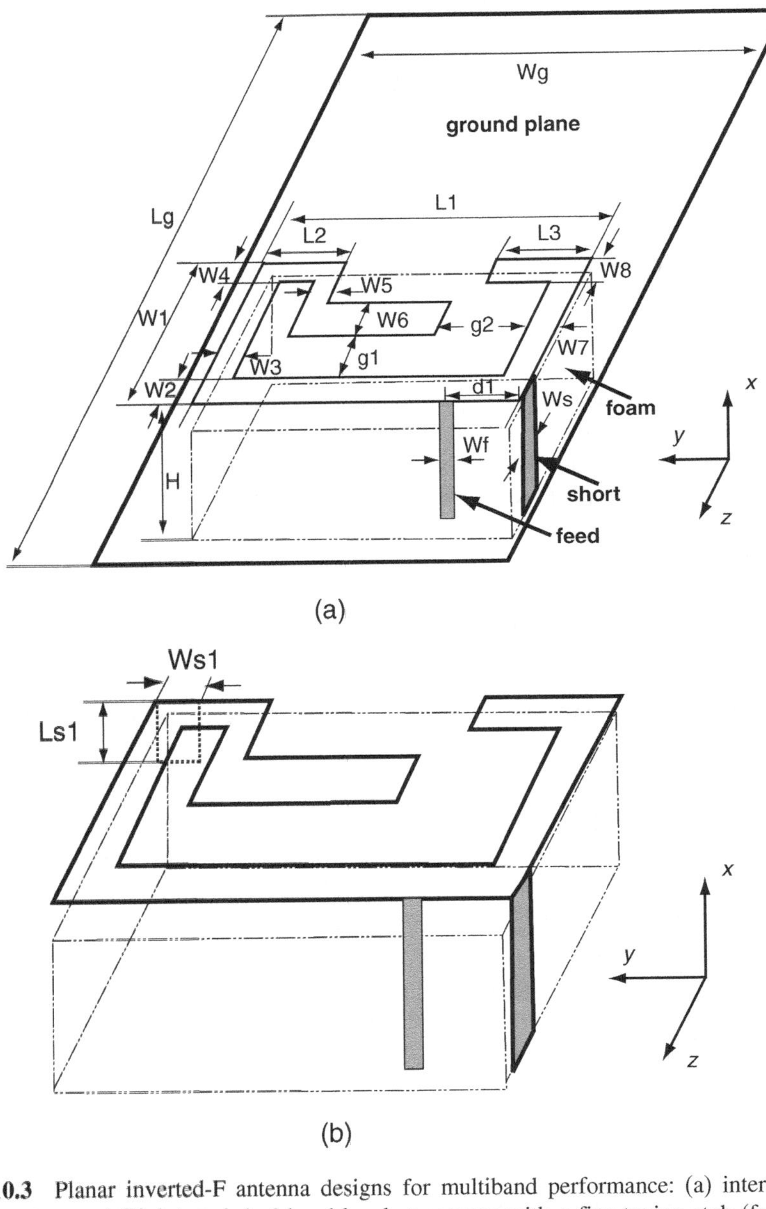

Figure 10.3 Planar inverted-F antenna designs for multiband performance: (a) internal dual-band handset antenna and (b) internal dual-band handset antenna with a fine-tuning stub (from Reference [10]. © 2004 IEEE. Courtesy of the IEEE).

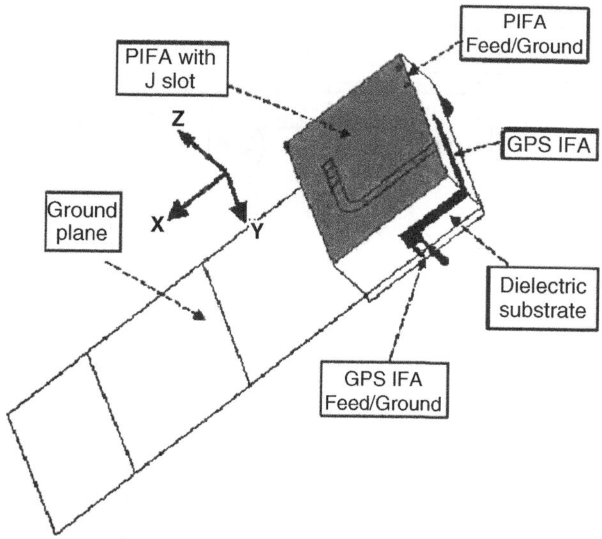

Figure 10.4 A multiantenna solution for handsets, including a planar inverted-F antenna and an inverted-F antenna suitable for operation in cellular telephone bands as well as the global positioning system (GPS) (from Reference [26]. © 2005 IEEE. Courtesy of the IEEE).

Related to the PIFA structure, two different shorted patch antenna designs have also been developed and evaluated for operation at 1800 MHz in a handset [19]. To provide insight into the effects of user proximity on antenna behaviour, simulations were performed with the antenna and handset alone and then head and hand models were added. It was found that the shorted patch antennas reduced the spatial peak SAR value in the head by 70 % relative to a monopole on the same package. This work highlights the advantages of planar antennas in operating scenarios where users are extremely close to the antenna: the additional directionality provided by printed antennas can serve to route energy more efficiently in directions where it can best contribute to the quality of the wireless communication link.

Another team has also developed a multiband antenna system based on PIFAs for a small handset, including an analysis of efficiency and power deposition in a user [26]. The competing needs of the antennas for the cellular telephone bands at 800 and 1900 MHz, as well as GPS (1575 MHz), were considered, including the selection of antenna feed points to reduce the coupling and increase the efficiency of the antennas. A general topology for this antenna system is shown in Figure 10.4 [26].

The other major platform for portable wireless communication is currently the laptop computer. While the package is much larger than that of a handset, there is still little available space in which to integrate an antenna. Moreover, the complex, large, and asymmetric ground-plane can adversely affect the antenna's radiation characteristics.

In Reference [27], a novel dual-band flat-plate antenna with a shorted parasitic element for inclusion into the screen of a laptop computer is presented. The antenna can be integrated into the supporting metal frame of the laptop's display and has a very low profile. The geometry and intended integration position for this antenna are shown in Figure 10.5 [27]. The antenna design operates over two frequency bands covering the 2.4 and 5.2–5.8 GHz wireless local area network bands. In Reference [28], a similar deployment scenario

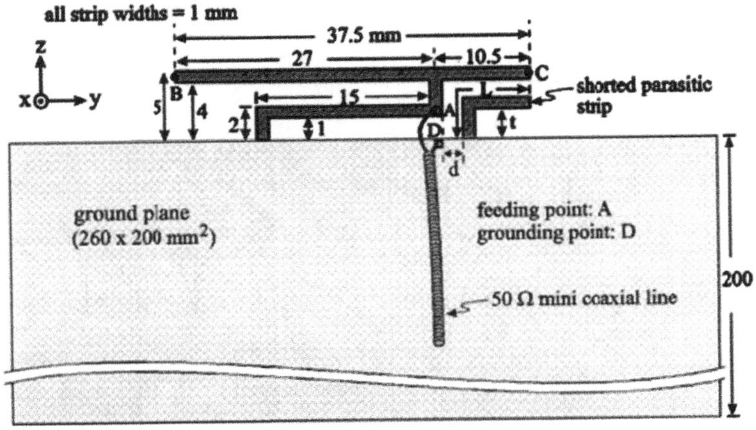

(a)

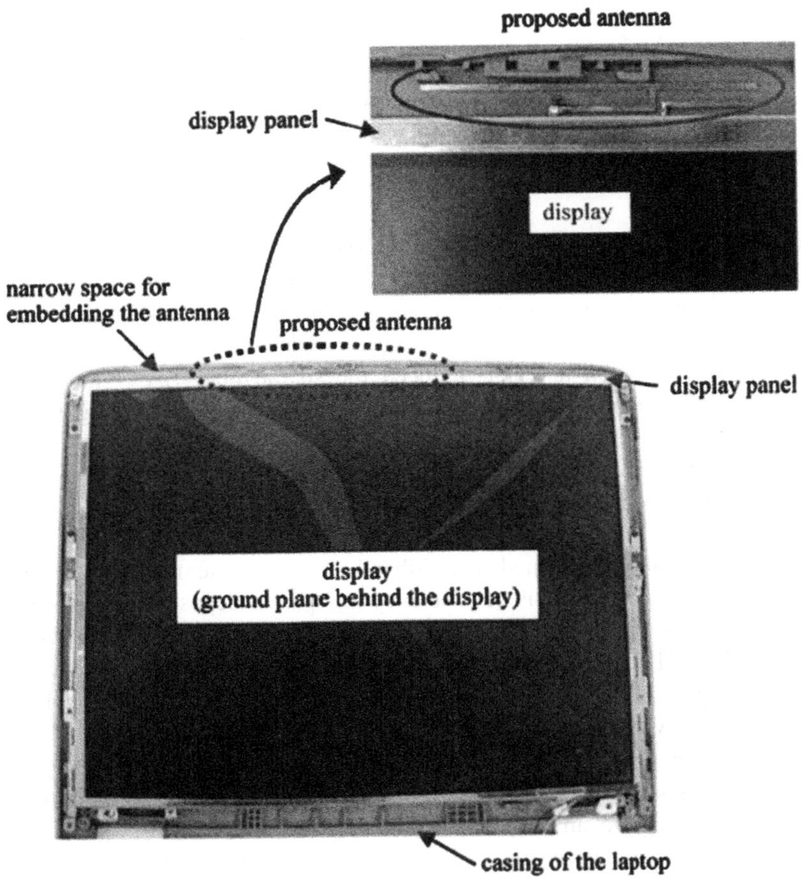

(b)

Figure 10.5 (a) Geometry of a dual-band flat-plate antenna with a shorted parasitic element for an internal laptop antenna for WLAN operation. (b) Picture showing the proposed antenna mounted on a laptop (the display portion only) for practical application (from Reference [27]. © 2005 IEEE. Courtesy of the IEEE).

is used for a planar inverted-F antenna with injected LCP material for structural support that covers the 2.4 and 4.9–5.9 GHz bands. In Reference [29], a dual-band PIFA is integrated into a very unique position in the centre back of the keyboard portion of a laptop. This position eliminates the need for a long coaxial cable connection to the radio. While simulated radiation patterns in both the 2.4 and 5.2 GHz bands show promise, the patterns at 5.2 GHz exhibit some significant lobe effects that are probably due to antenna – chassis interaction.

While most after-market wireless communication devices are designed for PCMCIA slots, an approach to the inclusion of an after-market antenna with an existing package is the vertically oriented antenna surface (VOAS) [30]. This is attached to a laptop and hangs freely from the back of the display to maintain a vertical position that can provide antenna operation independent of the display panel angle.

10.4 DESIGN METHODS FOR ANTENNAS AND ANTENNA PLACEMENT IN PACKAGES

Given the complexity and variability of the system, it is difficult to formulate any general design methodology that can be used to design antennas for particular packages. However, a number of researchers have developed some techniques for the inclusion of antennas into packages in situations where the antenna and the package are both comparable in size to an operating wavelength.

Liu *et al.* [31] outlines a design methodology for developing antennas intended for integration laptop computers. The first step initiates a simple antenna design that will fit into the prescribed integration location on the package. This activity can be conducted using mainly full-wave simulations, and then once a reasonable size and level of performance is achieved, the second step is fabrication of the antenna and measurement of the antenna in its intended environment. Once in the integration position, shifts in operating frequency or radiation characteristics can be observed and the antenna design can be adjusted in the third and final step [31].

In Reference [11], performance of a mobile telephone handset antenna–chassis combination is analysed. Using a resonant circuit model for the entire system, the work indicates that the mobile handsets in the range of approximately 0.25λ–1λ can be described by treating the system of antenna and chassis as a combination of the separate wavemodes of the antenna element and the chassis. By matching the resonant frequencies of the antenna and the chassis, very large bandwidths can be obtained and very small and practically nonradiating coupling elements can be used to replace traditional antenna elements for lower frequency operation [11].

Another related example is the use of 'electromagnetic visibility' to determine placements for antennas on wireless devices [17, 32]. An electromagnetic visibility study (EVS) can be considered as a shortcut to a full modal and diffraction analysis of the device and was first considered in Reference [32] and expanded in Reference [17] to aid in the integration process of antennas on to laptop computers. The EVS relies on the fundamental relationship between the induced surface currents on the host device generated from incoming electromagnetic fields and their impact on the antenna's performance. The study works by locating suitable regions on the host chassis with the highest steady-state current, and translates them into regions of visibility, from high to low, for the antenna. The process begins with the host surface (e.g. a laptop computer) defined by $S_{\text{Object}}(S_{\text{Object}} = S_{\text{Laptop}})$ and that of the antenna

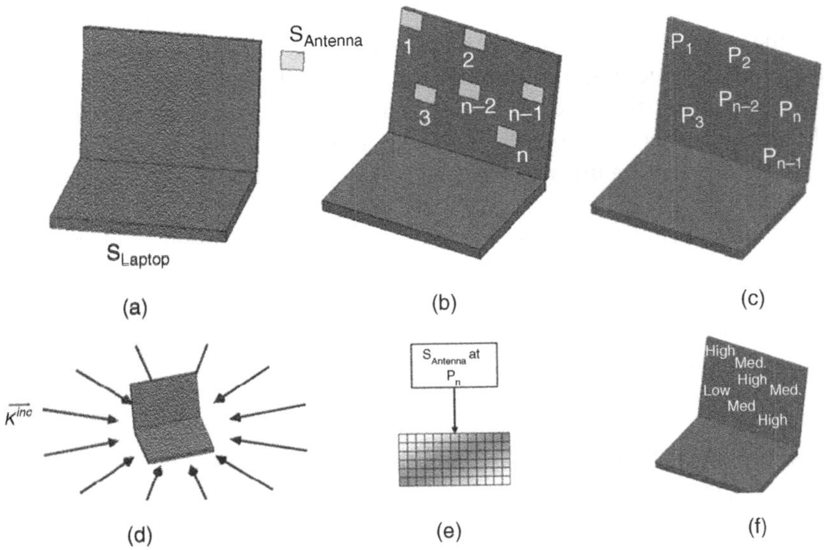

Figure 10.6 Electromagnetic visibility study (EVS). (a) Laptop surface and antenna surface; (b) realizable positions of the antenna on a laptop; (c) point representation of the integrated position for book-keeping; (d) representation of plane waves illuminating the object; (e) representation of surface currents on the laptop surface confined by the antenna's surface; and (f) ranking of position based on evaluation of the EVS (from Reference [17]. © 2004 IEEE. Courtesy of the IEEE).

defined by S_{Antenna} (Figure 10.6(a) [17]). A set of guidelines is then followed for the conformal integration scheme [17]:

1. For conformal integration, the antenna surface, S_{Antenna}, must satisfy the physical restraints for placement, and conformally occupy a surface on S_{Laptop} (Figure 10.6(b) [17]):

$$S_{\text{Antenna}} \in S_{\text{Laptop}} \tag{10.1}$$

2. Regions satisfying step 1 create a set of N candidate locations P (Figure 10.6(c) [17]):

$$\{P_n\}_{n=1}^{N} = \{P_1, P_2, \ldots, P_N\} \tag{10.2}$$

   Practically, these positions are determined not only by the projected size of the antenna but also by the availability of internal space for connections, cables, etc., within very crowded packages.

3. The laptop is discretized and illuminated by plane waves incident from all angles in the direction(s) and/or orientation(s) most likely to receive/transmit data. For this study, the plane of incoming data is considered to be in the azimuthal plane of the laptop computer (Figure 10.6(d) [17]) in the direction $\boldsymbol{k}^{\text{inc}}$. In the simulation space (typically using a finite difference time domain (FDTD) or finite element (FE) full-wave solver), the laptop is illuminated by plane waves propagating in the azimuthal plane ($\theta = \pi/2$) at intervals of $\Delta\phi$ ($= \pi/6$ for this study), for $0 \leq \phi \leq 2\pi$. Doing so produces a total of M plane waves

(where $M = 2\pi/(\Delta\phi)$ and $m = 1, 2, \ldots, M$ in the following equation), for which the electric field polarization for these directions can be given as

$$\{E_m^{\text{Inc}}\}_1^M = \sum_{m=1}^{M} \left(\{[\cos(m\ \Delta\phi)\hat{x} + \sin(m\ \Delta\phi)\hat{y}]_{\text{Horizontal}} + [\hat{z}]_{\text{Vertical}}\} \right. \\ \left. e^{j\{\omega t - k[x\cos(m\ \Delta\phi) + y\sin(m\ \Delta\phi)]\}} \right) \tag{10.3}$$

4. The magnitude of the steady-state conduction current density $\{|J_m^{\text{Con}}|\}_1^M$ on S_{Laptop} generated by $\{E_m^{\text{Inc}}\}_1^M$ can then be determined. Once $\{|J_m^{\text{Con}}|\}_1^M$ is known for $0 \le \phi \le 2\pi$, S_{Antenna} is revisited on the set of possible locations $\{P_n\}_{n=1}^N$ residing on S_{Laptop}. In the FDTD solution space, this is realized as the surface of the laptop with discritized spatial dimensions $(\Delta x, \Delta y, \Delta z)$ confined by S_{Antenna} (Figure 10.6(e) [17]).
5. With the information generated in step 4, the conduction current density in the region(s) confined by S_{Antenna}, u (with U being the total number of cells $\{u\}_1^U \in S_{\text{Antenna}}$), can then be used to find the average conduction current density, $\overline{J}_C$, over all incoming angles, M, and its standard deviation, σ:

$$\overline{J}_C = \frac{1}{MU} \sum_{j=1}^{M} \sum_{i=1}^{U} |J_{ij}^{\text{Con}}| \quad \text{A/m}^2 \tag{10.4}$$

$$\sigma = \sqrt{\frac{\sum_{j=1}^{M} \left[\left(\frac{1}{U} \sum_{i=1}^{U} |J_{ij}^{\text{Con}}| \right) - \overline{J}_C \right]^2}{M}} \quad \text{A/m}^2 \tag{10.5}$$

6. Using Equations (10.4) and (10.5), the areas of interest $\{P_n\}_{n=1}^N$ can be evaluated and ranked using the average steady-state conduction current densities as well as the standard deviation within the areas confined by S_{Antenna} (Figure 10.6(e) [17]).

Extensive simulation and experimental studies indicate that, in general, positions with higher levels of electromagnetic visibility (higher values of $\overline{J}_C$ and lower values of σ) will support antenna operation closer to that of the free space [17]. Results from other independent studies (e.g. Reference [33]) also confirm this.

A related but more rigorous methodology that provides more detailed information about antenna position on packages is that of characteristic modes. While these methods have been widely applied to situations where the package is much larger than an operating wavelength [34, 35] (e.g. airplanes, ground vehicles, etc.), use of characteristic modes for antennas and packages that are comparable in size has also proved to be helpful in explaining variations in packaged impedance and radiation characteristics [11, 36–38].

10.5 DIRECTIONS FOR FUTURE WORK IN THIS AREA

While much is now known about integration of printed antennas into packages, a number of challenges remain. The first, of course, is one of miniaturization. The trend for wireless devices to become smaller and smaller while supporting more and more functions necessarily puts pressure on designers to reduce the occupied volume of the antenna further. Reduction in size, however, has significant effects on antenna efficiency, which has a direct impact on battery life and communication range.

Several research groups around the world are tackling this problem. Recently, the use of magnetic materials has been proposed to miniaturize antennas in handsets and other portable devices [39, 40]. Size reductions of 50 % and more may be realized, but are accompanied by reductions in efficiency caused by the nonzero losses incurred in the magnetic materials. Another group has also used artificial magnetic conductors to reduce the thickness and size of internal printed antennas [41]. There remains, however, much more work to be done in this area. Another possible solution is to simply use the chassis of the device as the main antenna, but this requires careful coordination of the design team throughout product development and may not be feasible in the short term with existing design processes.

A second major challenge for the integration of printed antennas in packages is the development of multifunction devices that require more than one antenna. If a device has multiple internal antennas, they generally share the ground-plane presented by the chassis or internal circuit board of the wireless device. To minimize coupling between antennas, for use of separate systems that share a frequency band (such as 2.4 GHz wireless local area networks and Bluetooth), to isolate antennas on sensitive systems such as the GPS [26], or to isolate antennas intended for multiple input–multiple output (MIMO) systems, one group has implemented artificial magnetic conductors (AMCs) as part of the chassis [42]. A detailed depiction of the AMC used in Reference [42] is shown in Figure 10.7 [42]. The AMC is used as an edge treatment, as shown in Figure 10.8 [42]. As shown in Figure 10.9 [42], this isolation technique can achieve up to 45 dB of isolation between antennas that share the laptop ground-plane. This technique, however, may not be appropriate for smaller devices that do not have the available space to contain the AMC.

Finally, another challenge for antenna packaging is the inclusion of tunable, adaptable, or reconfigurable antennas into packages [17, 43]. While the ability to adjust the antenna to

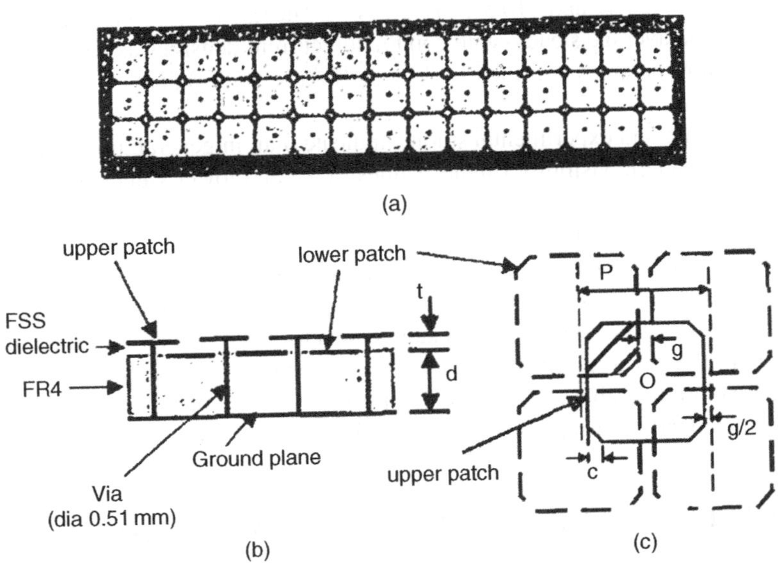

Figure 10.7 (a) AMC edge treatment; (b) stack-up of a typical AMC ($t = 0.051$ mm, $d = 2.49$ mm); (c) drawing of the overlapping solid copper patches in a capacitive FSS ($P = 3.96$ mm, $g = 0.25$ mm, $c = 0.64$ mm) (from Reference [42]. © 2003 IEEE. Courtesy of the IEEE).

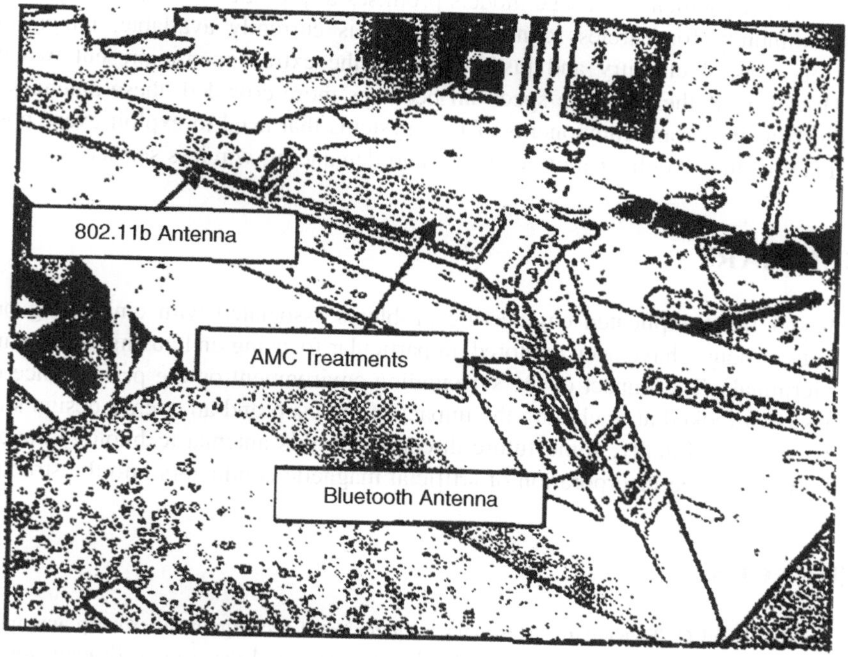

Figure 10.8 An 802.11b antenna and Bluetooth antenna placed around the screen of a surrogate laptop with an AMC edge treatment located between them (from Reference [42]. © 2003 IEEE. Courtesy of the IEEE).

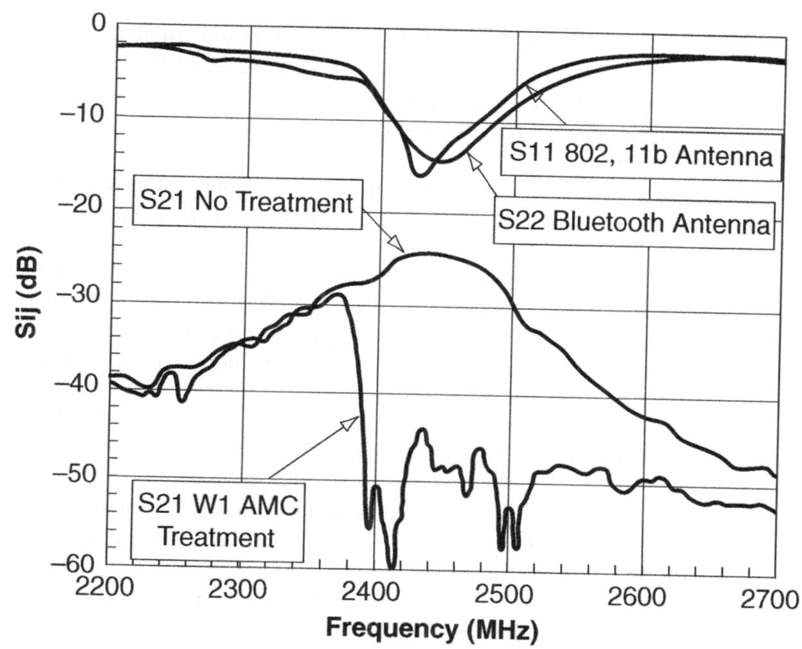

Figure 10.9 Return loss and mutual coupling data of the 802.11b antenna and Bluetooth antenna of Figure 10.8, with and without the AMC edge treatment (from Reference [42]. © 2003 IEEE. Courtesy of the IEEE).

new operating environments or usage models promises to provide more reliability, security, and noise immunity for wireless connections than is currently available, integration of the necessary control circuitry and justification of the extra expense present two major hurdles. However, as the available spectrum becomes more crowded, these antennas must be considered as important components of new systems that need to exploit every possible level of functionality to deliver more complex and demanding wireless services.

10.6 SUMMARY

In this chapter the complicated engineering problems associated with embedding printed antennas into packages have been explored, in particular focusing on housing the antennas in computer terminals. The impact of the surrounding environment on the performance of the antenna has been looked at, including the impact of the user, and an overall design strategy has also been given. Finally, some future directions in this antenna technology have been discussed, including the incorporation of artificial magnetic conductors into the structures.

REFERENCES

1. J. T. Bernhard and C. Tousignant, 'Resonant frequencies of rectangular microstrip antennas with flush and spaced dielectric superstrates', *IEEE Transactions on Antennas and Propagation*, **47**(2), 302–308, February 1999.
2. D. Guha and J. Y. Siddiqui, 'Resonant frequency of circular microstrip antenna covered with dielectric superstrate', *IEEE Transactions on Antennas and Propagation*, **51**(7), 1649–1652, July 2003.
3. R. Alzalzadeh and R. N. Karekar, 'Effect of dielectric protecting superstrate on radiation pattern of microstrip patch antenna', *Electronics Letters*, **27**(13), 1218–1219, 20 June 1991.
4. X.-H. Shen, P. Delmotte and G. A. E. Vandenbosch, 'Effect of superstrate on radiated field of probe fed microstrip patch antenna', *IEE Proceedings on Microwaves, Antennas and Propagation*, **148**(3), 141–146, June 2001.
5. Y. Hwang, Y. P. Zhang, T. K. C. Lo, K. M. Luk and E. K. N. Yung, 'Miniaturization on planar antennas with very high permittivity materials', in *Proceedings of the Asia-Pacific Microwave Conference*, Vol. 1, pp. 217–220, December 1997.
6. N. G. Alexopoulos and D. R. Jackson, 'Fundamental superstrate (cover) effects on printed circuit antennas', *IEEE Transactions on Antennas and Propagation*, **32**(8), 807–816, August 1984.
7. L. Bernard, R. Loison, R. Gillard and T. Lucidarme, 'High directivity multiple superstrate antennas with improved bandwidth', in *Proceedings of the IEEE International Symposium on Antennas and Propagation*, Vol. 2, pp. 522–525, 2002.
8. S. Nishizawa and O. Hashimoto, 'Effectiveness analysis of lossy dielectric shields for a three-layered human model', *IEEE Transactions on Microwave Theory and Techniques*, **47**(3), 277–283, March 1999.
9. K. H. Pan, J. T. Bernhard and T. Moore, 'Effects of lossy dielectric materials on microstrip antennas', in *Proceedings of the IEEE AP-S Conference on Antennas and Propagation for Wireless Communications*, pp. 39–42, November 2000.
10. Y.-X. Guo, M. Y. W. Chia and Z. N. Chen, 'Miniature built-in multiband antennas for mobile handsets', *IEEE Transactions on Antennas and Propagation*, **52**(8), 1936–1944, August 2004.
11. P. Vainikainen, J. Ollikainen, O. Kivekas and I. Kelander, 'Resonator-based analysis of the combination of mobile handset antenna and chassis', *IEEE Transactions on Antennas and Propagation*, **50**(10), 1433–1444, October 2002.

12. A. T. Arkko and E. A. Lehtola, 'Simulated impedance bandwidths, gains, radiation patterns and SAR values of a helical and a PIFA antenna on top of different ground planes', in *Proceedings of the Eleventh International Conference on Antennas and Propagation*, Vol. 2, pp. 651–654, 2001.
13. B. J. Herting, A. Perrotta and J. T. Bernhard, 'Finite ground plane packaging effects on a dual-band PIFA', in *Proceedings of the IEEE Topical Meeting on Electrical Performance of Electronic Packaging*, pp. 95–98, 2002.
14. M. Geissler, M. Gehrt, D. Heberling, P. Waldos and I. Wolff, 'Investigations on radiation *Q* of integrated handset antennas', in *Proceedings of the IEEE International Symposium on Antennas and Propagation*, Vol. 3, pp. 54–57, 2001.
15. G. Huff and J. T. Bernhard, 'Improvements in the performance of microstrip antennas on finite ground planes through ground plane edge serrations', *IEEE Microwave Wireless Composition Letters*, **12**, 308–310, August 2002.
16. S. Ponnapalli, 'Modeling and design of antennas for RF wireless systems', *IEEE Transactions on Composition, Packaging, and Manufacturing Technology, Part B: Advanced Packaging*, **19**(3), 487–502, August 1996.
17. G. H. Huff, J. Feng, S. Zhang and J. T. Bernhard, 'Directional reconfigurable antennas on laptop computers: simulation, measurement, and evaluation of candidate integration positions', *IEEE Transactions on Antennas and Propagation*, **52**(12), 3220–3227, December 2004.
18. M. A. Jensen and Y. Rahmat-Samii, 'EM interaction of handset antennas and a human in personal communications', *Proceedings of the IEEE*, **83**(1), 7–17, January 1995.
19. J. T. Rowley and R. B. Waterhouse, 'Performance of shorted microstrip patch antennas for mobile communications handsets at 1800 MHz', *IEEE Transactions on Antennas and Propagation*, **47**(5), 815–822, May 1999.
20. M. Hirose and M. Miyake, 'Gain improvement of a planar inverted F antenna on a handset by passive loading', in *Proceedings of the IEEE International Symposium on Antennas and Propagation*, Vol. 2, pp. 1128–1131, 1995.
21. G. Lazzi, J. Johnson, S. S. Pattnaik and O. P. Gandhi, 'Experimental study on compact, high-gain, low SAR single- and dual-band patch antenna for cellular telephones', in *Proceedings of the IEEE International Symposium on Antennas and Propagation*, Vol. 1, pp. 130–133, 1998.
22. J. Fuhl, P. Balducci, P. Nowak and E. Bonek, 'Internal broadband antenna for hand-held terminals with reduced gain in the direction of the user's head', in *Proceedings of the IEEE 45th Vehicular Technology Conference*, Vol. 2, pp. 848–852, 1995.
23. J. Fuhl, P. Nowak and E. Bonek, 'Improved internal antenna for hand-held terminals', *Electronics Letters*, **30**(22), 1816–1818, 27 October 1994.
24. G. Goussetis, J. C. Vardaxoglou and A. P. Feresidis, 'Handset antenna performance using flexible MEBG structures', in *Proceedings of the IEEE International Workshop on Antenna Technology: Small Antennas and Novel Metamaterials*, pp. 55–58, 2005.
25. K. L. Virga and Y. Rahmat-Samii, 'Low-profile enhanced-bandwidth PIFA antennas for wireless communications packaging', *IEEE Transactions on Microwave Theory and Techniques*, **45**(10, part 2), 1879–1888, October 1997.
26. Z. Li and Y. Rahmat-Samii, 'Optimization of PIFA-IFA combination in handset antenna designs', *IEEE Transactions on Antennas and Propagation*, **53**(5), 1770–1778, May 2005.
27. K.-L. Wong, L.-C. Chou and C.-M. Su, 'Dual-band flat-plate antenna with a shorted parasitic element for laptop applications', *IEEE Transactions on Antennas and Propagation*, **53**(1), 539–544, January 2005.
28. S. Rogers, J. Scott, J. Marsh and D. Lin, 'An embedded quad-band WLAN antenna for laptop computers and equivalent circuit model', in *Proceedings of the IEEE International Symposium on Antennas and Propagation*, Vol. 3, pp. 2588–2591, 2004.
29. K. Ito and T. Hosoe, 'Study on the characteristics of planar inverted F antenna mounted laptop computers for wireless LAN', in *Proceedings of the IEEE International Symposium on Antennas and Propagation*, Vol. 1, pp. 22–25, 2003.

30. D. A. Strohschein and J. T. Bernhard, 'Evaluation of a novel integrated antenna assembly for mobile data networks using laptop computers', in *Proceedings of the IEEE International Symposium on Antennas and Propagation*, pp. 1962–1965, 1998.
31. D. Liu, E. Flint and B. Gaucher, 'Integrated laptop antennas – design and evaluation', in *Proceedings of the IEEE International Symposium on Antennas and Propagation*, Vol. 4, pp. 56–59, 2002.
32. J. T. Bernhard, 'Analysis of integrated antenna positions on a laptop computer for mobile data communication', in *Proceedings of the IEEE International Symposium on Antennas and Propagation*, pp. 2210–2213, 1997.
33. R. D. Lyle and D. Liu, 'Numerical analysis of inverted-F antenna for WLAN applications', in *Proceedings of the IEEE International Symposium on Antennas and Propagation*, Vol. 2, pp. 934–937, 2003.
34. R. F. Harrington and J. R. Mautz, 'Theory of characteristic modes for conducting bodies', *IEEE Transactions on Antennas and Propagation*, Vol. 19(5), 633–639, September 1971.
35. E. Newman, 'Small antenna location synthesis using characteristic modes', *IEEE Transactions on Antennas and Propagation*, **27**(4), 530–531, July 1979.
36. B. A. Austin and K. P. Murray, 'Application of characteristic-mode techniques to vehicle-mounted NVIS antennas', *IEEE Antennas and Propagation Magazine*, **40**(1), 7–12, February 1998.
37. K. P. Murray and B. A. Austin, 'Synthesis of vehicular antenna NVIS radiation patterns using the method of characteristic modes', *IEE Proceedings on Microwaves, Antennas and Propagation*, **141**(3), 151–154, June 1994.
38. D. A. Strohschein, K. Sivaprasad and J. T. Bernhard, 'Application of characteristic modes to antenna placement on portable wireless devices', in *Proceedings of the IEEE International Symposium on Antennas and Propagation*, **URSI**, p. 222, 2001.
39. T. Tanaka, S. Hayashida, K. Imamura, H. Morishita and Y. Koyanagi, 'A study on miniaturization of a handset antenna utilizing magnetic materials', in *Proceedings of the 2004 Joint Conference of the 10th Asia-Pacific Conference on Communication and the 5th International Symposium on Multi-dimensional Mobile Communications*, Vol. 2, pp. 665–669, 2004.
40. K. Buell, H. Mosallaei and K. Sarabandi, 'Embedded-circuit magnetic metamaterial substrate performance for patch antennas', in *Proceedings of the IEEE International Symposium on Antennas and Propagation*, Vol. 2, pp. 1415–1418, 2004.
41. S. Rogers, J. Marsh, W. McKinzie and J. Scott, 'An AMC-based 802.11a/b antenna for laptop computers', in *Proceedings of the IEEE International Symposium on Antennas and Propagation*, Vol. 1, pp. 10–13, 2003.
42. S. Rogers, J. Marsh, W. McKinzie and G. Mendolia, 'AMC edge treatments enable high isolation between 802.11b and Bluetooth antennas on laptop computers', in *Proceedings of the IEEE International Symposium on Antennas and Propagation*, Vol. 1, pp. 38–41, 2003.
43. V. Stoiljkovic, F. Hamma, B. Winter, K. Simmons and J. Sullivan, 'Review of antenna design options for dual mode handsets', in *Proceedings of the IEE Colloquium on Adaptable and Multi-standard Mobile Radio Terminal*, No. 406, pp. 1–6, 1998.

Part III

Advanced Concepts and Applications in Wireless Systems

11

Printed Reflectarray Antennas

Marek E. Bialkowski[1], **Jose A. Encinar**[2], **J. A. Zornoza-Ramirez**[3] **and F. C.-E. Tsai**[4]
[1] *School of ITEE, University of Queensland, Brisbane, Queensland, Australia*
[2] *Polytechnic University of Madrid, ETSI Telecomunicación, Madrid, Spain*
[3] *Antenna Centre of Competence, EADS Astrium Limited, Stevenage, UK*
[4] *Wistron NeWeb Corporation, Taiwan, ROC*

11.1 INTRODUCTION

The conventional parabolic and shaped reflector antennas are widely used in both terrestrial and satellite communication systems. They feature high radiation efficiency and a large operational bandwidth, which is mainly governed by the bandwidth of the feeding structure, which is usually formed by a single-horn antenna or a cluster. However, their nonplanar format creates a challenge for manufacturers and is unwelcome in some applications. One limitation concerns an elaborate and expensive custom moulding production of these antennas. The other drawback is associated with transportation and deployment. Due to the three-dimensional shape they require a considerable storage volume, which is unwelcome, especially in satellite communication applications, because of the limited stowage space on board the satellite. Due to these reasons, efforts have been made to replace parabolic and shaped reflector antennas by their planar equivalents.

The advantage of planar antennas is that their aperture is easier to establish and maintain, especially when thin substrates are used for their construction. This is the case of light inflatable antennas, which are envisaged for future satellite communication applications. One example of a planar antenna, which replaces a parabolic reflector in many applications, is a printed microstrip patch array antenna [1]. This array antenna is formed by many planar antenna elements, arranged in a regular lattice, which are connected by sections of microstrip or strip-type transmission lines. It has a low profile, low weight, and is easy to manufacture using photolithographic techniques. However, it is limited to applications in which the required gain is in the order of less than 30 dB [2, 3]. The reason for this limitation is the loss of the circuit-type power combining network, which is formed by sections of microstrip or strip-type transmission lines. For large gains this circuit becomes quite long so the conduction and dielectric losses become excessively large. Such power loss leads

Printed Antennas for Wireless Communications Edited by R. Waterhouse

to the poor radiation efficiency of a microstrip patch array. The poor radiation efficiency of microstrip patch arrays can be overcome using alternative low-loss power combining networks. One option is to use waveguides instead of microstrip or striplines to form a feeding circuit. One such solution is a radial line slot array (RLSA) antenna [4–7], which uses a radial cavity or waveguide to feed an array of slots that forms a radiating aperture. This power combiner/divider is low loss and therefore makes the RLSA antenna more attractive than the microstrip patch array in high-gain applications. However, the use of a radial cavity as the power combiner comes at the expense of less flexibility in obtaining multipolarization and multibeam operation.

A compromise between a parabolic or shaped reflector and a microstrip patch array is a microstrip *reflectarray* [8–10], which is the focus of this article. This type of antenna, whose origin can be traced to the work described in References [11] and [12], is formed by a planar reflector constituted by many patches or printed dipoles arranged in a regular planar lattice with a feed element in the form of a horn antenna. Being a type of *planar reflector* antenna, the reflectarray combines positive features of the conventional reflector antennas and planar patch array technologies into one. Similar to the microstrip patch array, it can be developed using commonly available microstrip substrates and photoetching techniques. This leads to a very reliable and low-cost manufacturing. Because it employs a horn antenna as a feed, which acts as a spatial power combiner (with air as a combining medium), conduction and dielectric losses present in conventional patch arrays with a circuit-type feed network are avoided. The resulting low-loss power combining allows the creation of very large aperture antennas. By employing flexible membrane materials they can be folded or unfolded using a simple hinge-type mechanism [10, 13]. Due to the use of a horn feed and many elements forming the reflector, it offers flexibilities of multipolarization and multibeam operation similar to a parabolic reflector. The use of many elements forming the reflector structure offers many degrees of freedom in shaping a radiated beam. Because of a similar operation principle, its efficiency governed by spillover loss, aperture illumination loss, and insertion loss is similar to that of a parabolic reflector. However, its bandwidth is expected to be smaller than that of the parabolic reflector counterpart. The following factors, the element bandwidth, element spacing, and the spatial phase delay, are responsible for this phenomenon. Among them, the first factor is most crucial for a single pencil beam reflectarray of moderate size [14]. It can be overcome by the appropriate design of the phase-shifter element. However, for contoured beams or for very large reflectarrays, such as those used in space applications, the third factor is the most significant in limiting the operating bandwidth of the reflectarray. Parabolic reflector antennas use the physical curvature to equalize signals arriving at the feed. This is the true time signal compensating method, which in principle is frequency independent. In contrast, printed elements of the reflectarray offer, similar to microstrip phased arrays, only phase compensation. The phase adjustment acting as a signal delay mechanism is only valid over a limited frequency band. For large apertures, the compensated path length differences (with respect to the centre elements) can be in the order of many wavelengths for edge elements. In this case, the delay compensation obtainable via the phase compensation becomes valid only over a very narrow frequency band. The only counter measure to this adverse bandwidth narrowing is the introduction of a true time delay mechanism. This approach can be implemented using stubs with a length that varies in a range of several wavelengths. The inconvenience of this implementation is that a large space is required to accommodate the long delay lines. An alternative approach is to use a multilayer printed reflectarray with rectangular patches of variable size [15]. This technique gives more degrees of freedom and allows the patch dimensions to be optimized in order to match, not only the phase shift at the central frequency but also its variation with frequency. Using this technique, a 10 % bandwidth has been achieved for a 1 m reflectarray at 12 GHz [15–17].

11.2 PRINCIPLES OF OPERATION

The reflectarray operation can be explained using a ray-tracing approach with respect to its transmission mode. At a given frequency, a horn antenna having a centre or offset position produces a spherical wave, which is incident on to a planar reflector formed by many planar antenna elements, as shown in Figure 11.1. In order for the reflected wave to be converted to a plane wave travelling in a specified direction, the individual reflector elements need to provide a suitable phasing, as given by [8, 13]

$$\Phi_i = k_0 (R_i - \bar{r}_i \bullet \hat{r}_0) \tag{11.1}$$

Here Φ_i denotes the reflection phase for element i, k_0 is a wave number in free space, R_i is the distance from the phase centre of the feed to the element, $\bar{r}_i$ is the vector from the centre of the array to the element, and $\hat{r}_0$ is the unit vector in the main beam direction.

The resulting value of Φ_i depends on the physical size of the reflector. The difference in the values of Φ_i between the centre and edge elements can be in the order of many integer multiples of 2π. Assuming a narrowband operation of reflectarray, the phase values determined by formula (11.1) can be truncated to the $<0, 2\pi>$ range by applying the truncation rule $\Phi_i' = k_0 (R_i - \bar{r}_i \bullet \hat{r}_0) - 2\pi N$, where N is a suitably chosen integer number. Note that formula (11.1) only concerns the wavefront conversion from a spherical to a planar one. It does not cover other issues such as polarization or the shape of a radiated beam.

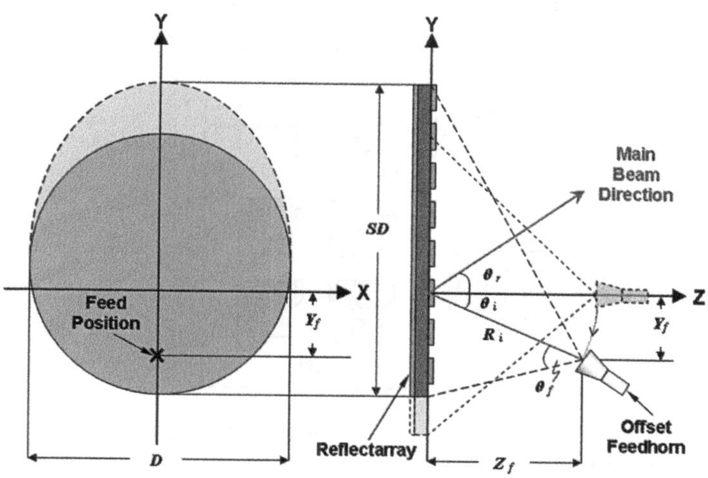

Figure 11.1 Configuration of the reflectarray with an offset horn.

11.3 PHASING METHODS

In order to obtain the required phase compensation, as given by the truncated version of formula (11.1), various phasing mechanisms can be applied. For a microstrip reflectarray they are illustrated in Figure 11.2. These include (a) identical patches with variable-length phase delay lines (in the form of open-circuit stubs) without [8, 10, 18, 19] and (b) with variable angular rotations [13, 20, 21], (c) variable-size patches [22, 26], (d) variable-size cross-dipoles [27], and (e) aperture-coupled patches with stubs of different lengths [28–30]. As required by the truncated version of phasing formula (11.1), each one of them should be capable of producing the reflected wave phase range of approximately 360°. The reflectarray design requires the availability of exact variations of this phase as a function of the variable parameter such as the patch size, stub length, or patch rotation angle. Such data can be obtained from measurements or computer simulations, in which a suitable EM field problem is solved. The designer has to have a full understanding of the working principles of the chosen phasing mechanisms and be aware of the accuracy limitations of the acquired phasing data.

11.3.1 Rectangular Patches with Stubs of Different Lengths

Firstly, an explanation is provided for the phasing mechanism involving fixed-size patches terminated in variable-length open-circuited stubs (Figure 11.2(a)). In this case, the spherical wave produced by the feed is assumed to be an approximately plane wave incident on a small

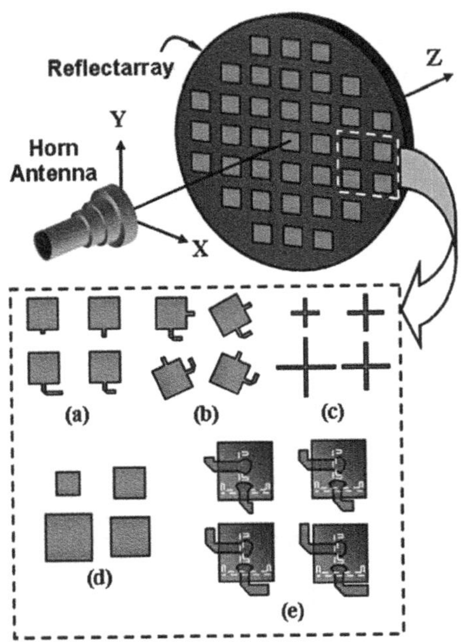

Figure 11.2 Printed microstrip reflectarray with various types of phasing elements: (a) identical patches with a variable-length phase delay lines without and (b) with variable angular rotations; (c) variable-size cross-dipoles; (d) variable-size patches; and (e) aperture-coupled patches with stubs of different lengths.

cluster of adjacent antenna elements in the reflector. This assumption is used in the remaining phasing methods shown in Figure 11.2. Because an arbitrary polarized wave can be resolved into two linearly polarized waves, only the case of a linearly polarized wave is considered. In general, this wave is obliquely incident, and therefore can be of H- (when the electric field is parallel to the plane of the reflector) or E-type (when the magnetic field is planar to the reflector plane). As a result, the reflected wave can depend on the angle and the type (E or H) of polarization of the incident wave. However, the usual situation for centred-fed reflectarrays is that the central elements of the reflector are the subject of an approximately normal incidence. As this group of elements is responsible for reflecting the largest fraction of the incident power, considerations only of the case of normal incidence may provide a good approximation to working out the reflection coefficient phase curves. This assumption is used in the following considerations; however, note that for offset reflectarrays the angle of incidence is larger, and this approach may not be accurate.

Assuming that the patch dimensions are suitably chosen so it resonates, the power intercepted by a patch is transferred to the stub. This transfer occurs without reflection if the patch input impedance is equal to the characteristic impedance of the stub. The signal travelling towards the stub's end is reflected and travels back to the patch, which radiates it. Compared with the incident wave, the radiated wave undergoes a phase shift, which is governed by the patch itself and the length of the open-circuited stub. Assuming that the plane wave is normally incident on two identical side-by-side configured patches, the differential phase shift can be estimated as $2\beta\Delta l$, where β is the propagation constant of the signal along the stub and Δl is the physical length difference between the stubs used in the two patches [19, 31]. For offset reflectarrays, where the angle of incidence is large, the previous law for a relative phase shift between elements will be valid if the angle of incidence in both elements is not very different. For this reason, it is better to avoid large angles of incidence. This simplified principle of operation of a resonating patch terminated in an open-circuited stub forms a simple design principle of a microstrip reflectarray with a variable-length delay line phasing mechanism [8, 10, 18, 19]. Prior to its use, the designer has to be aware of its shortcomings. Firstly, it does not take into account the effect of mutual coupling due to the presence of other elements. Because of mutual coupling the patch initially designed in isolation may be mismatched to the stub. As a result of this impedance mismatch the reflection coefficient at the connection point of the stub may not be given by the simple formula $\Gamma_{\text{stub}} = e^{j2\beta l}$, which is usually assumed in the design procedure. Another shortfall of this approach is that it does not fully take into account the presence of the conducting ground-plane supporting the patches. When the patches are sparsely distributed, the exposed portion of the ground-plane that supports the specular reflection may adversely affect the formation of the desired reflected wave [26]. These considerations show that the task of determining accurate phasing curves for a microstrip reflectarray employing fixed-size patches with stubs may require a more sophisticated electromagnetic field analysis which takes into account the presence of the remaining patches, at least from the neighbourhood and the ground [32]. Such an accurate approach has been developed for reflectarrays using the variable-size patch phasing mechanism. This is discussed later in this section.

Square patches with stubs can be used for single linear polarization when the stubs are connected to only one side of the patches, as in Figure 11.2(a), or for circular, dual-linear, or dual-circular polarization, when two delay lines are connected to orthogonal feeding points in each patch [18]. In any case, the stubs are usually folded to be accommodated in the space between patches, and the spurious radiation produced by the stubs increases the cross-polarization. A solution was proposed in Reference [18] to reduce the cross-polarization, which is based on folding the stubs in different directions, keeping a symmetry with respect to the principal planes of the reflectarray. As a result, the overall cross-polarization of the reflectarray was reduced.

The phasing mechanism to obtain circularly polarized microstrip reflectarrays using patches without and with rotation (Figure 11.2(b)) has been described in References [13], [20], and [21]. Circular polarization can be obtained using dual-polarized patches [18] and a circularly polarized feed, or by using a single linear polarized feed and sequentially rotated patches with dual stubs in References [13], [20], and [21]. The second approach exploits the *principle of sequential rotation*, which has been explained with respect to microstrip patch arrays in References [33] and [34]. The assessment of the two types of phasing was carried out in References [20] and [21]. Two types of circularly polarized reflectarrays of identical size, each producing about 42 dBi gain, were compared. The first type using identical patches with variable-length delay lines achieved an overall 69 % aperture efficiency. The second type with identical patches and stubs using variable rotation featured 60 % aperture efficiency. The patches without the rotation produced higher cross-polarization. This could be due to the choice of some lengths of the stubs that made them resonant. This problem was overcome in the second design because the stub lengths were fixed and resonant lengths were avoided.

A special case of fixed-size resonant patches with open-circuit stubs is when the patch width is small and equal to the stub width. This configuration becomes equivalent to the variable-length printed dipole configuration (Figure 11.2(c)). Its phasing properties cannot be explained by considering separately a resonant patch and a phasing stub. This extreme case shows that in order to obtain in an accurate manner the phasing properties of the patch terminated in an open-circuited stub the two should be considered as one entity. Because of the EM interactions between the patch and the stub, it is possible that the phase shift may not necessarily be a linear function of the stub length. This example shows the shortfall of the simplified explanation of the phasing mechanism involving a fixed-size patch and a variable-length stub. One remedy to this problem is to use a right-angle bend arrangement for a stub to minimize its EM interactions with the patch [32]. For this arrangement, the assumption that the phase shift is a linear function of the bent part of the stub may be more accurate. An undesired effect of this approach is an increased cross-polar radiation due to the bent stub. However, in the array configuration this problem can be overcome using the symmetrical feeding arrangement proposed in Reference [18]. The use of right-angle bent stubs and the symmetrical feeding method could be the reason for reducing phase errors and for achieving high-radiation efficiency with low cross-polarization for the microstrip reflectarray described in Reference [18]. One extra challenge in this design concerns a very high characteristic impedance of the phasing stub. As explained earlier, in order to meet the condition of reflectionless power transfer between the patch and the stub, the characteristic impedance of the stub has to be equal to the input impedance of the edge-fed patch. For thin substrates it is represented by a very large value, typically in the order of 150 Ω. This problem can be overcome using an inset patch feed arrangement. This is because moving the feed point closer to the centre of the patch reduces the input impedance.

11.3.2 Aperture-Coupled Patches with Stubs of Different Lengths

A similar mechanism to produce the required phase shift in the reflectarray elements using stubs of different lengths is based on the use of an aperture-coupled patch antenna

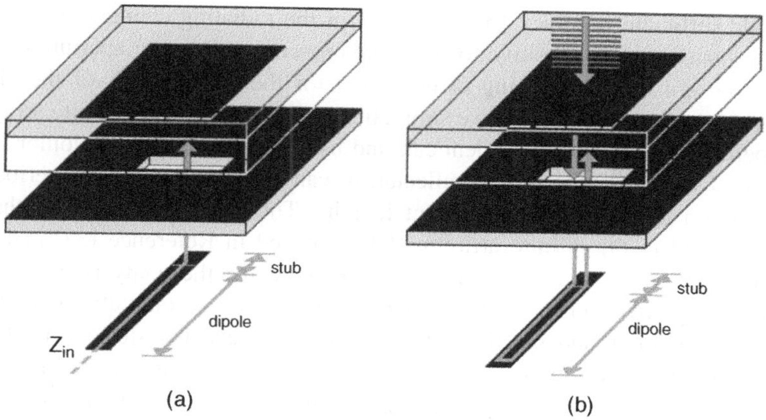

Figure 11.3 Aperture-coupled element: (a) design model of the radiating element and (b) model for computing the phase-shift curves.

(Figure 11.2(e)). In this configuration, each stub is formed by an open-ended length of microstrip line on the opposite side of the ground-plane, which is electromagnetically coupled to the radiating patch by an aperture in the ground-plane (Figure 11.3). The principle of operation is as follows: when the plane wave representing the field coming from the reflectarray feed antenna impinges on the resonant square patch, it is coupled to the microstrip line on the bottom (dipole) and propagates until reaching the open-circuit extreme, where it is reflected, coupled, and reradiated by the patch. This concept has been used for single and dual linear polarization reflectarrays [28–30]. In this configuration, as in that of the attached stubs, the phase of the reflected field is proportional to twice the length of the stub, and a real phase delay can be achieved. The range of phase delay can be larger than 360° and is only limited by the maximum length of the stub [29]. By printing the stubs in a different layer than the radiating elements, there is more room for the line, so the range in phase delay can be increased. The spurious radiation produced by the stubs is in the backward direction with respect to the reflectarray main beam and can be eliminated by placing an additional ground-plane. The main advantage of this configuration is that active elements or controllable phase-shifters can be included in the stub layer for reflectarrays with reconfigurable beams [30].

For the design of the phase-shifter, the microstrip line or the dipole underneath the aperture (see Figure 11.3) is considered as two segments seen from the aperture centre; the first one is a stub of a fixed length, which is adjusted to match the radiating element, whereas the second has variable length and is used to control the phase shift. Firstly, the aperture-coupled radiating element must be designed to achieve a good matching between the line and the patch, considering a microstrip feed line as in the case of conventional planar arrays (see Figure 11.3(a)). At this step, by using a full-wave simulation tool, all the geometrical parameters, such as dielectric materials, thickness, dimensions of the aperture, patch, and fixed stub, have to be determined. The matching of the radiating element is important to allow the energy coming from the feed to be coupled to the delay line, according to the reciprocity theorem. Once the aperture-coupled radiating element is designed, the curves of the phase shift are obtained by computing the phase of the reflection coefficient for different dipole lengths when a plane wave is incident on the patches (see Figure 11.3(b)). Some simple equivalent circuits have been proposed for the aperture-coupled patch [35, 36], though they do not provide sufficient accuracy for the design of the radiating element. This is because the input impedance is very sensitive to some parameters, such as the aperture dimensions. For both, the design of the radiating element and for the computation of the phase-shift curves, it is preferable to use a full-wave simulation tool assuming the 'infinite array model'. Since the apertures and patches are identical, and usually the reflectarray comprises a very large number of elements, the whole reflectarray can be considered as a periodic array, with the exception of the stubs of different lengths. The technique based on the method of moments applied to the infinite array model, proposed in Reference [37] and extended in Reference [38] for stacked patches, is very accurate for the analysis of the radiating element in the array environment. This is because it takes into account all possible coupling effects between apertures and patches. The coupling between the stubs is neglected, but this is less significant. For computing the phase-shift curves, the structure can be analysed as a multilayer periodic structure using a method of moments technique, as described in References [39] and [40]. This full-wave technique allows the phase shift to be computed for any angle of incidence and for an arbitrary polarization of the incident field, expressed as a combination of E- and H-type waves. The cross-polarization components are also computed and are used to predict the cross-polarization patterns of the reflectarray.

In principle the curves of phase shift versus stub length is expected to be linear. However, it was found in the simulations that the slope presents some variations. This is due to the resonances of the dipole. These variations can be minimized by an appropriate design of the radiating element. To achieve a large bandwidth of the phase-shifter element, the phase-shift curves calculated at different frequencies must be parallel, so that only a constant phase is added to all the elements in the reflectarray and the beam is not distorted when the frequency changes. In addition, a linear variation of the phase shift is desirable not only to improve the bandwidth but also to reduce the sensitivity to manufacturing tolerances.

For the experimental validation of the phase-shifter element, an aperture-coupled radiating element as defined in Figure 11.4 was designed to have input impedance from the microstrip approximately equal to 50 Ω in the frequency band 7.5–8.0 GHz. The resulting phase shift was measured for different lengths in a waveguide simulator. This is a well-known technique used to measure the reflection coefficient in infinite arrays by inserting a few radiating elements inside a waveguide, where an infinite array is emulated by the imaging action of the waveguide walls [41]. Two radiating elements with 10 mm × 10 mm patches and 6.61 mm × 1.23 mm apertures were inserted in a WR112 waveguide, being the ground-plane with the apertures fixed to the waveguide flange, so that the printed dipoles were out of the waveguide. The printed dipoles were 1.6 mm wide with a length of 2.06 mm for the fix stub (see Figure 11.3). The characteristics of all the dielectric layers are given in Table 11.1. An

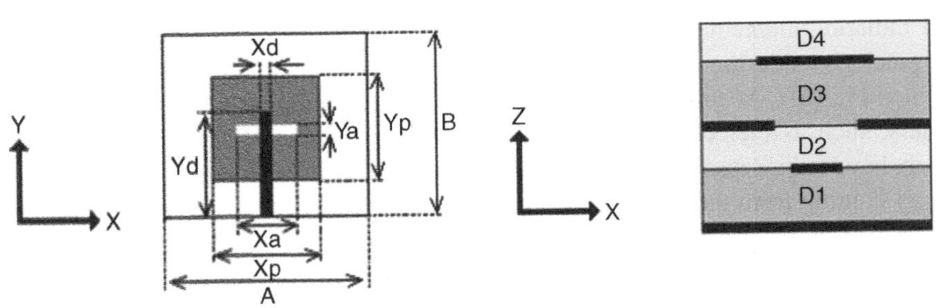

Figure 11.4 Definition of the phase-shifter cell.

Table 11.1 Dielectrics between layers of the built cell.

Dielectric	Thickness (mm)	ε_r	Tan δ
D4	0.79	3.20	0.0030
D3	2.00	1.05	0.0002
D2	0.508	3.38	0.0050
D1	9.67	1.05	0.0002

additional ground-plane separated a quarter-wavelength of the dipoles by the Rohacell layer D1 was added to eliminate the backward radiation produced by the apertures and dipole. The comparison between the calculated and measured phase of the reflection coefficient is shown in Figure 11.5(a) for 7.5 and 8.0 GHz as functions of the dipole length. The worst

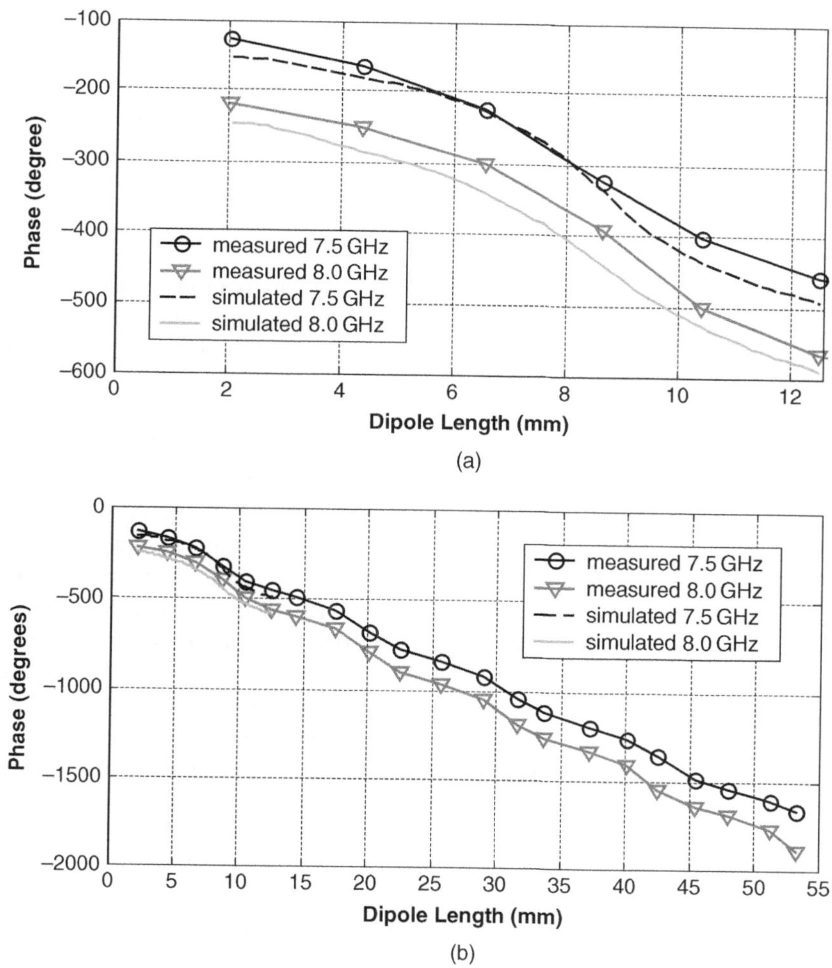

Figure 11.5 Phase of the reflection coefficient in the waveguide simulator (WGS): (a) comparison of simulations and measurements; (b) measurements in a large range of lengths.

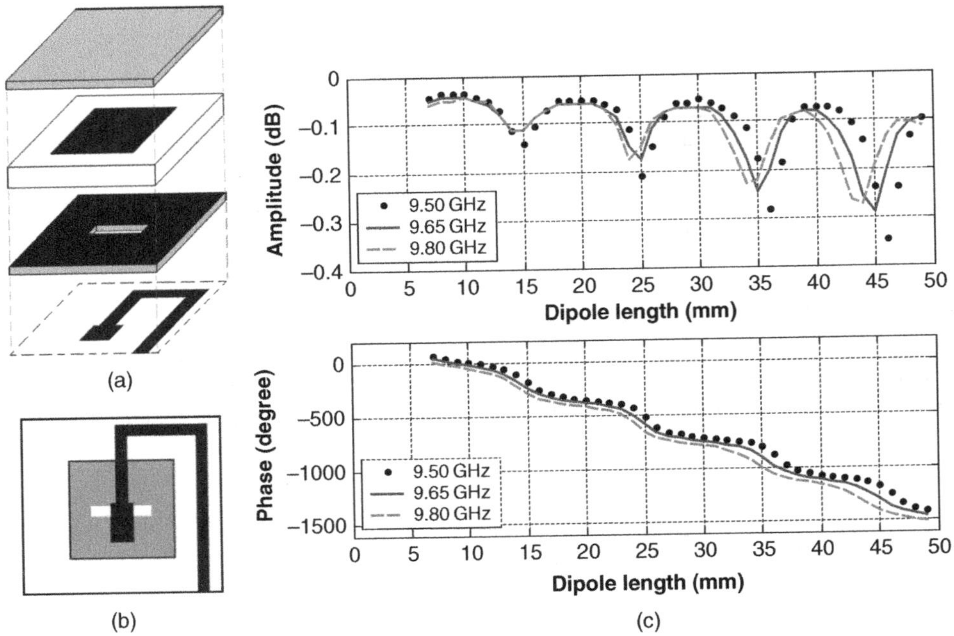

Figure 11.6 Unit cell with the dipole length increased (U-geometry): (a) expanded view; (b) top view; and (c) simulated amplitude and phase of the reflection coefficient.

difference between measurements and theoretical simulations is in the order of 20°, that is within an acceptable range. These discrepancies can be due to manufacturing tolerances, the finite dimensions of the two ground-planes, and the assumption of an infinite array of dipoles in the simulations. Figure 11.5(b) shows the measurement results for dipoles having dimensions in the order of 55 mm exceeding those of the waveguide dimensions. A very large range in phase delay is achieved.

As shown, the phase delay range depends on the dipole length. To avoid the stub going out of the unit cell, the dipole can be folded as shown in Figure 11.6. Using this configuration, a large phase delay range and the linearity of phase delay versus dipole length can be accomplished. The linearity of the phase delay as a function of the dipole length can be improved by modifying the matching stub as shown in Figure 11.6. The reflection losses and phase delay are shown in Figure 11.6(c) at the central and extreme frequencies, for a structure with the following parameters: square periodic cell of 20 mm, $X_p = Y_p = 10$ mm, $X_a = 6$ mm, $Y_a =$ 1 mm, $X_d = 2$ mm, $Y_d = 2.7$ mm + variable segment. The dielectric layers are the same as defined in Table 11.1, except for D2, which is equal to D4. The phase delay is greater than four times 360°. The amplitude of the reflection coefficient represents the dissipative losses in the different dielectric layers, since a ground plane is considered underneath the dipoles. These dissipative losses increase at the dipole resonances and are in the order of 0.1 dB.

11.3.3 Rectangular Patches or Crosses of Variable Size

The cross-dipole phasing illustrated in Figure 11.2(c) can be used to achieve dual-linear or circular polarization. In this case, the reflectarray polarization is governed by the polarization of the feed. The remaining issue, similar to that for the other phasing methods, concerns the phasing range the printed dipoles can offer. In order to provide the full beam-forming capability, the phasing elements should deliver a phasing range close to 360°. This phasing range is impossible to be fully achieved for a single-layer microstrip reflectarray. However, it can be easily met using stacked patch configurations.

The phasing mechanism exploiting variable-size patches has been described in References [24] to [26] and [42] to [44]. The analysis method for determining the phasing curves for a single-layer microstrip reflectarray was first presented in Reference [44]. Works shown in References [25] and [26] report on the extension of these analyses to the multilayer microstrip reflectarray configurations. The variable-size microstrip patch phasing mechanism relies on the fact that the phase of the reflection coefficient of a resonant patch varies over a large range for frequencies close to the resonant frequency. As a result, for a resonating patch a small change in its size leads to a wide range in phase variation of the reflected wave. This behaviour was recognized first in microstrip patch array antennas where small errors in the resonant length of a microstrip patch were found to cause large errors in the radiated field [45]. In order to obtain the phasing curves for a microstrip reflectarray formed by variable-size patches, solving the problem of scattering from an infinite array of identical microstrip patches has been postulated [24, 44]. The relevance of this canonical problem to the current problem is motivated by the fact that it should provide good approximation because of small differences in patch size of the actual reflectarray. This approach takes into account mutual coupling effects between the reflectarray elements, as well as the specular reflection from the ground-plane supporting the patches.

The necessary analysis can be performed with respect to the configuration shown in Figure 11.1. The incident electric field is assumed to be given by

$$\bar{E}_i = \bar{E}_0 e^{jk_0(xu_i + yv_i + z\cos\theta_i)} \tag{11.2}$$

where (θ_i, ϕ_i) is an arbitrary angle of incidence, $u_i = \sin\theta_i \cos\phi_i$, $v_i = \sin\theta_i \sin\phi_i$, and $\bar{E}_0 = E_{0\theta}\hat{a}_\theta + E_{0\phi}\hat{a}_\phi$.

Assuming that identical patches arranged in a rectangular lattice form the reflectarray, the total reflected field, being the sum of the field reflected from the ground dielectric substrate and the field scattered from the patch elements, can be expressed as

$$\bar{E}_t = \bar{E}_r + \bar{E}_s = (\bar{\bar{R}} + \bar{\bar{S}})\bar{E}_0 e^{jk_0(xu_i + yv_i - z\cos\theta_i)} \tag{11.3}$$

where

$$\bar{\bar{R}} = \begin{bmatrix} R_{\theta\theta} & 0 \\ 0 & R_{\phi\phi} \end{bmatrix}$$

are the component plane wave reflection coefficients for the dielectric substrate and

$$\bar{\bar{S}} = \begin{bmatrix} S_{\theta\theta} & S_{\theta\phi} \\ S_{\phi\theta} & S_{\phi\phi} \end{bmatrix}$$

are the component plane wave scattering coefficients for the patches.

Assuming that the given substrate and the element spacing do not introduce grating lobes or surface wave generation, and assuming lossless dielectrics, the magnitude of the total

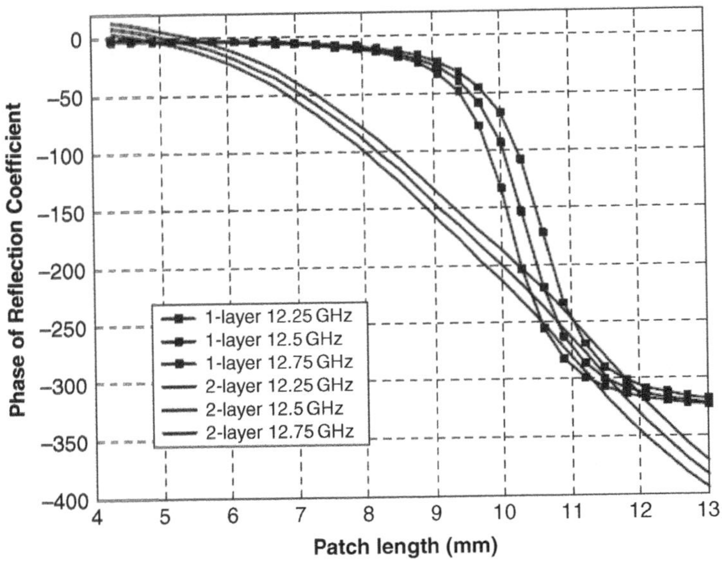

Figure 11.7 Unit cells for a single- and double-layer Ku-band microstrip reflectarray and the corresponding phasing curves reported in Reference [47].

reflection coefficient for the incident electric field $\bar{E}_0 = E_{0\theta}\hat{a}_\theta$ $((\theta_i, \phi_i) = (\theta_i, 0))$ is equal to 1: $|R_{t\theta\theta}| = |R_{\theta\theta} + S_{\theta\theta}| = 1$. As a result, only the total reflection coefficient phase becomes a function of the angle of incidence. The solution to this electromagnetic field problem is based on the Green function and Fourier transform approach and is similar in principle to the one described in Reference [46]. Typical results for the reflection coefficient phase as a function of the patch length for normal incidence $(\theta_i = 0°)$ over a narrow frequency band are shown in Figure 11.7.

The results presented in Figure 11.7 concern single- and double-layer Ku-band microstrip reflectarray designs, which were reported in Reference [47]. The single-layer design assumes a Kevlar (relative dielectric constant, $\varepsilon_r = 3.4$) layer 0.105 mm thick, the metal patch array, another 0.16 mm thick Kevlar followed by a 3 mm thick *honeycomb* layer ($\varepsilon_r = 1.067$), and the ground-plane. The double-layer design concerns a 0.105 mm thick Kevlar layer supporting the upper patch layer, followed by a 3 mm thick *honeycomb* layer, another 0.16 mm thick Kevlar layer supporting the lower patch layer, a 3 mm thick *honeycomb* layer, and a ground-plane. The two designs assume square-shaped patches and a unit cell of 14 mm × 14 mm. In the double-layer configuration, the size ratio for upper and lower square patches is 0.67.

The phasing curves seen in Figure 11.7 confirm a rapid variation of the phase around the patch resonance and the slow variation off-resonance for the single-layer microstrip configuration. Another property concerns an attainable phasing range. As observed in Figure 11.7, the achievable phasing range for the single-layer microstrip reflectarray is approximately $360° \times (1.0 - kh/\pi)$, where k is a wavenumber in the substrate and h is the substrate thickness. This formula shows that thinner substrates offer a greater phase range of typically more than 300°. However, this is achieved at the expense of a high nonlinear relationship between the reflection phase versus patch size, which makes the design prone to manufacturing errors. Another problem associated with a thin substrate is that it limits operation of a reflectarray to a narrow bandwidth. By increasing the substrate thickness, a much smoother phase variation and a larger bandwidth can be achieved. However, the resultant phase range becomes reduced to less than 300°, making practical designs unfeasible.

In order to obtain simultaneously a gently varying reflection phase curve, a larger phasing range, and a larger operational bandwidth, a multilayer structure involving stacked patches can be used in designing a microstrip reflectarray [25]. To avoid surface waves, low-loss low-dielectric constant materials (with properties similar to those of air) used as spacers between thin substrates on which patches are developed are assumed for these structures. To predict phasing characteristics for a multilayer patch structure an efficient electromagnetic analysis based on the generalized scattering matrix (GSM) and cascading the different matrices can be used [40]. This approach is similar in principle to the one described for a single-layer microstrip reflectarray [44]. This is because the reflection phase curves are obtained by considering an infinite periodic array, although this time formed by two layers of radiating elements. Note that due to the periodic arrangement of square patches, the analysis requires only a unit rectangular cell with appropriate periodic boundary conditions to be considered [40]. The phase of the reflection coefficient as a function of patch size is usually determined for an assumed ratio of patch size in the upper and lower layers. Similar to a single layer, the phase of the reflection coefficient can be dependent not only on the element size but also on the angle of incidence (θ_i, ϕ_i) and type (E or H) of polarization. However, it has to be noted [22] that if θ_i is less than about 40° the phase variation (considered as a function of the angle of incidence) is small. In turn, the use of incidence angles θ_i greater than 40° is not recommended. Hence to achieve proper behaviour, a microstrip reflectarray design is often restricted to scan angles smaller than 40°. Although for a more accurate design the angles of incidence must be taken into account, as pointed out in Reference [48], this restriction indicates that for well-behaved reflectarrays, the reflection phase curves can be estimated from the case of normal plane wave incidence on a periodic infinite array. Based on this assumption, the phasing curves can be generated by using the equivalent unit cell approach described in Reference [26]. By using the proper patch size ratio for upper (smaller patch) and lower (larger size patch) layers the phasing curves can be optimized. The ratio between 0.6 and 0.7 has been found both in References [25] and [26] to provide a gentle slope of phase as a function of patch size. This is illustrated both in Figures 11.7 and 11.8.

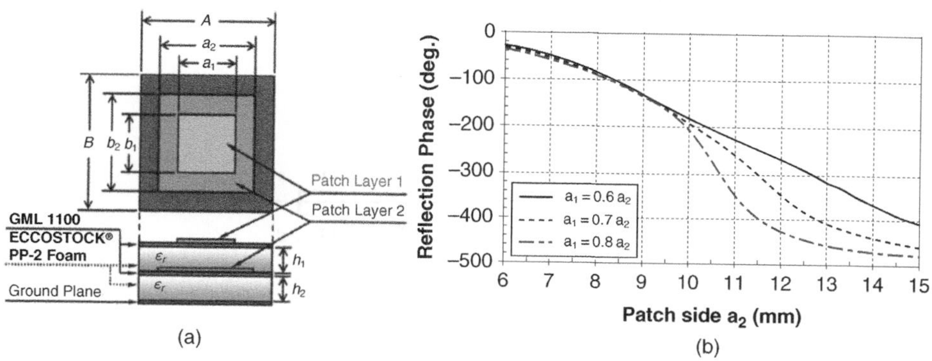

Figure 11.8 (a) Configuration of a unit cell of a two-layer microstrip reflectarray and (b) a simulated reflection phase at normal plane wave incidence versus square patch size, as obtained at 12.5 GHz [26].

In Figure 11.8, the two patch layers use 0.078 mm (0.0031 inch) thick GML 1100 thin copper laminate ($\varepsilon_r = 3.32$) manufactured by GIL Technologies®. Layers of ECCOSTOCK® PP-2 low loss foam ($\varepsilon_r = 1.03$) of 3.2 mm ($h_1 = h_2 = 3.2$ mm) in thickness are stacked beneath the patch layers for bandwidth and phase range enhancement. As observed in Figures 11.7 and 11.8, not only is the slope of the phase curve gentler in comparison with the single-layer one but also the phase range is increased to more than the required 360°. The two layers of patches can serve not only the purpose of increasing the phasing range over an increased operational bandwidth, as shown in Figure 11.7, but can also be used for the purpose of obtaining a dual-frequency operation [49].

The phase curves for a unit cell comprised of single or stacked patch configurations shown in Figures 11.7 and 11.8 have been generated using own-developed software described in References [25] and [26]. An alternative method is to apply a modern EM simulation software package such as the Ansoft High Frequency Simulator [26]. In this case, the newer versions of this software, which have capabilities of handling periodic boundary conditions, have to be used.

11.4 SINGLE-BEAM REFLECTARRAY DESIGN

The design of a single-beam microstrip reflectarray can be accomplished using the guidelines for designing conventional parabolic reflector antennas in conjunction with the element phasing information, which is obtained via suitable computer simulations. The results of these simulations are the geometrical parameters of the phase-shifter element (stub length or dimensions of the patches), which realize the required phase distribution in the reflectarray surface given by expression (11.1). Figure 11.9 shows two centred-fed configurations of (a) a prime focus parabolic reflector and (b) a reflectarray whose operation can be compared.

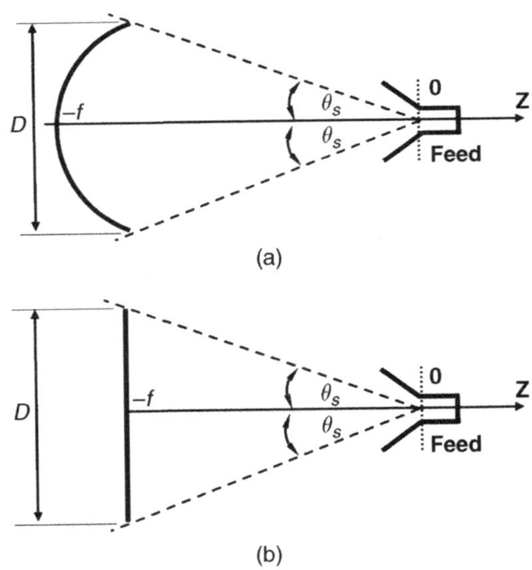

Figure 11.9 Configurations of (a) a prime focus parabolic reflector and (b) a reflectarray including the associated subtended angles.

Both antennas have the same subtended (half) angle θ_s but different f/D ratios. For the reflectarray, the relation between θ_s and f/D is given by

$$\theta_s = \tan^{-1}\frac{D}{2f} \tag{11.4}$$

This expression differs from the one for the parabola: $\theta_s = 2\tan^{-1}(0.25D/f)$. The comparison between the two expressions shows that for the same subtended angle and aperture size, the reflectarray will have smaller depth, which can be of advantage in some applications.

The prime focus configuration of Figure 11.9 creates aperture blockage. An offset configuration, shown in Figure 11.1, can provide a remedy to this problem. However, as the projected aperture is reduced by the $\cos\theta$ factor so is the directivity. In order to minimize beam squinting versus frequency the angle of the main beam has to be chosen close to the natural specular reflection angle [24]. Given the value of the subtended angle θ_s a suitable choice of a feed antenna has to be made to maximize the aperture efficiency. This is defined as the product of spillover and taper efficiencies. Assuming an axis-symmetric feed-power pattern of the form

$$G_f(\theta, \phi) = \cos^n \theta \quad \text{where } n = 2, 4, 6, \ldots \tag{11.5}$$

and a circular array, the spillover efficiency is given by the closed-form expression

$$\eta_S = 1 - \cos^{n+1}\theta_s \tag{11.6}$$

Similarly, the amplitude taper efficiency is given by

$$\eta_t = \frac{2n}{\tan^2\theta_s}\frac{(1-\cos^{n/2-1}\theta_s)^2}{(n/2-1)^2(1-\cos^n\theta_s)} \tag{11.7}$$

It has been shown [24] that for $n=2$ an optimal aperture efficiency is obtained for a subtended angle θ_{Sopt} of about 60°, for $n=4$, $\theta_{Sopt} \approx 50°$, and for $n=6$, $\theta_{Sopt} \approx 45°$. The corresponding maximum aperture efficiency is about 72 % for $n=6$ and is reduced to about 70 % for $n=2$. These aperture efficiency values are slightly inferior in comparison with the parabolic reflector because of a slightly lower taper efficiency.

The obtained results for aperture efficiencies assume that the feed phase centre is known prior to the reflectarray manufacturing. This assumption is usually fulfilled, since the phase centre can be accurately predicted with commercial software or measured, but in any case small adjustments of the feed position can be made once the reflector is developed in order to obtain the maximum measured gain.

In a reflectarray, the radiated field is composed of the scattered field by each element and the field reflected by the substrate and conducting ground, as stated in Equation (11.3). The scattered field due to individual antennas is computed by considering the infinite array approach. The infinite periodic structures are characterized by the Floquet theorem, which postulates the following. The scattered field due to a plane wave impinging on an infinite periodic structure is defined by a discrete and infinite number of plane waves that only depend on the incident wave and the periodicity of the structure. These plane waves that define the scattered field are named Floquet harmonics [50]. In order to reflect and radiate the field in a desired direction, the reflected field must be locally a plane wave when there are not higher order Floquet modes propagating. The condition that must be fulfilled to ensure that the only propagation of the fundamental mode occurs is given by

$$\frac{A}{\lambda_0}, \frac{B}{\lambda_0} \leq \frac{1}{1+|\sin(\theta_{\text{max inc}})|} \tag{11.8}$$

where A and B are the periodic cell dimensions, λ_0 the free-space wavelength at the design frequency, and $\theta_{\text{max inc}}$ the maximum angle of incidence, defined in the geometry of 0 as $\tan(\theta_{\text{max } inc}) = \left(SD - D/2 + |Y_f|\right)/Z_f$. The use of incidence angles greater than 40° is not recommended, as pointed out in Section 11.3.

Equation (11.8) closely relates the feed position with the periodic cell dimensions and therefore is a key condition to proper design of a reflectarray antenna. In order to avoid grating lobe generation, patch spacing in the range of 0.6–0.8 free-space wavelength is used in practical designs. In addition, to minimize the blockage due to the feed and reduce beam squint [23] an offset feed for illuminating the reflectarray is recommended.

Having chosen the reflectarray geometry, the next step concerns the design of the patch array. It includes determining the dimensions of the patches (or stubs) to achieve at a given frequency the phase shift distribution defined by Equation (11.1) for a pencil beam, or any other phase distribution for a contoured beam. Sometimes the adjustment of the phase-shifter elements is made in an approximate way using tables for phase shift obtained for normal or oblique incidence. In this case, some approximations are made since the real variation of phase shift with the angle of incidence and polarization of the field is not taken into account. It must be noted that as the incidence is oblique in each element, the phases for E- and H-polarizations may be different. For the design of reflectarrays based on stacked patches with variable size, a more accurate process has been proposed in Reference [25]. In this technique, the design of the reflectarray is carried out element by element, first at the central frequency, assuming a fixed relative size of the stacked patches. In this stage, the dimensions of the stacked rectangular patches are adjusted in each cell to match the objective phase by using a zero finding routine that iteratively calls the analysis routine. The real angles of incidence θ_i, ϕ_i and the polarization of the field coming from the feed as a combination of E- and H-type waves are considered. Therefore, when the reflectarray is designed for dual-linear or circular polarization, the two dimensions of the patches are adjusted in order to obtain the required phase-shift distribution for the two orthogonal polarizations, with the electric field in the X and Y directions (E_x and E_y). In addition, this technique permits the real amplitude and phase of the field incident on each reflectarray element to be considered, which is available from simulations or measurements of the feed horn [51]. This leads to producing the required phase distribution in a more accurate manner than from geometrical optics, as offered by Equation (11.1). In addition, optimization of patch dimensions can be used to overcome the frequency band limitation in large reflectarrays [15].

The design of a reflectarray in a frequency band is carried out using the following steps. Firstly, the required phase delay $\phi_{\text{di}}(f)$, which is the one given in Equation (11.1) without truncation, is determined at the central and extreme frequencies in order to achieve a reflected field with a progressive phase distribution in a predetermined frequency band. The difference of phase delay at the extreme frequencies, $D_{\text{di}}(f_2, f_1) = \phi_{\text{di}}(f_2) - \phi_{\text{di}}(f_1)$, depends on the element location, and is larger near the centre of the reflectarray. Secondly, using the patch dimensions obtained for the design at the central frequency, an optimization routine based on the Fletcher–Powell algorithm is applied to adjust all the dimensions of the stacked patches to match both the required phase shift ϕ_i and phase delay difference $D_{\text{di}}(f_2, f_1)$.

It has to be noted that using two stacked patches there are not enough degrees of freedom to carry out the optimizations in a frequency band; however, using three array layers the optimization can be successfully achieved. This has been demonstrated in the design of a 1 m reflectarray for 10 % bandwidth at 12 GHz [15].

Having completed the reflectarray design, its radiation pattern can be determined next. This task can be accomplished using the information about the tangential component of the reflected electric field over its aperture. This approach uses the EM field equivalence principle in which the residual backward field due to feed is neglected. In practice, the radiation pattern calculations are performed in a different manner when the two phasing mechanisms, one employing the fixed-size patch with variable stubs and the other using the variable size patches, are assumed. For patches with variable-length stubs, without or with rotations, the preferable approach is to treat the reflectarray as a planar phased array of patches [18, 20, 21]. In this case, mutual coupling effects between individual elements are neglected, although they might be taken into account during the procedure of determining the phasing curves. For the without-rotation case, when all the patches are copolarized, the far-field pattern can be written [18, 20] as

$$E(R_0, \theta_0, \phi_0) = \frac{\mathrm{e}^{-\mathrm{j}k_0 R_0}}{R_0} \sum_m \sum_n E_{mn} \mathrm{e}^{-\mathrm{j}k_0(r_{mn} - z + \bar{\boldsymbol{\rho}} \cdot R_0) + \mathrm{j}\Phi_{mn}} \tag{11.9}$$

where $E_{mn} = (F_{mn}/r_{mn})A(\hat{\boldsymbol{r}}_{mn} \cdot \hat{\boldsymbol{r}}_c)A(\hat{\boldsymbol{r}}_c \cdot \hat{z})$, F_{mn} is the pattern function of the feed to the mnth patch, $A(\cdot)$ is the pattern function of the mnth patch, $\hat{\boldsymbol{r}}_c$ is the unit vector from the feed to the centre of reflectarray, $\hat{\boldsymbol{r}}_{mn}$ is the unit vector from the feed to the mnth patch, r_{mn} is the distance between the feed and the mnth patch, $\bar{\boldsymbol{\rho}}$ is the position vector of the projected aperture on the $z = 0$ plane, and Φ_{mn} is the phase delay due to the open-circuit stub connected to the mnth patch. Note that expression (11.9) neglects the field terms concerning the specular reflection from the ground and due to the diffraction at the edges of the ground-plane [21].

For the reflectarrays designed using variable-size patches, the radiation pattern is obtained by treating the array as an assembly of unit cells [24, 25]. In this case, the far-field expression is given by

$$\begin{aligned} E(R_0, \theta_0, \phi_0) = \frac{\mathrm{e}^{-\mathrm{j}k_0 R_0}}{R_0} \sum_m \sum_n \bar{\bar{Q}}(\theta_0, \phi_0, \theta_{mn}, \phi_{mn}) \Big[\bar{\bar{R}}(\theta_{mn}, \phi_{mn}) \\ + \bar{\bar{S}}(\theta_{mn}, \phi_{mn}) \Big] E_f(\theta_{mn}, \phi_{mn}) \mathrm{e}^{\mathrm{j}k_0(x_{mn} \sin\theta_0 \cos\phi_0 + y_{mn} \sin\theta_0 \sin\phi_0)} \end{aligned} \tag{11.10}$$

where x_{mn}, y_{mn} are the coordinates of the centre of the mnth patch, θ_{mn}, ϕ_{mn} are the angles of incidence from the feed to the mnth patch, and $E_f(\theta_{mn}, \phi_{mn})$ is the field pattern of the feed including the $\mathrm{e}^{-\mathrm{j}k_0 R_{mn}}/R_{mn}$ dependence. The dyadics $\bar{\bar{R}}(\theta_{mn}, \phi_{mn})$ and $\bar{\bar{S}}(\theta_{mn}, \phi_{mn})$ have already been introduced in Equation (11.3). The dyad $\bar{\bar{Q}}(\theta_{mn}, \phi_{mn})$ accounts for transformation from a plane wave field to a spherical wave and is related to the radar cross-section of a rectangular plate [24]:

$$\bar{\bar{Q}} = \frac{-\mathrm{j}AB}{\lambda_0^2} \hat{\boldsymbol{e}}_\mathrm{r} \hat{\boldsymbol{e}}_\mathrm{i} (\hat{z} \cdot \hat{\boldsymbol{e}}_\mathrm{r} \times \hat{\boldsymbol{h}}_\mathrm{i}) \tag{11.11}$$

where A, B are the dimensions of the unit cell, $\hat{\boldsymbol{e}}_\mathrm{r}, \hat{\boldsymbol{e}}_\mathrm{i}$ are unit vectors representing the polarizations of the incident and reflected waves, and $\hat{\boldsymbol{h}}_\mathrm{i}$ represents a unit vector in the

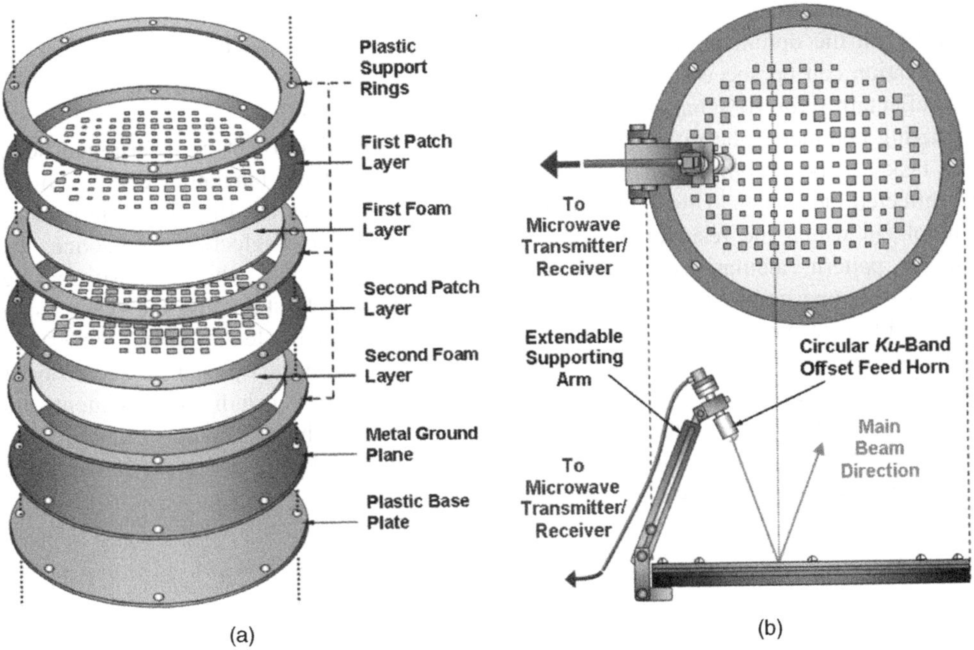

Figure 11.10 A 161-element two-layer printed microstrip reflectarray with variable-size patches: (a) multilayer structure construction and (b) mechanical layout [26].

direction of the incident magnetic field. The expression (11.10) includes both co- and cross-polar components of the radiated field. Note that in contrast to (11.9), expression (11.10) takes into account the specular reflection from the ground.

Usually, the theoretically predicted radiation patterns are compared with the measured ones [18, 19, 24, 25]. A relatively good agreement between simulated and measured radiation patterns obtained by many researchers gives the confidence in the developed theoretical tools for designing and assessing the performance of reflectarrays with various types of phasing mechanism.

The following figures show examples of single-beam microstrip reflectarrays, which have been reported by the authors and other researchers in the antenna literature. Figure 11.10 shows the construction details of a small dual-polarized reflectarray for operation at the Ku-band, which was reported in Reference [26]. In this design, the horn is pointing directly at the centre element of the reflectarray offset by an incident angle, $\theta_i = 20°$, away from the broadside direction. The 161-element microstrip reflectarray is of elliptical shape and is aimed for dual-linear polarization with an offset feed to scan the main beam in the direction $\theta_r = 20°$.

Figure 11.11 shows a breadboard of 406 mm diameter of a two-layer reflectarray with patches of variable size. The reflectarray was designed to radiate a pencil beam in the direction $\theta_0 = 25°$, $\Phi_0 = 0°$ at 11.95 GHz for two orthogonal linear polarizations. A commercial horn is used with its phase centre located at coordinates $X_f = -150$, $Y_f = 0$, $Z_f = 300$ mm with respect the reflectarray centre. Figure 11.12 shows the co- and cross-polar patterns in the XZ plane measured at 11.95 GHz for X polarization (with the tangential electric field on the reflectarray surface in the x direction). The simulated patterns are also represented, which

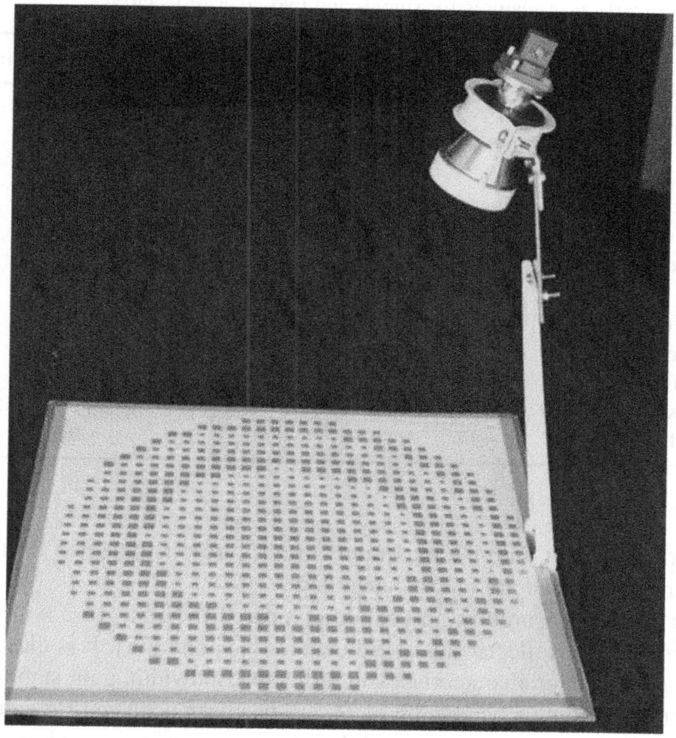

Figure 11.11 Two-layer breadboard for dual polarization. (From Reference [25]. © 2001 IEEE. Reproduced by permission of the IEEE.)

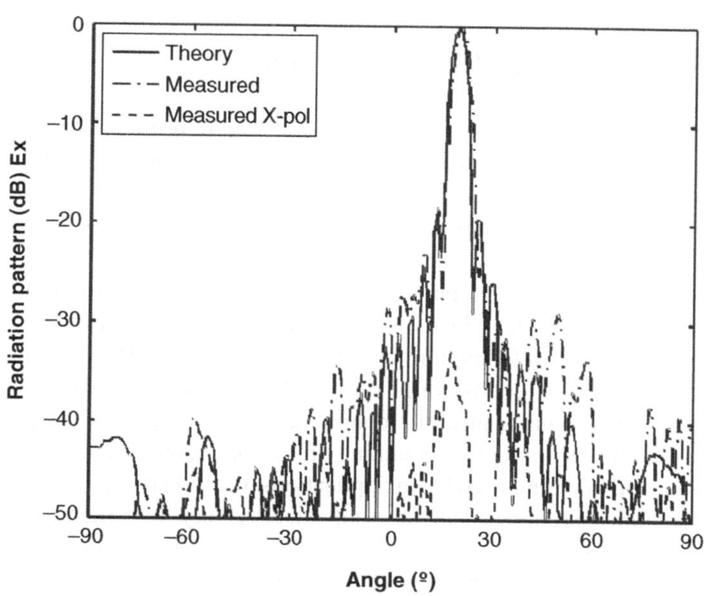

Figure 11.12 Comparison between theoretical and measured patterns. (From Reference [25]. © 2001 IEEE. Reproduced by permission of the IEEE.)

Table 11.2 Comparison of features for single- and two-layer reflectarrays.

	Reflectarray [52]	Reflectarray [18]	Two-layer prototype [25]	
Central frequency (GHz)	9.75	9.075	11.95	
Band (GHz)	9.5–10.0	8.85–9.5	11.5–12.4	
Band (%)	5.1	7.2	11.5–12.4	16.7
Polarization	Linear	Dual linear	Dual linear	
F/D	0.9	0.87	0.84	
Radiation angle (deg)	25	30	19	
Surface/λ^2	261.6	383.0	205.5	
Maximum directivity (dBi)	34.7	36.2	33.8	
Measured directivity (dBi)	—	—	32.47	
Measured gain (dBi)	30.5	33	31	
Gain variations (dB)	1.5	3	0.3	1.5
Cross-polar (dB)	≤ -17	≤ -25	≤ -26	

are in very good agreement with the measurements. The results obtained for the orthogonal polarization were similar and are not included here.

The features of the measured prototype are compared in Table 11.2 with the measured results of the single-layer reflectarrays in References [18] and [52]. It is clear that the two-layer prototype offers better characteristics in terms of bandwidth, gain stability, and cross-polarization. For this design, a bandwidth of 16.7 % has been achieved for gain variations of only 1.5 dB (±0.75 dB). It has been proved that the maximum measured cross-polarization (−26 dB) corresponds to that generated by the commercial feed, where the cross-polarization produced by the reflectarray is in the order of −30 dB.

11.5 SHAPED-BEAM REFLECTARRAYS

So far, the considerations have focused on the design of single-beam microstrip reflectarrays featuring single, dual or circular polarization using single- or multilayer patch/substrates. The design procedure follows the guidelines for conventional parabolic reflector antennas accompanied by reflection coefficient phase simulations of the elements forming the planar reflector.

The possibility of an independent phase adjustment for each reflectarray printed element can be used to another advantage, which concerns multibeam or shaped-beam operations. In particular, such antennas are welcome for direct broadcast satellite (DBS) application. In this application, a large-size antenna positioned on board of a stationary satellite is required to produce a number of high-gain beams, or a contoured beam, towards highly populated areas and low-density radiation towards the remaining areas, for example representing other parts of a continent or ocean. For contoured beam applications, the radiation patterns have to follow the prescribed radiation outlines, usually accompanied by very demanding specifications in terms of gain, cross-polarization discrimination, and isolation. One solution to this problem is to use a parabolic reflector antenna with a cluster of feed elements [53]. Hundreds of waveguide feeds can be employed for this purpose. This is the case of the Intelsat VI satellite [53] where two sets of 146 C-band dual-polarized horns are used for transmit and receive purposes. The use of clusters of feeds has advantages such as in-built redundancy and frequency reuse. Also, providing that the feeds are phased, the beam shape can be reconfigurable. An alternative method for producing a shaped beam is via the use of a single feed illuminating a shaped reflector. In this case, the reflector has to be shaped (deformed in a continuous manner) so that the radiated beam pattern meets the specified contour specifications. The task of obtaining the desired reflector shape is not trivial. However, it can be facilitated with the use of modern antenna design software packages such as COBRAW and POS, as offered by TICRA [54]. This software is based on physical optics and uses a minimax optimization engine driving a reflector analysis program to arrive at a smoothly curved reflector shape to meet target directivity within a given geographical contour. The output of the program is defined either in the form of Zernike polynomials or spline coefficients, which describe the reflector surface.

An already accomplished design of a shaped reflector can form the basis for designing a shaped beam microstrip reflectarray. This approach has been demonstrated in Reference [55] where the authors had an access to the parameters of the Ku-band shaped reflector, which produced a specified contour beam for the European continent. To obtain the same beam pattern with a reflectarray, they applied geometrical optics to translate the aperture phase, which was available for the shaped reflector, to the patch reflection phase. The remaining design step was realization of the phase data across the planar aperture by using variable-size patches. This step is analogous to the one for realizing a single-beam reflectarray, which has already been covered in an earlier section of this article. In Reference [55], the undertaken approach has been fully demonstrated by manufacturing and testing an elliptically shaped (110 cm $\times$ 90 cm) single-layer microstrip reflectarray illuminated by a linearly polarized horn antenna. It has been found that the manufactured reflectarray provided illumination of 99 % of the specified coverage area with the minimum desired directivity of 23 dB in the 14 GHz frequency band. In their conclusions, the authors of Reference [55] indicated that a better method would be to find the required reflectarray excitations by applying the power pattern array synthesis directly, without the intermediate step of shaped reflector synthesis. This alternative approach to designing shaped-beam reflectarrays has been demonstrated in References [56] to [60].

In contrast to Reference [55] the work in References [56] to [60] has concentrated on dual-polarization and multiple-layer microstrip reflectarrays. The use of two- and triple-layer configurations used in these designs allowed for more degrees of freedom in order to obtain the specified beam contours over a given frequency band [59]. Using two layers of patch arrays, the bandwidth of the phase-shifter element is drastically improved, as pointed out during the discussion of the various phasing mechanisms, and this is sufficient for designing moderate-size reflectarrays with an increased operational bandwidth. On the other hand, three-layer structures provide more flexibility for achieving the required reflection phase values at different frequencies, i.e. to compensate the phase delay in a frequency band. Therefore, a three-layer configuration is preferred for large reflectarrays, and particularly in contoured beam reflectarrays, to reduce the contour distortions in the operational frequency band of the antenna, as demonstrated in Reference [59].

The direct synthesis of a reflectarray amplitude pattern creates a significant challenge due to the highly nonlinear nature of the problem, which leads to the use of iterative numerical techniques to obtain the desired solution. Under the previously stated assumption that the amplitude of the reflection coefficient of each reflectarray element is unity, the amplitude of the reflections from individual cells forming the planar aperture becomes fixed by the radiation pattern of the feed. Thus, the only variables to be optimized in Equation (11.10) are the reflection coefficient phases of the reflectarray printed elements.

A phase-only synthesis technique must be applied. The power pattern synthesis method utilized in References [58] to [60] relies on the alternating projections technique, which was first applied to array pattern synthesis in Reference [61]. In this technique, two sets are considered: the available radiation patterns associated with the reflectarray geometry (R) and the radiation patterns that fulfil the required specifications with an arbitrary shape (M), both defined at the design frequency. The synthesis consists of minimizing the distance between the two sets by projecting from R to M and vice versa successively. To accomplish that task with flexibility (introduction of constraints in the excitations) and efficiency (fast algorithm even for large number of elements), suitable definitions of the corresponding projector operators are required, as described in References [60] and [61]. The choice of the starting pattern belonging to set R becomes very important in order to avoid the chance that the iterative synthesis process converges to a nondesired solution (represented by a local minimum). The more the number of elements (variables), the higher the degrees of freedom in the synthesis and, consequently, the more local minima make it difficult to achieve good convergence of the method. Two different general choices are considered. The first one takes as a starting phase distribution the one that is associated with the radiation pattern of several pencil beams that approximate the desired shaped beam [56]. The second one, simpler than the first, considers an out-of-focus beam being properly directed. Then, the initial radiation pattern is projected into set M, by modifying the gain up to the specified levels. From this *modified* radiation pattern that fulfils the requirements, both amplitude and phase distributions are obtained by using a two-dimensional fast fourier transform (FFT) algorithm. The projection into the R set is performed by changing the amplitude at each element to the value imposed by the feed. The process is repeated until the algorithm converges to a minimum distance between the two sets.

When applying the synthesis method to a reflectarray with several hundreds or thousands of elements (DBS antennas), the convergence to a local minimum where the shaped pattern is synthesized with narrow holes inside it is very common. To overcome this problem, the optimization process is carried out in several stages where the illumination on the reflectarray's edge is gradually increased from one stage to the next one [60]. The flowchart associated with the stepped pattern synthesis is shown in Figure 11.13. The feed-power radiation pattern

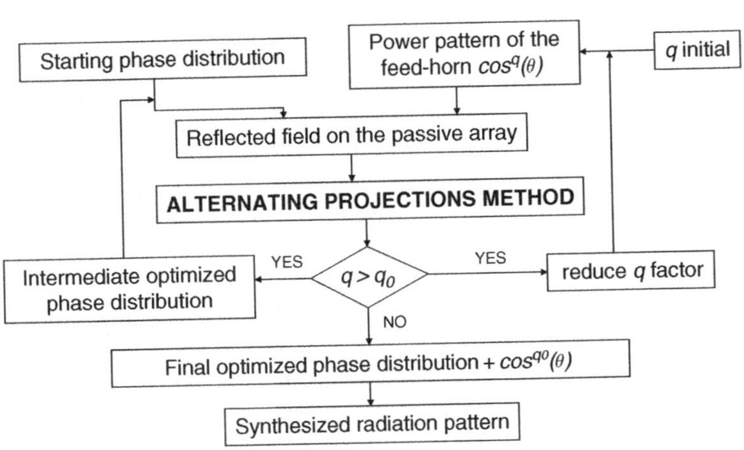

Figure 11.13 Phase-only synthesis technique consists of a several-stage process applied to reflectarrays.

is typically modelled as a cosine function to the exponent q, as in Equation (11.5), which facilitates the variation of the taper of the amplitude distribution. The higher the taper, i.e. the q factor, the smaller is the reflectarray's edge illumination. In the first stage a very high taper is usually chosen so that the illumination level near the border is very low, and hence only the central part of the reflectarray surface contributes significantly to the resulting radiation pattern. This is equivalent to reducing the size of the reflectarray and thus the number of variables, since the outer elements undergo a negligible amplitude excitation. As a result of the reduction of the number of variables, there is a smaller number of local minima. Therefore, an improvement in the convergence is expected. The phase-only synthesis based on the alternating projections technique is applied for a taper high enough that no holes appear in the coverage region. In the following stages, the taper level is gradually decreased (the level at the border increased) and the phase distribution obtained in the previous stage is used as a starting point for the synthesis method. The procedure finishes with the phase-only synthesis for the specified taper defined for q_0 due to other antenna requirements, as for instance spillover reduction, high illumination efficiency, or a compromise between the two (maximum aperture efficiency). In case the field pattern of the feed is known, a final stage of the stepped synthesis process can be realized where the cosine model is substituted by the real field pattern, obtained from simulations or measurements. This last stage does not lead to significant variations of the synthesized phase distribution if the feed antenna exhibits a quasi-axisymmetrical radiation pattern, i.e. stepped circular, Potter, and corrugated horns, as the modelled pattern does.

The above outlined procedure has been applied to designing a number of shaped beam reflectarrays, which are presented here. The first design concerns a direct broadcast satellite (DBS) application with Australia and New Zealand coverage [47, 58]. The information concerning the required contours was obtained by accessing the information about Optus B1 geostationary satellites. The available EIRP contours were translated into directivity by subtracting 17 dBW due to the transponder power and then defined by applying the student version of the program COBRAW, as offered by TICRA [54]. They are drawn in Figure 11.14(a) with a thick grey line superimposed on to the world map expressed in u and v coordinates referring to the reflectarray reference system (a translation from the satellite coordinate system into the reflectarray one has been applied). It has to be mentioned that the negative sign in the u coordinate is used to reflect the fact that the patterns are seen from the satellite. The required directivity levels are 34 dBi for New Zealand, 27 dBi for the coastal region including cities like Brisbane, Sydney, Melbourne, and Adelaide, 27 dBi for Perth, and 27 dBi for Broome, and not less than 23 dBi for other parts of the Australian continent. The specifications so defined are suitable for use in the reflectarray pattern synthesis summarized in Figure 11.13.

To achieve a contoured beam similar to that of the actual 2.4 m × 1.8 m reflector antenna used on Optus B1 satellites, an elliptical planar reflectarray with axes of 1.9 m × 1.8 m was assumed in the design. Note that Optus B1 uses a 38-element feed array while the reflectarray only assumed a single feed. In order to achieve dual-polarization operation, a periodic square cell structure forming the reflectarray was employed. With respect to the selection of substrates and spacers, a two-layer structure formed by a Kevlar layer 0.105 mm thick covering a patch array supported over another Kevlar layer 0.16 mm thick was assumed. Below this sandwiched structure a 3 mm-thick *honeycomb* layer was included. These layers were repeated and the complete two-layer planar structure was supported by a conducting ground-plane. Patch dimensions in the upper layer were assumed to be smaller by a factor of 0.67 in comparison with the patches belonging to the lower layer. The thickness of

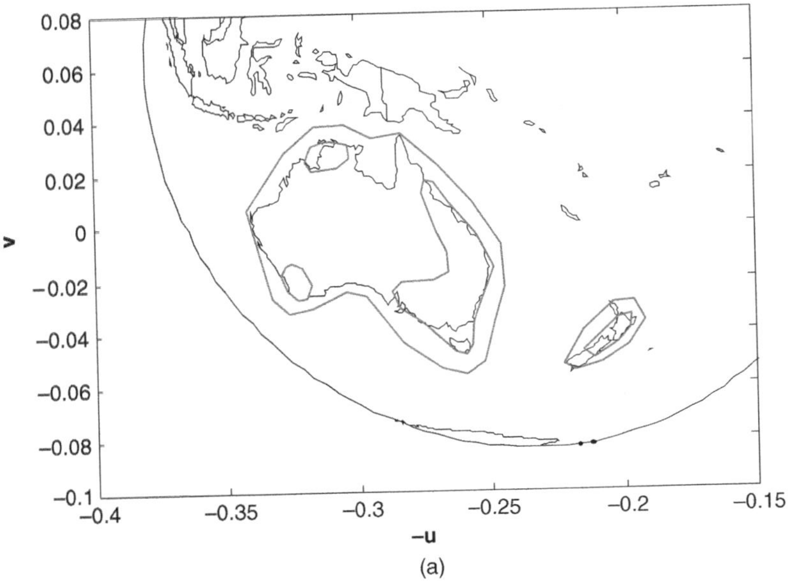

(a)

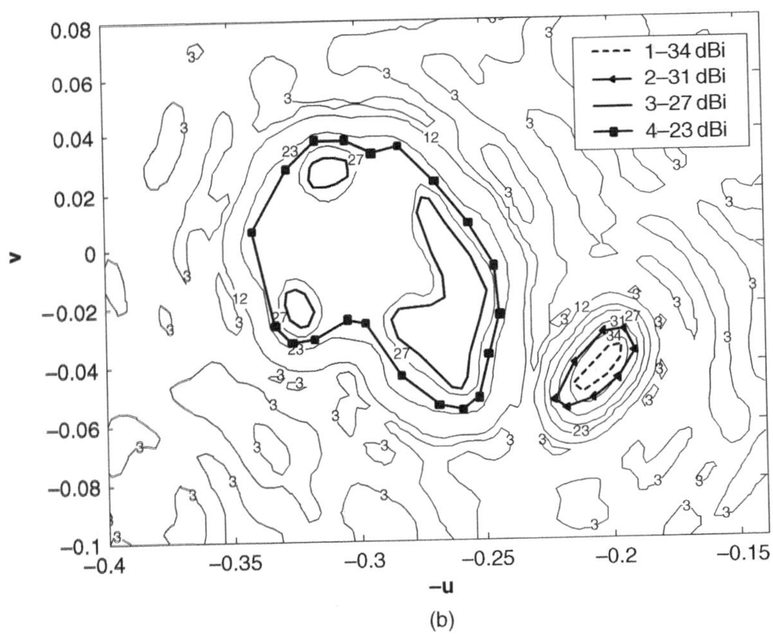

(b)

Figure 11.14 Optus B1 DBS coverage: (a) required shaped-beam contours over the world map and (b) directivity pattern of the double patch layer reflectarray at 12.5 GHz superimposed on to the specified polygons.

the whole multilayer structure was approximately 6.5 mm. To allow propagation of only the fundamental Floquet harmonic at the centre frequency of 12.5 GHz a cell size was chosen of 14 mm, which is 0.58λ at this frequency. This notion led to 13 460 elements forming the reflectarray. With respect to the feed, an offset feed position was assumed 1.775 m away from the reflectarray and 0.0 m (X direction) and −0.061 m (Y direction) with regard to its centre. The simulations were performed using the developed software described in Reference [25] and Matlab routines for displaying the obtained results.

To complete the assessment of performance of the two-layer reflectarray design, the gain radiation patterns at 12.25 and 12.75 GHz were determined. Figure 11.14(b) shows the simulated radiation patterns for X-polarization (electric field parallel to the horizon). This pattern was found to be in excellent agreement with the synthesized pattern [58]. Similar results were achieved for the Y-polarization.

11.6 MULTIBEAM REFLECTARRAYS

The design procedure of a contour beam antenna is analogous to that used for multibeam antennas. Multiple-beam reflectarrays can be designed using either single-feed or multifeed arrangements. In the latter case, the antenna configuration and the design process is similar to the one used in multifeed reflectors. Achieving several beams with a single-feed reflectarray is possible due to the flexibility that is offered by the many individual phasing elements forming a reflectarray. This task can be accomplished using multilayer patch configurations and an accurate analysis and design method described in Section 11.4, which is demonstrated as follows.

In order to radiate several pencil beams in the directions (θ_n, ϕ_n), where n is an integer number from 1 to the number of beams N, by a single reflectarray surface, the total tangential field radiated can be considered as the superposition of the individual fields associated with each beam, as given by

$$E_R(x_i, y_i) = A_R(x_i, y_i)\ \mathrm{e}^{\mathrm{j}\Phi_R(x_i, y_i)} = \sum_{n=1}^{N} A_n(x_i, y_i)\ \mathrm{e}^{\mathrm{j}\psi_n(x_i, y_i)} \tag{11.12}$$

where (x_i, y_i) are the coordinates of the centre of element i in the reflectarray (see Figure 11.15(a)), $A_n(x_i, y_i)$ and $\psi_n(x_i, y_i)$ are, respectively, the amplitude and phase on element i to radiate a beam in the direction (θ_n, ϕ_n). The phase distribution is a progressive phase, given by

$$\psi_n\,(x_i, y_i) = -k_0 \sin\theta_n \cos\phi_n x_i - k_0 \sin\theta_n \sin\phi_n y_i \tag{11.13}$$

After performing the summation of the complex field distributions, the overall required amplitude and phase distributions $A_R(x_i, y_i)$ and $\psi_R(x_i, y_i)$ in Equation (11.12) are obtained. The phase distribution corresponding to the multiple beams can be implemented by simply adjusting the phase shift at each element, as in the case of a single-beam reflectarray. When the reflectarray is designed to produce the multibeam phase distribution, but with a smooth amplitude variation imposed by the radiation pattern of the feed, normally the beams are achieved in the required directions, but with high sidelobes produced by the difference in amplitude. However, in some particular cases, as in the following example, a good multibeam performance can be achieved by applying an appropriate design method.

Consider two symmetrical beams in the YZ plane at $+27.5°$ and $-27.5°$ from the Z axis at 11.95 GHz that have to be obtained with a single feed placed in the XZ plane, as shown in Figure 11.15(a). The required field in the reflectarray surface, given by expression (11.12), becomes

$$E_R(x_i, y_i) = 2A(x_i, y_i) \cos\left[k_0 \sin(27.5°) y_i\right] \tag{11.14}$$

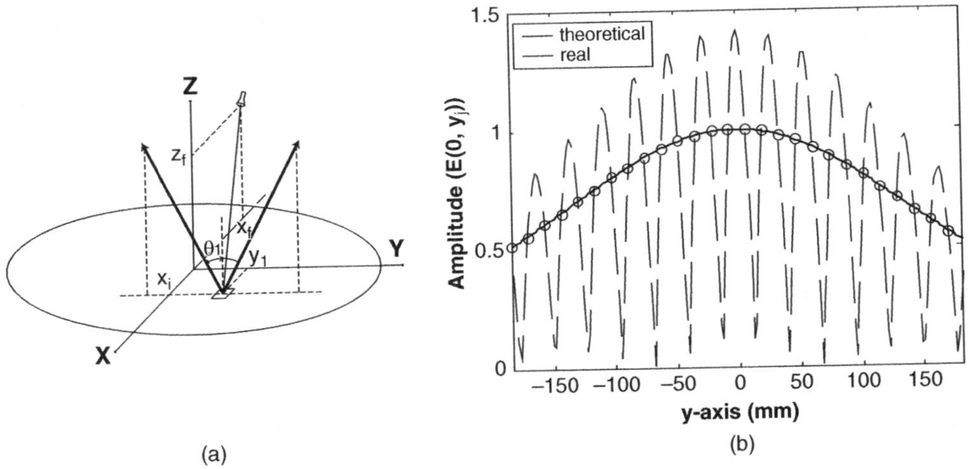

Figure 11.15 Two-beam reflectarray: (a) antenna geometry and (b) amplitude distribution in the *y* direction.

which is real-valued as a result of the symmetrical beams. The absolute value of this expression presents the periodical variations in the *y* direction associated with the cosine function, as shown in the discontinuous line of Figure 11.15(b). The required amplitude for the points separated by a semi-period (13.6 mm) is marked with circles.

For the reflectarray design the periodic cell is chosen as 13.6 mm × 13.6 mm (half a period of the oscillations in amplitude), and the radiation pattern and position of the horn are defined so that the amplitude of the incident field on the reflectarray, shown as a continuous line in Figure 11.15(b), matches the values marked with circles. Once the phase distribution and the feed position are defined, a two-layer reflectarray is designed, manufactured, and tested. The design assumes dual-linear polarization. The structure is formed by two layers of metallic rectangular patches printed on Cuclad ($\varepsilon_r = 3.4$) substrate 250 μm thick and a ground-plane, separated from each other by 3 mm Rohacell ($\varepsilon_r = 1.067$) layers. The breadboard is shown in Figure 11.16(a). The measured radiation pattern for E-polarization (tangential electric field on the reflectarray in the *x* direction) is given in Figure 11.16(b), which agrees very well with the simulated patterns. The patterns are also simulated and measured at 11.7 and 12.2 GHz, showing a deviation in the beam directions of $\pm 0.7°$ and a reduction in gain less than 0.4 dB, in the whole frequency band. Similar patterns were obtained for H-polarization but are not shown here. The dissipative losses measured in the reflectarray are 0.35 dB.

For the multifeed arrangement, a reflectarray can be designed in a similar way as in multifeed reflectors for multiple-beam applications with frequency reuse. The reflectarray can be designed to generate a beam associated with each feed. The design simulates the electrical behaviour of a parabolic reflector that generates a collimated beam in the direction (θ_0, φ_0) for a feed located at the focal point. When other feeds are located in the vicinity of the focus, several beams are generated in the directions (θ_n, ϕ_n), with subindex *n* varying from 1 to the number of feeds. Figure 11.17(a) shows the radiation patterns at 12 GHz for a 994 mm × 938 mm elliptical reflectarray with five feeds. One feed is placed at the focus and the others are located in a plane perpendicular to the line from the focus to the centre of the reflectarray. Some aberration effects can be seen in the radiation patterns for the four surrounding beams due to the feed defocusing. This problem has been studied in

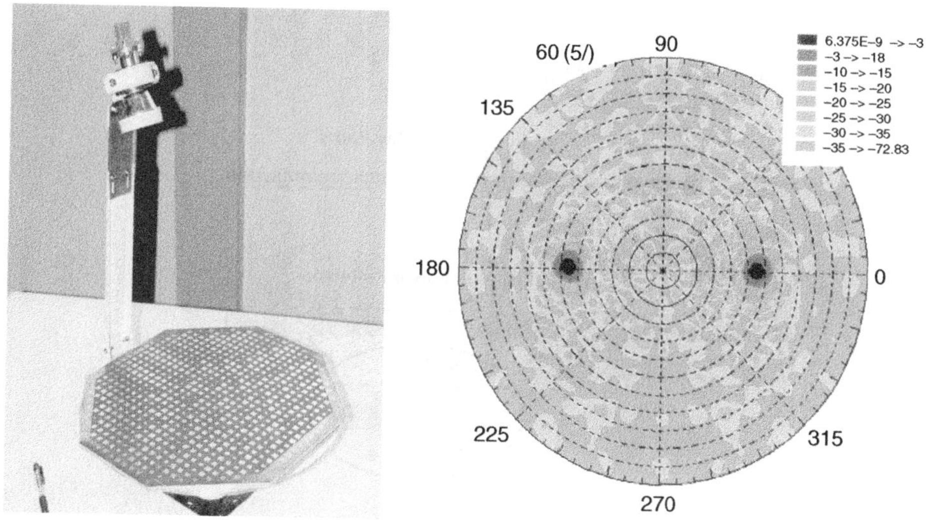

(a) (b)

Figure 11.16 Two-beam reflectarray: (a) breadboard and (b) measured radiation pattern at 11.95 GHz. (From Reference [62]. © 2003 IEEE. Reproduced by permission of the IEEE.)

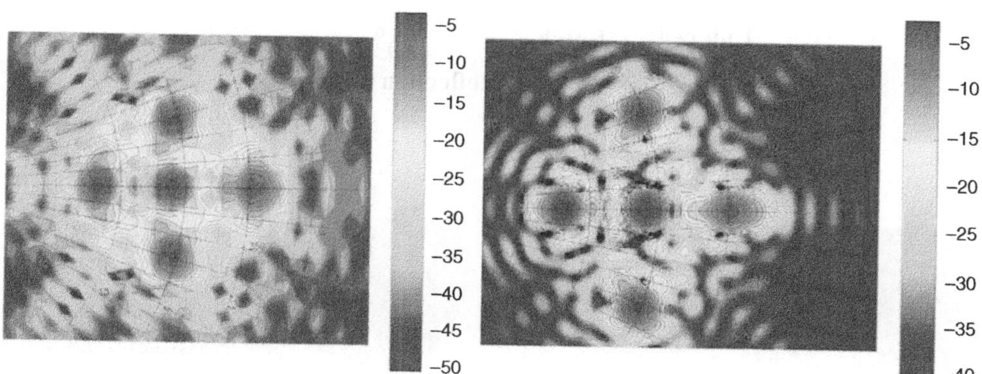

(a) (b)

Figure 11.17 Radiation patterns for a multifeed multibeam reflectarray: (a) without optimizations and (b) for optimized feed positions.

parabolic reflectors [63], and optimal feed positions can be found to minimize the aberration. Figure 11.17(b) shows the radiation patterns for the same reflectarray with the feeds located at the optimal positions [63], where an improvement on the beam contour and lower sidelobe levels are achieved. For further improvements, not only the feed positions but also the patch dimensions in the reflectarray can be optimized.

11.7 FOLDED REFLECTOR ANTENNAS

The configurations of microstrip reflectarrays that have been considered so far are similar in operation to conventional parabolic reflector antennas. This is because, like their conventional

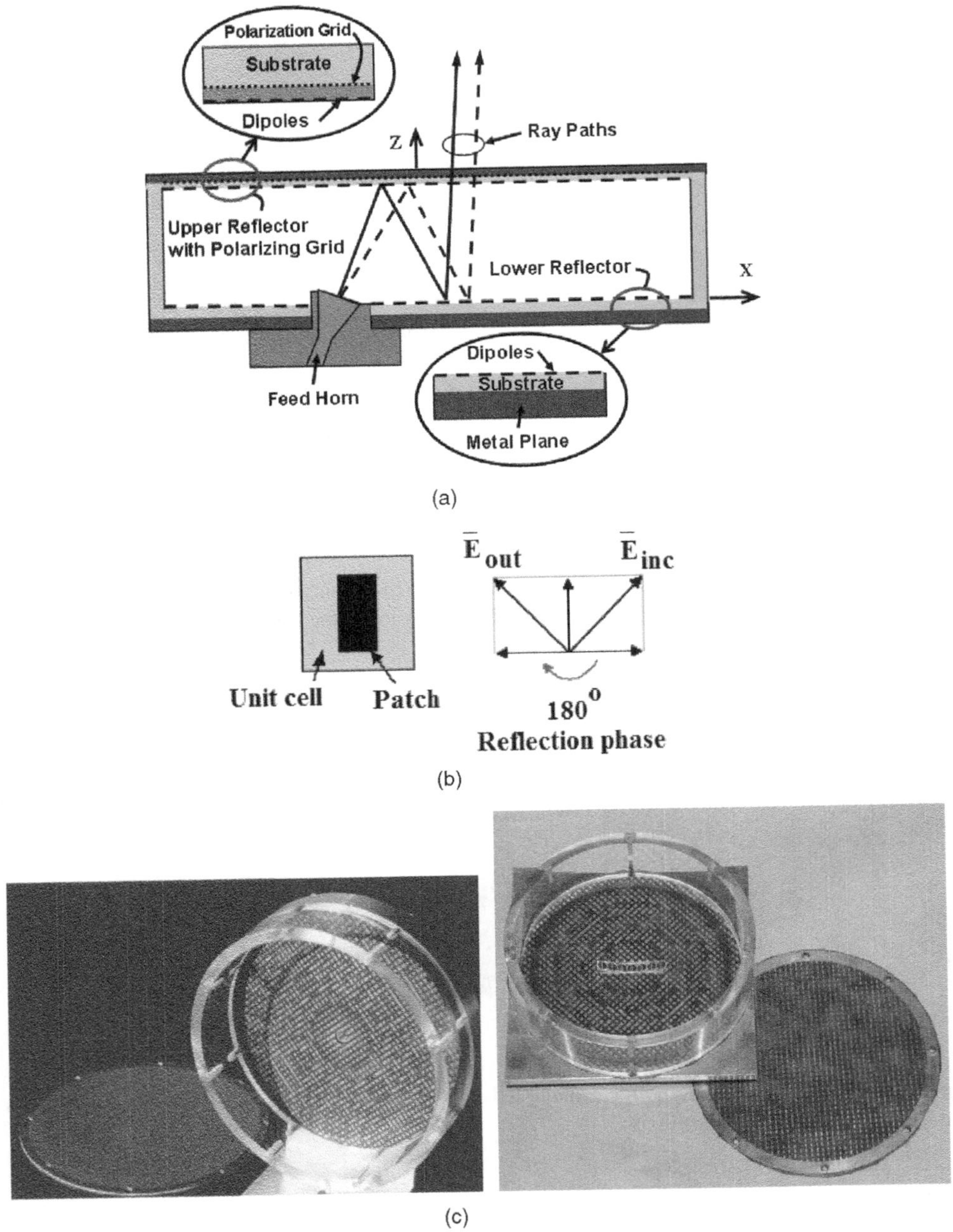

Figure 11.18 (a) Configuration of the folded reflector antenna; (b) illustration of the vector decomposition of the incident and reflected electric field for 180° of the reflection phase-angle difference; and (c) photographs of the developed prototypes for single- and multibeam operation. (From Reference [59]. © 2004 IEEE. Reproduced by permission of the IEEE.)

counterparts, they use a feed element and a reflector to produce a pencil or shaped beam. The main difference concerns the shape of the reflector. The location of the feed away from the reflector, in both antennas, is unwelcome in applications, in which the feed can be subject to an accidental damage. This is the case of radar systems for automotive applications [64]. In order to overcome this problem, a folded reflector antenna structure, which employs a reflectarray and a polarizing grid acting as an additional reflector and a shield, has been proposed in Reference [65]. The configuration of this antenna is shown in Figure 11.18.

The antenna consists of a feed located within a printed reflectarray and a planar polarization filter. As seen in Figure 11.18, the feed is a cylindrical waveguide feed horn, but a planar array antenna may equally be used. The polarization filter is formed by a grid or a resonant slot array, printed on a dielectric substrate, which acts as a radome. The detailed operation of this antenna has been explained in Reference [66].

11.8 SUMMARY

This article has covered the principles of operation and design guidelines of microstrip reflectarrays, which have been intensively researched over the last two decades at various parts of the world. The reflectarrays that have been described here are able to produce single- and multibeam operation with linear, dual-linear, or circular polarization. For the target gain of up to 40 dB they are capable of providing high radiation efficiency that is comparable with their parabolic reflector counterparts. The use of many phasing elements in the reflectarray aperture allows a contoured beam operation to be obtained that is required in direct broadcast satellite systems. Designs of such reflectarrays based on a flexible phase-only synthesis procedure, accompanied by examples of working prototypes, have been covered in detail. Another interesting application, which has been described here, deals with multibeam reflectarrays, either single-fed or multi-fed spot and/or shaped beams. The design principles accompanied by several examples of such antenna have been presented. Finally, it has been shown that the reflectarray concept can also be utilized in folded reflector antennas for automotive radar and communication applications.

ACKNOWLEDGEMENTS

The authors acknowledge the financial support of the Australian Research Council, the European Space Agency, and the Spanish Commission of Science and Technology CICYT under Project TEC-2004-02155. The authors express their gratitude to E. Carrasco and M. Arrebola for providing some results presented in this chapter. Special thanks are expressed to Dr J. Huang for supplying the photographs of inflatable reflectarrays and Prof. W. Menzel for making available pictures of his folded reflector antennas.

REFERENCES

1. K. R. Carver and J. W. Mink, 'Microstrip antenna technology', *IEEE Transactions on Antennas and Propagation*, **AP-29**(1), 2–24, January 1981.
2. P. S. Hall and C. M. Hall, 'Coplanar corporate feed effects in microstrip patch array design', *Proceedings of the IEE, Part H*, **135**, 180–186, June 1988.

3. H. J. Song and M. E. Bialkowski, 'Ku-band 16 × 16 array of aperture coupled microstrip patches', *IEEE Antennas and Propagation Magazine*, **40**(5), 25–29, 1998.
4. F. J. Goebels and K. C. Kelly, 'Arbitrary polarization from annular slot planar antennas', *IRE Transactions on Antennas and Propagation*, **AP-9**, 342–349, July 1961.
5. N. Goto and M. Yamamoto, 'Circularly polarized radial line slot antennas', *IECE Technical Report* (in Japanese), **AP80-57**, 43, August 1980.
6. M. Ando, T. Numata, J. Takada and N. Goto, 'A linearly polarised radial line slot antenna', *IEEE Transactions on Antennas and Propagation*, **AP-36**(12), 1675–1680, December 1988.
7. P. W. Davis and M. E. Bialkowski, 'Experimental investigations into a linearly polarized radial slot antenna for DBS TV in Australia', *IEEE Transactions on Antennas and Propagation*, **AP-45**(7), 1123–1129, July 1997.
8. J. Huang, 'Microstrip reflectarray', in *IEEE Antennas and Propagation Society Symposium Digest*, London, Ontario, Canada, pp. 612–615, June 1991.
9. R. E. Munson, H. A. Haddad and J. Hanlen, 'Microstrip reflectarray antenna for satellite communication and radar cross-section enhancement or reduction', US Patent 4684952, August 1987.
10. J. Huang, 'Capabilities of printed reflectarray antennas', in *Proceedings of the International Symposium on Phased Arrays*, Boston, Massachusetts, pp. 131–134, 1996; http://techreports.jpl.nasa.gov/1996/96-0862.pdf
11. D. G. Berry, R. G. Malech and W. A. Kennedy, 'The reflectarray antenna', *IEEE Transactions on Antennas and Propagation*, **AP-11**, 645–651, November 1963.
12. C. S. Malagasi, 'Microstrip disc element reflectarray', in *Electronics in Aerospace Systems Convention*, September 1978.
13. J. Huang, 'The development of inflatable array antennas', *IEEE Antennas and Propagation Magazine*, **43**(4), 44–50, August 2001.
14. D. M. Pozar, 'Bandwidth of reflectarrays', *Electronics Letters*, **39**(21), 1490–1491, October 2003.
15. J. A. Encinar and J. A. Zornoza, 'Broadband design of three-layer printed reflectarrays', *IEEE Transactions on Antennas and Propagation*, **51**(7), 1662–1664, July 2003.
16. J. A. Encinar, J. A. Zornoza, M. Arrebola, L. Datashvili, H. Baier, H. Legay and G. Toso, 'Design of a three-layer printed reflectarray for dual polarisation and dual coverage', in *27th ESA Antenna Workshop on Innovative Periodic Antennas*, Santiago de Compostela, Spain, pp. 273–280, March 2004.
17. J. A. Encinar, L. Datashvili, H. Baier, M. Arrebola, M. Sierra-Castaner, J. L. Besada, H. Legay and G. Toso, 'Breadboard of a three-layer printed reflectarray for dual polarisation and dual coverage', in *28th ESA Antenna Workshop on Space Antenna Systems and Technologies*, Noordwijk, The Netherlands, June 2005.
18. D.-C. Chang and M.-C. Huang, 'Multiple-polarization microstrip reflectarray antenna with high efficiency and low cross-polarization', *IEEE Transactions on Antennas and Propagation*, **43**(8), 829–834, August 1995.
19. R. D. Javor, X.-D. Wu and K. Chang, 'Design and performance of a microstrip reflectarray antenna', *IEEE Transactions on Antennas and Propagation*, **43**(9), 932–938, September 1995.
20. J. Huang and R. J. Pogorzelski, 'Microstrip reflectarray with elements having variable rotation angles', in *IEEE Antennas and Propagation Society Symposium Digest*, pp. 1280–1283, 1997.
21. J. Huang and R. J. Pogorzelski, 'A *Ka*-band microstrip reflectarray with elements having variable rotation angles', *IEEE Transactions on Antennas and Propagation*, **46**, 650–656, May 1998.
22. S. D. Targonski and D. M. Pozar, 'Analysis and design of a microstrip reflectarray using patches of variable size', in *IEEE Antennas and Propagation Society Symposium Digest*, pp. 1820–1823, 1994.
23. S. D. Targonski and D. M. Pozar, 'Minimization of beam squint in microstrip reflectarrays using an offset feed', in *IEEE Antennas and Propagation Society Symposium Digest*, Baltimore, Maryland, pp. 1326–1329, July 1996.

24. D. M. Pozar, S. D. Targonski and H. D. Syrigos, 'Design of millimeter wave microstrip reflectarray', *IEEE Transactions on Antennas and Propagation*, **45**, 287–295, February 1997.
25. J. A. Encinar, 'Design of two-layer printed reflectarrays using patches of variable size', *IEEE Transactions on Antennas and Propagation*, **49**(10), 1403–1410, October 2001.
26. F.-C. E. Tsai and M. E. Bialkowski, 'Designing a 161-element Ku-band microstrip reflectarray of variable size patches using an equivalent unit cell waveguide approach', *IEEE Transactions on Antennas and Propagation*, **51**(10), 2953–2962, October 2003.
27. A. Kelkar, 'FLAPS: conformal phased reflecting surfaces', in *Proceedings of the IEEE National Radar Conference*, Los Angeles, California, pp. 58–62, March 1991.
28. A. W. Robinson, M. E. Bialkowski and H. J. Song, 'A passive reflect-array with dual-feed microstrip patch elements', *Microwave and Optical Technology Letters*, **23**(5), 295–299, 5 December 1999.
29. E. Carrasco, B. Alfageme and J. A. Encinar, 'Design of a multilayer aperture-coupled cell used as phase-shifter in reflectarrays', in *13e Journées Internationales de Nice sur les Antennes (JINA)*, Nice, France, pp. 194–195, November 2004.
30. A. W. Robinson, M. E. Bialkowski and H. J. Song, 'An 137-element active reflect-array with dual-feed microstrip patch elements', *Microwave and Optical Technology Letters*, **26**(3), 147–151, August 2000.
31. M. E. Bialkowski, H. J. Song, K. M. Luk and C. H. Chan, 'Theory of an active transmit/reflect array of patch antennas operating as a spatial power combiner', in *IEEE Antennas and Propagation Society Symposium Digest*, Boston, Massachusetts, Vol. IV, pp. 764–767, July 2001.
32. F.-C. E. Tsai and M. E. Bialkowski, 'Investigation into phasing of dual-feed microstrip patch elements for use in a reflectarray', in *Proceedings of MIKON2002*, Gdansk, Poland, pp. 805–808, May 2002.
33. T. Teshirogi, M. Tanaka and W. Chujo, 'Wideband circularly polarized array antenna with sequential rotations and phase shift of elements', in *Proceedings of the International Symposium on Antennas and Propagation, ISAP*, Japan, pp. 117–120, 1985.
34. P. S. Hall, J. Huang, E. Rammos and A. Roederer, 'Gain of circularly polarized arrays composed of linearly polarised elements', *Electronics Letters*, **25**(2), 124–125, January 1989.
35. J. R. James and P. S. Hall, *Handbook of Microstrip Antennas*, Peter Peregrinus Ltd., London, pp. 330–337, 1989.
36. D. M. Pozar, 'Microstrip antennas', *Proceedings of the IEEE*, **80**, 79–91, January 1992.
37. D. M. Pozar, 'Analysis of an infinite phased array of aperture coupled microstrip patches', *IEEE Transactions on Antennas and Propagation*, **37**(4), 418–425, April 1989.
38. P. Gay-Balmaz, J. A. Encinar and J. R. Mosig, 'Analysis of a multilayer printed arrays by a modular approach based on the generalized scattering matrix', *IEEE Transactions on Antennas and Propagation*, **48**(1), 26–34, January 2000.
39. R. Mittra, C. H. Chan and T. Cwik, 'Techniques for analysing frequency selective surfaces – a review', *Proceedings of the IEEE*, **76**(12), 1593–1615, December 1988.
40. C. Wan and J. A. Encinar, 'Efficient computation of generalized scattering matrix for analyzing multilayered periodic structures', *IEEE Transactions on Antennas and Propagation*, **43**, 1233–1242, November 1995.
41. P. W. Hannan and M. A. Balfour, 'Simulations of phased-array antennas in waveguides', *IEEE Transactions on Antennas and Propagation*, **AP-13**, 342–353, May 1965.
42. D. Pilz and W. Menzel, 'Full wave analysis of a planar reflector antenna', in *Proceedings of the Asia-Pacific Microwave Conference*, Hong Kong, pp. 225–227, December 1997.
43. K. W. Lam, 'On the analysis and design of microstrip reflectarrays', PhD Dissertation, City University of Hong Kong, October 2001.
44. D. M. Pozar and T. A. Metzler, 'Analysis of a reflectarray antenna using microstrip patches of variable size', *Electronics Letters*, **29**(8), 657–658, 15 April 1993.
45. D. M. Pozar and B. Kaufman, 'Design considerations for low-sidelobe microstrip arrays', *IEEE Transactions on Antennas and Propagation*, **38**(8), 1176–1185, 1990.

46. D. M. Pozar, 'Radiation and scattering from a microstrip patch on a uniaxial substrate', *IEEE Transactions on Antennas and Propagation*, **35**(6), 613–621, 1987.
47. J. A. Zornoza and M. E. Bialkowski, 'Australia and New Zealand satellite coverage using a microstrip patch reflectarray', *Microwave and Optical Technology Letters*, **37**(5), 321–325, June 2003.
48. D. M. Pozar, 'Microstrip reflectarrays, myths and realities', in *13ᵉ Journ´ees Internationales de Nice sur les Antennes (JINA)*, pp. 175–179, November 2004.
49. J. A. Encinar, 'Design of a dual frequency reflectarray using microstrip stacked patches of variable size', *Electronics Letters*, **32**(12), 1049–1050, June 1996.
50. N. Amitay, V. Galindo and C. P. Wu, *Theory and Analysis of Phased Array Antennas*, Wiley-Interscience, New York, pp. 37–44, 1972.
51. J. A. Zornoza, R. Leberer, J. A. Encinar and W. Menzel, 'Folded multi-layer microstrip reflectarray with shaped pattern', *IEEE Transactions on Antennas and Propagation*, **54**(2, Part 2), 510–518, February 2006.
52. M. Patel and J. Thraves, 'Design and development of a low cost, electronically steerable, X-band reflectarray using planar dipoles', in *Proceedings of Military Microwaves*, London, pp. 174–179, 1994.
53. D. Roddy, *Satellite Communications*, Prentice-Hall, Englewood Cliffs, New Jersey, 1989.
54. http://www.ticra.com.
55. D. M. Pozar, S. D. Targonski and R. Pokuls, 'A shaped-beam microstrip patch reflectarray', *IEEE Transactions on Antennas and Propagation*, **47**(7), 1167–1173, July 1999.
56. J. A. Zornoza and J. A. Encinar, 'Multi-layer printed reflectarrays as an alternative to shaped reflectors', in *Proceedings of the 24th ESTEC Antenna Workshop on Innovative Periodic Antennas*, Noordwijk, Holland, pp. 243–247, June 2001.
57. J. A. Zornoza and J. A. Encinar, 'Design of shaped beam reflectarrays for direct broadcast satellites', in *12ᵉ Journées Internationales de Nice sur les Antennes (JINA)*, Nice, France, pp. 367–370, November 2002.
58. J. A. Zornoza, J. A. Encinar and M. E. Bialkowski, 'A double-layer microstrip reflectarray design to obtain Australia and New Zealand footprint', in *IEEE Antennas and Propagation Society Symposium Digest*, Vol. 3, pp. 310–313, June 2003.
59. J. A. Encinar and J. A. Zornoza, 'Three-layer printed reflectarrays for contoured beam space applications', *IEEE Transactions on Antennas and Propagation*, **52**(5), 1138–1148, May 2004.
60. J. A. Zornoza and J. A. Encinar, 'Efficient phase-only synthesis of contoured beam patterns for very large reflectarrays', *International Journal of RF and Microwave Computer-Aided Engineering*, 415–423, September 2004.
61. O. M. Bucci, G. Franceschetti, G. Mazzarella and G. Panariello, 'Intersection approach to array pattern synthesis', *IEE Proceedings*, **137**(6), 349–357, December 1990.
62. J. A. Zornoza and M. E. Bialkowski, 'Design of a shaped beam multi-layer microstrip reflectarray with Australia and New Zealand Coverage pattern', *IEEE phased Array Systems and Technology Symposium*, Boston, USA, October 2003.
63. C. J. Sletten, *Reflector and Lens Antennas*, Chapter 7, Artech House, Norwood, Massachusetts, 1988.
64. W. Menzel, D. Pilz and R. Leberer, 'A 77-GHz FM/CW radar front-end with a low-profile, low-loss printed antenna', *IEEE Transactions on Microwave Theory and Technology*, **47**(12), 2237–2241, December 1999.
65. D. Pilz and W. Menzel, 'Folded reflectarray antenna', *Electronics Letters*, 832–833, April 1998.
66. W. Menzel, D. Pilz and M. Al-Tikriti, 'Millimeter-wave folded reflector antennas with high gain, low loss, and low profile', *IEEE Antennas and Propagation Magazine*, **44**(3), 24–29, June 2002.

12

Antennas on Artificial Magnetic Conductors and Other Impedance Surfaces

Daniel F. Sievenpiper
HRL Laboratories LLC, Malibu, California, USA

12.1 INTRODUCTION

Surface textures have been developed that can alter the electromagnetic boundary condition of a metal surface to provide a variety of useful properties [1, 2]. Known as high-impedance surfaces, or artificial magnetic conductors, these materials are typically built as subwavelength mushroom-shaped metal protrusions. They can be analysed as resonant LC circuits, in which the capacitance is provided by the proximity of the metal plates, and the inductance is related to the thickness of the structure. Near their LC resonance frequency, these materials provide a high-impedance boundary condition, and reflect incoming waves with a phase shift of 0, rather than π as with an ordinary electric conductor. They are sometimes known as artificial magnetic conductors because the tangential magnetic field is zero at the surface, rather than the electric field, as with an electric conductor.

In addition to their unusual reflection phase properties, these materials have a surface wave bandgap, within which they do not support bound surface waves of either transverse magnetic (TM) or transverse electric (TE) polarization. They may be considered as a kind of electromagnetic bandgap structure, or photonic crystal [3, 4] for surface waves [5]. Although bound surface waves are not supported, leaky TE waves can propagate within the bandgap, which can be useful for certain applications.

Engineered electromagnetic surface textures can be used to provide several important properties for antenna applications. Because of their unusual boundary condition, high-impedance surfaces can be used to build new kinds of low-profile antennas and to control diffraction from metal structures. These materials can also be electronically tuned by incorporating varactor diodes between the capacitive plates. Tunable impedance surfaces can be

Printed Antennas for Wireless Communications Edited by R. Waterhouse

used as simple, low-profile electronically steerable antennas. Finally, borrowing concepts from holography, impedance surfaces can be patterned to provide arbitrary radiation patterns from conformal surfaces.

12.2 HIGH-IMPEDANCE SURFACES

High-impedance surfaces typically consist of an array of metal protrusions on a flat metal sheet. The protrusions are arranged in a two-dimensional lattice, and can be visualized as mushrooms or thumbtacks protruding from the surface. High-impedance surfaces are typically constructed as printed circuit boards, where the bottom side is a solid metal ground-plane and the top contains an array of small ($<< \lambda$) metal patches, as shown in Figure 12.1. The plates are connected to the ground-plane by metal-plated vias to form a continuous conductive metal texture. For greater capacitance, multilayer circuit boards with overlapping plates can be used.

When the period is small compared to the wavelength of interest, it is possible to analyse the material as an effective medium, with its surface impedance defined by effective lumped-element circuit parameters that are determined by the geometry of the surface texture. A wave impinging on the material causes electric fields to span the narrow gaps between the neighbouring metal patches, which can be described as an effective sheet capacitance C. As currents oscillate between the neighbouring patches, the conducting paths through the vias and the ground-plane provide a sheet inductance L. These form a parallel resonant circuit that dictates the electromagnetic behaviour of the material. Its surface impedance is given by the following expression:

$$Z_s = \frac{j\omega L}{1 - \omega^2 LC} \tag{12.1}$$

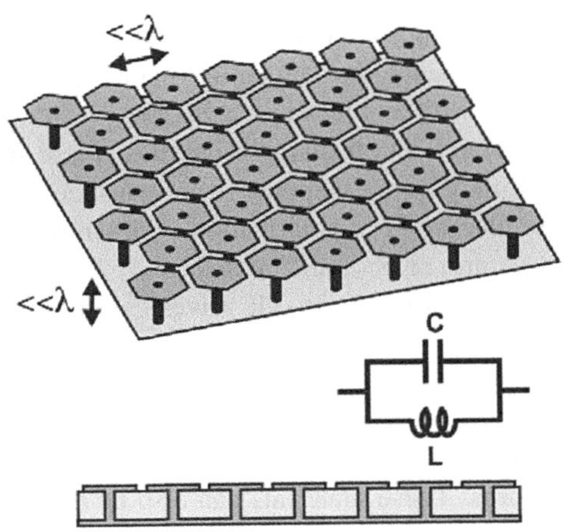

Figure 12.1 A high-impedance surface is built as a thin two-dimensional lattice of plates attached to a ground-plane by metal-plated vias. The plates provide capacitance and the thickness of the surface provides inductance. It may be modelled as an effective impedance surface, with the impedance defined by a parallel resonant LC circuit. It has high electromagnetic impedance near its LC resonance frequency.

The resonance frequency of the circuit is given by

$$\omega_0 = \frac{1}{\sqrt{LC}} \tag{12.2}$$

Below resonance the surface is inductive and above resonance the surface is capacitive. Near ω_0, the surface impedance is much higher than the impedance of free space.

12.3 SURFACE WAVE BANDS

The general behaviour of surface waves on an impedance surface is derived in several electromagnetics textbooks [6]. The derivation proceeds by assuming a surface having an impedance Z_s and a wave that decays exponentially away from the surface with the decay constant α, as shown in Figure 12.2.

It can be shown that TM waves occur on an inductive surface, in which the surface impedance is given by the following expression:

$$Z_s = \frac{j\alpha}{\omega\varepsilon} \tag{12.3}$$

Conversely, TE waves can occur on a capacitive surface, with the following impedance:

$$Z_s = \frac{-j\omega\mu}{\alpha} \tag{12.4}$$

In the above expressions, ε and μ are the permittivity and permeability of the space surrounding the surface, which may be a vacuum, and ω is the angular frequency of the wave.

Many of the important properties of the high-impedance surface can be explained using the effective surface impedance model. The surface is assigned an impedance equal to that of a parallel resonant LC circuit, as described above. The use of lumped circuit parameters to describe electromagnetic structures is valid as long as the wavelength is much longer than the size of the individual features. The effective surface impedance model can predict the reflection properties and some features of the surface wave band structure, but not the bandgap itself, which by definition must extend to large wave vectors.

The wave vector, k, is related to the spatial decay constant, α, into the surrounding space, and the frequency, ω, by the following dispersion relation:

$$k^2 = \mu_0\varepsilon_0\omega^2 + \alpha^2 \tag{12.5}$$

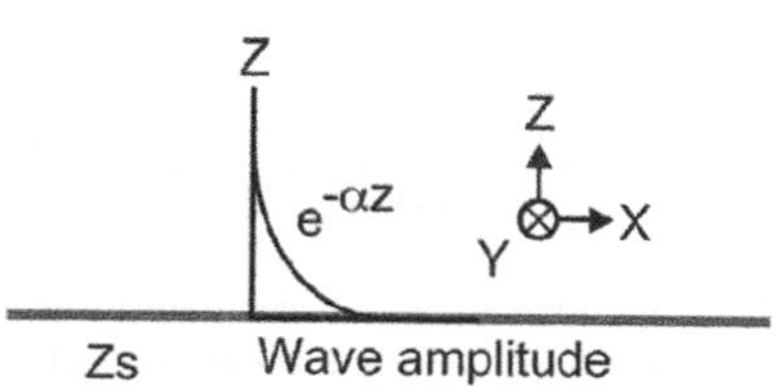

Figure 12.2 A surface wave is a wave that is bound to a surface and decays into the surrounding space.

For TM waves Equation (12.5) can be combined with Equation (12.1) to find the following expression for k as a function of ω, in which μ is the impedance of free space and c is the speed of light in vacuum:

$$k = \frac{\omega}{c}\sqrt{1 - \frac{Z_s^2}{\eta^2}} \qquad (12.6)$$

A similar expression can be found for TE waves by combining Equations (12.5) and (12.2):

$$k = \frac{\omega}{c}\sqrt{1 - \frac{\eta^2}{Z_s^2}} \qquad (12.7)$$

Inserting Equation (12.3) into Equations (12.6) and (12.7), the dispersion diagram can be plotted for surface waves, in the context of the effective surface impedance model. An example of the complete dispersion diagram, calculated using the effective medium model, is shown in Figure 12.3.

Below resonance, TM surface waves are supported. At low frequencies, they lie very near the light line, indicated in Figure 12.3 by the dotted line with a slope equal to the speed of light, c. The fields extend many wavelengths beyond the surface, as they do on a flat metal sheet. Near the resonant frequency, the surface waves are tightly bound to the surface, and have a very low group velocity. The dispersion curve is bent over away from the light line. In the effective surface impedance limit, there is no Brillouin zone [7] boundary and the TM dispersion curve approaches the resonance frequency asymptotically. Thus, this approximation does not predict the bandgap.

Above the resonance frequency, the surface is capacitive and TE waves are supported. The lower end of the dispersion curve is close to the light line and the waves are weakly bound to the surface, extending far into the surrounding space. As the frequency is increased, the curve bends away from the light line and the waves are more tightly bound to the surface. The slope of the dispersion curve indicates that the waves feel an effective index of refraction that is greater than unity. This is because a significant portion of the electric field is concentrated in the capacitors.

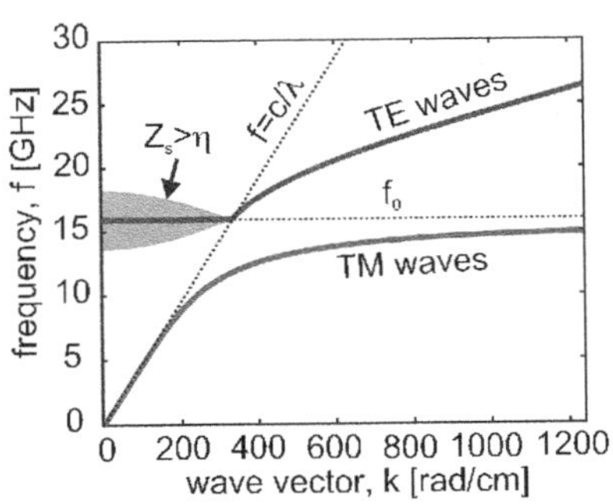

Figure 12.3 The effective surface impedance model can determine many of the properties of the high-impedance surface, including the shape and polarization of the surface wave bands. This is the predicted surface wave dispersion diagram for a surface with sheet capacitance of 0.05 pF and sheet inductance of 2 nH. The surface supports TM waves below the resonance frequency and TE waves at higher frequencies. This model does not predict the bandgap, but it does predict a region of radiative loss.

The TE waves that lie to the left of the light line exist as leaky waves that are damped by radiation, which can be modelled as a resistor in parallel with the high-impedance surface. The damping resistance is the impedance of free space, projected on to the surface at the angle of radiation. This blurs the resonance frequency, so the leaky waves actually radiate within a finite bandwidth. Small wave vectors represent radiation perpendicular to the surface, while wave vectors near the light line represent radiation at grazing angles. In place of a band gap, the effective surface impedance model predicts a frequency band characterized by radiation damping.

In the effective impedance surface model described above, the properties of the textured surface are summarized into a single parameter – the surface impedance. This model correctly predicts the shape and polarization of the surface wave bands, and also the reflection phase, to be described later. However, it does not predict the bandgap itself. For a more accurate picture of the surface wave properties, a finite element numerical model can be used. The metal and dielectric regions are discretized on a grid, and the electric field at all points on the grid is described in terms of an eigenvalue equation, which may be solved numerically. A single unit cell is simulated and Bloch boundary conditions [8] are used. The calculation yields the allowed frequencies for each wave vector. An example high-impedance surface is shown in Figure 12.4, along with the calculated dispersion diagram. The lowest band is TM, the second band is TE, and both have a similar shape to that predicted by the effective surface impedance model. A bandgap, within which the surface does not support bound

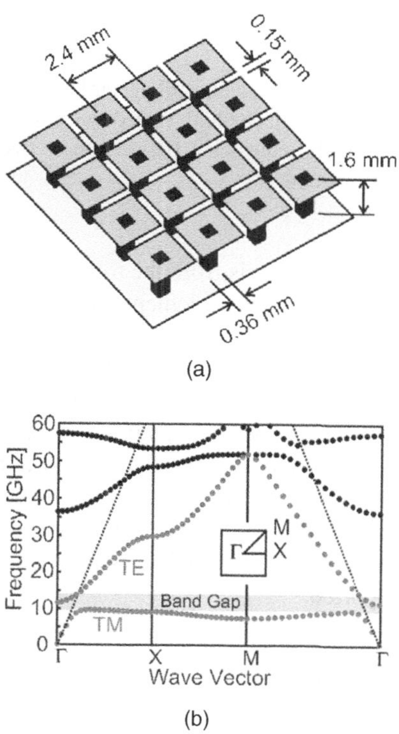

Figure 12.4 (a) The complete dispersion diagram can be obtained accurately using numerical methods, and square lattices are often easier to simulate. The substrate, which is not shown, has a relative dielectric constant of 2.2. (b) The lowest bands are qualitatively similar to that of the effective surface impedance model. The finite element model also predicts a bandgap where bound surface waves of neither polarization are supported, between the first two bands. It also predicts several higher bands.

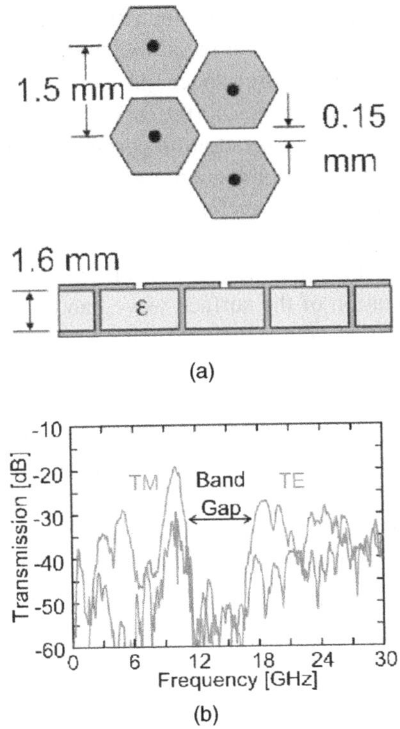

Figure 12.5 (a) Measurements were performed on a triangular lattice of hexagons, built on a substrate with a relative dielectric constant of 2.2. (b) The high-impedance surface supports TM surface waves at low frequencies and TE waves at high frequencies. Between these two bands is a gap, within which waves of neither polarization are supported.

surface waves of either polarization, extends from the top of the TM band to the point where the TE band crosses the light line. The finite element model also predicts additional higher order bands that are not predicted by the simple effective surface impedance model.

Surface wave modes can be measured by recording the transmission between a pair of small coaxial probes placed near the surface. Depending on their orientation, the probes will excite surface waves with TM, TE, or both polarizations. An example of a high-impedance surface and the measured surface wave transmission across a 12 cm sample is shown in Figure 12.5 for both TM and TE polarizations. As predicted by both models described above, TM waves are supported at low frequencies and TE waves are supported at high frequencies. The TM and TE bands are separated by a bandgap within which bound surface waves of either polarization are not supported.

12.4 REFLECTION PHASE

The surface impedance defines the boundary condition at the surface for the standing wave formed by incident and reflected waves. For a low-impedance surface, such as an electric conductor, the ratio of electric field to magnetic field is small. The electric field has a node at the surface and the magnetic field has an antinode. Conversely, for a high-impedance

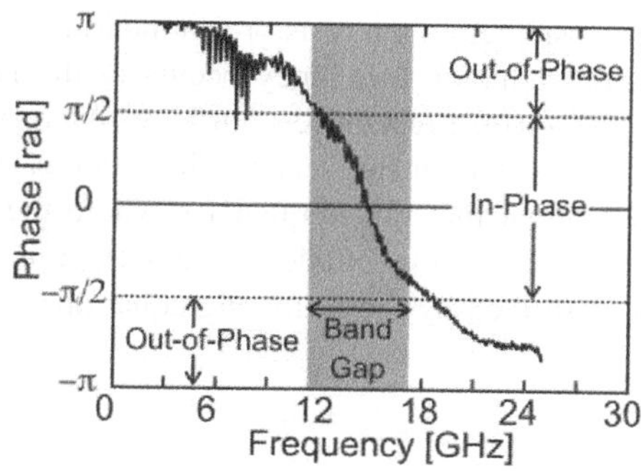

Figure 12.6 The reflection phase was measured for the same surface as shown in Figure 12.5. The phase is zero at the resonance frequency, but it approaches π for frequencies far from the resonance. The phase crosses through $\pi/2$ and $-\pi/2$ near the edges of the surface wave bandgap.

surface the electric field has an antinode at the surface while the magnetic field has a node. Another term for such a surface is again an artificial magnetic conductor. Using the effective surface impedance model described previously, the reflection phase can be determined for the resonant textured surface described above. For a normally incident wave, the reflection phase of the surface is given as follows:

$$\Phi = \mathrm{Im}\left\{ \ln\left(\frac{Z_s - \eta}{Z_s + \eta} \right) \right\} \tag{12.8}$$

In the above expression, Z_s is given by Equation (12.1) and η is the impedance of free space. At very low frequencies, the reflection phase is π and the structure behaves like a smooth metal surface. At higher frequencies, the reflection phase slopes downward and eventually crosses through zero at the resonance frequency, where it behaves as a magnetic conductor. Above the resonance frequency, the phase returns to $-\pi$. The phase falls within $\pi/2$ and $-\pi/2$ when the magnitude of the surface impedance exceeds the impedance of free space. The behaviour of the reflection phase predicted by Equation (12.8) is identical to the measured result shown in Figure 12.6. This reflection phase curve was measured using the surface shown in Figure 12.5. It is worth noting that for a wide range of geometries, the edges of the surface wave bandgap occur at the same frequencies where the reflection phase crosses through $\pi/2$ and $-\pi/2$.

12.5 BANDWIDTH

An antenna lying parallel to the textured surface will see the impedance of free space on one side and the impedance of the surface on the other side. Where the textured surface has low impedance, far from the resonance frequency, the antenna current is mirrored by an opposing current in the surface. Since the antenna is shorted-out by the nearby conductor, the radiation efficiency is very low. Within the bandgap near resonance, the textured surface has a much higher impedance than that of free space, so the antenna is not shorted-out. In this range of frequencies, the radiation efficiency is high.

The textured surface is modelled as an LC circuit in parallel with the antenna, and the radiation into free space is modelled as a resistor with a value of the impedance of free space. The amount of power dissipated in the resistor is a measure of the radiation efficiency of the antenna. The maximum radiation efficiency occurs at the LC resonance frequency of the ground-plane, where the surface reactance is infinite. At very low frequencies, or at very high frequencies, currents in the surface cancel the antenna current and the radiated power is reduced. It can be shown that the frequencies where the radiation drops to half of its maximum value occur when the magnitude of the surface impedance is equal to the impedance of free space, as described by the following equation:

$$|Z_s| = \eta \tag{12.9}$$

In the above expression, η is the impedance of free space, and is given by the following:

$$\eta = \sqrt{\mu_0/\varepsilon_0} \tag{12.10}$$

Substituting Z_s from Equation (12.1) into (12.9), allows ω to be solved:

$$\omega^2 = \frac{1}{LC} + \frac{1}{2\eta^2 C^2} \pm \frac{1}{\eta C}\sqrt{\frac{1}{LC} + \frac{1}{4\eta^2 C^2}} \tag{12.11}$$

The terms in $1/(\eta C)^2$ are typically small compared to the terms in $1/(LC)$, so will be neglected. This approximation yields the following equation for the edges of the operating band:

$$\omega = \omega_0\sqrt{1 \pm \frac{Z_0}{\eta}} \approx \omega_0\left(1 \pm \frac{1}{2}\frac{Z_0}{\eta}\right) \tag{12.12}$$

In the above expression, Z_0 can be considered as a kind of characteristic impedance of the surface:

$$Z_0 = \sqrt{L/C} \tag{12.13}$$

The two frequencies designated by the $\pm$ signs in Equation (12.12) delimit the range over which an antenna would radiate efficiently on such a surface. The total bandwidth (BW) is roughly equal to the characteristic impedance of the surface divided by the impedance of free space:

$$\text{BW} = \frac{\Delta\omega}{\omega_0} \approx \frac{Z_0}{\eta} = \frac{\sqrt{L/C}}{\sqrt{\mu_0/\varepsilon_0}} \tag{12.14}$$

This is the bandwidth over which the phase of the reflection coefficient falls between $\pi/2$ and $-\pi/2$, and image currents are more in-phase than out-of-phase. As noted in the previous section, this range often coincides with the surface wave bandgap. It also represents the maximum usable bandwidth of a flush-mounted antenna on a resonant surface of this type.

It can be shown that the inductance of the surface L is equal to the product of the permeability μ and the thickness t. Using Equation (12.14) and substituting for L in Equation (12.13), a more useful expression can be obtained for the bandwidth of a thin ($t << \lambda$) nonmagnetic ($\mu = \mu_0$) resonant textured ground-plane:

$$\mathrm{BW} = \frac{2\pi}{\lambda_0} t \tag{12.15}$$

In the above expression, λ_0 is the free-space wavelength at the resonance frequency. This result is significant because it proves that the bandwidth is determined entirely by the thickness of the surface, with respect to the operating wavelength. Note that the dielectric constant of the substrate has no direct effect on the bandwidth, and dielectric loading cannot be used to reduce the thickness, except at the expense of the bandwidth. A similar limitation exists for all small antennas, and their bandwidth is determined by a relation analogous to Equation (12.15), for three-dimensional problems [9–11].

12.6 ANTENNAS ON HIGH-IMPEDANCE SURFACES

In the frequency range where the surface impedance is very high, the tangential magnetic field is small, even with a large electric field along the surface. Such a structure is sometimes described as an artificial magnetic conductor. Because of this unusual boundary condition, the high-impedance surface can function as a new type of ground-plane for low-profile antennas. The image currents in the ground-plane are in-phase with the antenna current, rather than out-of-phase, allowing radiating elements to lie directly adjacent to the surface, while still radiating efficiently. For example, a wire antenna lying flat against a high-impedance ground-plane is not shorted as it would be on an ordinary metal sheet.

An example of a horizontal low-profile antenna on a high-impedance ground-plane is shown in Figure 12.7(a). A simple wire antenna is fed through the back of the surface by a coaxial cable, in a manner similar to the monopole, and is bent over across the surface. The wire is typically about a half-wavelength long at the resonance frequency of the surface. On a flat metal ground-plane, a horizontal wire is shorted-out, and most of the power transmitted to the feed is reflected back. However, on the high-impedance surface, a horizontal wire antenna is well matched if operated within the bandgap, as shown by the return loss in Figure 12.7(b). The radiation pattern in Figure 12.7(c) indicates that the antenna produces significant gain, despite being roughly 1 mm above the ground-plane. This is because the reflection phase of the surface is 0, rather than π, as with an ordinary conductor. Thus, currents in the high-impedance surface reinforce the currents in the wire, instead of cancelling them as a smooth metal surface does. This effect can be used to build a variety of low-profile antennas that can lie directly adjacent to the artificial magnetic ground-plane, such as antennas with various polarizations, including circular, as well as various directive radiation patterns.

12.7 DIFFRACTION CONTROL

Surface textures such as artificial impedance surfaces can be used to control the diffraction properties of a metal object for one or both polarizations. One way to visualize this is through the support or suppression of surface waves. The high-impedance surface described

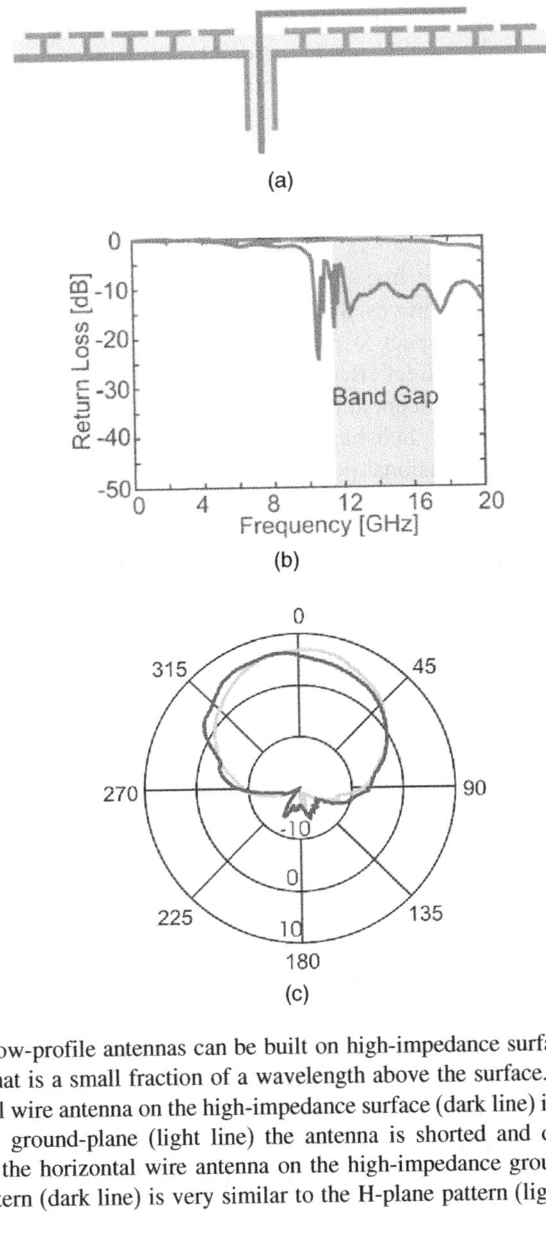

Figure 12.7 (a) Low-profile antennas can be built on high-impedance surfaces, such as a horizontal bent wire antenna that is a small fraction of a wavelength above the surface. (b) The measured return loss of the horizontal wire antenna on the high-impedance surface (dark line) is low within the bandgap. On a smooth metal ground-plane (light line) the antenna is shorted and does not radiate. (c) The radiation pattern of the horizontal wire antenna on the high-impedance ground-plane is symmetrical, and the E-plane pattern (dark line) is very similar to the H-plane pattern (light line).

above suppresses TM, TE, or both polarizations, depending on the resonance frequency of the surface and the bandgap, in relation to the frequency of interest.

One example of where diffraction control can be useful is in the generation of circular polarization. The task of designing an antenna that transmits or receives in circular polarization over a wide range of angles is often complicated by the presence of the metallic structures on which the antenna is mounted. For example, antennas on a flat metal ground-plane will tend to emit in vertical polarization at low angles, because horizontal fields are shorted by the metal surface while vertical fields can propagate along the metal. The polarization purity of a wave is often described in terms of its axial ratio, which is the ratio of the major axis to the minor axis of the polarization ellipse.

A simple aperture antenna, shown in Figure 12.8, demonstrates the use of the high-impedance surface as a means of improving the symmetry of the radiation pattern. The aperture was the open end of a standard Ku-band rectangular waveguide, which was attached to a similarly sized rectangular hole in the centre of a high-impedance ground-plane. In this particular case, the surface has a bandgap that spans from 12 to 18 GHz. The results from this surface can be compared to an identical aperture antenna with a metal ground-plane of the same size.

While the radiation pattern is somewhat affected by the shape of the aperture, it is primarily determined by the geometry of the surrounding ground-plane and the electromagnetic boundary condition of that surface. The flat metal ground-plane supports the propagation of TM polarized waves, because in these waves the electric field is perpendicular to the metal surface. Waves with this polarization can propagate for long distances in close proximity to a metal surface. For this reason, the E-plane radiation pattern in Figure 12.9(a) is quite broad. TE waves, on the other hand, cannot propagate at grazing angles to a metal surface because their transverse electric field is shorted by the conducting surface. The H-plane is therefore much narrower. This is the expected radiation pattern for an aperture antenna in a conducting ground-plane.

On the textured ground-plane, the pattern is much more symmetrical, as shown in Figure 12.9(b). This can be attributed to the suppression of both TM and TE surface waves near the resonance frequency. The gain is also higher in the forward direction, which can be attributed to standing waves that occur at the resonance frequency and surround the aperture, and cause a slight increase in the effective aperture area. The radiation patterns shown here are for 13 GHz, within the bandgap. The antenna produces a similar pattern throughout most of the bandgap region. However, as the frequency is increased towards the upper edge where leaky TE waves are supported, the H-plane actually becomes broader than the E-plane, as shown in Figure 12.9(c). If a ground-plane were to behave as a magnetic conductor, a broad H-plane and a narrow E-plane would be expected – the opposite of the electric conductor. Thus, in this way the textured surface mimics a magnetic conductor.

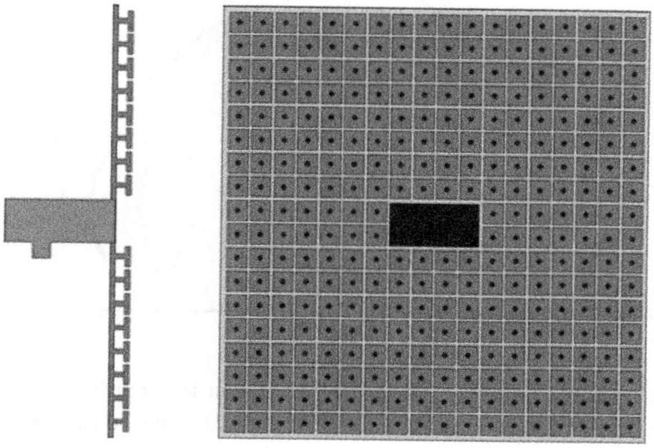

Figure 12.8 The side view (left) and front view (right) of our aperture antenna in a high-impedance surface. The unit cells of the high-impedance surface measure 3.7 mm and the size of the ground-plane (not shown to scale) is 12.7 cm. The aperture is fed by a coax to Ku-band rectangular waveguide transition.

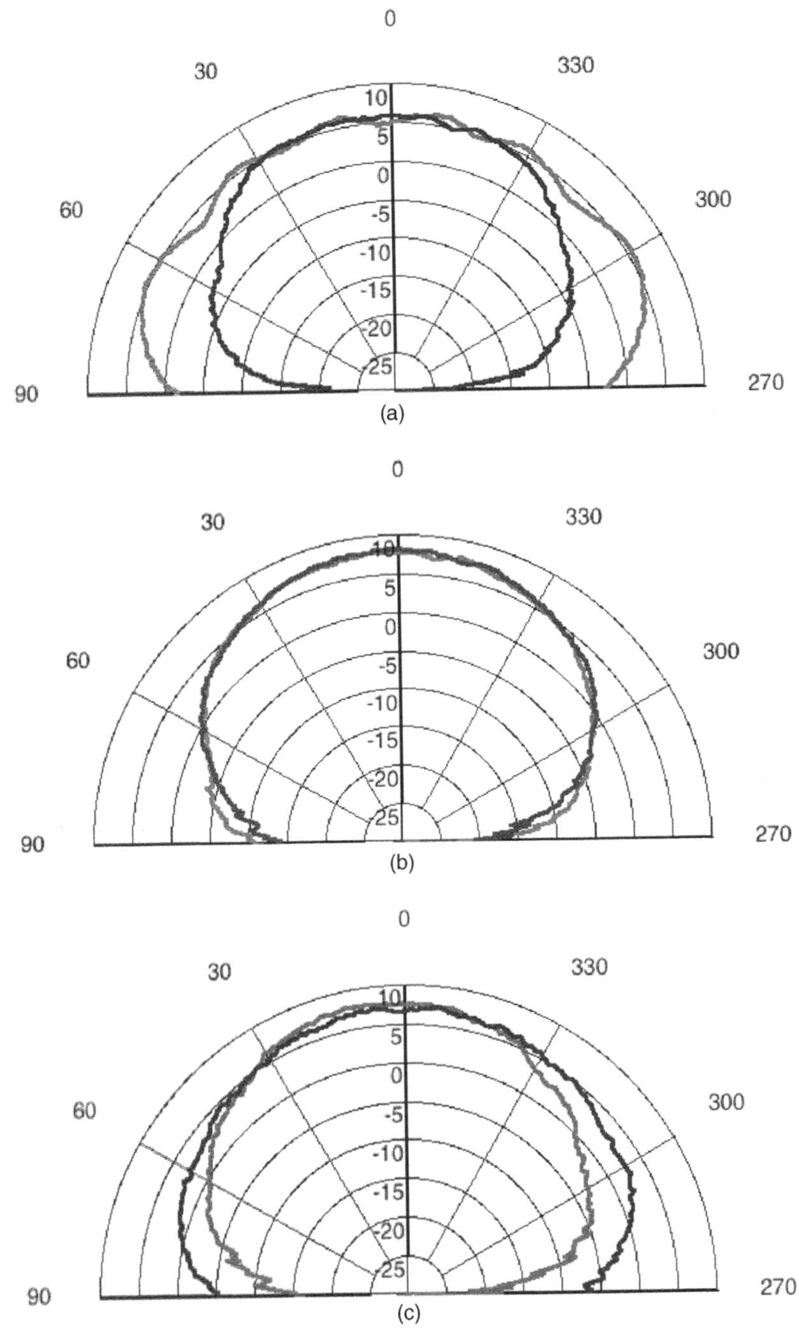

Figure 12.9 (a) The radiation pattern of an aperture antenna in a conventional metal ground-plane. The E-plane (light line) has a much broader pattern than the H-plane (dark line) because horizontally polarized waves are shorted out by the metal surface, while vertically polarized waves are allowed. (b) The radiation pattern of the aperture antenna in the high-impedance surface. The E-plane and H-plane are nearly identical at the resonance frequency of the surface, where waves of both polarization are suppressed. (c) The radiation pattern of the aperture antenna in the high impedance ground-plane near the edge of the TE band. The H-plane is broader than the E-plane, as would be expected if the surface were a magnetic conductor.

12.8 TUNABLE IMPEDANCE SURFACES

The resonance frequency and the reflection phase of a high-impedance surface can be tuned by changing its effective capacitance by adding tunable lumped capacitors. Because the reflection phase is determined by the frequency of the incoming wave with respect to the resonance frequency, such a surface can perform as a distributed phase-shifter, and can be used to build simple electronically steerable antennas.

An electrically tunable impedance surface can be built by connecting neighbouring cells with varactor diodes. Changing the bias voltage on the diodes adjusts their capacitance and tunes the resonance frequency. In order to supply the required voltage to all of the varactors, half of the cells are alternately biased and the other half are ground in a checkerboard pattern, as shown in Figure 12.10. At the centre of each biased cell, a metal via passes through a hole in the ground-plane and connects to a control line located on a separate circuit layer on the back of the surface. The varactors are oriented in opposite directions in each alternate row, so that when a positive voltage is applied to the control lines, all the diodes are reverse-biased. By individually addressing each cell, the reflection phase can be programmed as a function of position across the surface. By programming the reflection phase as a function of position, the tunable impedance surface can be used as an electronically steerable reflector [12].

While a reflector geometry is the simplest way of building electronically steerable antenna with the tunable impedance surface, it is only useful for applications where a space feed can be used. For applications requiring conformal printed antennas, the tunable impedance surface can be used as the basis for an electronically steerable [13] leaky wave antenna [14–17]. In this case, a surface wave is excited directly in the surface and then radiates energy into the surrounding space as it propagates. This method involves programming the surface with a periodic impedance function that scatters the surface wave into free space. The period of the surface impedance can be varied to change the phase-matching condition

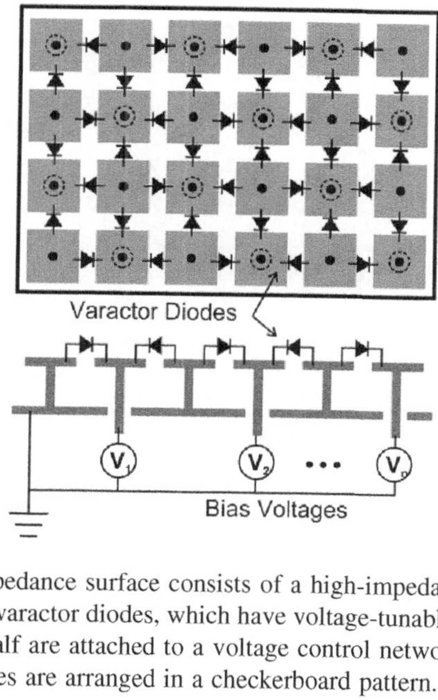

Figure 12.10 A tunable impedance surface consists of a high-impedance surface in which adjacent cells have been connected by varactor diodes, which have voltage-tunable capacitance. Half of the vias are grounded, but the other half are attached to a voltage control network on the back of the surface. The grounded and biased plates are arranged in a checkerboard pattern.

between the surface wave and the space wave, and thus steer the radiated wave. The beam can be electronically steered over a wide range in both the forward and backward directions. The decay rate of the surface waves can also be controlled independently of the beam angle to allow adjustment of the aperture profile.

To build a steerable leaky wave antenna, a feed structure is integrated into the tunable surface, such as the flared notch antenna shown in Figure 12.11. It can be as close as a small fraction of a wavelength from the surface, but it should not be close enough to detune the capacitance between the plates below it. A flared notch antenna will generate TE waves or a wire antenna can be used for TM waves.

A periodic pattern of voltages is applied to the tunable surface to create a periodic surface impedance function. When waves propagate across the surface, they are scattered by the nonuniform surface impedance. The scattered energy radiates at an angle determined by the wave vector of the surface wave and the periodicity of the surface impedance. The radiation angle may be determined by assuming that a wave launched into the surface feels an effective refractive index of n_{eff}. Its wave vector is $k_0 n_{\mathrm{eff}}$, where $k_0 = 2\pi/\lambda$ is the free-space wave vector. The surface impedance has period p, corresponding to a wave vector $k_p = 2\pi/p$. The scattered radiation in free space must have a total wave vector of k_0 and phase matching requires that it have a component parallel to the surface, which is equal to the sum of the wave vectors of the surface wave and the surface impedance function. As illustrated in Figure 12.12, the radiation is scattered into the forward direction if $k_p < k_0 n_{\mathrm{eff}}$ and is scattered backward if $k_p > k_0 n_{\mathrm{eff}}$. In general, the radiation angle is given by the following expression:

$$\theta = \sin^{-1}\left(\frac{k_0 n_{\mathrm{eff}} - k_p}{k_0}\right) \tag{12.16}$$

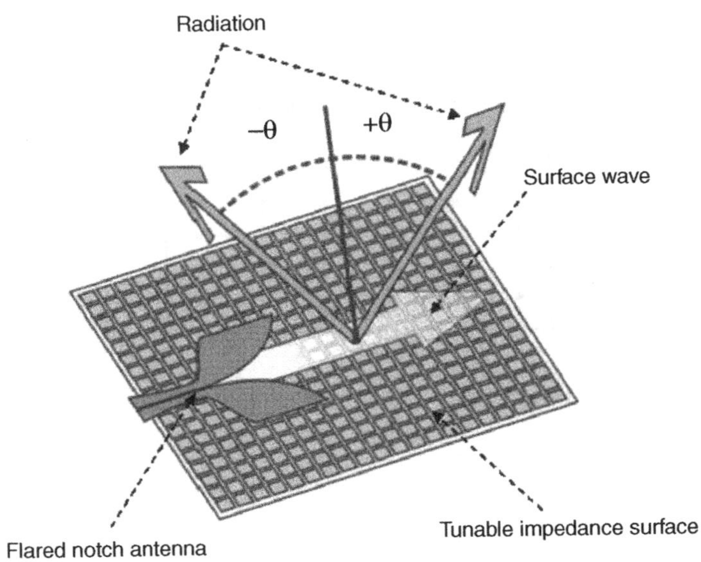

Figure 12.11 The tunable impedance surface can be used as an electronically steerable leaky wave antenna by incorporating a conformal feed, such as a flared notch antenna. The surface wave propagates away from the antenna, but the radiation can propagate in either the forward or backward direction, depending on the phase-matching condition at the surface.

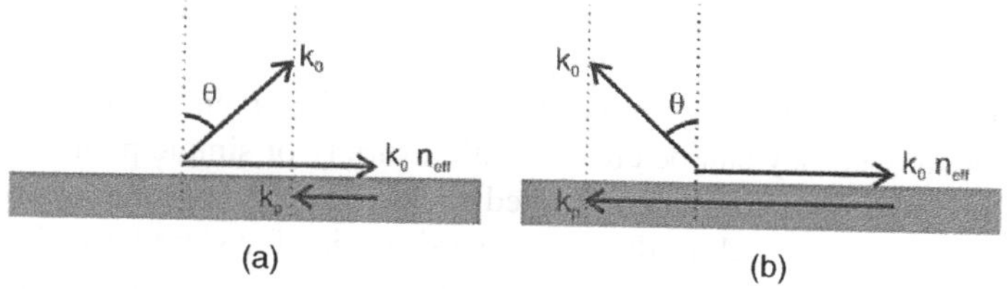

Figure 12.12 The direction of radiation is determined by phase matching at the surface. The tangential component of the wave vector of the space wave must match the difference between that of the surface wave and that of the periodic surface impedance. (a) Forward leaky waves are generated when the surface impedance has a period that is greater than the wavelength of the surface waves, corresponding to a shorter wave vector. (b) Backward leaky waves are generated when the period of the surface impedance is shorter than the wavelength of the surface wave.

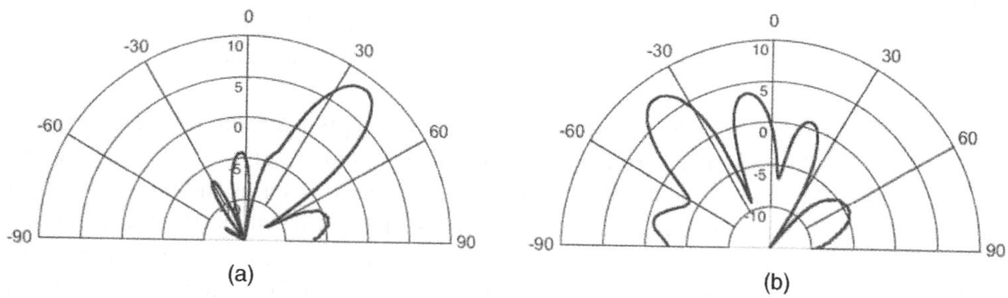

Figure 12.13 The surface can be configured for either forward or backward leaky wave radiation. This shows examples of the leaky wave radiation patterns when the surface is programmed for (a) forward radiation and (b) backward radiation.

For backward leaky waves, the energy still travels outward from the feed, so its group velocity is in the forward direction, but its phase velocity, which determines the radiation angle, is in the backward direction. Leaky wave structures capable of backward or broadside radiation have been studied extensively [18–20], but tunable impedance surfaces are novel because they can be electronically reconfigured to steer continuously from the forward to the backward direction at a single frequency. Figure 12.13 shows examples of radiation in both the forward and backward directions.

12.9 HOLOGRAPHIC ARTIFICIAL IMPEDANCE SURFACES

A common problem for designers of planar antennas is their integration into complex metallic objects while maintaining the desired radiation pattern. Simple shadowing by the surrounding structure can result in deep, wide nulls in the radiation pattern, so it can be nearly impossible to achieve appreciable radiation to some angles. Furthermore, the antenna excites currents in the surrounding structure, and these currents contribute to the radiation pattern in unexpected ways. Artificial impedance surfaces can provide a way to control the electromagnetic scattering and the propagation of surface currents around complex metallic shapes, to simultaneously enable greater flexibility in the design and position of conformal antennas and more control over the radiation pattern.

Artificial impedance surfaces do not need to contain all elements of the high-impedance surfaces discussed above; they can be created with a variety of simple planar metal patterns. The material used here consists of a grounded dielectric layer covered with a pattern of square conductive patches, avoiding the use of vertical vias for simplicity. The patches are small compared to the wavelength of interest, so their scattering properties can be described in terms of their effective surface impedance. By varying the geometry of the patches, the surface impedance can be controlled as a function of position.

The surface impedance is designed using a method that is analogous to optical holography, as the interference pattern between the expected antenna currents and the desired radiation fields. Currents from the antenna are scattered by the modulated surface impedance to produce the desired radiation pattern. The surface impedance can be designed so that surface currents propagate over a desired distance before being scattered into free space, so the radiation can reach regions that are not in the direct line of sight of the antenna. This method decouples the performance of the antenna from the shape of the body to which it is attached and its location on that body, thus allowing new kinds of conformal antennas that were not previously possible.

Holographic artificial impedance surfaces combine concepts from three fields: (1) leaky waves on modulated impedance surfaces, (2) holographic antennas, and (3) artificial electromagnetic media. The study of leaky wave antennas dates back as far as the mid 1950s [21]. Many early leaky wave antennas involved dielectric structures in which the thickness was modulated [22, 23] or tapered [24] to vary the effective surface impedance. Oliner and Hessel [25] developed a comprehensive model of leaky waves on modulated impedance surfaces. This model describes how the propagation and radiation of leaky waves are controlled by the surface reactance, modulation depth, and period, and it serves as a foundation for the thin printed leaky wave antennas described here.

12.9.1 Generation of Microwave Holograms

The surface impedance is designed using the concept of holography. Originally demonstrated at optical frequencies, holography involves producing an interference pattern using two waves, one representing the object to be imaged, as illuminated by a laser beam, and the other representing a reference plane wave from the same laser. The interference pattern formed by these two waves is recorded on photographic film. When the reference wave illuminates the developed film, it is scattered to produce a copy of the original image wave.

Microwave holography [26, 27] works in much the same way. A source antenna produces the reference wave and the desired radiation pattern corresponds to the image wave. The microwave hologram is a collection of scatterers arranged on the interference pattern between these two waves. Like the optical analog, the microwave hologram scatters the reference wave to produce the image wave. The reference wave can either be a surface wave [28] or a plane wave propagating parallel [29] or perpendicular [30] to the surface. Most traditional holographic antennas involve metal strips or dipoles [31, 32] that are printed on a dielectric layer. However, silicon-based steerable holographic antennas have also been studied [33, 34]. Recent work shows that holographic antennas are capable of high efficiency if the surface and feed are properly designed [35]. All of these previous structures can be considered as binary holograms – they consist of discrete scatterers or alternating regions of metal and dielectric.

Figure 12.14 A microwave hologram generated as the interference pattern between a point source and a plane wave propagating at 60° to the surface.

Artificial impedance surfaces allow a continuous variation in the surface impedance, which provides greater control over the scattering properties of the surface.

An example of a microwave hologram is shown in Figure 12.14. The interference pattern that is formed by a point source in a two-dimensional plane and a plane wave propagating 60° from normal to that plane appear as a pattern of ellipses.

12.9.2 Characterization of Artificial Impedance Surfaces

Traditional holographic antennas are built using discrete scatterers, usually either metallic strips or dipoles. Designs based on alternating regions of electric and magnetic conductors have also been studied numerically [36]. Artificial impedance surfaces provide a continuously variable surface impedance. Implemented as a lattice of subwavelength metallic patterns, they produce an effective surface impedance that depends on the geometry of the metal shapes. Compared to the previously studied binary holograms, artificial impedance surfaces allow greater control over the radiation properties of the antenna.

Various kinds of artificial impedance surfaces have been studied in the past, such as pin-bed structures [37] and high-impedance surfaces [2]. These materials can be combined with the concept of tunable impedance surfaces [12] to allow for completely arbitrary electronic reconfigurability of the surface scattering properties. Because the metallic patterns in these structures are small compared to the wavelengths of interest, they can be described in terms of effective medium properties such as the effective surface impedance. This permits the use of fast numerical methods that can analyse structures that are many wavelengths in size, without modelling the detailed structure of the material, which would render the problem impractically large.

One implementation of an artificial impedance surface is a lattice of subwavelength metal patches on a grounded dielectric substrate. Figure 12.15 shows an example of a single unit cell of such a surface, and the surface impedance range as a function of the gap between

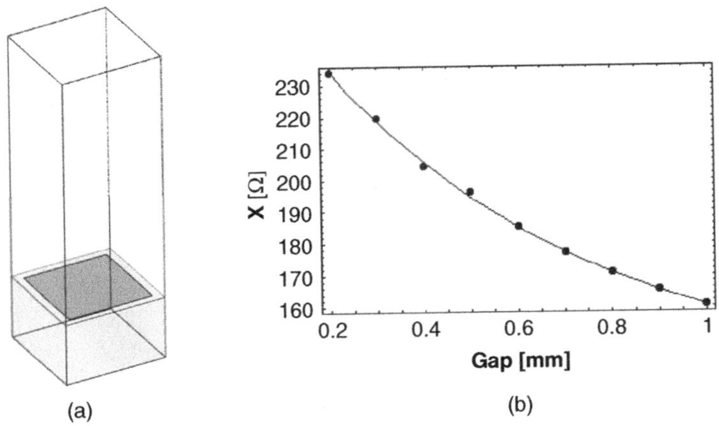

Figure 12.15 (a) The effective surface impedance of the lattice of metal regions is simulated with a single unit cell, consisting of a small metal patch on a grounded dielectric slab. Periodic boundary conditions are imposed on the opposing pairs of vertical walls. (b) Calculated reactance of the square unit cell with a period of 3 mm on a 1.57 mm thick slab of dielectric constant 2.2, as a function of the gap between neighbouring unit cells.

the metal patches, as calculated by Ansoft HFSS electromagnetic software. The impedance range is plotted for the case of a phase difference of 72° between neighbouring cells. The impedance is defined here as the ratio of the average electric to magnetic field above the surface, for the eigenmode that satisfies the periodic boundary conditions on the cell walls. For gaps ranging from 1 to 0.2 mm, the effective impedance varied from 161j to 234j Ω at a frequency of about 17 GHz.

12.9.3 Holographic Artificial Impedance Surfaces

The holographic pattern shown in Figure 12.14 is combined with the impedance versus geometry data shown in Figure 12.15 to determine the required geometry versus position on the surface, and generated a pattern of squares. A small section of the nonuniform pattern of metal squares is shown in Figure 12.16.

The surface is fed by a small monopole antenna inserted from the back, at the focus of the ellipses shown in Figure 12.14. Currents generated by the monopole are scattered by the holographic impedance surface to produce a narrow pencil beam at 60°, as shown in Figure 12.17, with a gain of about 20 dBi. Thus, the holographic pattern has transformed the initial wave (the currents from the monopole) into the desired wave (a plane wave at 60°). For comparison, a monopole on a similar smooth metal surface produces the expected pattern, shown on the same plot.

12.9.4 Propagation Around Curved Surfaces

Holographic artificial impedance surfaces can be used to design antennas on curved shapes to have radiation properties that would ordinarily be impossible. To illustrate this concept, a holographic pattern was designed to enable a monopole antenna on a conducting cylinder to produce a narrow beam on the other side of the cylinder, towards a direction that is

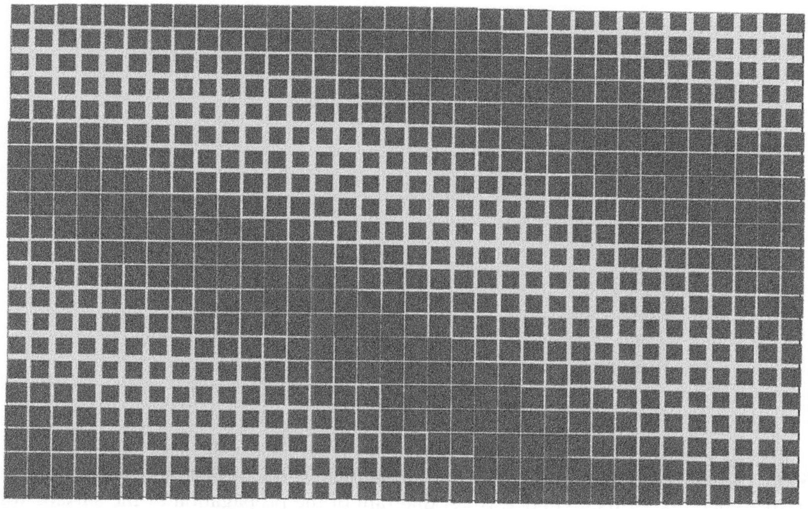

Figure 12.16 A small section of the artificial impedance surface showing a nonuniform lattice of metal squares. The squares are small compared to the wavelength of interest, so the wave only sees the effective surface impedance of the artificial medium.

otherwise shadowed. The monopole antenna generates surface currents that propagate around the cylinder, and a holographic impedance surface was designed using the interference pattern formed by those surface currents and a plane wave at 135° on the opposite side of the cylinder, as shown in Figure 12.18. As the surface currents propagate around the cylinder, they are scattered by the modulated impedance surface. The scattered radiation experiences constructive interference to produce a narrow beam at 135° on the other side of the cylinder, the angle for which the surface was designed. This is shown in Figure 12.19, along with the radiation pattern of a similar monopole antenna on an ordinary conducting cylinder.

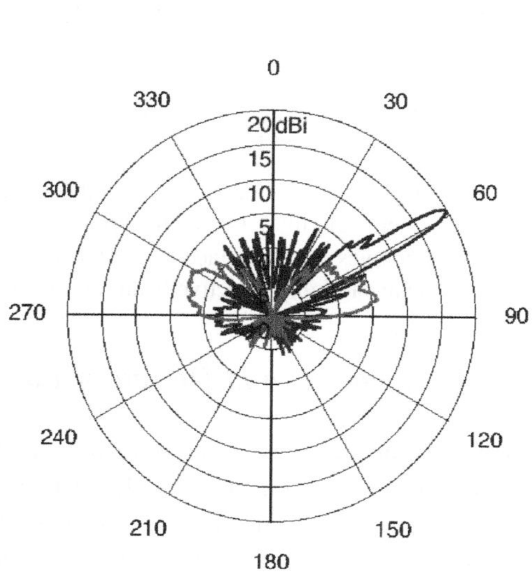

Figure 12.17 Radiation pattern of a monopole antenna on the holographic artificial impedance surface (dark line) and on a flat conducting surface of the same size (light line). The patterned surface transforms the monopole into a highly directive antenna.

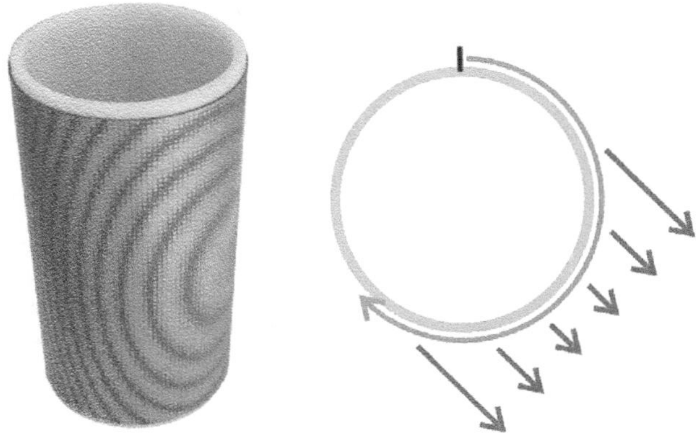

Figure 12.18 (a) A cylinder covered with an artificial impedance surface that is designed to produce radiation on the other side of the cylinder. (b) A diagram of the propagation of the surface wave (light line) and the radiation produced by the surface wave (dark lines).

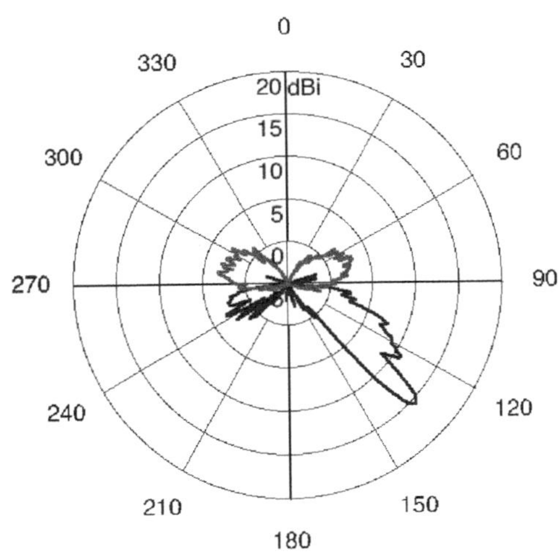

Figure 12.19 Measured radiation pattern of a point dipole antenna above the patterned cylinder (dark line) and on a smooth conducting cylinder (light line). On the holographic artificial impedance surface, radiation from the monopole antenna couples into a surface wave that wraps around the cylinder and radiates in a controlled manner to produce a highly directive beam on the other side, at 135°.

12.9.5 Numerical Tools for Analysis and Design of Large Surfaces

Radiation patterns from artificial impedance surfaces have been computed using software such as the FastScat electromagnetic scattering code [38, 39]. FastScat is a frequency-domain boundary integral code using high-order Nyström discretizations and the fast multipole method. Impedance boundary conditions for varying impedance surfaces are handled by a suitable extension of the usual boundary integral operators; e.g. the extended combined field integral equation has two additional integral operator terms that include the effects of arbitrary surface impedance profiles. Surface impedance profiles are specified by functional

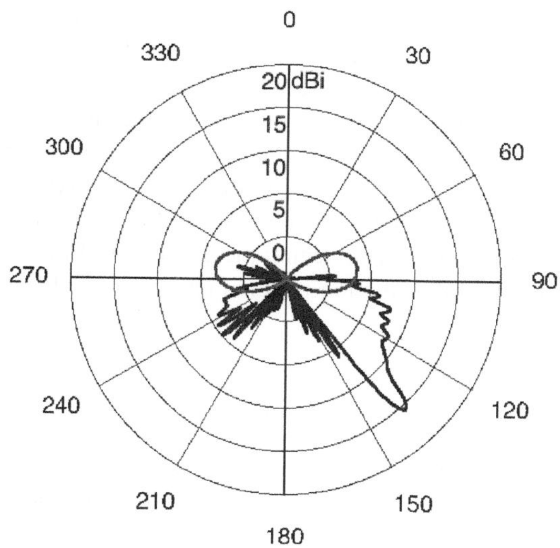

Figure 12.20 Calculated radiation pattern of a point dipole antenna above the patterned cylinder (dark line) and on a smooth conducting cylinder (light line). The calculated result agrees with the measurement shown in Figure 12.19.

form; hence, a holographic impedance surface is fully characterized by the functional forms of the desired radiation and surface waves. The modelled surfaces are then the smoothly varying approximations (as shown in Figure 12.14) to the constructed patched surfaces (as shown in Figure 12.16). The FastScat results for the cylinder shown in Figure 12.18 are plotted in Figure 12.20. A simple point dipole above the cylindrical surface is used to model the monopole excitation. Good agreement is found between the simulated and measured results for both holographic impedance and conducting cylinders. This is important for two reasons: (1) it shows that the effective surface impedance is sufficient to describe the surface properties to allow for accurate prediction of the antenna performance and (2) because this method does not require the simulation of the detailed structure of the surface, it can be used to simulate large objects with practical computing resources. Thus, the holographic artificial impedance surface method can be used to design thin printed antennas on large complicated objects with a high degree of confidence.

12.10 SUMMARY

Subwavelength textures can be applied to metal surfaces to change their electromagnetic properties. Thin coatings containing lattices of grounded metal plates can behave as a high-impedance surface and can be analysed using a simple lumped circuit parameter model. These surfaces have two important properties: (1) they suppress the propagation of surface waves within a bandgap and (2) they provide a reflection phase of zero at the resonance frequency. The bandwidth of these properties is related to the thickness of the surface. High-impedance surfaces can be used for a variety of antenna applications, such as to build new kinds of low-profile antennas or to control scattering by the nearby ground-plane.

Electronically tunable impedance surfaces can be built by incorporating varactor diodes into the lattice. These can be used with a conformal feed to build electronically steerable leaky wave antennas. Using this technique, the radiation pattern can be steered over a wide range of angles in both the forward and backward directions. Because it involves only printed circuit boards, varactor diodes, and low-frequency control electronics, tunable impedance surfaces can be used as a simple, low-cost alternative to traditional phased arrays.

The concept of holography can also be applied to artificial impedance surfaces to allow antennas with novel radiation properties to be formed into curved shapes. Some of the important capabilities of these antennas include the ability to achieve high gain towards angles that are otherwise shadowed. This technique can be used to build high-performance conformal antennas with arbitrary radiation patterns on complex objects.

REFERENCES

1. D. Sievenpiper, 'High-impedance electromagnetic surfaces', PhD Dissertation, Department of Electrical Engineering, University of California, Los Angeles, California, 1999.
2. D. Sievenpiper, L. Zhang, R. Broas, N. Alexopolous and E. Yablonovitch, 'High-impedance electromagnetic surfaces with a forbidden frequency band', *IEEE Transactions on Microwave Theory and Techniques*, **47**, 2059–2074, November 1999.
3. E. Yablonovitch, 'Inhibited spontaneous emission in solid-state physics and electronics', *Physical Review Letters*, **58**, 2059–2062, May 1987.
4. J. Joannopoulos, R. Meade and J. Winn, *Photonic Crystals: Molding the Flow of Light*, Princeton University Press, Princeton, New Jersey, 1995.
5. W. Barnes, T. Priest, S. Kitson and J. Sambles, 'Physical origin of photonic energy gaps in the propagation of surface plasmons on gratings', *Physical Review B*, **54**, 6227–6244, September 1996.
6. S. Ramo, J. Whinnery and T. Van Duzer, *Fields and Waves in Communication Electronics*, 2nd edn, John Wiley and Sons, Inc., New York, 1984.
7. L. Brillouin, *Wave Propagation in Periodic Structures*, McGraw-Hill, New York, 1946.
8. N. Ashcroft and N. Mermin, *Solid State Physics*, Saunders College Publishing, Orlando, Florida, 1976.
9. H. Wheeler, 'Fundamental limitations of small antennas', *Proceedings of the IRE*, **35**, 1479–1484, December 1947.
10. L. Chu, 'Physical limitations of omni-directional antennas', *Journal of Applied Physics*, **19**, 1163–1175, December 1948.
11. J. McLean, 'A re-examination of the fundamental limits on the radiation Q of electrically small antennas', *IEEE Transactions on Antennas and Propagation*, **44**, 672–675, May 1996.
12. D. Sievenpiper, J. Schaffner, H. J. Song, R. Loo and G. Tangonan, 'Two-dimensional beam steering reflector using an electrically tunable impedance surface', *IEEE Transactions on Antennas and Propagation*, **51**, 2713–2722, October 2003.
13. D. Sievenpiper, 'Forward and backward leaky wave radiation with large effective aperture from an electronically tunable textured surface', *IEEE Transactions on Antennas and Propagation*, **53**, 236–247, January 2005.
14. G. Broussaud, 'Un nouveau type d'antenne de structure plane', *Annales de Radioelectricite*, **11**, 70–88, January 1956.
15. J. W. Lee, J. J. Eom, K. H. Park and W. J. Chun, 'TM-wave radiation from grooves in a dielectric-covered ground plane', *IEEE Transactions on Antennas and Propagation*, **49**(1), 104–105, January 2001.
16. C.-N. Hu and C.-K. C. Tzuang, 'Analysis and design of large leaky-mode array employing the coupled-mode approach', *IEEE Transactions on Microwave Theory and Techniques*, **49**, 629–636, April 2001.
17. P. W. Chen, C. S. Lee and V. Nalbandian, 'Planar double-layer leaky wave microstrip antenna', *IEEE Transactions on Antennas and Propagation*, **50**, 832–835, June 2002.

18. T. Tamer and F. Kou, 'Varieties of leaky waves and their excitation along multilayered structures', *IEEE Journal of Quantum Electronics*, **22**, 544–551, April 1986.
19. M. Guglielmi and D. Jackson, 'Broadside radiation from periodic leaky-wave antennas', *IEEE Transactions on Antennas and Propagation*, **41**, 31–37, January 1993.
20. S. -G. Mao and M. -Y. Chen, 'Propagation characteristics of finite-width conductor-backed coplanar waveguides with periodic electromagnetic bandgap cells', *IEEE Transactions on Microwave Theory and Techniques*, **50**, 2624–2628, November 2002.
21. J. Simon and V. Biggi, 'Un noveau type d'aerien et son applicatoin a la transimssion de television a grand distance', *L'Onde Electrique*, **332**, 1–16, November 1954.
22. A. Thomas and F. J. Zucker, 'Radiation from modulated surface wave structures I', 1957 IRE National Convention Record, Part 1, pp. 153–160, March 1957.
23. R. Pease, 'Radiation from modulated surface wave structures II', 1957 IRE National Convention Record, Part 2, pp. 161–165, March 1957.
24. L. Felsen, 'Radiation from a tapered surface wave antenna', *IRE Transactions on Antennas and Propagation*, **8**(6), 577–586, November 1960.
25. A. Oliner and A. Hessel, 'Guided waves on sinusoidally-modulated reactance surfaces', *IRE Transactions on Antennas and Propagation*, **7**(5), 201–208, December 1959.
26. R. Dooley, 'X-band holography', *Proceedings of the IEEE*, **53**(11), 1733–1735, November 1965.
27. W. Kock, 'Microwave holography', *Microwaves*, **7**(11), 46–54, November 1968.
28. P. Checcacci, V. Russo and A. Scheggi, 'Holographic antennas', *IEEE Transactions on Antennas and Propagation*, **18**(6), 811–813, November 1970.
29. K. Iizuka, M. Mizusawa, S. Urasaki and H. Ushigome, 'Volume-type holographic antenna', *IEEE Transactions on Antennas and Propagation*, **23**(6), 807–810, November 1975.
30. D. Sazonov, 'Computer aided design of holographic antennas', in *Antennas and Propagation Society International Symposium*, Vol. 2, pp. 738–741, July 1999.
31. K. Levis, A. Ittipiboon, A. Petosa, L. Roy and P. Berini, '*Ka*-band dipole holographic antennas', *IEE Proceedings of Microwaves Antennas and Propagation*, **148**(2), 129–132, April 2001.
32. M. ElSherbiny, A. Fathy, A. Rosen, G. Ayers and S. Perlow, 'Holographic antenna concept, analysis, and parameters', *IEEE Transactions on Antennas and Propagation*, **52**(3), 830–839, March 2004.
33. N. Farhat, 'Holographically steered millimeter wave antennas', *IEEE Transactions on Antennas and Propagation*, **28**(4), 476–480, July 1980.
34. A. Fathy, A. Rosen, H. Owen, F. McGinty, D. McGee, G. Taylor, R. Amantea, P. Swain, S. Perlow and M. ElSherbiny, 'Silicon-based reconfigurable antennas – concepts, analysis, implementation and feasibility', *IEEE Transactions on Microwave Theory and Techniques*, **51**(6), 1650–1661, June 2003.
35. S. Thingvold, A. Ittipiboon, A. Sebak and A. Petosa, 'Holographic antenna efficiency', in *Antennas and Propagation Society International Symposium*, Vol. 3, pp. 721–724, June 2003.
36. V. Sanchez and R. Diaz, unpublished work.
37. R. King, D. Thiel and K. Park, 'The synthesis of surface reactance using an artificial dielectric', *IEEE Transactions on Antennas and Propagation*, **31**(3), 471–476, May 1983.
38. L. F. Canino, J. J. Ottusch, M. A. Stalzer, J. L. Visher and S. M. Wandzura, 'Numerical solution of the helmholtz equation in 2D and 3D using a high-order Nystrom discretization', *Journal of Computational Physics*, **146**, 627–663, 1998.
39. J. J. Ottusch, M. A. Stalzer, J. L. Visher and S. M. Wandzura, 'Scalable electromagnetic scattering calculations on the SGI Origin 2000', in *Proceedings of the Supercomputing Conference*, 1999.

13

Transceiver Architectures for Highly Integrated RF Front-Ends

Sanjay Raman[1] **and Richard M. Svitek**[2]

[1] *Bradley Department of Electrical and Computer Engineering, Virginia Polytechnic Institute and State University, Blacksburg, Virginia, USA*

[2] *Renaissance Wireless, Pittsburgh, Pennsylvania, USA*

13.1 INTRODUCTION

The twenty-first century has been witnessing a tremendous demand for wireless (*untethered*) communications and information services, such as Personal Communications Services (PCS – 3G, 4G, and beyond), wireless data networks and internet access, position location, navigation, roadway informatics, and wireless sensor networks. While cost and form factor (size and weight) have been primary drivers in the commercial arena, military systems must also be highly reliable, rapidly reconfigurable, aware of their user/location/environment, and robust in the quality of service under a wide range of hostile environmental conditions. Of course, in both arenas, by definition, *mobility demands low power dissipation.*

The necessity for low-cost and high-efficiency *system* implementations for these untethered communications and sensor capabilities has generated an explosion in the development of integrated circuit and packaging technologies, in both commercial and military sectors. For example, radio frequency integrated circuits (RFICs) and/or monolithic microwave (or mm-wave) integrated circuits (MMICs) are generally packaged together with very large scale integrated (VLSI) digital signal processing (DSP) and microprocessor (μP) control chips in advanced multichip modules (MCMs) or system-in-package/system-on-package (SiP/SoP) implementations. Figure 13.1(a) shows a conceptual diagram of such an MCM for an RF wireless communications node. The figure shows the various functional chips fabricated in different integrated circuit (IC) technologies (e.g. GaAs HBT for power amplifiers, SiGe BiCMOS for low-noise receivers, and Si CMOS for baseband and digital processing), connected to the system using conventional wirebonding or solder ball grid arrays (BGAs). In addition, discrete high-quality passives (or embedded passives in the MCM carrier), and even antennas, can be provided as well. A key advantage of this approach is that the

Printed Antennas for Wireless Communications Edited by R. Waterhouse

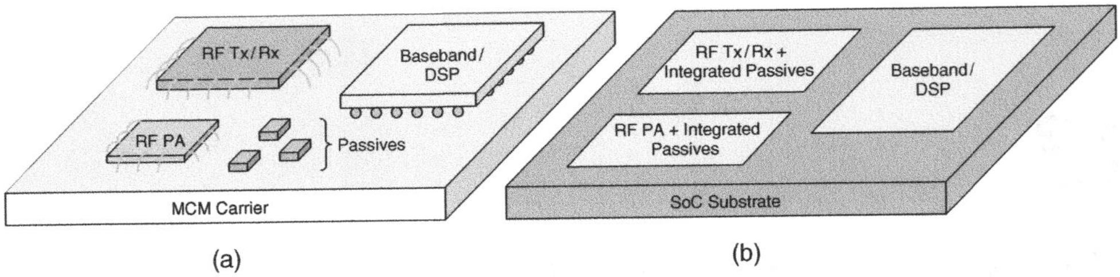

Figure 13.1 Conceptual diagrams of (a) multichip module (MCM) and (b) system-on-a-chip (SoC) implementations of an RF wireless communications node.

optimum technology for each system function can be selected (e.g. GaAs HBT of RF power amplifiers, Si CMOS for mixed-signal/DSP, off-chip passives for high-Q). On the other hand, the MCM approach leads to greater manufacturing costs due to the requirements for handling, mounting, and connecting the different components. In addition, this approach may suffer from lower reliability and lower manufacturing yields.

Meanwhile, significant strides have been made in the development of *mixed-signal* integrated circuits combining RF, and low-frequency analog and digital functions on the same chip, now commonly referred to as system-on-a-chip (SoC) implementations [1, 2]. The development of such SoCs is motivated by lower packaging and handling costs, greater reliability, reduced size of the overall electronic system, reduced parasitic reactances and flexibility in impedance matching, the ability to incorporate on-chip digital-domain filtering, frequency synthesis, etc. [3], and the ability to provide increased functionality/multimode/multiband products (e.g. for interoperability between different wireless standards and services) at low cost.

Figure 13.1(b) shows a conceptual diagram of an SoC implementation of an RF wireless communications node. However, the SoC approach does present some potential drawbacks. Firstly, SoC fundamentally limits technology flexibility, since by definition all functions will be implemented in one given technology (e.g. SiGe BiCMOS or Si CMOS) – in other words, the *best* technology is not necessarily being used for every function. For example, on-chip passive components tend to have lower quality factors in comparison to their off-chip counterparts. On the other hand, device engineers and IC designers continue to push IC technologies (primarily silicon-based) and system architectures, respectively, such that high-performance SoC implementations are becoming a reality [4]. Other issues with the SoC approach include longer design cycle times and significant mask redesign costs, mixed-mode noise/coupling issues, and thermal dissipation issues.

This chapter focuses on transceiver architectures that can lead to highly integrated RF front-ends, particularly from the point of view of silicon SoC integration. However, the architectures and design trade-offs to be discussed are also relevant to RF transceiver module/system-in-package/system-on-package (SiP/SoP) implementations as well. The chapter will also briefly discuss the implications of such architectures on the design of the RF front-end–antenna interface.

13.2 WIRELESS SYSTEM-ON-A-CHIP

Figure 13.2 shows a conceptual block diagram of a wireless system-on-a-chip design. From the antenna and RF transceiver to the microprocessor and baseband input/output, all compo-

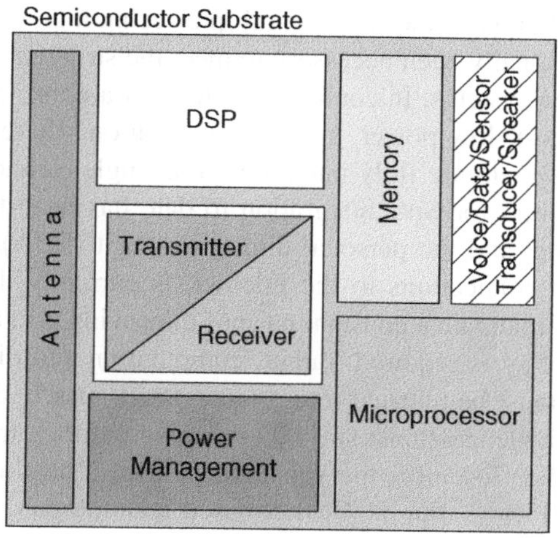

Figure 13.2 Block diagram of a generic, SoC-integrated, wireless transceiver.

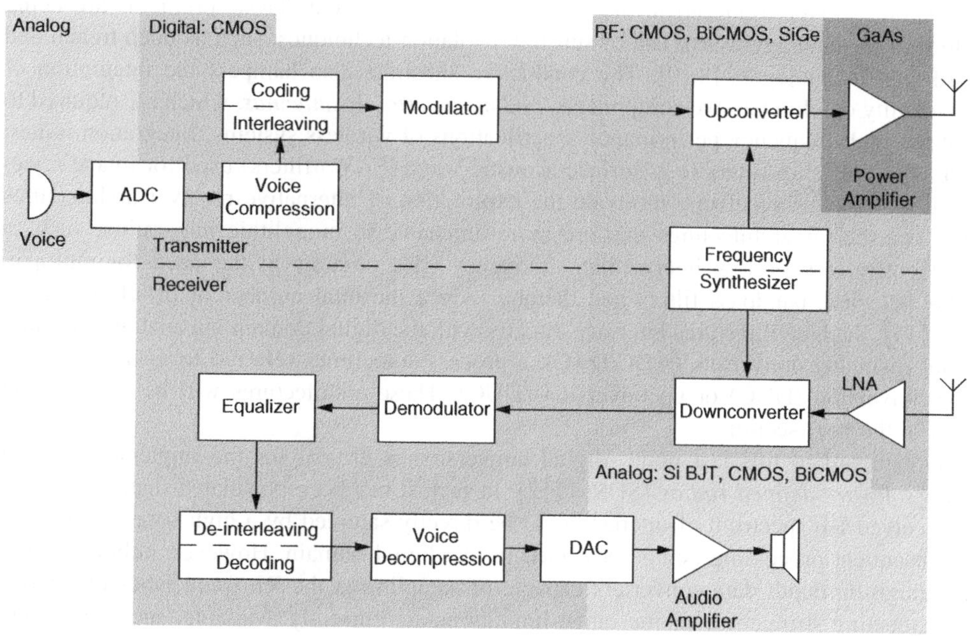

Figure 13.3 Block diagram of a generic digital wireless transceiver, i.e. the 'transmitter/receiver' block from Figure 13.2. Blocks are categorized by circuit type and process (after Reference [6]).

nents are integrated in a single semiconductor technology on a common substrate. Focusing on just one part of the entire system, Figure 13.3 shows the block diagram of a generic RF transceiver (transmitter and receiver) categorized by integrated circuit technology and circuit type (RF, analog, or digital). While the digital components are implemented in Si CMOS, the RF sections of the receiver and transmitter, along with the frequency synthesizer, can be realized with different technology options. The optimum choice of semiconductor technology is a complicated issue, involving performance, wafer cost, level of integration, and time-to-market [5]. Although the use of multiple technologies ultimately may not be cost effective, it has traditionally been necessary to meet the stringent performance standards of wireless communication systems. In contrast, 'system-on-a-chip' (SoC) integration offers the possibility of a low-cost, low-power, single-chip solution where all the functions of a wireless communication system are fully integrated in a single technology on a single IC.

One of the biggest drivers for SoC integration to date has been the mobile phone. The merging of the 'cell phone' and the personal digital assistant (PDA), along with the desire for pervasive, high-speed connections to the internet, is currently driving the commercial marketplace. However, to maintain a constant pace of improving cost and performance, more than simple transistor scaling is required; rather, computation and digital signal processing and analog functionality must be married into an integrated, mixed-signal system [7].

A number of unique challenges must still be overcome before a true single-chip solution is commercially viable. For instance, the substrate of the IC has a significant impact on the performance of an RF-SoC. For Si CMOS-based technologies, the relatively conductive substrate typically employed in commercially available processes can couple undesired spurious signals into the sensitive receiver front-end and corrupt signal reception. Likewise, switching noise from high-speed digital circuitry can leak through the substrate into the RF/analog receiver front-end in the mixed-signal environment. These problems are being addressed through aggressive use of substrate isolation techniques, such as deep trench and guard ring technologies [8, 9]. The conductive substrate also hampers the integration of high-quality on-chip passive components, *and most certainly antennas*, which are required to meet the high-frequency performance specifications of wireless systems. Integration issues, such as the off-chip filters (e.g. surface acoustic wave (SAW) filters) used for image rejection in heterodyne receivers, motivate the exploration of alternative receiver architectures [10]. Transceiver architectures that are more amenable to monolithic integration, such as direct conversion or low intermediate frequency (IF), circumvent the need for multiple, highly selective bandpass filters and thereby reduce the total number of off-chip components [11]. Such architectures interface directly with the digital domain via analog-to-digital/digital-to-analog converters (ADC/DAC), a process sometimes referred to as direct digital downconversion (DDC) or upconversion (DUC). These architectures will be discussed in detail in the next section.

With regard to this last point, digital conversion is critical for the implementation of future *software-defined radios* (SDRs) [12]. In fact, it has been postulated that, ultimately, the received RF spectrum of interest could be directly sampled by a high-speed ADC with all subsequent processing being performed in the digital domain. However, although high-speed/high-bit-depth data converters capable of sampling at the RF frequencies of interest while meeting stringent dynamic range limitations are currently available, such converters have extremely high power consumptions, which rules out their implementation in mobile communication transceivers for the foreseeable future.

13.3 RECEIVER ARCHITECTURES

The ability of RF wireless receivers to distinguish a desired, faint signal from a spectrum of undesired and potentially interfering signals is a remarkable feat in electrical engineering. For more than 70 years, this reception of transmitted information was accomplished primarily with Edwin Armstrong's superheterodyne receiver topology. However, recent advances in IC technology have generated a great deal of interest in alternative receiver architectures that can provide higher levels of integration. The following sections give an overview of the dominant receiver architectures in use today.

13.3.1 Superheterodyne Receivers

Traditionally favoured for their high selectivity and sensitivity, heterodyne receivers downconvert an RF signal to sequentially lower intermediate frequencies (IFs) for amplification and filtering prior to baseband processing [13]. The block diagram of a generic heterodyne receiver chain is shown in Figure 13.4 and illustrates its operation: the RF spectrum received by the antenna is first filtered (1) to select the overall frequency band of the communications system of interest; a low-noise amplifier (2) follows to boost the weak, desired signal and offset the typically high noise figure of the subsequent filter (3) and mixer (4); the mixer performs the frequency downconversion, translating the RF spectrum to an intermediate frequency determined by the local oscillator (LO); a channel selection filter (5) is then employed to remove nearby, in-band interferers; finally, a series of high-gain stages (6) amplify the signal for envelope detection, demodulation, or conversion to the digital domain. In practice, multiple IF stages are used as discussed below.

The process of downconverting the signal to a lower, intermediate frequency provides at least two important advantages: selectivity and tunability. The first is related to the use of an intermediate frequency; by translating the desired signal to a lower frequency, the need for prohibitively high filter Qs is alleviated. As illustrated in Figure 13.4, sharp cutoff filters ('high-Q') are needed at some point in the receiver chain to attenuate any nearby interferers that can corrupt the reception of the desired signal. Since typical filters exhibit a trade-off between their insertion loss and the Q of their component resonators [6], placing a highly selective filter for channel selection at the front of the receiver introduces a significant

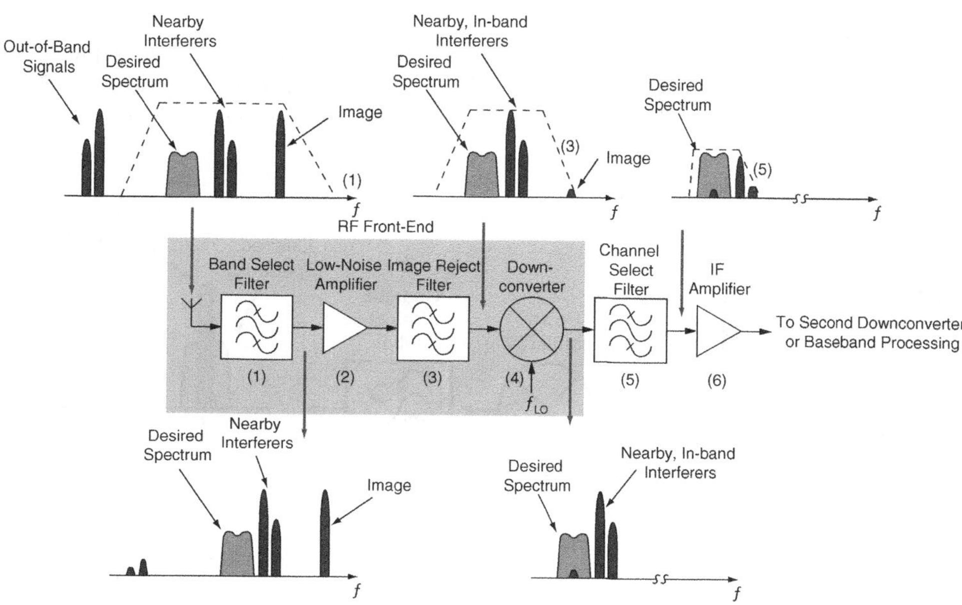

Figure 13.4 Block diagram of the superheterodyne receiver topology (after Reference [6]).

amount of loss. This loss requires a large amount of gain from the first stage of amplification and significantly degrades the overall receiver noise figure since the first stage sets the minimum noise figure in a cascade of components. In other words, the ability of the receiver to detect very low power signals, known as sensitivity, is degraded by the increase in noise.

A second advantage of the superheterodyne is in the tunability of the receiver. Since the IF is determined by the frequency of the LO, the centre frequencies of the filtering components after the mixer can remain fixed and the LO tuned to give a fixed IF for a range of RF frequencies. LO generation is typically done with a voltage-controlled oscillator (VCO) (often embedded within a phase-locked loop). In practice, VCO tunability is considerably easier than retuning the bandpass filter used for channel selection each time the RF frequency to be received changes.

A major drawback to the superheterodyne topology arises from the problem of *image frequencies*. In the mixing process, the intermediate frequency is defined as

$$f_{\rm IF} = |f_{\rm RF} - f_{\rm LO}| \tag{13.1}$$

where $f_{\rm RF}$ and $f_{\rm LO}$ are the frequency of the RF and local oscillator signals, respectively. After mixing the RF and LO signals, the polarity of the difference between the two signal frequencies is not maintained; thus, $f_{\rm RF} - f_{\rm LO}$ yields the same IF as $f_{\rm LO} - f_{\rm RF}$. In other words, the superheterodyne architecture downconverts the bands located an equal Δf on either side of the LO to the same IF frequency (Figure 13.5). In terms of the IF, the image frequency ($f_{\rm IM}$) is located at

$$f_{\rm IM} = f_{\rm LO} - f_{\rm IF} = 2f_{\rm LO} - f_{\rm RF} \tag{13.2}$$

for low-side injection (i.e. the LO is lower in frequency than the RF) and

$$f_{\rm IM} = f_{\rm LO} + f_{\rm IF} = 2f_{\rm LO} - f_{\rm RF} \tag{13.3}$$

for high-side injection (i.e. the LO is higher in frequency than the RF, as depicted in Figure 13.5). The desired RF signal and the image are separated in frequency by $2f_{\rm IF}$. With the image and the RF signal translated to the same IF, the image signal represents

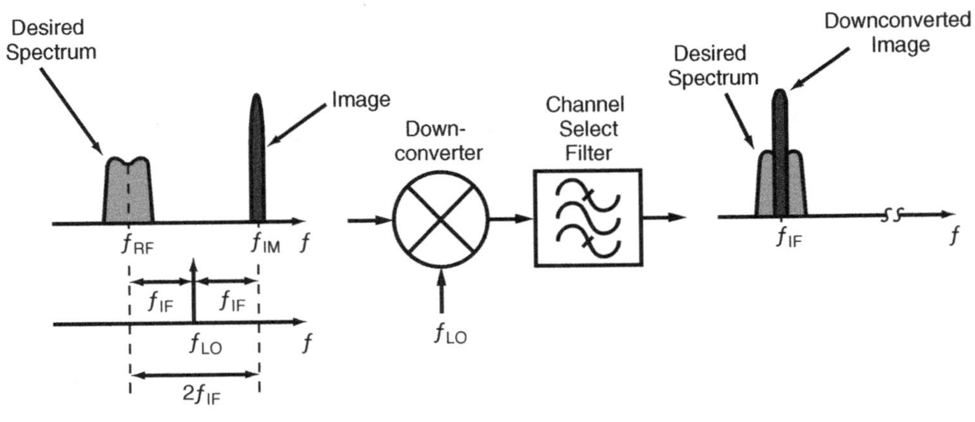

Figure 13.5 The image problem in heterodyne receivers.

an unwanted interference component corrupting the desired RF spectrum. Furthermore, at a significantly high power level the image acts like an in-band interferer, potentially desensitizing the receiver front-end from the desired signal due to gain compression. To mitigate these effects, an image-reject filter (IRF) must be placed before the mixer to attenuate the image sufficiently. This filter is designed to have minimal loss for the desired channel and large suppression of the image.

The choice of the IF frequency determines the image frequency and impacts the rejection of the image signal. A low IF allows for high selectivity and suppression of nearby interferers since the channel selection occurs at a low frequency. With an IF close to baseband, highly selective, active, analog filters, which are straightforwardly integrated on-chip, can perform the channel filtering. However, a low IF places the image close to the desired RF signal and limits the image rejection that can be achieved with practical filters at the RF frequency. Therefore, alternative methods of image rejection (to be discussed below) must be employed to reduce the image; otherwise, since the image cannot be attenuated without affecting the desired signal, greater image corruption must be tolerated within the communication system. On the other hand, a high IF results in substantial image rejection with an RF filter, since the image is far away in frequency from the desired RF spectrum and thus well into the rejection band of the (properly designed) image reject filter. However, with such a high IF, the receiver selectivity suffers because at a higher frequency, highly selective – but higher insertion loss – filters cannot be employed for channel selection and thus nearby interferers are poorly attenuated. Basically, heterodyne receivers exhibit a trade-off between sensitivity (image rejection) and selectivity.

To ease the trade-off between selectivity and image rejection, additional stages of down-conversion with successively lower IFs are used. The sequence of IFs allows for partial channel selection with increasing filter Qs at each lower IF and thus provides greater selectivity without requiring highly selective, lossy filters. In modern heterodyne receivers, the image-reject filter is typically a surface acoustic wave (SAW) device that is relatively large and expensive, sensitive to impedance mismatches, and currently not suitable for on-chip integration (although there is significant ongoing research on integrating such devices on chip, for example using microelectromechanical system (MEMS) technology). Given the push for higher levels of integration, this disadvantage has motivated designers to look at alternative receiver architectures.

13.3.2 Low-IF Image-Reject Receivers

The low-IF receiver architecture is an extension of the superheterodyne approach, where the intermediate frequency is very near baseband [14]. As mentioned in the previous section, the choice of a lower IF allows for the substitution of the RF image rejection filter with an IC-based approach for removing the image, for example an analog technique such as the Hartley or Weaver architectures [6] or a 'mixer-free' digital technique like subsampling with delay processing [15]. The performance of an image-rejection methodology can be quantified with the image-rejection ratio (IRR), the ratio between the power of the desired signal, P_{sig}, and the power of the image, P_{im}:

$$\mathrm{IRR\ (dB)} = 10\log(P_{sig}/P_{im}) \tag{13.4}$$

An IRR in the order of 70–80 dB is typically required in modern RF applications.

In a Hartley receiver, the desired RF signal is first downconverted in quadrature (i.e. into parallel 'I' and 'Q' signal paths) to a low IF and then low-pass filtered, typically with an integrated active filter [16]. The filtered Q channel signal is then shifted by 90° relative to the I channel, and the two channel signals are then summed. The 90° shift operation on the Q branch reverses the polarity of the image component such that when the I and Q signals are summed together, the images in each branch cancel; meanwhile, the desired signal simply experiences a 90° shift in both paths and its spectrum is theoretically downconverted without corruption by the image.

Alternatively, the Weaver architecture replaces the in-line 90° shift operation with an additional set of quadrature mixers at the IF to achieve the same result. Basically, the second I/Q mixing stage provides the additional 90° phase shift needed to result in image subtraction. Figure 13.6 shows the block diagram of a low-IF receiver with quadrature outputs using the Weaver architecture for image rejection. It should be noted that in this figure, analogue-to-digital (A/D) conversion occurs after the image-rejection processing occurs in the analog domain. However, it is also equally possible to perform the A/D conversion on the I and Q signal paths after the first stage of (reasonably) low-IF mixing, and then perform filtering, second-stage quadrature multiplication, and subsequent processing in the digital domain. In this latter case, the first stage IF selected will drive the ADC speed requirements, and consequently the power consumption of the receiver architecture.

However, since the second set of mixers downconverts the signal to a second IF, a second image signal exists and must be accounted for. Similar to the first image, and relative to the first IF, the second image is located at

$$f_{IM2} = 2f_{LO2} - f_{IF1} = 2f_{LO2} - f_{RF} + f_{LO1} = f_{IF1} - 2f_{IF2} \tag{13.5}$$

for low-side injection and

$$f_{IM2} = 2f_{LO2} - f_{IF1} = 2f_{LO2} - f_{LO1} + f_{RF} = f_{IF1} + 2f_{IF2} \tag{13.6}$$

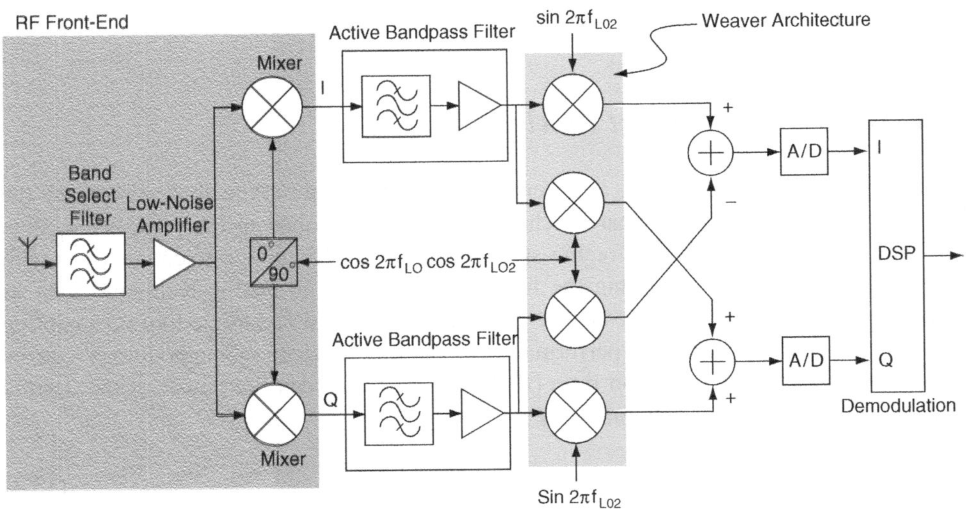

Figure 13.6 Block diagram of the low-IF receiver with the Weaver image-rejection architecture.

for high-side injection. To suppress this image, the signal after the first downconversion is passed through a bandpass filter rather than a low-pass filter. Further suppression can be achieved by choosing the first and second IFs so that the location of the second image frequency before any downconversion, which is given by

$$f_{IM2} = 2f_{LO2} - f_{RF} + 2f_{LO1} \tag{13.7}$$

is in an unused band that is not expected to have a significant amount of energy.

The downside of analog image-rejection architectures is their sensitivity to amplitude (ε) and phase ($\Delta\phi$) errors in the signal paths. In the case of the Hartley architecture, the realization of a constant-gain, frequency-independent 90° phase-shifter is extremely difficult; for the Weaver architecture, the LO signals must be exactly in quadrature, or image suppression degrades significantly. The image-rejection ratio can be expressed in terms of ε and $\Delta\phi$ to show the impact of such errors:

$$\text{IRR (dB)} = 10\log\left[4/(\varepsilon^2 + \Delta\phi^2)\right] \tag{13.8}$$

For example, a gain mismatch of 0.1 % and a phase imbalance of 1° results in an image rejection of ~ 41 dB – far less than what is normally required. Thus, in practice, neither the Weaver nor the Hartley architectures alone are sufficient. Other methods, such as additional filtering or digital correction/calibration in DSP, are necessary. Nevertheless, since modern ADCs and DSP can easily operate at low megahertz frequencies, low-IF architectures are now an extremely popular alternative to the superheterodyne architecture, although power consumption issues must be addressed.

13.3.3 Direct Conversion Receivers

Direct conversion receivers (DCRs), also known as zero-IF or homodyne receivers, are a special case of the heterodyne approach. Instead of down converting to a finite IF, DCRs translate the desired RF spectrum directly to DC, or 0 Hz, using an LO frequency exactly equal to the RF. The simplicity of this architecture affords two major advantages over the classical superheterodyne topology. Firstly, because the IF frequency is zero, the image to the desired RF signal is the desired signal itself. In other words, the RF bands on both sides of the LO signal contain transmitted information and need to be detected (see Section 13.3.3.4 below on I/Q mismatch). Thus, the bulky, off-chip, front-end image-reject filters required in the superheterodyne topology can be eliminated [17]. Second, with the desired spectrum downconverted directly to the baseband, channel selection can be performed either with a simple low-pass filter or by converting to the digital domain with an ADC and digitally performing channel selection in DSP. The capability of using DSP raises the possibility of having a universal RF front-end, where any type of wireless standard (e.g. GSM, CDMA, WiFi, WiMax, etc.) can be received with the same analog blocks and decoded using digital methods. A fundamental, unchanging RF front-end is one of the keys in realizing multimode, multistandard software radios. The block diagram of a DCR is shown in Figure 13.7.

Consequently, the direct conversion architecture is highly attractive for integrated RF receivers. The reduction in off-chip components results in higher levels of integration and lower costs. Furthermore, the elimination of off-chip filters simplifies interstage matching, since the need to match the input and output impedances of the off-chip filter is eliminated.

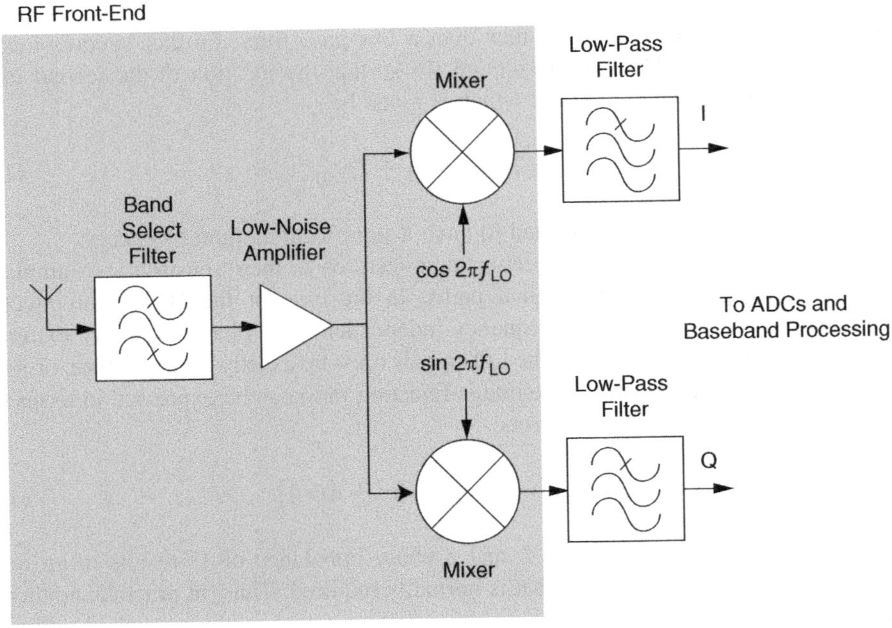

Figure 13.7 Block diagram of the direct conversion receiver.

By filtering at baseband frequencies, device parasitics are less severe and less current is needed for amplification; thus, DCRs also offer the potential for low-power consumption. However, DCRs suffer from a number of unique issues, making their actual implementation less than straightforward.

13.3.3.1 DC Offsets

RF band selection is the only filtering performed in the receive chain of a DCR before the signal is downconverted to the baseband. Therefore, strong, nearby signals, including the receiver's own LO, can mix with themselves (self-mixing) down to zero IF, generating DC levels (potentially time-varying) that appear as interference at the centre of the desired band. Figure 13.8 shows three potential self-mixing mechanisms for creating DC offsets.

Figure 13.8(a) shows self-mixing due to the finite isolation between the LO and RF ports of a mixer in one branch of the DCR. For an integrated mixer, port-to-port isolation is limited by substrate coupling, bondwire radiation, and capacitive and magnetic coupling [18]. Since the LO is typically a strong signal, in order to provide sufficient drive for the mixer switching core, the LO can leak through these unintended paths with sufficiently high amplitude back to the low-noise amplifier (LNA). There, the LO signal can reflect off the LNA output back into the mixer input and mix with itself, generating a static DC level. Furthermore, if the LO signal leaks back to the LNA input, the situation is exacerbated since the LO is amplified. The LO signal can also be reradiated by the antenna, reflect off obstructions such as a building or a moving vehicle, and be recaptured by the front-end. With fading and multipath reception, the received power level can therefore rapidly vary, resulting in a time-varying or dynamic DC offset (Figure 13.8(b)). The reradiation of the

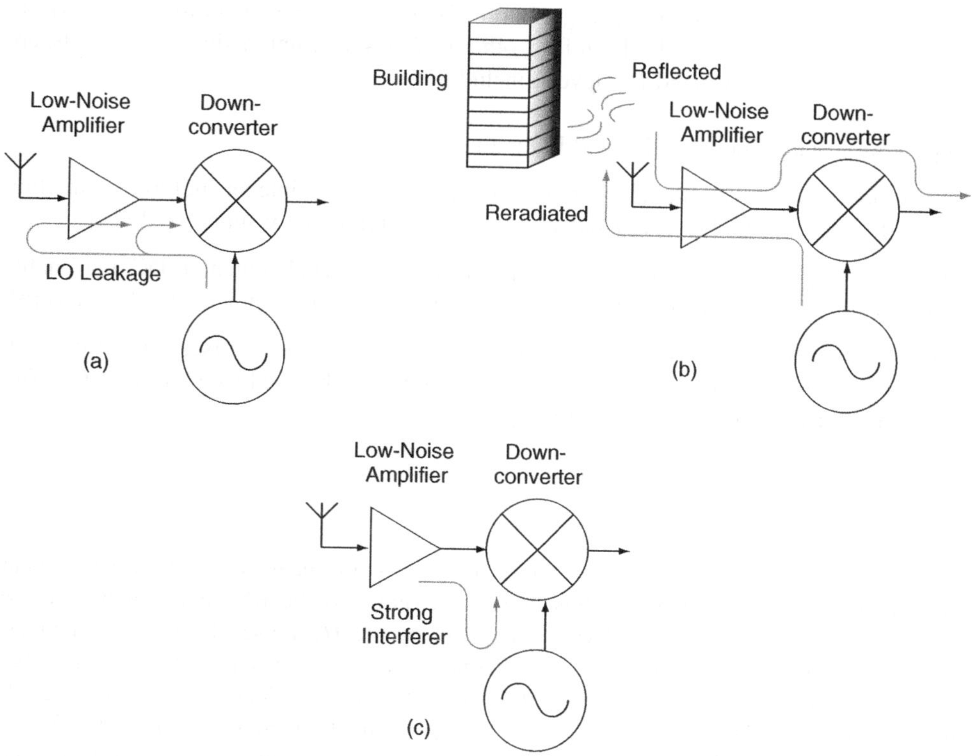

Figure 13.8 DC offset mechanisms: (a) LO leakage, (b) LO reradiation; and (c) strong in-band interferer (after Reference [18]).

LO can also cause problems for other users in the same receive band. Since the LO is at the same frequency as the RF, it appears in the passband of the RF band select filter of other users in the system. Thus, as shown in Figure 13.8(c), a strong nearby interferer such as another user's LO, can also generate DC offsets by finding a path to the mixer LO port and mixing with itself.

The impact of such DC offsets on a communication system depends on the type of modulation being employed. For example, Gaussian minimum phase-shift keying (GMSK) modulation, used in the Global System for Mobile Communications (GSM), has significant signal content near DC. Thus, a time-varying DC offset can lead to severe signal corruption. On the other hand, quadrature phase-shift keying (QPSK), used in code division multiple access (CDMA) schemes, is less susceptible to DC offsets [17]. Nonetheless, DC offsets appearing at the centre of the downconverted spectrum can dominate the output of the mixer. If large enough, the DC levels can saturate or desensitize the baseband components after the mixer, degrading amplification of the desired signal and thereby reducing sensitivity.

While the simplest solution for removing DC offsets is to high-pass-filter the spectrum with a series AC coupling capacitor, this method is not always feasible. As mentioned above, GMSK (and many other modulation schemes) has a significant signal content at DC; capacitively coupling the mixer output can lead to a loss of information and higher bit error rates (BERs). Furthermore, the corner frequency of the high-pass filter has to be extremely low, in the order of tens of hertz; the required on-chip area for a series capacitor at these frequencies is prohibitively large. Large-size capacitors also tend to react slowly to dynamically changing DC levels. Therefore, other methods, or combinations of methods, are employed to mitigate unwanted DC levels, including:

- Improving isolation between RF and LO paths.
- Subharmonic or superharmonic mixing. With an LO at some integer multiple or fraction of the RF, DC offsets due to LO self-mixing can be significantly suppressed [19].
- DC-free coding of the transmitted data. By removing the signal content at DC, reasonable high-pass filtering can be employed, particularly for wideband channels (e.g. CDMA) [6].
- Offset calibration techniques. Digital sample-and-hold feedback, offset estimation with corrective feedback at the baseband, and servo loops, can be used to track the DC offset and remove it from the desired spectrum [17].

13.3.3.2 Flicker (1/f) Noise

Flicker ($1/f$) noise is a significant problem in RF systems in general because nonlinear devices and circuits can upconvert this noise spectrum into the RF range, such as in an RF oscillator where the close-in phase noise is set by the $1/f$ noise of the active devices. However, in the direct conversion case, low-frequency noise presents an additional challenge as it can appear directly in the downconverted band and therefore degrade the overall signal-to-noise ratio SNR. The $1/f$ corner frequency f_α, is the frequency at which the flicker noise and the receiver thermal noise floor are equal and is often used to specify device noise performance. The value of f_α is dependent on the semiconductor process: for bipolar devices, a corner frequency as low as $\sim$8 kHz can be achieved, while for metal oxide silicon field-effect transistors (MOSFETs), f_α can range from tens of kilohertz to a few megahertz [17].

Like DC offsets, the impact of $1/f$ noise depends on the modulation scheme. For narrowband systems like GSM, high corner frequencies can effectively raise the noise figure and degrade the SNR of the received signal; on the other hand, in wideband systems, such as CDMA, $1/f$ noise has a more minimal effect.

13.3.3.3 Even-Order Distortion

Intermodulation distortion (IMD) can have deleterious effects on the system performance of a radio transceiver. IMD products result from the interaction of large signals with circuit nonlinearities, resulting in signal components arising at frequencies in or near the passband of a receiver or at baseband frequencies. IMD products can cause an increase in bit errors, suppress receiver sensitivity, and/or generate interference to other users or systems. Products due to odd-order distortion (in particular third-order) are a significant problem in the design of all receivers as third-order intermodulation products (e.g. IM3) fall directly within the band of interest and cannot be filtered easily. For direct conversion receivers, *even-order* distortions must also be considered as they lead to second-order intermodulation products (e.g. IM2) that either fall within the downconverted baseband spectrum [20] or, in the case of a large single-tone interferer, generate a time-varying DC offset [21]. Second-order nonlinearities are also problematic for signals with strong envelope variations, such as a

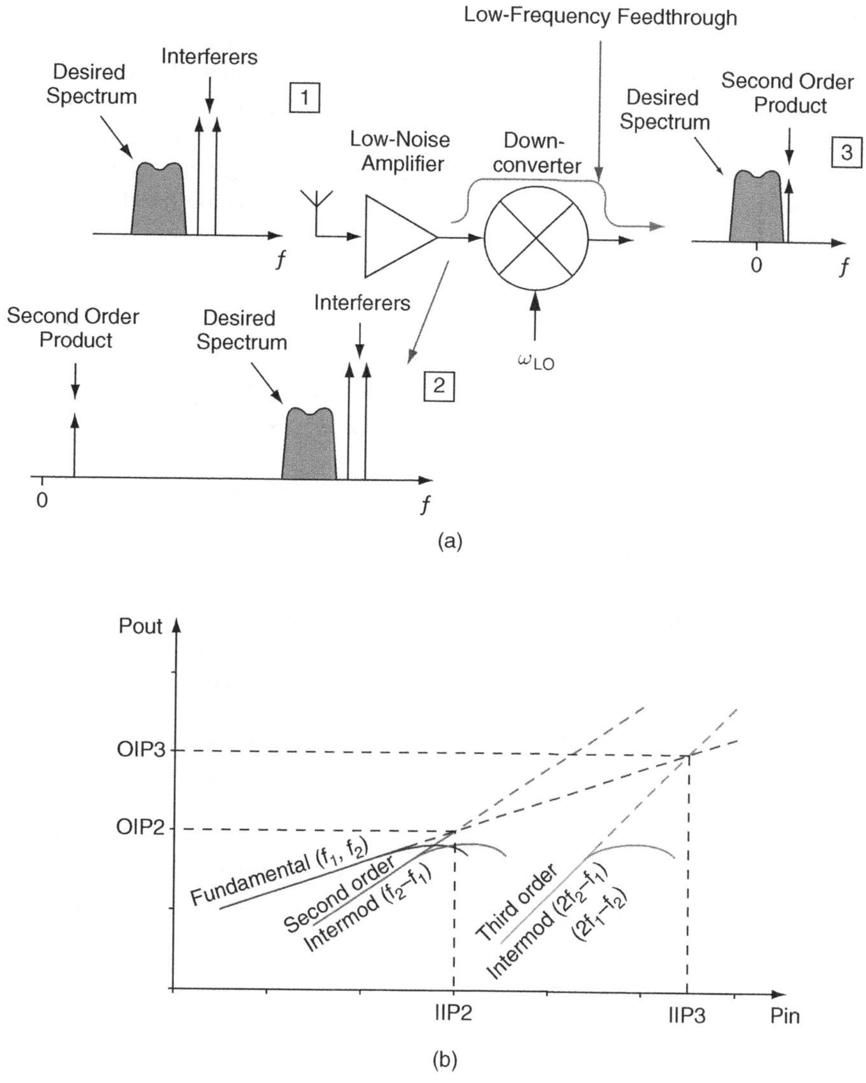

Figure 13.9 Intermodulation distortion: (a) second-order effect on strong interferers (after Reference [22]) and (b) extrapolation of tones to find the second- and third-order intercept points, IP2 and IP3, respectively.

signal that undergoes fading during propagation, since these amplitude modulations (AM) are essentially demodulated into baseband components by even-order nonlinearities [22].

To see the effects of second order distortion on a direct conversion receiver, Figure 13.9(a) shows the front-end of a generic DCR that experiences a second-order nonlinearity within the LNA. For analysis purposes, the nonlinear response of the LNA can be approximated with a power series approximation:

$$y(t) = k_1 x(t) + k_2 x^2(t) + k_3 x^3(t) + \cdots \tag{13.9}$$

Applying two tones of equal power ($A_1 = A_2 = A$) at the LNA input, i.e. the input is given by $x(t) = A\cos 2\pi f_1 t + A\cos 2\pi f_2 t$, the resulting output spectrum can be found by substituting $x(t)$ into Equation (13.9). Using trigonometric identities, the resulting equation can be simplified and leads to the second-order product $k_2\cos[2\pi(f_1 - f_2)t]$. This frequency component can then leak through the mixer (due to finite RF-to-IF isolation) and appear at the baseband, as shown at point 3 of Figure 13.9(a). Though not depicted here, the mixer can also have second-order nonlinearities and further degrade the desired baseband signal.

In the case of a single-tone applied at the LNA input, where $x(t) = A\cos 2\pi ft$, the output is given by:

$$\begin{aligned} y(t) &= k_1 A\cos(2\pi ft) + k_2 A^2\cos^2(2\pi ft) \\ &= k_1 A\cos(2\pi ft) + \tfrac{1}{2}k_2 A^2\{1 + \cos[2\pi(2f)t]\} \\ &= \tfrac{1}{2}k^2 A_2 + k_1 A\cos(2\pi ft) + \tfrac{1}{2}k_2 A^2\cos[2\pi(2f)t] \end{aligned} \tag{13.10}$$

The first term in Equation (13.10) is a DC component related to the second-order nonlinearity of the circuit, k_2, and is magnified by the square of the input signal amplitude. Therefore, any large signal within the passband of a direct conversion receiver can generate a large DC offset if the even-order performance of the receiver is poor (i.e. if k_2 is relatively large).

A common method for characterizing intermodulation distortion is the well-known two-tone test. Here, two (typically) equal-amplitude tones, slightly offset in frequency, are applied to the input of a circuit. The power levels of the second- and third-order IM products and the fundamental tones are measured at the circuit output. The measured data are linearly extrapolated beyond compression until the IM curves intersect with the curve of the fundamental. These intersections, which are in reality nonphysical since they lie beyond the compression point, are known as the intercept points (IPs) for the different IM products. Figure 13.9(b) shows the locations of the input second-order intercept point (IIP2) and the input third-order intercept point (IIP3). Assuming the slope of the second-order IM product is roughly 2 and the slope of the third-order IM product is roughly 3, the intercept points can be calculated from a two-tone test at a single-input power level (P_{in}) using the following equations:

$$\text{IIP2} = 2P_{\text{in}} - \text{IM2} \tag{13.11}$$

$$\text{IIP3} = \frac{3}{2}P_{\text{in}} - \text{IM3}/2 \tag{13.12}$$

The intercept points give the absolute power level of the IM products for a given input power level. From these, the relative suppression (RS) of the unwanted products with respect to the fundamental can be determined:

$$\text{RS2} = P_{\text{in}} - \text{IIP2} \tag{13.13}$$

$$\text{RS3} = 2(P_{\text{in}} - \text{IIP3}) \tag{13.14}$$

The required relative suppression for each block in a receiver chain is determined by the overall system design.

Two common ways to combat even-order distortion include the use of well-balanced, or differential, circuitry and the use of feedback. Balanced circuitry is effective because ideally common-mode signals are rejected and thus second-order nonlinearities become a nonissue. Although device mismatches make the actual nonlinearities nonzero, balanced circuitry is still extremely effective. The downside of this approach is the increased power consumption and die area required for essentially twice as much circuitry. For an LNA in particular, a differential topology requires two times the power of its single-ended counterpart to achieve the same noise figure.

13.3.3.4 I/Q Mismatch

The translation of the desired RF spectrum to DC mandates the use of quadrature downconversion to recover the negative-frequency portion of the spectrum [13]. Because the LO is at the same frequency as the RF, the negative-frequency spectrum, an equal but mirrored image of the desired spectrum, is also converted to DC, making it impossible to differentiate between the two spectra. Therefore, analog I/Q demodulation (which is subject to mismatches that lead to nonzero gain and phase imbalances, as mentioned above in the context of image IRR) is necessary to properly receive the desired signal. These imbalances introduce distortions into the received I/Q constellation and result in an increased BER [23]. An example of distorted I/Q constellations is shown in Figure 13.10.

Figure 13.11 illustrates the use of quadrature downconversion in an ideal DCR to remove the negative frequency portion of the spectrum. Here, the I and Q branches are assumed to have perfect amplitude and phase balance and the system is assumed to be noise free. For the in-phase channel, which by convention denotes the side multiplied by $\cos(2\pi f_{LO}t)$, an equal but mirrored image of the desired spectrum is centred at $-f_{RF}$; after downconversion, this negative spectrum falls at DC on top of the desired signal. Without a quadrature channel, the resulting IF spectrum is a mixture of the two halves of the bandwidth on either side of the LO, each half indistinguishable from the other, as in Fig. 13.11(c.i). On the other hand, with an orthogonal quadrature channel, where the input signal is multiplied by $\sin(2\pi f_{LO}t)$, the

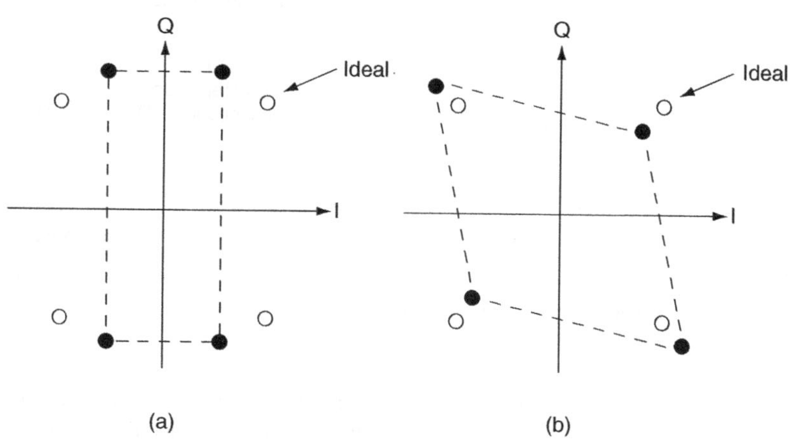

Figure 13.10 Effect of I/Q mismatch on signal constellation: (a) gain error and (b) phase error (after Reference [6]).

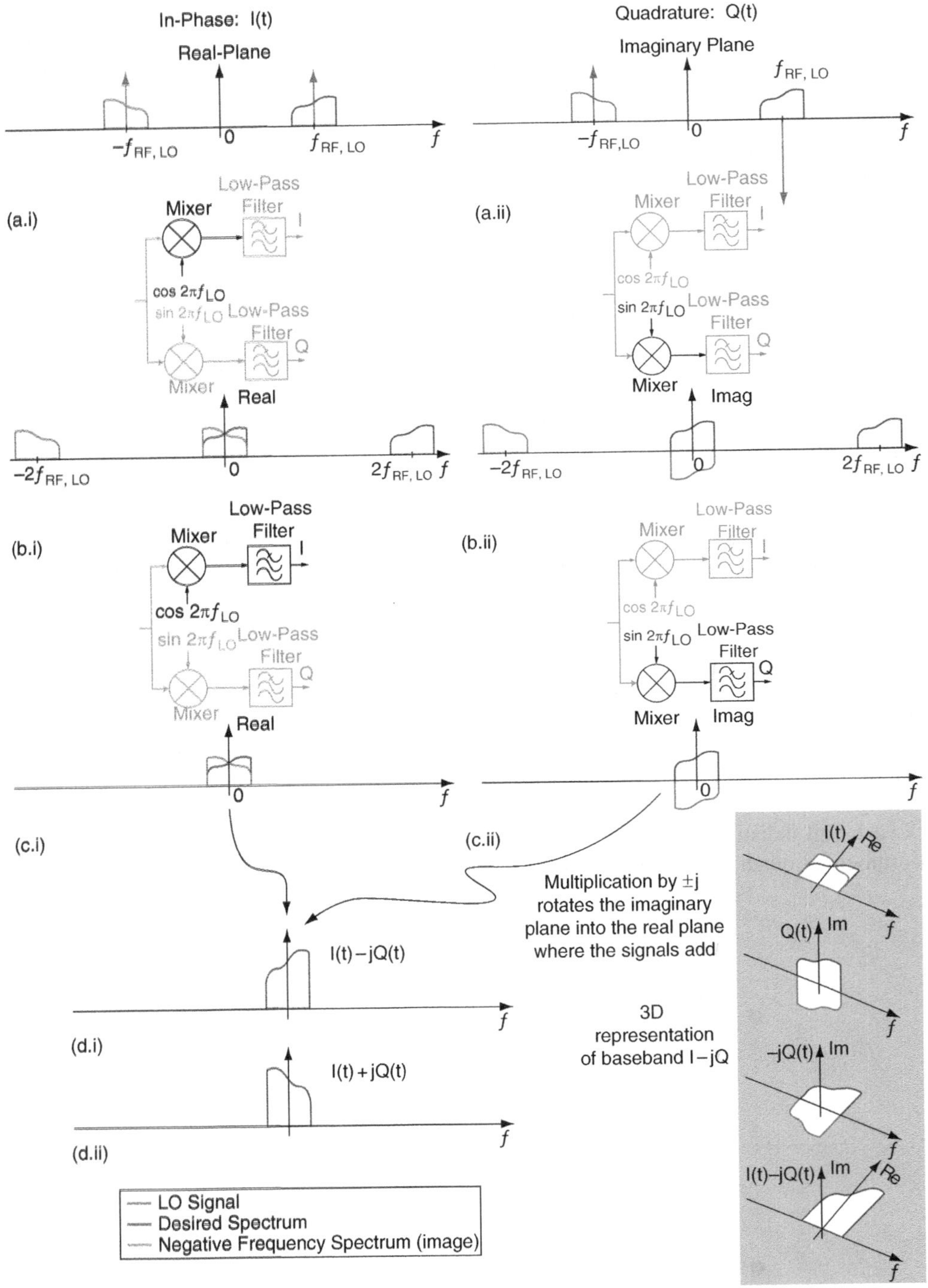

Figure 13.11 Illustration of quadrature downconversion and the removal of the negative 'image' frequency. Note that only the magnitude of the spectrum is shown in the picture. Both magnitude and phase portions lead to the removal of the negative frequency components. Gain and phase imbalances result in less rejection of the negative frequency and increases in the BER.

negative frequency portion of the spectrum can be removed, as shown in Figure 13.11(d.i) or (d.ii). The inset of Figure 13.11 shows how the I and Q channels interact to remove the image.

The 90° separation between the I and Q paths is typically accomplished in the LO path of the downconversion mixers, as shown in Figure 13.11. Quadrature signals can be generated with one of several different methods: RC polyphase filters [24], quadrature VCOs [25], frequency divider stages [26], and ring oscillators [27]. To remove the negative frequency portion of the spectrum completely, the amplitude and balance of the I and Q branches must be perfect; otherwise, imbalances between either the I and Q amplitudes or the phase (or both) result in greater corruption of the desired signal by the negative frequency spectrum. Although all components in the receive path contribute to gain and phase errors, the balance of the LO is critical. Negative frequency rejection (NFR) is one way to characterize the effects of I/Q amplitude (ε) and phase ($\Delta\phi$) imbalances. Given in dBc, the NFR for a receiver is approximately

$$\text{NFR} \approx 10 \log (\varepsilon^2 + \tan^2 \Delta\phi) \tag{13.15}$$

For an amplitude and phase imbalance of 0.1 dB and 1°, respectively, the associated NFR is ~ 35 dBc. With careful design techniques, gain and phase imbalances can be minimized. Furthermore, since the imbalances tend to be static, both analog and digital (DSP) calibration and adaptation methods can be used to correct the mismatches [28, 29].

13.4 TRANSMITTER ARCHITECTURES

When selecting an RF transmitter architecture, the type of modulation that will be employed should first be considered. For example, the familiar analog frequency modulation (FM) (or a simple digital frequency shift keying (FSK) modulation, for that matter) can be generated by tuning the frequency of the transmit VCO (either by directly tuning the varactor control voltage of the VCO or through programmable control of a phase locked loop (PLL)/frequency synthesizer). However, most modern wireless communication systems employ some type of digital phase modulated signalling scheme, which in turn typically necessitates I and Q signal paths in the transmitter. In principle, the I and Q baseband data can be directly upconverted in quadrature to RF; however, in practice, pulse shaping must be performed on the data streams using a combination of analog and digital techniques in order to minimize intersymbol interference [6]. Also, as with the direct conversion and low-IF receiver architectures described above, I/Q transmitters are sensitive to gain and phase mismatches in the I and Q signal paths.

This section will discuss two transmitter architectures that are currently in use for such systems.

13.4.1 Direct Conversion Transmitters

In direct conversion (also known as direct modulation) transmitters, the transmitted carrier frequency is equal to the local oscillator frequency; in other words, modulation and upconversion occur simultaneously [6]. This is basically analogous to the direct conversion receiver architecture described above. The basic implementation of this architecture with I and Q

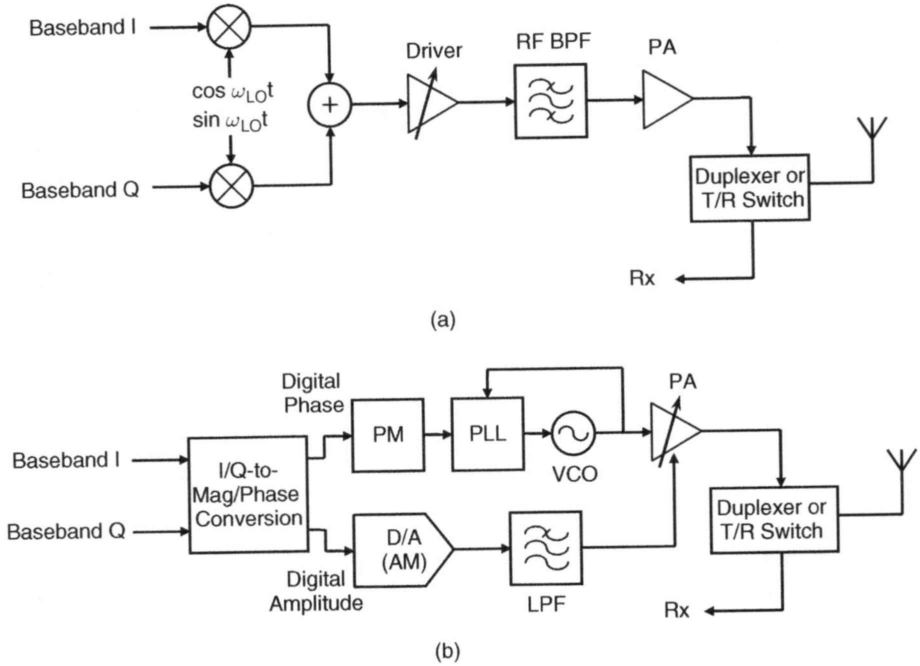

Figure 13.12 Block diagram of (a) direct conversion transmitter architecture and (b) open-loop polar modulation transmitter.

signal paths is shown in Figure 13.12(a). As can be seen, the I and Q baseband data streams are upconverted to RF, combined, and then amplified prior for transmission from the antenna. A bandpass filter is typically required between the driver amplifier and the PA in order to suppress transmitted noise outside the transmit band.

A major issue with this architecture is PA output leakage feeding back to the VCO. Although the RF carrier frequency is at the same frequency as the LO, the PA output is a 'noisy' modulated spectrum around that LO frequency [6]. Energy located *near* this centre frequency, which leaks back to the VCO, can result in 'injection pulling' or 'injection locking' of the VCO frequency, which in turn will cause the transmitted RF spectrum to be at an incorrect centre frequency. This effect can be mitigated somewhat through shielding strategies, including simply placing the PA and related circuitry in a separate module, rather than in the same environment as the transmit VCO. Another approach is to generate the transmitter carrier frequency by mixing two 'offset' VCOs whose individual oscillation frequencies are well away from the desired transmit centre frequency, which significantly reduces the potential for injection pulling or locking to occur. Alternatively, a more conventional two-step transmitter (analogous to the traditional superheterodyne receiver) can be used, which avoids VCO pulling problems through suitable selection of intermediate frequencies. Furthermore, sensitivity to I/Q mismatches is reduced since quadrature modulation occurs at lower frequencies [6]. However, significant high-Q filtering is required to suppress transmitted noise and spurious frequencies, which increases component requirements (in some cases off-chip) and degrades transmitter power efficiency. Single-sideband (SSB) mixer topologies can be used to suppress unwanted sidebands, but at the expense of further increased circuit size, complexity, and power consumption.

13.4.2 Polar Modulation Transmitters

Meanwhile, emerging higher data rate digital wireless communications standards (such as EDGE, enhanced data rates for GSM evolution) have led to implementations employing higher-order modulation schemes (such as a modified 8-PSK in the case of EDGE). At the same time, existing voice and other lower data rate services must also be supported, leading to modern multimode radio architectures. Of course, such higher-order modulation schemes typically require higher levels of PA linearity due to their higher peak-to-average voltage ratios. An increasingly popular approach to this problem is the polar modulation transmitter [30].

In the polar modulation transmitter architecture, the I/Q baseband data streams are converted into magnitude/phase data, using a digital processing algorithm such as the CORDIC (coordinate rotation digital computer) scheme. The phase data are then modulated on the transmit VCO phase, while the magnitude data are modulated directly on the PA bias. This is advantageous in that AM–AM and AM–PM conversion problems in nonlinear PAs can be largely mitigated, allowing the use of more power-efficient PA topologies to be used.

A basic diagram of a polar modulation transmitter is shown in Figure 13.12(b). The architecture shown is an *open-loop* polar modulation transmitter. However, this approach imposes stringent requirements on the linearity of the magnitude signal input to the PA bias, and typically requires some predistortion in order to transmit the amplitude information accurately over the full range of PA output voltage swings [30]. In addition, some predistortion or correction to the phase input may also be required. Such predistortion and correction schemes must also account for variations in temperature, supply voltage, etc. An alternative approach is a *closed-loop* polar modulation transmitter (also known as a polar-loop transmitter), wherein closed-loop feedback control is applied to both the amplitude and phase control paths based on the sensed PA output signal [30]. This architecture can inherently correct for nonlinearities in the amplitude and phase control paths, and obviate requirements for calibration and predistortion. However, stability of the closed control loops can be an issue and may impose operational bandwidth limitations on the system.

13.5 FRONT-END–ANTENNA INTERFACE AND INTEGRATION ISSUES

This chapter has focused so far on transceiver architectures that can lead to highly integrated RF front-ends, particularly from the point of view of silicon SoC integration. As described above, direct conversion/low-IF receivers and direct modulation transmitters have emerged as the architectures of choice for modern wireless communications transceivers due to their high level of integration, direct digital interface, and suitability for implementation of multimode/multiband products. However, the aforementioned receiver architectures are susceptible to a number of impairments, including I/Q mismatches, second-order nonlinearities, $1/f$ noise, and static/dynamic DC offsets (DCR only). The transmitter architectures can also be impaired by I/Q mismatch errors, as well as PA leakage and harmonic issues. Nonetheless, the significant advantages of these architectures for low-cost, multiband, multimode architectures has led to the development of various device, circuit, and system level strategies to overcome such impairments.

An important design approach in modern integrated transceivers is the use of well-balanced differential circuits in both I and Q signal paths. Such differential circuits mitigate second-order distortion effects, as well as offer common-mode noise rejection. However, this in turn requires careful consideration of the single-ended-to-differential transformation process where required. Of course, the simplest case is that of a balanced antenna, such as a printed dipole or bow-tie antenna, which could in principle be directly coupled to differential transmit or receive circuitry. For example, in Reference [31], 2 mm long dipole antennas were integrated on standard bipolar-grade (20 Ω cm) silicon substrates with integrated transmit and receive circuits, and interchip communications at 24 GHz for up to ~5 m ranges was demonstrated. Naturally the gain of the link was quite low owing to the lossy silicon substrate, but was sufficient for the short ranges in question. Similar antenna structures were integrated into silicon ICs for intrachip wireless clock distribution around 10 GHz over distances of 2.2 cm [32]. However, these single-chip transceiver nodes did not include front-end band-select filtering or duplexing functions, required for most modern wireless communication transceiver architectures, as shown in Fig. 13.12. The latter situation is complicated by the need to support multiple transmit and receive bands, which is currently addressed by using multiple filters (e.g. SAWs) and switch bank modules ('switchplexers'), which are not typically available in differential form. Therefore, the current approach is to employ single-ended antennas off-chip, single-ended filter/duplexer structures, and perform single-ended-to-differential conversion at the LNA or PA. This conversion can be performed with either (1) passive balanced-to-unbalanced (balun) structures or (2) single-ended-to-differential active circuit topologies in the LNA and/or transmit amplifier chains. The first approach is undesirable since it adds additional off-chip components and may introduce additional front-end losses in the low-noise receive and high-power transmit paths; the second approach may introduce differential path imbalances and impedance matching issues. Integration of *on-chip* balun structures is generally impractical for RF frequencies below ~10 GHz due to their large die area consumption and losses on typical silicon substrates, although it may become practical at higher (millimetre-wave) frequencies on semi-insulating substrates and/or by exploiting higher interconnect metal layers.

Additionally, while integrated antennas with reasonable on-chip dimensions are practical at higher microwave and millimetre-wave frequencies (>30 GHz), it is a significant challenge to integrate antennas on-chip at the lower RF range frequencies typically used in mobile/wireless communication systems. For example, in Reference [33] a 'chip-scale' integrated patch antenna was demonstrated for the 5–6 GHz range in a commercially available GaAs MESFET (metal semiconductor field effect transistor) power amplifier process (Figure 13.13). The antenna was fabricated on the *backside* of the chip through backside lithography. An integrated class-A tuned MESFET PA fed the antenna through a backside via and was output-matched directly to the antenna driving point impedance. The size of the antenna was ~4.2 mm × 4.2 mm, which is ~1/13 of λ_0 at ~5.6 GHz. The structure was mounted with the antenna facing up and is compatible with flip-chip technology. The combined active antenna had a measured gain on boresight of −3 dBi, ~10 dB greater than the passive (no PA) comparison antenna. This level of performance is of course contingent on the semi-insulating GaAs substrate, and a comparable approach on standard low-resistivity silicon would be impractical (although RF/mixed-signal IC processes employing high-resistivity silicon substrates are now becoming available).

The competing alternative to complete SoC integration is a module-based approach, also referred to as system-in-package (SiP) or system-on-package (SoP) solutions. As discussed in the introduction, a key advantage of module-based implementations is that the optimum technology for each system function can be selected. With respect to the specific issue of off-chip high-Q passives and efficient, low-loss antennas, a module-based approach enables

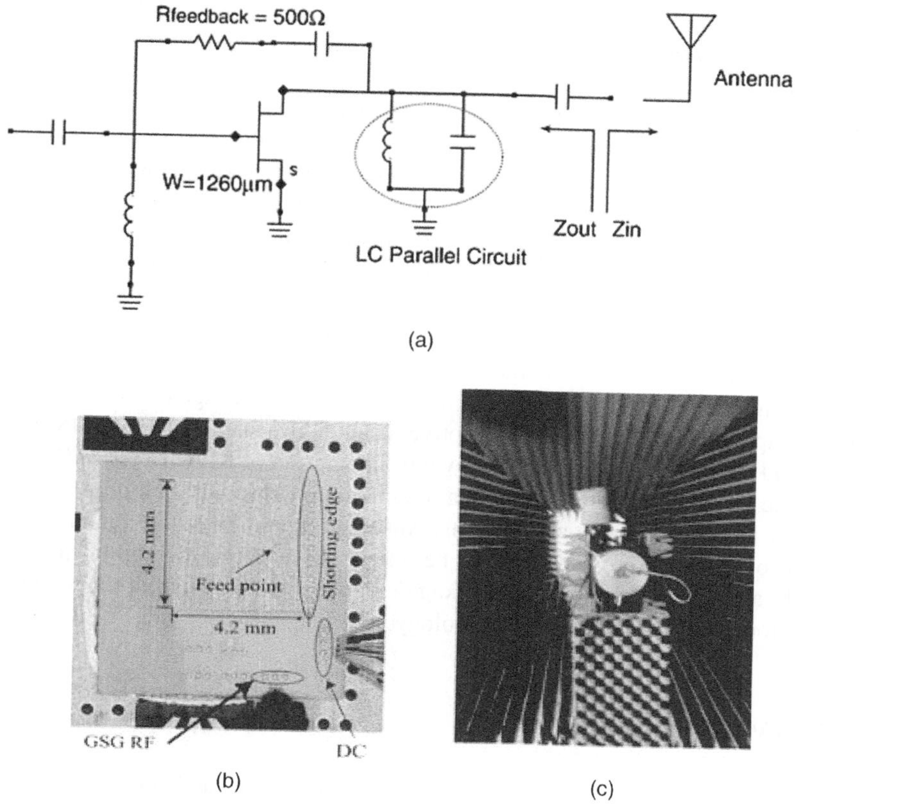

Figure 13.13 (a) Simplified circuit diagram of the integrated power amplifier; (b) active antenna (backside view); and (c) antenna mounted in an anechoic chamber for gain and pattern measurements.

such structures to be mounted on or directly fabricated in a low-loss module material (such as a low-temperature co-fire ceramic (LTCC) or liquid crystal polymer (LCP)). The larger nonsemiconductor area associated with the module or package structure allows for planar and multilayer antenna (and balun and filter) integration without sacrificing the relatively precious semiconductor die area. For example, in Reference [34], a circularly polarized slotted-patch antenna was implemented in a low-loss microwave laminate material, which was in turn used as the top shield of a multichip module, for 5.25 GHz WLAN applications. A three-dimensional feeding structure was used to couple the front-end circuits on the lower part of the MCM to the antenna feed points. In this case, the antenna size was much larger than what could economically be integrated at the IC level (2.3×2.3 cm^2), but with commensurate gains in performance. The module antenna demonstrated a measured gain of 6 dB in the boresight direction and a measured efficiency of 70 %. In Reference [35], LTCC substrates were stacked with micro-BGAs used for vertical interconnects between the substrates. Stepped cavities fabricated in the LTCC substrates provided space for embedding RF devices and ICs (and potentially MEMS), which are wirebond or flip-chip interconnected with the LTCC boards. Low-loss distributed and passive components, and antennas, could also be directly integrated into the LTCC structures.

Another advantage of a module-based approach is PA isolation. As discussed above, in direct conversion transmitters, PA output leakage can lead to undesired VCO injection pulling or locking. Therefore, location of the PA chip(s) in a separate module, perhaps along with transmit/receive/band-select switches and low-loss filter components, is currently the approach of choice.

Of course, the module-based approach potentially leads to greater manufacturing costs due to the requirements for handling, mounting, and connecting the different components, and may also suffer from lower reliability and lower manufacturing yields. However, based on the discussion in Reference [36], SoP *currently* tends to be the lower-cost solution overall, particularly when integration of high-Q passive elements and antennas is considered. However, such cost-benefits are not realized if the various RF, baseband, and data-conversion/DSP functional blocks are implemented as individual ICs within the module or package. Therefore, in conclusion, a continued push is expected for high-level integration of RF/analog/mixed-signal functions in an SoC environment, while high-Q/low-loss passive components are likely to continue to be integrated off-chip in the module or package environment for high-performance applications. Potentially disruptive technologies such as RF MEMS filters may ultimately lead to further advances in the level of integration of RF/wireless SoCs; however, based on basic physical limitations, antenna integration on-chip will be limited to particular short-range or low-performance applications. Antenna integration at the module or package level, on the other hand, is very likely to be an attractive option of highly integrated RF front-ends. In any case, careful system/package/module/chip co-design is required to obtain optimum performance from the chosen topology.

13.6 SUMMARY

In this chapter transceiver architectures have been presented that can lead to highly integrated RF front-ends. An overview has been given of wireless systems on a chip and both transmitter and receiver architectures have been looked at that can be utilized in this important wireless application. The focus has been directed to the point of view of silicon SoC integration, but the architectures and design trade-offs discussed are also relevant to RF transceiver module/system-in-package/system-on-package (SiP/SoP) implementations.

REFERENCES

1. D. Y. C. Lie, X. Yuan, L. E. Larson, Y. H. Wang, A. Senior and J. Mecke, ' "RFSoC": low-power single-chip radio design using Si/SiGe BiCMOS technology', in *Proceedings of the 2002 3rd International Conference on Microwave and Millimeter Wave Technology*, pp. 30–37, August 2002.
2. L. M. Franca-Neto, P. Pardy, M. P. Ly, R. Rangel, S. Suthar, T. Syed, B. Bloechel, S. Lee, C. Bumett, D. Cho, D. Kau, A. Fazio and K. Soumyanath, 'Enabling high-performance mixed-signal system-on-a-chip (SoC) in high performance logic CMOS technology', in *2002 Symposium on VLSI Circuits Digest of Technical Papers*, pp. 164–167, June 2002.
3. J.-F. Luy and G. E. Ponchak, 'Introduction to the special issue on microwave circuits on silicon substrates', *IEEE Transactions on Microwave Theory and Techniques*, **46**(5), 569–571, May 1998.
4. J. Sevenhans, F. Op't Eynde and P. Reusens, 'The silicon radio decade', *IEEE Transactions on Microwave Theory and Techniques*, **50**(1), 235–244, January 2002.
5. L. E. Larson, 'Integrated circuit technology options for RFICs – present status and future directions', *IEEE Journal of Solid-State Circuits*, **33**(3), 387–399, March 1998.
6. B. Razavi, *RF Microelectronics*, Prentice-Hall, Upper Saddle River, New Jersey, 1997.

7. A. Chatterjee, J. D. Cressler, D. B. M. Klaassen, A. Matsuzawa and H. Shichijo, 'Forward special issue on device integration technology for mixed-signal SOC', *IEEE Transactions on Electron Devices*, **50**(3), 543–545, March 2003.
8. S. M. Sinaga, A. Polyakov, M. Bartek and J. N. Burghartz, 'Circuit partitioning and RF isolation by through-substrate trenches', in *Proceedings of the 2004 Electronic Components and Technology Conference*, pp. 1519–1523, June 2004.
9. T. Blalack, Y. Leclercq and P. Yue, 'On-chip RF isolation techniques', in *Proceedings of the 2002 Bipolar/BiCMOS Circuits and Technology Meeting (BCTM)*, pp. 205–211, October 2002.
10. A. Matsuzawa, 'RF-SoC – expectations and required conditions', *IEEE Transactions on Microwave Theory and Techniques*, **50**(1), 245–384, January 2002.
11. L. E. Larson, 'Silicon technology tradeoffs for radio-frequency/mixed-signal systems-on-a-chip', *IEEE Transactions on Electron Devices*, **50**(3), 683–699, March 2003.
12. T. Hentschel, M. Henker and G. Fettweis, 'The digital front-end of software radio terminals', *IEEE Personal Communications*, 40–46, August 1999.
13. A. A. Abidi, 'Direct-conversion radio transceivers for digital communications', *IEEE Journal of Solid-State Circuits*, **30**(12), 1399–1310, December 1995.
14. J. Crols and M. S. J. Steyaert, 'Low-IF topologies for high-performance analog front ends of fully integrated receivers', *IEEE Transactions on Circuits and Systems – II Analog and Digital Signal Processing*, **45**(3), 269–282, March 1998.
15. M. Valkama and M. Renfors, 'A novel image rejection architecture for quadrature radio receivers', *IEEE Transactions on Circuits and Systems – II Analog and Digital Signal Processing*, **51**(2), 61–68, February 2004.
16. S. J. Roome, 'Analysis of quadrature detectors using complex envelope notation', *IEE Proceedings of Radar, Sonar, and Navigation*, **136**(2), 95–100, April 1989.
17. A. Loke and F. Ali, 'Direct conversion radio for digital mobile phones – design issues, status, and trends', *IEEE Transactions on Microwave Theory and Techniques*, **50**(11), 2422–2435, November 2002.
18. A. Mashhour, W. Domino and N. Beamish, 'On the direct conversion receiver – a tutorial', *Microwave Journal*, **44**(6), 113–128, June 2001.
19. R. Svitek and S. Raman, 'A SiGe active sub-harmonic front-end for 5-6 GHz direct conversion receiver applications', in *2003 IEEE RFIC Symposium Digest*, Philadelphia, Pennsylvania, pp. 675–678, 8–13 June 2003.
20. A. A. Abidi, 'General relations between IP2, IP3 and offsets in differential circuits and the effects of feedback', *IEEE Transactions on Microwave Theory and Techniques*, **51**(5), 1610–1612, May 2003.
21. R. Magoon, A. Molnar, J. Zachan, G. Hatcher and W. Rhee, 'A single-chip quad-band (850/900/1800/1900 MHz) direct conversion GSM/GPRS RF transceiver with integrated VCOs and fractional-*N* synthesizer', *IEEE Journal of Solid-State Circuits*, **37**(12), 1710–1720, December 2002.
22. B. Razavi, 'Design considerations for direct-conversion receivers', *IEEE Transactions on Circuits and Systems – II Analog and Digital Signal Processing*, **44**(6), 428–435, June 1997.
23. C. L. Liu, 'Impacts of I/Q imbalance on QPSK-OFDM-QAM detection', *IEEE Transactions on Consumer Electronics*, **44**(3), 984–989, August 1998.
24. F. Behbahani, Y. Kishigami, J. Leete and A.A. Abidi, 'CMOS mixers and polyphase filters for large image rejection', *IEEE Journal of Solid-State Circuits*, **36**(6), 873–886, June 2001.
25. P. Andreani, A. Bonfani, L. Romano and C. Samori, 'Analysis and Design of a 1.8 GHz CMOS LC Quadrature VCO', *IEEE Journal of Solid-State Circuits*, **37**(12), 1737–1747, December 2002.
26. A. Mazzanti *et al.*, 'Analysis and design of injection-locked LC dividers for quadrature generation', *IEEE J. Solid-State Circuits*, **39**(9), 1325–1333, September 2004.
27. L. Dai and R. Harjani, 'A low-phase-noise CMOS ring oscillator with differential control and quadrature outputs', in *Proceedings of the 2001 13th Annual IEEE International ASIC/SOC Conference*, pp. 134–138, September 2001.

28. Y.-H. Hsieh *et al.*, 'An auto-I/Q calibrated CMOS transceiver for 802.11g', *IEEE J. Solid-State Circuits*, **40**(11), 2187–2192, November 2005.
29. J. Chabloz *et al.*, 'A novel I/Q mismatch compensation scheme for a low-IF receiver front-end', in *IEEE International Symposium on Circuits and Systems*, Vol. 4, pp. 453–456, May 2004.
30. T. Sowlati *et al.*, 'Quad-Band GSM/GPRS/EDGE polar loop transmitter', *IEEE J. Solid-State Circuits*, **39**(12), 2179–2189, December 2004.
31. J.-J. Lin, L. Gao, A. Sugavanam, X. Gao, R. Li, J.E. Brewer and K. K. O, 'Integrated antennas on silicon substrates for communication over free space', *IEEE Electron Device Letters*, **25**(4), 196–198, April 2004.
32. K. K. O, K. Kim, B. A. Floyd, J. I. Mehta, H. Yoon, C.-M. Hung, D. Bravo, T. O. Dickson, X. Guo, R. Li, N. Trichy, J. Caserta, W. R. Boosted, J. Branch, J. Bohorquez, E. Seok, L. Gao, A. Sugavanam, J.-J. Lin, J. Chen and J. E. Brewer, 'On-chip antennas in silicon ICs and their application', *IEEE Transactions on Electron Devices*, **52**(7), 1312–1323, July 2005.
33. J. Zhao and S. Raman, 'An integrated chip-scale active antenna for 5–6 GHz wireless microsystems', in *2002 IEEE Radio and Wireless Conference (RAWCON) Digest*, Boston, Massachusetts, pp. 139–152, 11–13 August 2002.
34. S. Brebels, J. Ryckaert, B. Come, S. Donnay, W. De Raedt and R. P. Mertens, 'SOP integration and codesign of antennas', *IEEE Transactions on Advanced Packaging*, **27**(2), 341–351, May 2004.
35. S. Pinel, S. Chakraborty, M. Roellig, R. Kunze, S. Mandal, H. Liang, C.-H. Lee, R. Li, K. Lim, G. White, M. Tentzeris and J. Laskar, '3D integrated LTCC module using μBGA technology for compact C-band RF front-end module', in *2002 International Microwave Symposium Digest*, pp. 1553–1556, 2002.
36. L.-R. Zheng, X. Duo, M. Shen, W. Michielsen and H. Tenhunen, 'Cost and performance tradeoff analysis in radio and mixed-signal system-on-package design', *IEEE Transactions on Advanced Packaging*, **27**(2), 364–375, May 2004.

14

Frequency-Agile Antennas for a Software-Defined and Cognitive Radio

J. T. Aberle, S.-H. Oh and G. A. Taylor
Department of Electrical Engineering, Arizona State University, Tempe, Arizona, USA

14.1 INTRODUCTION

At the present time, there is tremendous demand for antennas with high efficiencies in very small form factors that fit inside ever-shrinking portable wireless devices like handsets and personal digital assistants. These antennas must cover a variety of frequency bands and support different wireless standards. In the past, antenna engineers have attempted to design antennas for these applications using computationally intensive full-wave electromagnetic simulations followed by long hours in the lab tweaking performance. It has become increasingly apparent that the requirements of the software-defined radio (SDR) and its proposed successor, the cognitive radio, will render this approach untenable. In this chapter, a new approach is described for allowing electrically small radiating structures to be used in multiband, multimode radio transceivers. In this approach, the narrow instantaneous bandwidth of the radiator is automatically tuned over a much wider frequency range. The approach described in this chapter is becoming increasingly attractive as low-loss, highly linear control devices such as RF MEMS switches become more widely available at low cost.

The frequency selectivity inherent in electrically small antennas (ESAs) can greatly simplify the RF front-end design for multiband, multimode radios by reducing the requirements for analog filters. This reduction in filter requirements leads to a decrease in the size, cost, and power requirements of these radios. The frequency-agile antenna can be tuned to a specific narrow bandwidth over a much wider frequency range by reconfiguring its geometry using appropriate control elements such as RF switches or varactors. In principle, the control elements can be configured in an open-loop fashion based on a digital word received from the radio's baseband processor. However, open-loop tuning cannot be used in practice because of

Printed Antennas for Wireless Communications Edited by R. Waterhouse

limitations on the values of commercially available components, tolerances associated with physical components, and variations in the antenna's response due to environmental changes. These factors can cause the exact centre of the antenna's narrow instantaneous bandwidth to fluctuate in an unpredictable way. Furthermore, open-loop tuning is not compatible with the ultimate goals of SDR and cognitive radio, where the radio can be reprogrammed to cover new frequency bands and modes of operation simply by downloading new software. Thus, the implementation of a closed-loop automatic antenna tuning unit (AATU) is imperative. It will be shown that the AATU makes the tunable antenna fully compatible with the goals of the SDR and cognitive radio.

The basic block diagram of the proposed AATU system is shown in Figure 14.1. The closed-loop system for antenna matching ensures that a narrowband antenna is automatically matched to any desired frequency under all environmental conditions using practical component values and tolerances. However, as the block diagram reveals, the antenna system is no longer simply an electromagnetic transducer, but a mixed-signal system that involves several microelectronic circuits as well as appropriate software algorithms running on one or more programmable logic devices (PLDs) such as field programmable gate arrays (FPGAs) or digital signal processors (DSPs). The central component in realizing the AATU is a tunable RF matching network that is configured into one of some number of possible states by a command from the antenna control unit (ACU).

In implementing the tunable matching network, a pi-network impedance synthesizer is used that consists of variable reactive components. Variable capacitors of the impedance synthesizer are realized using a bank of switched capacitors [1]. The control elements are RF switches such as PIN diodes, FET switches, or RF MEMS switches. To provide feedback data to the ACU, the incident and reflected powers at the input to the impedance synthesizer are measured using a three-line directional coupler and RF power detectors.

The role of the ACU is to reconfigure the impedance synthesizer such that the matching state is optimum by generating the required switch control signals based on the acquired incident and reflected power levels. The ACU determines the optimum state for the RF

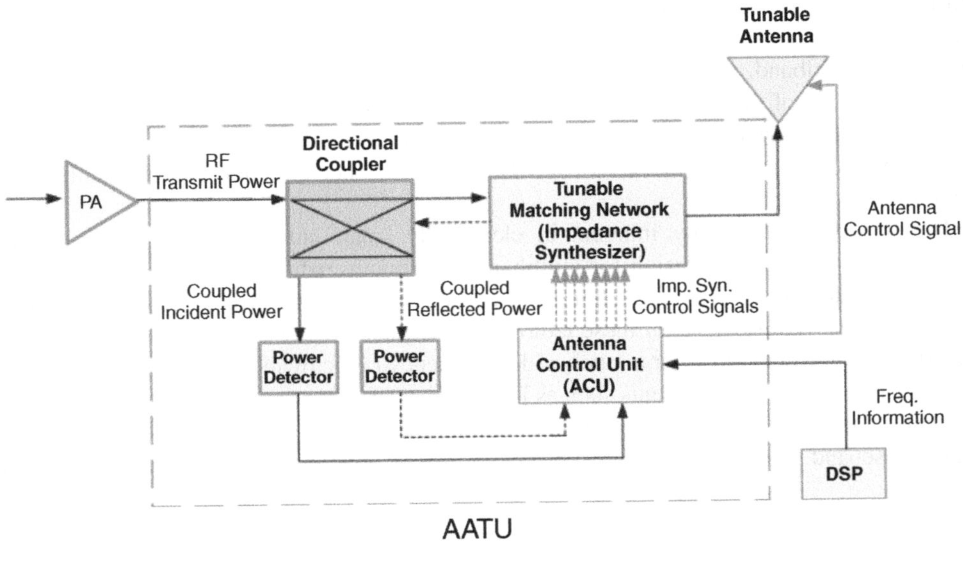

Figure 14.1 Block diagram of the closed-loop automatic antenna tuning unit (AATU).

switches by running a search algorithm. Employing a PLD for the ACU provides the end users with compete control over the implementation of the impedance synthesizer and antenna.

In this chapter a number of practical issues concerning on AATU is studied, focusing on their tunable matching networks and concomitant control circuitry and algorithms. In Section 14.2, a new theoretical tool is discussed for characterizing antennas from a system point of view, allowing their frequency selectivity to be incorporated into the RF front-end design. This new tool involves modelling the antenna as a two-port device. In Section 14.3 tunable matching networks are discussed as well as appropriate analog and mixed-signal control circuitry to allow frequency-agile antennas to be reconfigured on demand by the radio's baseband processor. In Section 14.4, the implementation of an AATU prototype is discussed.

14.2 TWO-PORT REPRESENTATION OF AN ANTENNA

In many situations it is desirable to model an antenna as a two-port network. Such a model can be used in circuit simulations to compute the overall efficiency of the antenna with a matching network. Furthermore, such a representation of the antenna could be quite useful in overall link simulations that include a channel model. In this section a procedure is presented for deriving a rigorous two-port model for an antenna [2], and its use with a matching network is demonstrated.

The approach described here produces an exact model of the antenna from the given data in the form of two-port network parameters that can be readily imported into a commercial circuit simulator. The idea presented here is remarkably simple and profoundly useful and can be applied to any antenna (with a single feed point), and does not require the assumption of a particular equivalent circuit.

14.2.1 Two-Port Antenna Model

Given the input impedance and radiation efficiency of an antenna at a specified frequency (from either simulation or measurements), a two-port representation of the antenna can be derived as follows. Let the complex input impedance of the antenna be denoted by Z_a and the radiation efficiency (as a dimensionless quantity between 0 and 1) be denoted by e_{cd}. Then the equivalent circuit (valid at that specific frequency) is shown in Figure 14.2, where

$$\begin{aligned} Z_a &= R_a + jX_a = R_r + R_l + jX_a \\ R_r &= e_{cd}\, R_a = \text{radiation resistance} \\ R_l &= (1 - e_{cd})\, R_a = \text{dissipative loss resistance} \\ X_a &= \text{antenna reactance} \end{aligned} \tag{14.1}$$

Since the radiation resistance represents power that is 'delivered' by the antenna to the rest of the universe, the radiation resistance can be replaced with a transformer to the impedance of free space or, more conveniently, to any port impedance that is required (such as 50 Ω). The turns ratio of the transformer is given by

$$N = \sqrt{\frac{R_r}{Z_0}} \tag{14.2}$$

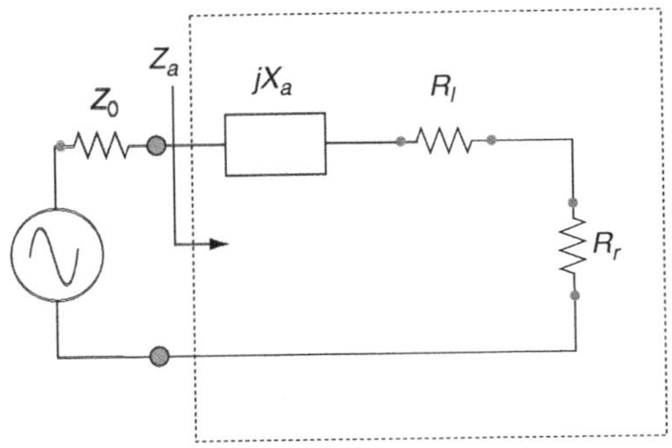

Figure 14.2 Equivalent circuit of an antenna (valid at a single frequency).

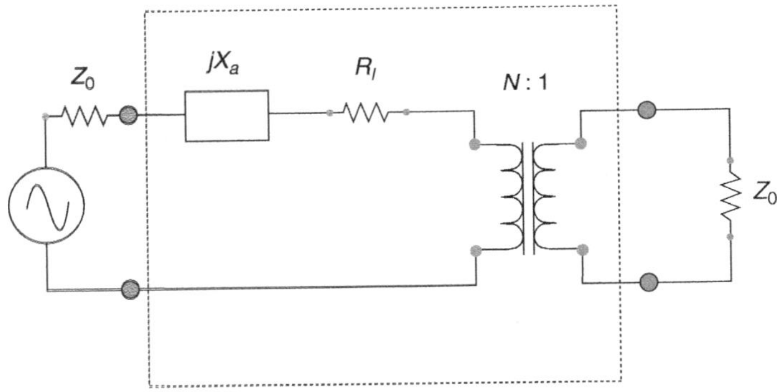

Figure 14.3 Two-port representation of the antenna (valid at a single frequency).

where Z_0 is the desired port impedance. The resulting two-port representation of the antenna is shown in Figure 14.3. At each frequency, a two-port representation of the form shown in Figure 14.3 can be constructed, and the two-port scattering matrix evaluated and written into an appropriate file format (such as Touchstone) for use in a circuit simulator. Note that when port 2 of the two-port network shown in Figure 14.3 is terminated by the proper port impedance, the antenna's input impedance is obtained as

$$Z_{\mathrm{a}} = Z_0 \frac{1 + S_{11}}{1 - S_{11}} \tag{14.3}$$

and its total efficiency is obtained as

$$e_{\mathrm{tot}} = |S_{21}|^2 = (1 - |S_{11}|^2)\ e_{\mathrm{cd}} \tag{14.4}$$

14.2.2 Use of the Proposed Model with Matching Networks

In this subsection, the efficacy of the proposed two-port model developed above will be demonstrated. To do so, the performance of an electrically small antenna with a passive matching network will be determined. Particular interest will be taken in the overall radiation efficiency of the antenna with the passive matching network. The two-port model of the antenna presented here allows these parameters to be evaluated readily.

Consider the input impedance of a shorted patch antenna (SPA) designed for operation at 1.8 GHz. Any number of passive matching circuits can be used to provide a (theoretical) perfect match to the SPA at a specified frequency. One of the most common ways to match such an antenna is to use an L-section consisting of two reactive elements as shown in Figure 14.4 (however, a Π-matching network topology is used for the design of an AATU, as will be demonstrated in Section 14.3). Using readily available design formulas for the L-section (e.g. from Chapter 5 of Reference [3]), the following values for the inductor and capacitor are obtained for a perfect match at 1800 MHz where the measured antenna input impendence is $9.3 + j96.6\ \Omega$:

$$L_1 = 2.24\ \text{nH}, \qquad C_1 = 0.81\ \text{nH}$$

The major disadvantage of using a passive matching network with an electrically small antenna is that any dissipative losses in the components of the matching network reduce the overall radiation efficiency. To examine this effect, assume that the inductor and capacitor have a Q of 50 and infinity, respectively, and that the antenna radiation efficiency e_{cd} is equal to 1. The combination of the matching network and two-port model of the antenna can be analysed using an appropriate circuit simulator. Here the HP-Advanced Design System (ADS) is used. The schematic of the antenna along with its matching network captured from the ADS is shown in Figure 14.4. The computed return loss looking into the input of the matching network is shown in Figure 14.5, and the total efficiency of the antenna together with its matching network is shown in Figure 14.6. Of course, the return loss result could have been obtained readily without the proposed two-port model of the antenna. However, without the use of a rigorous two-port model of the antenna, the total efficiency result would

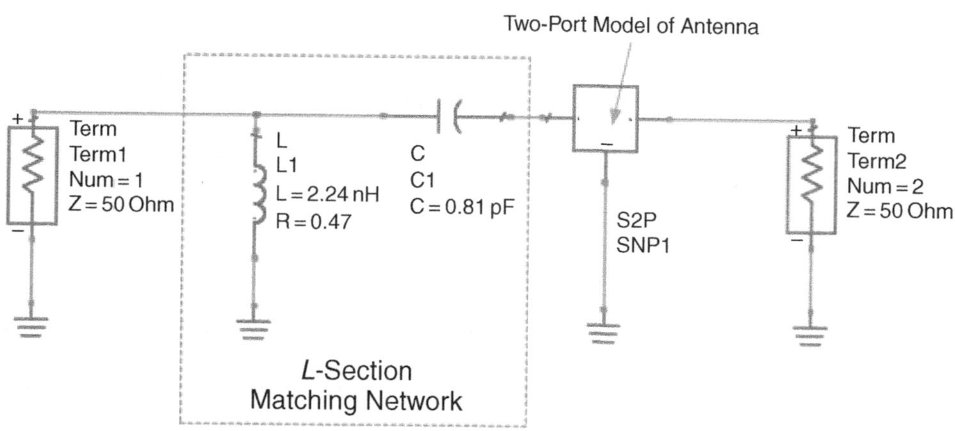

Figure 14.4 Schematic of an SPA with the L-section matching network for analysis in the Agilent ADS.

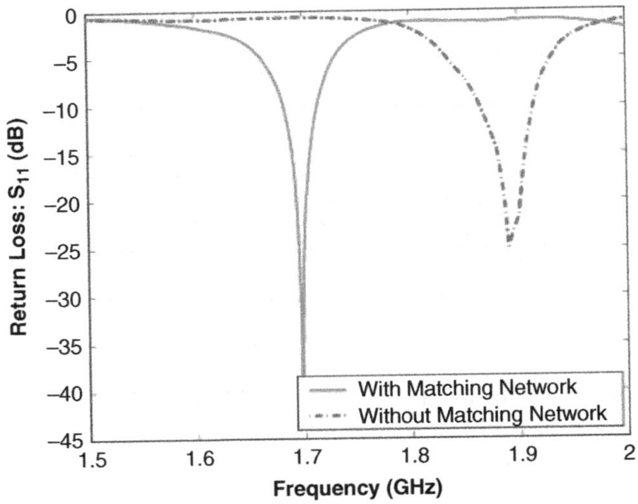

Figure 14.5 Return loss at the input of the *L*-section matching network computed using the Agilent ADS.

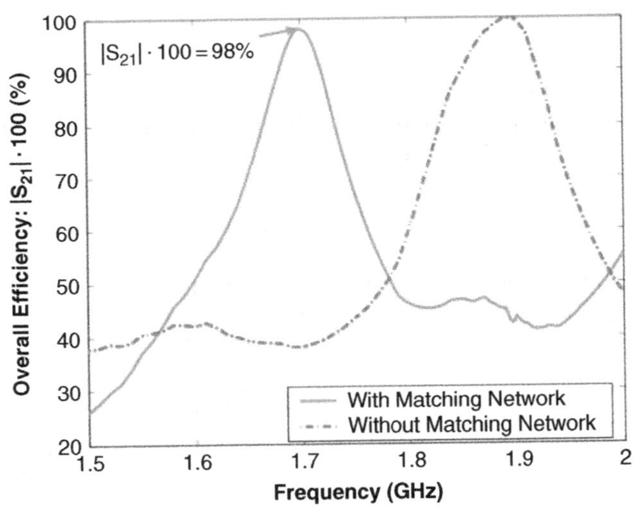

Figure 14.6 Overall efficiency (in percent) of the matching network and antenna computed using the Agilent ADS.

have to be calculated outside the circuit simulator. With the use of the proposed two-port model for the antenna, it becomes possible, for example, to use the circuit simulator's built-in optimization tools to maximize the overall radiation efficiency over commercially available inductor values.

In the following section, the usefulness of the proposed two-port antenna model will be used to evaluate and optimize the overall radiation and AATU system efficiencies.

14.3 AUTOMATIC ANTENNA TUNING UNIT (AATU)

14.3.1 Introduction

In Section 14.2, reconfigurable antennas were presented as a frequency-agile selective component to simplify RF front-end design for multiband, multimode radios. The bandwidth limitation and environmental effects of electrically small tunable antennas for hand-held devices were then discussed. In this section, an automatic antenna tuning unit (AATU) will be introduced, which, as will be seen, makes the tunable antenna fully compatible with the goals of the SDR and cognitive radio.

AATUs can be designed using both closed-loop and open-loop control schemes. In the open-loop control scheme, the ACU shown in Figure 14.7 reconfigures the tunable antenna without feedback. The ACU receives operating frequency information from the radio's baseband processor and refers to a look-up table to generate required switch control signals to reconfigure the tunable antenna. However, the narrowband tunable antenna is very sensitive to component tolerances and environmental factors, and its characteristics vary over the operating frequency range. Furthermore, the instantaneous bandwidth of the antenna tuned by the open-loop control is not usually wide enough to cover all channels in an operating band for current RF wireless communication systems. In order to overcome these limitations, a closed-loop matching network is implemented. In order to meet the requirements of a current RF wireless system such as frequency hopping and multiband operation, high-speed tuning techniques for the closed-loop matching system are used.

In this section, the major issues in AATU design are studied from the perspective of current RF communication systems. These include the design, fabrication, and measurement of the tunable matching network, RF power sensor, and analog and mixed-signal control circuitry.

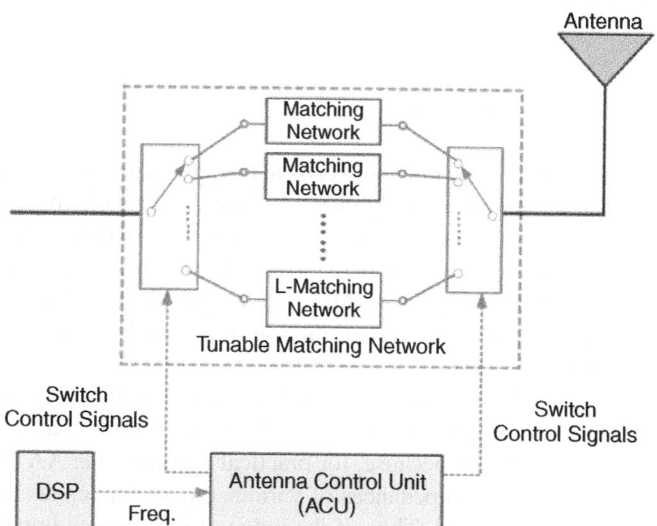

Figure 14.7 Block diagram of the open-loop tuning unit.

14.3.2 Closed-Loop Antenna Tuning System

In this subsection general considerations related to the design and performance of a closed-loop antenna tuning system are discussed. A closed-loop system is one that uses feedback from the output to control the input [4]. As shown in Figure 14.1, the inputs of the closed-loop ACU are the incident and reflected power levels, which are measured by a directional coupler and power detectors. The output of the ACU is a set of digital (control) signals used to reconfigure the impedance synthesizer. The details of the closed-loop control scheme are studied in Section 14.4.

One of the benefits of the closed-loop AATU system is that it does not require accurate antenna impedance measurements, which are usually very hard to achieve in a constantly changing environment. Moreover, the AATU system will be able to match almost any kind of antenna automatically. Also, by automatically reducing the mismatches at the output of the power module, the system can eliminate the need for an isolator, which is an expensive and bulky component.

In the following subsections, major issues in AATU design are discussed. These include the analysis, design, fabrication, and measurement of the impedance synthesizer, directional coupler, RF power detector, and the ACU.

14.3.3 Impedance Synthesizer

To provide a complex-conjugate matching capability for a wide range of impedances and a variety of antennas that might be used under changing environments, an impedance synthesizer will be introduced. Firstly, the basic topology of a tunable pi-matching network will be described, and then a discussion is provided realizing the tunability of the network using PIN diode switches controlled by a digital microcontroller (ACU). In later subsections, the practical design issues are discussed for the fabrication of the impedance synthesizer and its measurement methodology.

14.3.3.1 Tunable Pi-Matching Network

In order to design a tunable matching network, a lowpass-type pi-matching network is considered due to its harmonic rejection capability and wide matchable impedance range [5]. Figure 14.8 illustrates the basic topology of a lowpass-type pi-matching network configuration. With a fixed inductor and two variable capacitors, the matching network is capable of generating a wide range of tuning possibilities. Although the matching network provides better tunability if a variable inductor is used, the use of a tunable inductor is eliminated in order to simplify the AATU design.

Note that, for the AATU design, there is more interest in a matchable domain at a given frequency for given ranges of component values (C_1, C_2, and L), rather than specific solutions for a given load impedance. This is because, for practical purposes, the AATU is required to be able to match a wide range of impedances of various types of antennas under constantly changing environment conditions [6]. Thus, if the range of the antenna impedance is within the matchable domain, the antenna can always be matched by tuning the component values. Methods of determining the matchable domain for given ranges of component values are

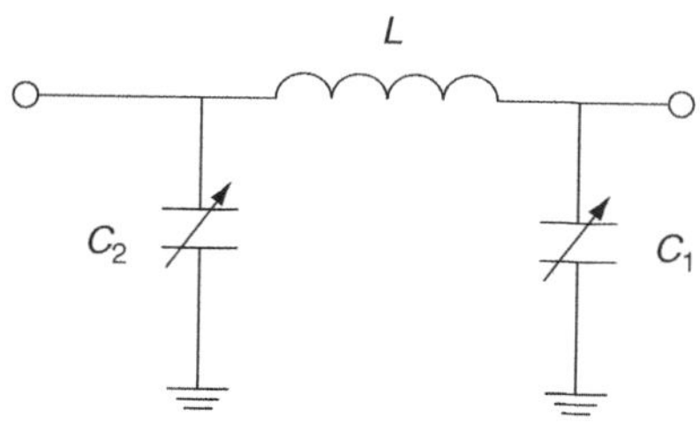

Figure 14.8 Tunable pi-matching network configuration.

described in detail in References [7] to [9]; for completeness these are also discussed in the following subsections.

14.3.3.2 Impedance Transformation and Conjugate Matching

The object of the tunable pi-matching network is to transform an antenna impedance to a specified system impedance such as 50 Ω. Figure 14.9 shows a pi-matching network placed between an arbitrary antenna impedance Z_A and a system impedance Z_0. The impedance seen looking into the matching network Z_{out} at frequency f_0 is expressed by

$$Z_{out} = \frac{1}{\dfrac{1}{\dfrac{1}{(1/Z_0 + jB_2)} + jX_L} + jB_1} \tag{14.5}$$

where

$$X_L = 2\pi f_0 L$$
$$B_1 = 2\pi f_0 C_1$$
$$B_2 = 2\pi f_0 C_2$$

Notice that the impedance Z_{out} is easily obtained by a series or parallel circuit combination.

The maximum power transfer from the input of the matching network to the antenna will occur when the antenna input impendence is conjugate-matched to the impedance seen looking into the matching network; i.e.

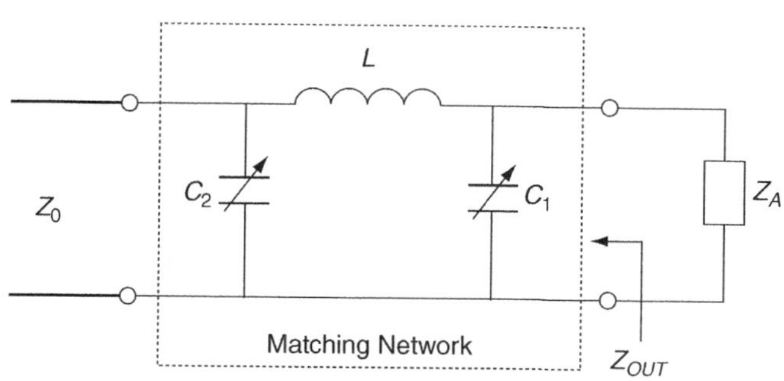

Figure 14.9 A pi-matching network transforms an arbitrary antenna impedance Z_A to a system impedance Z_0.

$$Z_A = Z_{out}^* \tag{14.6}$$

Substituting Equation (14.5) into Equation (14.6) gives

$$Z_A = \frac{1}{\dfrac{1}{\dfrac{1}{(1/Z_0 - jB_2)} - jX_L} - jB_1} \tag{14.7}$$

which represents the antenna impedance to be matched. Provided that the component values (C_1, C_2, L) are adjusted from zero to infinity, the pi-matching network is able to synthesize any antenna impedance. However, the ranges of component values of the matching network that can be realized in practice are limited.

14.3.3.3 Binary Capacitor Array

In this subsection a means of realizing the variable capacitors (C_1, C_2) shown in Figure 14.9 are discussed. To realize variable capacitors, a binary capacitor array [8] is considered, and its configuration is shown in Figure 14.10. The capacitance of the variable capacitor C is expressed as

$$C^1 = (a_0 2^0 + a_1 2^1 + a_2 2^2 + \cdots + a_{N-1} 2^{N-1}) C_{min} \tag{14.8}$$

where the coefficients are either 1 or 0 representing the switch status (i.e. 1 when the switch is on and 0 when the switch is off) associated with each capacitor, N is the total number of capacitor elements used to replace the variable capacitor C, and C_{min} is the lowest capacitor value of the array that determines the minimum increment of the capacitance.

By switching the binary array of the fixed capacitors on and off, the capacitance C varies from zero (when $a_0 = a_1 = a_2 = \cdots = a_{N-1} = 0$) to $(2N - 1)C_{min}$ (when $a_0 = a_1 = a_2 = \cdots = a_{N-1} = 1$) in steps of C_{min}. In other words, 2^N uniformly distributed capacitance values can be generated by using N capacitors and N switching elements. Later it will be shown that the lowest capacitor value C_{min} determines the resolution of its matchable domain, and the number of capacitors N determines the highest capacitor value of the array and the range of the matchable impedances. The most important characteristic of the binary capacitor array is that it can interface with a microprocessor as long as the switching elements can be controlled by a digital signal.

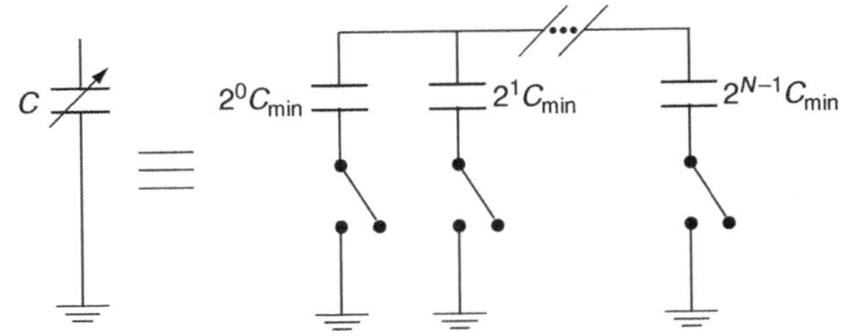

Figure 14.10 Binary switched capacitor array as a variable capacitor.

There are several types of RF switches that can implement the impedance synthesizer in practice, depending on whether the switching time is fast enough and whether the losses and parasitic effects are appreciable or not. Among these RF switches are PIN diodes, FET switches, and RF MEMS switches. Here the application of PIN diodes as the switching elements is considered in detail as these elements are currently the most appropriate low-cost devices available for the frequency range considered here. Fortuitously, as RF MEMS switches become more widely available at low cost, the deleterious effects of control element parasitics and nonlinearity will become less important.

In this design for the binary capacitor array, a single-pole, single-throw (SPST) series PIN diode switch is chosen, which is commonly used when a minimum insertion loss is required over a broad frequency range [10]. Figure 14.11 shows a binary capacitor array structure combined with PIN diode switches. Notice that additional RF chokes are used to provide a bias voltage (i.e. digital control signal). Practical issues involved in implementing the PIN diode switches, such as losses and parasitic effects, are discussed in detail in a later subsection.

14.3.3.4 Impedance Synthesizer Configuration

The tunable pi-matching network is configured with a fixed inductor and two binary capacitor arrays as shown in Figure 14.12. Notice that the DC biasing circuits (refer to Figure 14.11) are not included in the figure.

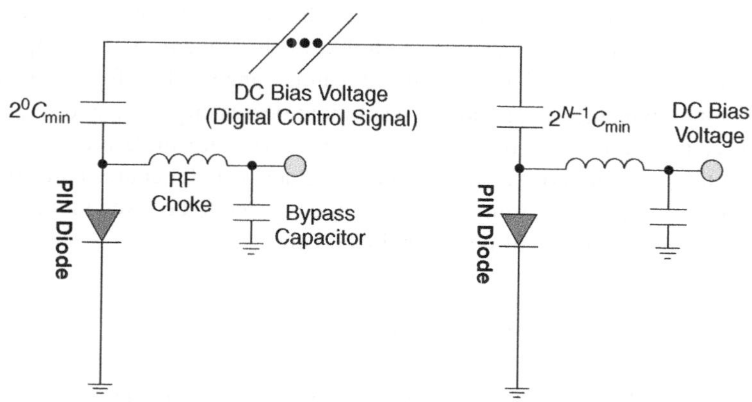

Figure 14.11 Binary capacitor array structure combined with PIN diode switches.

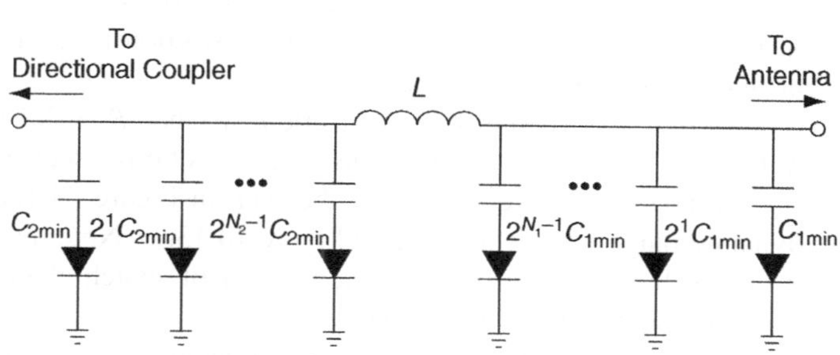

Figure 14.12 Impedance synthesizer configuration.

The variable capacitors C_1 and C_2 shown in Figure 14.8 are replaced with binary capacitor arrays, as discussed in the previous subsection. The values of L, $C_{1\min}$ and $C_{2\min}$, and the numbers (N_1 and N_2) of capacitors in the configuration are determined by the desired operating frequency and matchable region, discussed in the following subsection. Using Equation (14.8), the capacitance values are expressed as

$$C_1 = (a_0 2^0 + a_1 2^1 + a_2 2^2 + \cdots + a_{N_1-1} 2^{N_1-1}) C_{1\min} \tag{14.9}$$

$$C_2 = (b_0 2^0 + b_1 2^1 + b_2 2^2 + \cdots + b_{N_2-1} 2^{N_2-1}) C_{2\min} \tag{14.10}$$

The total number $N(N = N_1 + N_2)$ of capacitors composing the two binary capacitor arrays generates $2N$ different matching states. For the rest of the analysis, it is assumed that the number of capacitors composing each binary capacitor array are the same, i.e. $N_1 = N_2$.

14.3.3.5 Matchable Domain

With the preceding study about the tunable matching network, it is now pertinent to discuss the matchable impedance domain based on its operating frequency and component values. It is important to note that any type of matching network is capable of producing a perfect match as long as its component values are variable from zero to infinity. However, the practical component values that can be realized are limited. Component values will define the frequency range, dynamic range, and resolution of the domain. For example, the total number of capacitors, N, will determine the dynamic range. Also, for a set of N, $C_{1\min}$ and $C_{2\min}$ affect the dynamic range and the resolution; decreasing $C_{1\min}$ and $C_{2\min}$ increases the resolution at the expense of the dynamic range. Therefore there are many trade-offs associated with the realization of the tunable matching network. Of course parasitic elements associated with the devices (for example the PIN diodes) need to be taken into consideration, but for the sake of brevity this will not be included here.

14.3.3.6 Matchable Domain with Parasitics Effects

The effects of the PIN diode parasitics on the matchable domain of the impedance synthesizer are discussed in this section. Figure 14.13(a) shows the simulated matchable domain for a 12-capacitor impedance synthesizer at 900 MHz when ideal switches are used. The values of ideal reactive components of L, $C_{1\min}$, and $C_{2\min}$ are 3 nH, 0.5 pF, and 1 pF, respectively. Now an observation is made on how the deterioration and distortion in the dynamic range and resolution occur due to the PIN diode parasitics.

Figure 14.13(b) shows the case when a series parasitic resistance $R_s = 2.3\ \Omega$ is included in the simulation. When the PIN diode switch is ON, the parasitic resistance R_s generates loss to the impedance synthesizer. Therefore, as can be seen in Figure 14.13(b), there is a contraction of the dynamic range when compared to Figure 14.13(a). Note that losses of the impedance synthesizer will become significant as the number of switch elements increases mainly due to the series resistance of the PIN diodes.

Figure 14.13(c) shows the case when a series parasitic inductance $L_s = 2.8$ nH is included in the simulation. When the PIN diode switch is ON, the parasitic inductance is in series

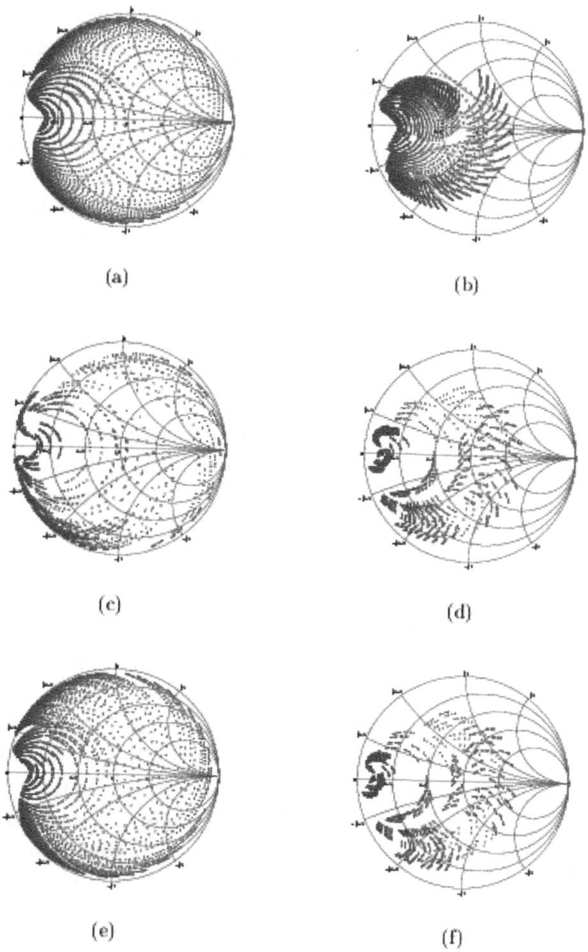

Figure 14.13 Simulated matchable domain at $f = 900$ MHz with 12 capacitors ($2^{12} = 4096$ states). Values of ideal components of L, C_{1min}, and C_{2min} are 3 nH, 0.5 pF, and 1 pF, respectively: (a) ideal PIN diode; (b) PIN diode with only series parasitic resistance ($R_s = 2.3\ \Omega$); (c) PIN diode with only series parasitic inductance ($L_s = 2.8$ nH); (d) PIN diode with series parasitic resistance and inductance; (e) PIN diode with only parallel parasitic capacitance ($C_p = 0.18$ pF); and (e) PIN diode with both on-state and off-state parasitic effects.

with each one of the capacitors of the impedance synthesizer, which introduces loss to the impedance synthesizer and fails to distribute the matchable impedances uniformly. It should be noted that, as the operating frequency is increased, the impedance of the inductor increases, so the effects of the parasitic inductor are significant.

Figure 14.13(d) shows the case when both the series parasitic inductance ($L_s = 2.8$ nH) and resistance ($R_s = 2.3\ \Omega$) are included in the simulation. As can be seen, the dynamic range is contracted and distorted due to the parasitic resistance and inductance. Notice that the matchable impedances are no longer distributed uniformly on the Smith chart.

When the PIN diode is OFF, the parasitic capacitance $C_p = 0.18$ pF is added in series with each one of the capacitors of the impedance synthesizer. This limits the minimum capacitance values (C_{1min} and C_{2min}) that can be realized in practice, and hence deteriorates the resolution of the matchable domain. This is because, as discussed earlier, the resolution is determined by C_{1min} and C_{2min}. Figure 14.13(e) shows the matchable domain when only the parasitic capacitance is included.

Figure 14.13(f) depicts the matchable impedance domain for the impedance synthesizer when both the on-state and off-state parasitic effects are included in the simulation. In later sections, it will be shown that the parasitic effects of the PIN diodes ultimately set the limit on the impedance synthesizer, in terms of efficiency, matching domain, and its control methodologies.

It should be mentioned that when the parasitics are involved in the impedance synthesizer, the matchable impedances plotted on the Smith chart no longer represent impedances that can be perfectly matched to the system impedance, but rather represent impedances that can be matched with a reasonable return loss.

A dramatic performance improvement of the impedance synthesizer can be achieved by using low-loss, low-parasitic switches. The most promising technology for implementing these switches is considered to be MEMS. However, MEMS switches are not yet commercially practical due to size and cost. The PIN diode based impedance synthesizer can be refined with RF MEMS switches when possible.

14.3.3.7 Impedance Synthesizer Prototype Design and Fabrication

The compromise between large matchable domains and small losses led to consideration of an impedance synthesizer with six switches ($2^6 = 64$ states). In general, variations of antenna impedance due to the environmental conditions are limited to a certain region of the Smith chart. Thus, it is not necessary for the dynamic range of an impedance synthesizer to cover all the Smith chart area, but be sufficient to generate the complex conjugates of the antenna impedances confined to a certain region on the Smith chart. In other words, fabricating an impedance synthesizer for the purpose of matching a *particular* antenna (as opposed to *any* antenna) will require only a small tuning range of capacitors. Therefore, engineering trade-offs are necessary to determine the optimal number of matching states with an acceptable dynamic range.

To determine the required values for L_s, C_{1min}, and C_{2min}, a series of simulations was performed with a circuit model designed with HP-ADS software. As shown in Figure 14.14, all the parasitic effects (i.e. $R_s = 2.3\ \Omega$, $L_s = 2.8$ nH, $C_p = 0.18$ pF, and $R_p = 1.92$ k Ω) of the PIN diodes are included in the circuit model, as well as the transmission line effects of the PCB on which the components of the impedance synthesizer are mounted. The vales of L, C_{1min}, and C_{2min} are determined to be 1.0 nH, 4.7 pF, and 2.2 pF, respectively. It should be noted that the matchable domains presented here are only a sample of many simulations performed. Other values of L, C_{1min}, C_{2min}, and N could possibly be better in terms of both dynamic range and matching capability.

A 1 nH SMD-type inductor (0402CS-1N0XJLW Chip Inductor, Coilcraft, Inc.) and thin-film capacitors (Accu-P® Thin-Film RF/Microwave Capacitors, AVX Corp.) were used to build the impedance synthesizer. The commercially available values for the design and part numbers are shown in Table 14.1.

The photograph of the impedance synthesizer prototype fabricated on FR4 board is shown in Figure 14.15. The simulation results with the commercially available values are shown in Figure 14.16.

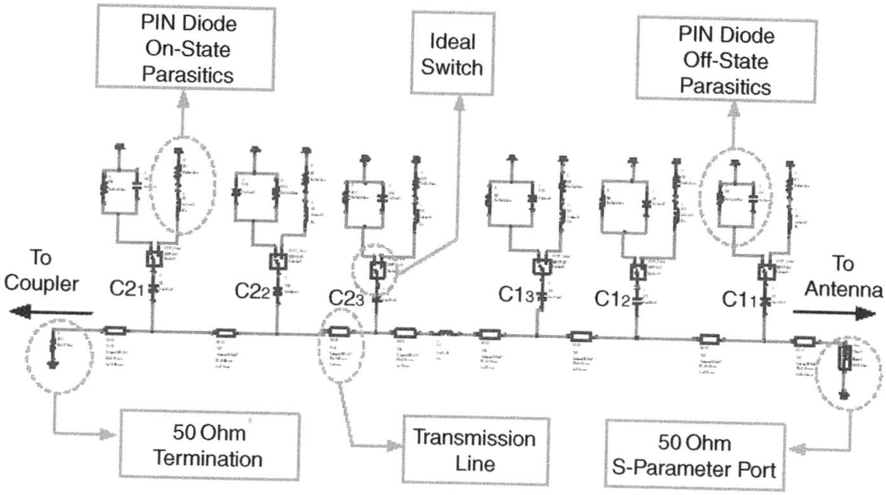

Figure 14.14 ADS system simulation schematic layout.

Table 14.1 The calculated and commercially available values of capacitors for the fabrication of the impedance synthesizer.

	Calculated values (pF)	Commercially available values (pF)	Part numbers
$C_{1_1} = C_{1\min}$	4.7	4.7	06035J4R7BBTTR
C_{1_2}	$4.7 \times 2^1 = 9.4$	10	06035J100GBTTR
C_{1_3}	$4.7 \times 2^2 = 18.8$	18	06035J180GBTTR
$C_{2_1} = C_{2\min}$	2.2	2.2	06035J2R2BBTTR
C_{2_2}	$2.2 \times 2^1 = 4.4$	4.7	06035J4R7BBTTR
C_{2_3}	$2.2 \times 2^2 = 8.8$	8.2	06035J8R2BBTTR

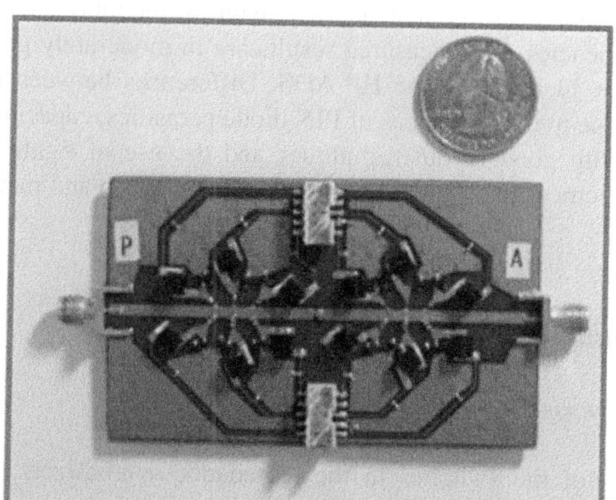

Figure 14.15 Photograph of the fabricated impedance synthesizer.

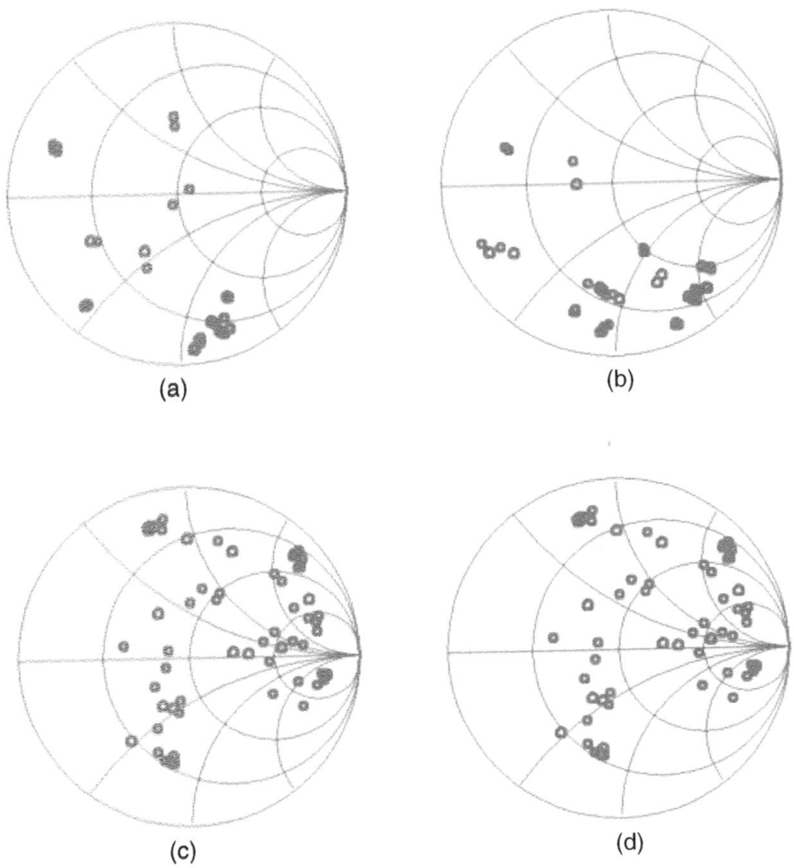

Figure 14.16 ADS simulated matchable domain results at (a) 800 MHz, (b) 900 MHz, (c) 1800 MHz, and (d) 1900 MHz.

Figure 14.17 shows the measured matchable domains of the fabricated impedance synthesizer at different frequencies. The measured results are in moderately good agreement with the results (see Figure 14.16) from the HP-ADS. Differences between the simulation and measured results are due to the tolerances of PIN diode parasitics, capacitors, and RF chokes. It is expected that, with good layout techniques and the use of tighter tolerance components, even better agreement between the results from simulation and measurement could be achieved.

14.3.4 Other components

14.3.4.1 RF Power Sensor

To control the status of the switches in the impedance synthesizer, the AATU requires feedback data that allow it to determine the impedance mismatch corresponding to a given state. In certain applications (CDMA, for example), the incident power level is constantly being adjusted to optimize signal-to-interference plus noise ratio for all users. Hence, it is

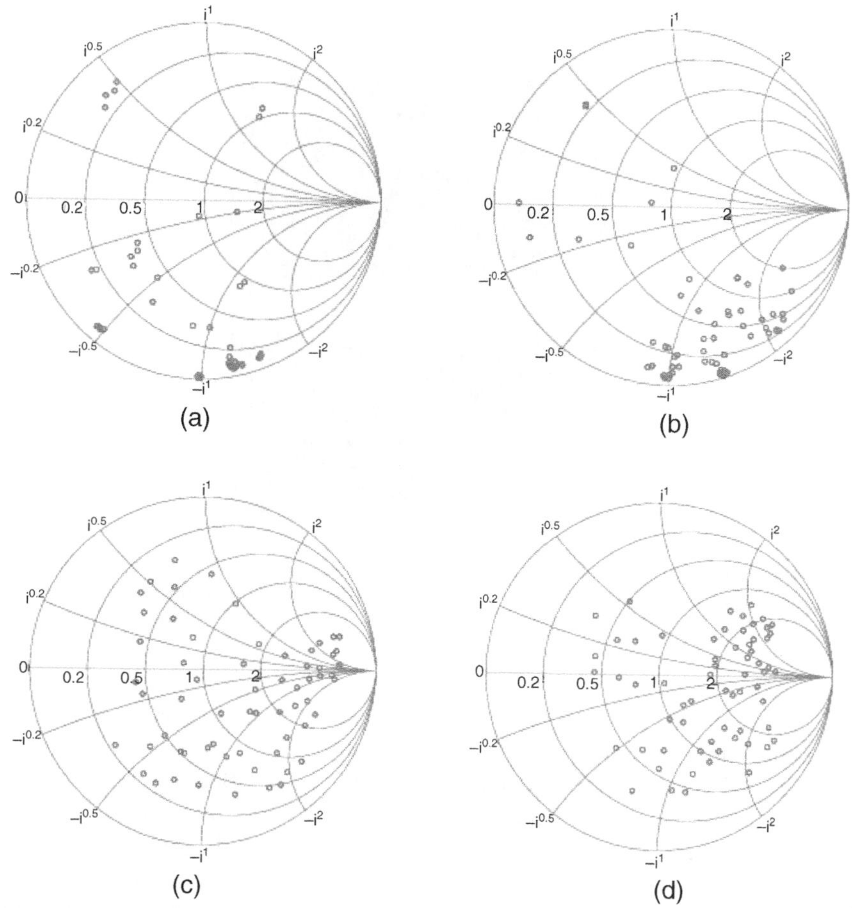

Figure 14.17 Measured matchable domain of the fabricated impedance synthesizer with six capacitors at (a) 800 MHz, (b) 900 MHz, (c) 1800 MHz, and (d) 1900 MHz.

generally necessary to measure incident power levels as well as reflected power levels in order to be able to determine the input reflection coefficient. In the next subsections, the RF power sensors used to detect the incident and reflected power levels at the input of the impedance synthesizer are considered. Implementation of these sensors includes the design and fabrication of a three-line directional coupler, RF power detectors, and ADCs. The design of the coupler and power detectors use standard RF design principles and therefore for the sake of brevity will not be included here.

14.3.4.2 Analog-to-Digital Converter (ADC) and Data Acquisition

After detecting the incident and reflected power level with the RF power detectors, it is necessary to convert the analog signal to a digital signal for further digital signal processing of the ACU. The National Semiconductor's ADC1173 15 Msps ADC is chosen to digitize the analog voltage signals from the outputs of the RF power detectors. An evaluation board (ADC1175EVAL Evaluation Board, National Semiconductor Corp.) was purchased and is

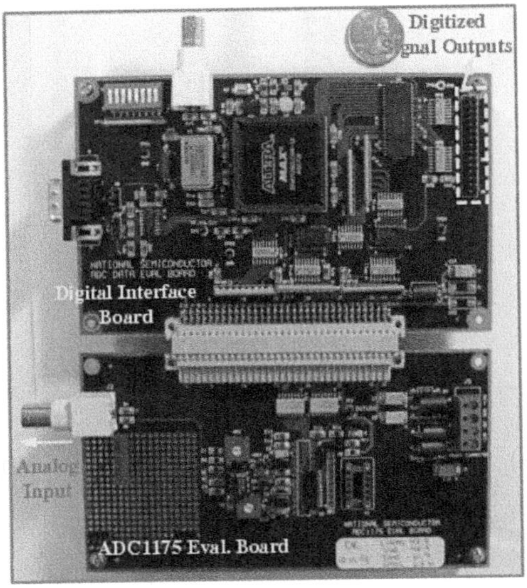

Figure 14.18 Photograph of the ADC evaluation board (ADC1175EVAL Evaluation Board, National Semiconductor Corp.) and digital interface board (WaveVison™ Digital Interface Board, National Semiconductor Corp.).

shown in Figure 14.18. The ADC1173 is a low-power, 15 Msps converter that digitizes signals to 8-bits while consuming just 33 mW of power.

The 8-bit digital output from the ADC1175EVAL Evaluation Board is provided in parallel format. To capture the digitized signal and feed the data into a microprocessor for further signal processing, the National Semiconductor's WaveVison™ Digital Interface Board is chosen and connected to the ADC1175EVAL Evaluation Board, as shown in Figure 14.18. The WaveVison™ software is used to evaluate the performance of the ADC and eliminate the need for a logic analyser.

14.3.5 Antenna Control Unit (ACU)

The reconfigurability of the antenna system, i.e. a tunable antenna and impedance synthesizer, is achieved by altering both the geometry of the antenna and the state of the impedance synthesizer by an antenna control unit (ACU). The ACU employs an open-loop and closed-loop system for the tunable antenna, as discussed before, and the impedance synthesizer (discussed in Section 14.3.3), respectively.

14.3.5.1 Closed-Loop ACU for the Impedance Synthesizer

After the electronically tunable antenna is reconfigured by the open-loop ACU, the AATU adopts a closed-loop system to ensure that the coarsely tuned antenna is automatically matched to any given frequency under all environmental conditions. Based on the coupled incident and reflected signal levels (refer to Section 14.4), a search algorithm running on the ACU tries to minimize the impedance mismatch of the antenna.

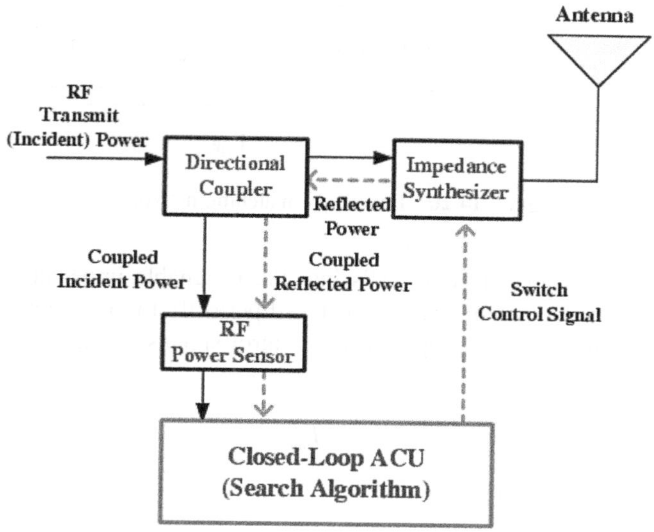

Figure 14.19 Block diagram outlining the functionality of the closed-loop ACU.

The block diagram of the closed-loop ACU is illustrated in Figure 14.19. The dashed arrow lines indicate the feedback signals of the closed-loop control system. The ACU receives the feedback from the RF power sensors, i.e. the directional coupler, RF power detectors, and ADCs, to measure the ratio between the incident and reflected power, and consequently control the impedance synthesizer. The ACU controls the switches of the impedance synthesizer via the digital output voltages from its output pins which generate TTL level voltages. In order to be compatible with current RF communication systems such as UMTS/W-CDMA, it is necessary to find an optimum matching state quickly and efficiently.

14.3.5.2 Tuning Method for the Impedance Synthesizer

The purpose of the closed-loop ACU is to generate proper switch control signals automatically which reconfigure the impedance synthesizer for an optimum matching state based on the incident and reflected power level information. The search algorithm running on the ACU will efficiently determine the switching status required to produce a conjugate matching between the impedance synthesizer and antenna at a desired operating frequency. The cost function that is used to optimize the impedance matching condition is the reflection coefficient, Γ_{imp}, seen looking into the input of the impedance synthesizer. Referring to Figure 14.20, Γ_{imp} can be obtained by [11]

$$\Gamma_{\mathrm{imp}} = S_{11} + \frac{S_{12}S_{21}\Gamma_{\mathrm{ant}}}{1 - S_{22}\Gamma_{\mathrm{ant}}} \tag{14.11}$$

where Γ_{ant} is the reflection coefficient seen looking toward the antenna, which can be calculated in terms of the antenna impedance, Z_{a}, and the characteristic impedance, Z_0, given by

$$\Gamma_{\mathrm{ant}} = \frac{Z_{\mathrm{a}} - Z_0}{Z_{\mathrm{a}} + Z_0} \tag{14.12}$$

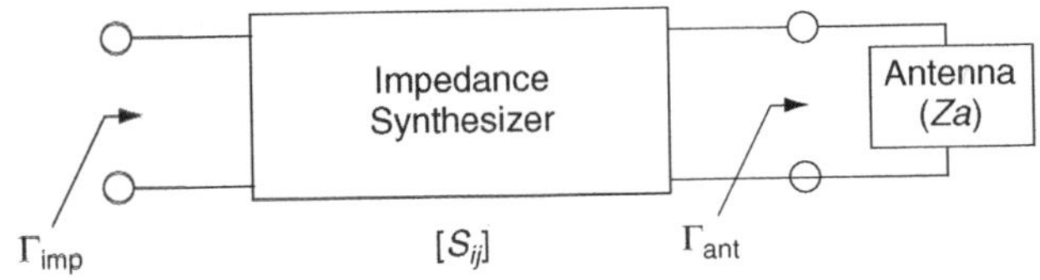

Figure 14.20 Impedance matching network.

As long as the antenna impedance Z_a falls into the matchable impedance domain of the impedance synthesizer, the antenna can be matched perfectly to the system impedance, Z_0, if the search algorithm finds the proper switch control signals required for the impedance synthesizer to produce

$$S_{22}^* = \Gamma_{ant} \tag{14.13}$$

which reduces Equation (14.11) to

$$\Gamma_{imp} = S_{11} + \frac{S_{12}S_{21}S_{22}^*}{1 - S_{22}S_{22}^*} = 0 \tag{14.14}$$

In practice, a perfect impedance match is not realizable due to various issues such as the parasitic effects of PIN diodes [12]. For this reason, the perfect matching condition of Equation (14.14) cannot be achieved in practice. In most cases, values of Γ_{imp} between −15 dB and −20 dB are considered acceptable.

14.3.5.3 Nearest Neighbor Search Algorithm

A number of different types of search algorithms can be implemented according to References [1], [6], and [8]. In Reference [1], several different algorithms such as the 'Hooke and Jeeves's Algorithm', 'Powell Algorithm', 'Simplex Method', and 'Single-Step Algorithm' were tested to compare their performance, and the 'Single-Step Algorithm' was implemented in an ASIC prototype. In Reference [12], a search method was developed based on genetic algorithms, and the implementation of the tuning algorithm using a microprocessor is considered.

In this section, the focus will be on a nearest neighbor search algorithm that is very similar to the 'Single-Step Algorithm' of Reference [1]. The proposed search algorithm has also been implemented in hardware using FPGA technology, and its implementation using DSP is also considered.

In order to find the switching status for the optimum matching condition, a two-dimensional search plane is considered, shown in Figure 14.21. Each axis of the plane corresponds to the decimal representations of N_i binary digital control signals. Note that, as discussed earlier, the impedance synthesizer consists of binary capacitor arrays and that the switching elements of the each array are controlled by N_i digital signals.

The search process starts from an arbitrary point on the two-dimensional plane. The algorithm compares the ratios of the incident and reflected power levels detected from this point and its eight nearest neighboring points. From among these nine points, the one that generates a minimum ratio of these power levels is selected as the new starting point. The search continues to iterate until it finds a starting point that produces a smaller mismatch than any of its nearest neighbors.

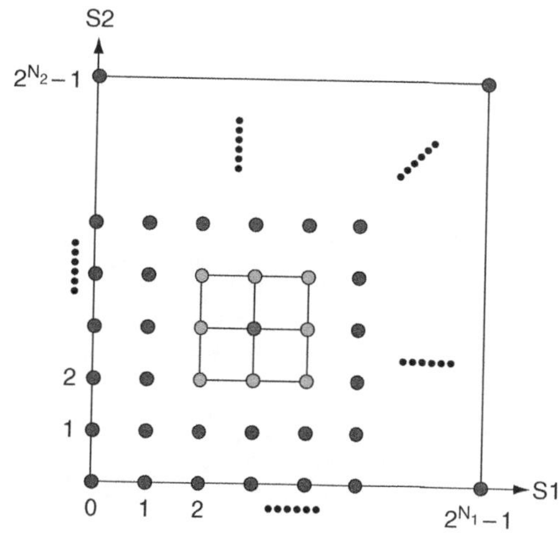

Figure 14.21 Topology of nearest neighbor search.

It has to be mentioned that the number of iterations required to find an optimum point depends on its initial search point. Therefore, occasionally, the algorithm will require many iterations before a satisfactory matched condition is reached. In order to reduce the search iteration, a look-up table can be used to determine a beginning search point that is close to the optimum point. The look-up table can be created by storing optimum points from previous operating frequencies [6]. The number of iterations required can also be reduced by ending the search process when an acceptable return loss has been achieved, even if it would be possible to improve the return loss further by additional iterations.

A significant drawback of the nearest neighbor search algorithm is that it requires the return loss (reflected power level) to converge monotonically. In other words, if other local minima exist, the algorithm might end up finding a local minimum point that does not provide a proper matching condition. To overcome this problem, random global searches such as generic algorithms [12] can be developed. A thorough examination was undertaken of the nearest neighbor search algorithm using MATLAB and it was concluded that the results suggest that to ensure a successful result, the nearest neighbor search algorithm requires either that only one local minimum exists in the search plane or that a look-up table should be used to determine a proper beginning search point.

14.3.5.4 Hardware Implementation of ACU

The system requirements are to design and implement tuning algorithms by using a low-cost platform. Since DSPs and FPGAs provide inexpensive and expandable platforms that allow rapid design prototyping, both approaches will be discussed in the following sections.

14.3.5.5 FPGA Implementation

The ACU is programmed and implemented using an FPGA device (Altera UP2 Education Board, Altera Corp.), shown in Figure 14.22. The FPGA is an array of logic gates that can

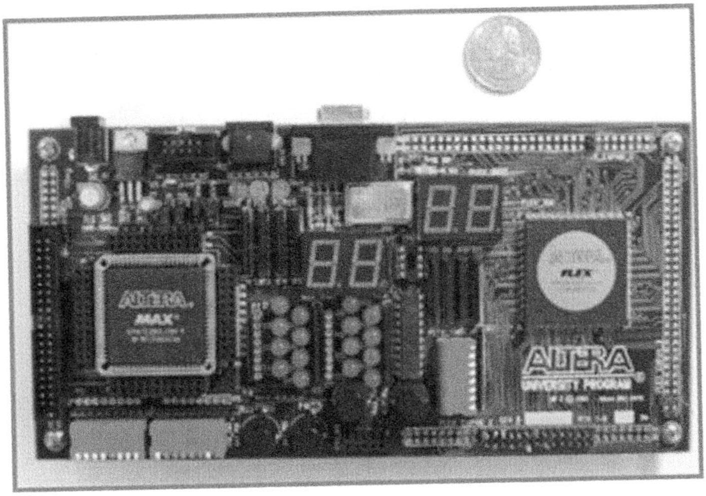

Figure 14.22 Photograph of the Altera UP2 Education Board.

be programmed dynamically in the field after it is manufactured and often used to prototype digital integrated circuits [13].

Both the open-loop and closed-loop algorithms are programmed using very high-speed integrated circuit hardware description language (VHDL) and compiled using the MAX+PLUS II Software. To run the algorithms, the EPF10K70 device on the Altera board is configured in-system with the ByteBlasterMV download cable, which is a hardware interface to a standard parallel port. More details about the board can be found in Reference [14].

Employing the FPGA for the ACU provides the end users with complete control over the tunability of the impedance synthesizer and antenna. For large-scale production, improved performance, and low-power consumption, the VHDL code can be retargeted to an application-specific integrated circuit (ASIC). In Reference [1], an ASIC unit is developed to implement a complex search algorithm.

14.3.5.6 DSP Implementation

DSPs are specialized microprocessors used to perform digital signal manipulations efficiently. Since a DSP chip is a very common component of almost all commercial wireless devices, the existing chip can be utilized to implement the ACU. The search algorithms can be written in high-level languages such as C and C++, which can be developed efficiently and easily [15]. Also by using a DSP platform, DSP commands can be simulated, implemented, and specified, e.g. operating frequency information required to reconfigure the tunable antenna and matching network on demand.

The ACU prototype is being developed using a DSP device (Sundance SMT8036 Software Defined Radio Kit, Sundance Digital Signal Processing, Inc.), which contains ADCs (AD6645-108)/DAC(AD9777) as well as a baseband DSP (TMS320C6416), and is shown in Figure 14.23. Code Composer Studio Software is used to develop, debug, and optimize C-code for the search algorithms and to configure the DSP chip on the board.

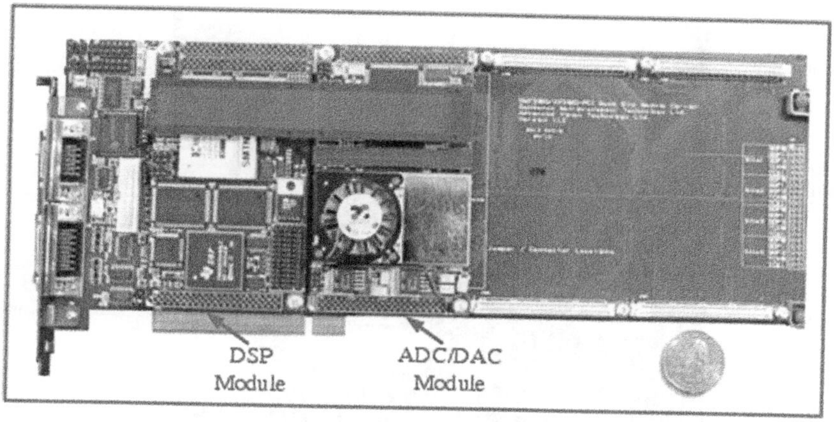

Figure 14.23 Photograph of the Sundance SMT8036 Software Defined Radio Platform.

14.4 AATU PROTOTYPE AND TEST SETUP

In the previous sections, the design issues associated with the development of the closed-loop AATU were discussed, with particular emphasis on the impedance synthesizer. To establish the feasibility of using the AATU in SDR and cognitive radio applications, an AATU system was constructed and demonstrated.

14.4.1 AATU Prototype

In order to verify the AATU functionality experimentally, the overall AATU prototype system shown in Figure 14.24 was assembled using the components discussed throughout Section 14.3. The AATU demonstration setup is based on the block diagram shown in Figure 14.1.

The impedance synthesizer discussed previously was placed next to the electronically tunable antenna. The power detector combined with the three-line direction coupler was inserted between the impedance synthesizer and an RF MEMS switch (Magfusion MagLatch™ RF Switch, Magfusion, Inc.). The RF MEMS switch is not part of the AATU, but employed to switch automatically between an RF signal generator (Agilent E4432B Signal Generator) and a VNA (HP-8510C Vector Network Analyser). The detailed photo and description of the current off-the-shelf (COTS) RF MEMS switch is shown in Figure 14.25.

It should be mentioned that, in this AATU prototype, the ACU reconfigures the impedance synthesizer based on the reflected power level only. A 50 Ω matched load is connected to the output of the incident power level detector. However, note that, as discussed in Section 4.4.1, it is necessary to detect both the incident and reflected power level to be compatible with the dynamic power variation scheme in an RF communication system such as the CDMA.

14.4.2 AATU Measurement Results

Based on the system setup discussed in the above subsection, the system return loss of the AATU will be measured. Note that in evaluating the AATU system performance, the most

Figure 14.24 Photograph of the AATU demonstration setup.

Figure 14.25 Photograph of the RF MEMS (Magfusion MagLatch™ RF Switch, Magfusion, Inc.) switch evaluation board.

important quantity is the return loss of the system. This is because the purpose of the AATU is to eliminate automatically the impedance mismatch between an antenna and a power module for the maximum power transfer.

Now consider the case where the frequency of an RF signal applied to the input of the directional coupler (output of the signal generator) is arbitrarily set to 1.87 GHz. Firstly, the tuning circuit of the SPA is controlled by the open-loop ACU discussed before. The solid line in Figure 14.26 presents the measured return loss of the SPA which is reconfigured based on the frequency information received by the ACU. Then, the nearest neighbor search algorithm running on the FPGA tries to minimize the reflected power level in order to eliminate the impedance mismatch between the antenna and impedance synthesizer. Figure 14.26 shows

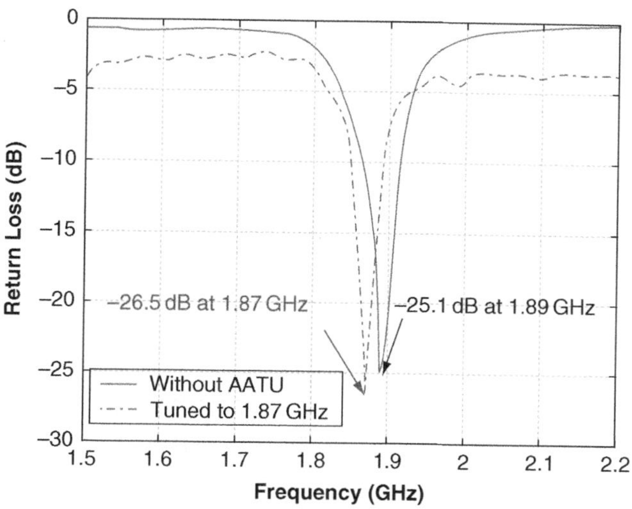

Figure 14.26 Narrowband tuning ability of the AATU. The SPA is tuned to 1890 MHz.

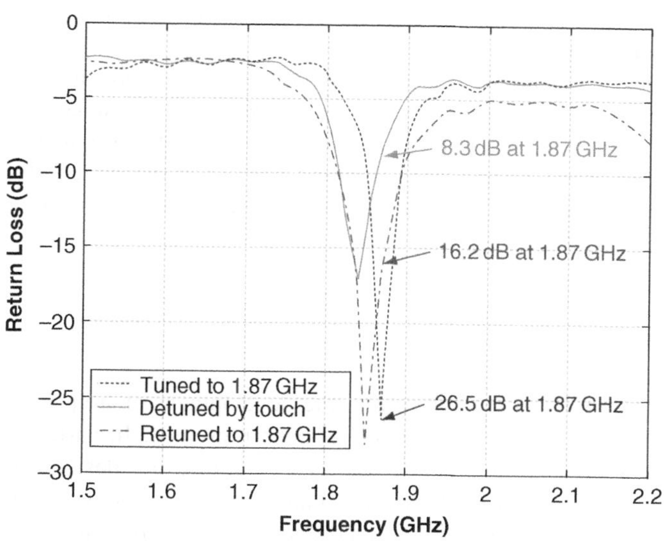

Figure 14.27 Automatic tuning ability of the AATU.

the results achieved by the prototype. As can be seen in the figure, the AATU enables very narrowband tunability.

Hand-held devices are generally used under constantly changing environment conditions. To demonstrate how well the AATU compensates for changing environmental conditions, the antenna is brought to close proximity of a human hand. Figure 14.27 shows the detuned response of the antenna when it is in contact with the human body. As can be seen, significant degradation of the antenna's performance occurs. Once the antenna is detuned, the search algorithm automatically reconfigures the impedance synthesizer to correct for this sudden

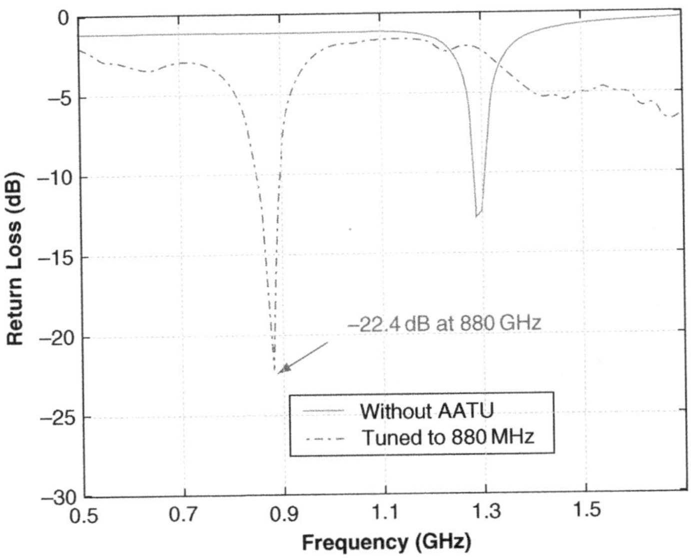

Figure 14.28 Narrowband tuning ability of the AATU. The SPA is tuned to 880 MHz.

environmental change. The retuned response, with the antenna still in contact with the hand, is shown in the same figure.

In many wireless applications it is necessary to change the operating frequency of the transmitter. To emulate this situation, the operating frequency is changed to 880 MHz from 1.87 GHz. Based on the frequency information, the ACU first reconfigures the SPA to work at a lower frequency band, as shown in Figure 14.28. Then the search algorithm finds an optimum switching configuration for the impedance synthesizer. After tuning, the measured response is shown in the same figure.

14.5 SUMMARY

In this chapter a thorough investigation into the development of frequency agile antennas for a software-defined and cognitive radio has been presented. In particular, attention has been focused on a number of practical issues concerning the design and modelling of electrically small antennas, as well as their tunable matching networks and concomitant control circuitry and algorithms. A new theoretical tool was presented for characterizing antennas from a system point of view, allowing their frequency selectivity to be incorporated into the RF front-end design. Tunable matching networks were also presented as well as appropriate analog and mixed-signal control circuitry to allow frequency-agile antennas to be reconfigured on demand by the radio's baseband processor. Finally, the implementation of an AATU prototype was discussed.

REFERENCES

1. J. Mingo, A. Valdovinos, A. Crespo, D. Navarro and P. Garcia, 'An RF electronically controlled impedance tuning network design and its application to an antenna input impedance automatic

matching system', *IEEE Transactions on Microwave Theory and Techniques*, **52**, 489–497, February 2004.

2. S. Rogers, J. Aberle and D. Auckland, 'Two-port model of an antenna for use in characterizing wireless communications systems obtained using efficiency measurements', *IEEE Antennas and Propagation Magazine*, **45**, 115–118, June 2003.
3. J. Aberle and S.-H. Oh, 'Reconfigurable antenna technology for VHF/UHF applications', Technical Report, General Dynamics Decision System, Inc., October 2004.
4. C. Phillips and R. Harbor, *Feedback Control Systems*, 4th edn, Prentice-Hall, Upper Saddle River, New Jersey, 1992.
5. Y. Sun and J. Fidler, 'Design of Π impedance matching networks', in *IEEE International Symposium on Circuits and Systems (ISCAS'94)*, Vol. 5, pp. 5–8, May 1994.
6. M. Thompson and J. Fidler, 'Frequency agile antenna tuning and matching', in *Eighth International Conference on HF Radio Systems and Techniques, 2000*, pp. 169–174, July 2000.
7. J. Fidler and Y. Sun, 'Computer-aided determination of impedance matching domain', in *IEE Twelfth Saraga Colloquium on Digital and Analogue Filters and Filtering Systems*, pp. 1/1–1/6, November 1992.
8. Y. Sun and J. Fidler, 'High-speed automatic antenna tuning units', in *Ninth International Conference on Antennas and Propagation (ICAP'95)*, Vol. 1, pp. 218–222, April 1995.
9. Y. Sun and J. Fidler, 'Practical considerations of impedance matching network design', in *Sixth International Conference on HF Radio Systems and Techniques, 1994*, pp. 229–233, July 1994.
10. Application Note (APN1002): 'Design with PIN diodes', Alpha Industries, Inc., Woburn, Massachusetts, 1999.
11. D. Pozar, *Microwave Engineering*, Second edn, John Wiley & Sons, Inc., New York, 1998.
12. Y. Sun and W. Lau, 'Evolutionary tuning method for automatic impedance matching in communication systems', in *IEEE International Conference on Electronics, Circuits and Systems, 1998*, Vol. 3, pp. 73–77, September 1998.
13. V. Nelson, H. Nagle, B. Carroll and J. Irwin, *Digital Logic Circuit Analysis and Design*. Prentice-Hall, Englewood Cliffs, New Jersey, 1995.
14. *University Program Design Laboratory Package User Guide*, Altera Corporation, San Jose, California, 2001.
15. P. Burns, *Software Defined Radio for 3G*, Artech House, Boston, Massachusetts, 2003.

15

Antenna Evaluation for Communications with Diversity/MIMO

Rodney Vaughan
School of Computing Science, Simon Fraser University, Burnaby, British Columbia, Canada

15.1 INTRODUCTION

This chapter addresses the evaluation of antennas for communications. Most antenna applications concern a digital communications link. The antenna gain has a direct impact on the link performance including the spectral efficiency. The classical directive gain and its measurement are reviewed with the natural progression to the distributed gain and the diversity gain for antennas designed for multipath situations. With most links operating in multipath, multielement antennas with high distributed gain and good diversity performance are required. In developing compact antennas, multielement antennas with statistical performance measures require convenient experimental evaluation techniques.

15.1.1 Background: Digital Communications Link Performance

Most antenna applications are for the transport of digital information. The link performance is measured by a statistic – the throughput of correctly detected bits per bandwidth, or *capacity efficiency*. For transmission, the digital data are typically encoded into a finite alphabet of analogue symbols comprising several bits of information. If the bit error rate (BER) can be kept to less than about 10^{-4}, then forward error correction coding can reduce the coded bit error rate to a negligible value. For most applications, a negligible error rate is not necessary, and an acceptable uncoded BER for wireless communications can be as low as or even 10^{-2}. A direct evaluation of the capacity of a link, or its digital outage, etc., is relatively straightforward, involving digital test sequence analysis. However, such an evaluation, dealing with only the data channel, does not reveal the mechanisms causing the bit

Printed Antennas for Wireless Communications Edited by R. Waterhouse

errors, or where link improvements could be made within the system. While fractions of a dB improvement are a cause for celebration in data coding research, much larger improvements, especially in most personal and mobile links, can be made at the antennas, through better matching (spatial, polarization, and impedance), leading to higher gain designs. Figure 15.1 depicts the channels of a communications link.

Independent of the encoding and alphabet size, the primary parameter that defines the average uncoded BER is the average energy per bit over the noise energy density, denoted ϵ_b/N_0, where the energy averaging is over the alphabet of symbols. The ϵ_b/N_0 relates directly to the signal-to-noise ratio (SNR) at the receiver (see Equation (15.1) below). The instantaneous SNR is calculated by time averaging over several RF cycles. The noise is normally dominated by the receiver thermal noise, which is otherwise independent of the wireless link. In a changing link, such as in mobile communications, the SNR becomes a random variable, and its distribution is estimated over either a fixed duration or over a set of environments for link characterization.

The Shannon capacity efficiency, in C bits per second, over the bandwidth used, B in Hz, is related to ϵ_b/N_0 by Shannon's law

$$\frac{C}{B} = \log_2\left(1 + \frac{C}{B}\frac{\epsilon_B}{N_0}\right) = \log_2(1 + \mathrm{SNR}) \tag{15.1}$$

This leads to the limiting wireless link capacity (and also the achievable or *practicable capacity*) and the capacity efficiency, being directly related to the signal energy at the receiver. The signal energy for a given signal design and regulated transmitter amplification level is determined by the path loss and antenna gains. The importance of the capacity efficiency metric lies with the need to use wisely the radio spectrum, a finite and shared resource.

With communications signal processing techniques such as equalization, rake reception, and error correction coding, wireless digital communications systems can now operate within

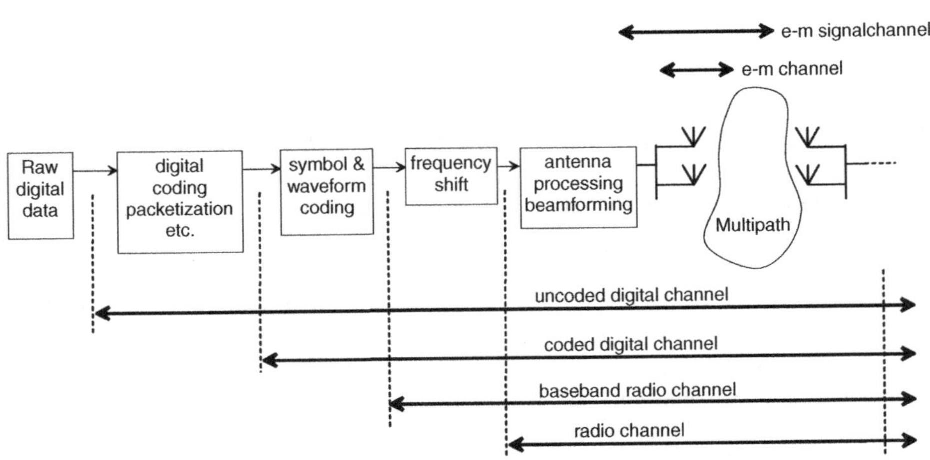

Figure 15.1 The channels of a digital link. Evaluation of the link is simplest using digital data sequences, but this reveals little about the RF performance and how it may be changed to improve the link. With the communications signal processing capability having reached its limits of capacity improvement for a channel, it is the performance of the antenna, and in particular the multielement antenna, that remains the source of potential capacity improvement.

fractions of a dB (in the SNR) of the Shannon capacity limit. In this sense, the signal processing capability has progressed to achieve effective saturation of the Shannon capacity efficiency.

In the meantime, the antenna gain has not only maintained its fundamental role in the wireless link but has now become the only capacity-limiting factor left within the control of the link designer. If the antenna's physical aperture can be allowed to increase in order to increase the antenna gain, then the capacity efficiency can increase. Such a capacity efficiency increase is unbounded, at least in a mathematical sense. The increased aperture can be achieved using multielement antennas, and the associated signal processing technology, together called multiple-input, multiple output (MIMO), has opened up relatively new possibilities for the capacity efficiency. Consequently, multielement antennas are key for high-capacity efficiency systems.

To illustrate the power of multielement antennas, the Shannon capacity efficiency for M_{tx} combined transmit antennas and $M_{rx} \geq M_{tx}$ receive antennas can be written in an approximate form as

$$\begin{aligned}\frac{C}{B} &\approx M_{tx} \log_2 \left(1 + \tfrac{M_{rx}}{M_{tx}} \mathrm{SNR}\right) \\ &\approx M \log_2(1 + \mathrm{SNR}), \qquad M_{rx} = M_{tx} = M\end{aligned} \tag{15.2}$$

This is a parallel-channels capacity formula [1]. It relates to Equation (15.1) in that there are M_{tx} independent (i.e. parallel) channels with the total transmit power divided evenly between the channels. The channels each have the same mean SNR, which is now SNR/M_{tx}, and there is also a receiving array gain of M_{rx} for each channel. The simplification in Equation (15.2) follows from having the same number of elements at each end of the link. Equation (15.2) can be considered as an upper limit to the Shannon capacity for MIMO links, both for the known channel (at the transmitter) case and the unknown channel case.

In an MIMO link, the antenna signals should be mutually uncorrelated over the fading from the multipath propagation. Relative to capacity efficiency limits for MIMO operating in multipath, the parallel channels equation above is optimistic by a few bits/s Hz for modest antenna numbers, say $M < 12$. However, the approximation is highly illuminating: there is a striking comparison between the single-channel case of Equation (15.1) and the MIMO case of Equation (15.2). While the receiver SNR still has the same logarithmic impact on the capacity efficiency as in the classical Shannon relationship, the linear dependence on M illustrates the unbounded (except by practicable considerations) growth of capacity efficiency with antenna dimension. For a fixed spectral cost, i.e. bandwidth and transmit power, the capacity can grow by simply adding antennas. The associated signal processing required for this is currently under development for commercially viable systems.

If there is unobstructed line-of-sight (LOS) between the M-element transmit and receive arrays, then the capacity efficiency will exceed that of the multipath case. In this case, conventional, directional beamforming would be used.

15.1.2 Interference-limited links

Most applications occupy a spectrum that is shared, intentionally or inadvertently, and the signals from other users, or interference, dominates the thermal noise as the unwanted signal. The primary parameter for communications capacity, instead of being the SNR, now becomes the signal-to-interference plus noise ratio (SINR).

With multielement antennas, array signal processing techniques can also be applied to suppress the interference power, which of course increases the SINR. This is a compounding motivation for using multielement antennas in communications.

In the following, techniques are developed for the evaluation of antennas that can provide the signals suitable for realization of high-capacity efficiencies.

15.2 BASIC ANTENNA PARAMETERS AND EVALUATION WITH IDEAL POINT-TO-POINT

Classically, antenna performance is gauged by the antenna pattern in a well-defined direction. This section introduces the classical parameters of antenna performance [2]. The principle of reciprocity is invoked throughout so that transmit and receive results can be interchanged except where stated otherwise.

The far-field pattern of an antenna is denoted here by

$$\boldsymbol{h}(\theta, \phi) = h_\theta(\theta, \phi)\hat{\theta} + h_\phi(\theta, \phi)\hat{\phi} \tag{15.3}$$

By convention, the antenna pattern refers to the electric field of the transmit pattern, and the unit of $\boldsymbol{h}$ is length, in metres. The transmit pattern is the same as the receive pattern except for a conjugation that stems from a reversal of the wave direction. Therefore the power patterns, e.g. $|h_\theta(\theta, \phi)|^2$, for transmit and receive, are the same. For linearly polarized antennas, the conjugate makes no difference to the polarization.

The open-circuit voltage at the receiving antenna is the product of the incident electric field, $E_{\mathrm{inc}}(\theta, \phi)$, in volts/m, and the receiving pattern, here written with the frequency dependence made explicit for both the antenna and field:

$$\begin{aligned} V_{\mathrm{o}}(\omega) &= \boldsymbol{E}_{\mathrm{inc}}(\theta, \phi; \omega) \cdot \boldsymbol{h}^*(\theta, \phi; \omega) \\ &= E_{\mathrm{inc}\theta}(\theta, \phi; \omega) h_\theta^*(\theta, \phi; \omega) + E_{\mathrm{inc}\phi}(\theta, \phi; \omega) h_\phi^*(\theta, \phi; \omega) \end{aligned} \tag{15.4}$$

This defines the pattern. Here the vertical $(\hat{\theta})$ and horizontal $(\hat{\phi})$ have been used for the polarizations, but any convenient orthogonal pair will do. In satellite communications, the right-hand $(\hat{\boldsymbol{R}})$ and left-hand $(\hat{\boldsymbol{L}})$ circular polarizations (RHCP and LHCP) are often used. RHCP means that the tip of the radiating transverse electric field vector follows a right-handed helical locus as it moves through space. Seen from the receiver, the electric field vector is turning anticlockwise. The unit vectors for circular polarizations (CPs) are

$$\hat{\boldsymbol{R}} = 1/\sqrt{2}(\hat{\theta} - \mathrm{j}\hat{\phi})\mathrm{e}^{-\mathrm{j}\phi}; \qquad \hat{\boldsymbol{L}} = 1/\sqrt{2}(\hat{\theta} - \mathrm{j}\hat{\phi})\mathrm{e}^{\mathrm{j}\phi} \tag{15.5}$$

and the $\mathrm{e}^{\mathrm{j}\phi}$ term is not required for magnitude-only patterns. Note that the transmit pattern is of the opposite hand to the receive pattern, and confusion here is a common pitfall in CP antenna and link design. Here it is assumed that the time dependency is $\mathrm{e}^{\mathrm{j}\omega t}$; for a time dependency of $\mathrm{e}^{-\mathrm{j}\omega t}$, the RHCP and LHCP are interchanged. The *copolar pattern* refers to the pattern in the wanted polarization and the cross-polar is the orthogonal, or unwanted, polarization.

15.2.1 Directive Gain

The classical performance measure for an antenna is the *antenna gain*, which is the normalized total (both polarizations) radiated power density in the direction of its maximum, denoted (θ_0, ϕ_0). The total radiated power from pattern $h(\theta, \phi)$ is represented by

$$P_{\text{rad}} = \frac{1}{4\pi} \int_{\phi=0}^{2\pi} \int_{\theta=0}^{\pi} \left(|h_\theta(\theta, \phi)|^2 + |h_\phi(\theta, \phi)|^2 \right) \sin\theta \, \mathrm{d}\theta \, \mathrm{d}\phi \tag{15.6}$$

and the *gain* of the antenna is

$$G(\theta_0, \phi_0) = \eta_{\text{ant}} \frac{|h_\theta(\theta_0, \phi_0)|^2 + |h_\phi(\theta_0, \phi_0)|^2}{P_{\text{rad}}} = \eta_{\text{ant}} D(\theta_0, \phi_0) \tag{15.7}$$

Here, the *directivity*, $D(\theta_0, \phi_0)$, is a purely total-power pattern shape parameter, and all the ohmic losses (metallic and dielectric) of the antenna are represented by the antenna efficiency factor, η_{ant}.

It is convenient to deal with normalized, unitless patterns, here denoted $g(\theta, \phi)$, meaning that $P_{\text{rad}} = 1$ when $g(\theta, \phi)$ is used instead of $h(\theta, \phi)$ in Equations (15.6) and (15.7); therefore

$$\int_{\phi=0}^{2\pi} \int_{\theta=0}^{\pi} \left(|g_\theta(\theta, \phi)|^2 + |g_\phi(\theta, \phi)|^2 \right) \sin\theta \, \mathrm{d}\theta \, \mathrm{d}\phi = 4\pi \tag{15.8}$$

For antenna evaluation, the gain is separated into the polarized gains, written here for normalized patterns as

$$G(\theta, \phi) = G_\theta(\theta, \phi) + G_\phi(\theta, \phi) = \eta_{\text{ant}} |g_\theta(\theta, \phi)|^2 + \eta_{\text{ant}} |g_\phi(\theta, \phi)|^2 \tag{15.9}$$

Therefore, for evaluating either the copolar gain or cross-polar gain, patterns of both polarizations are needed, and there is also a need to be able to calculate the total radiated power. Figure 15.2 gives the basic spherical coordinate system showing linear polarizations and the relation between the electric far-field and the normalized pattern, developed below.

15.2.2 Antenna Efficiency Terms

The concept of efficiency in antennas is illuminated by a basic circuit model (Figure 15.3) with an excitation current at the RF carrier frequency, $I = I_0 \mathrm{e}^{\mathrm{j}\omega_c t}$, running through series resistances whose dissipated powers respectively represent the radiated power and the antenna ohmic losses. Here the antenna impedance mismatch loss is included as well.

The link between the physical situation and the circuit model is the radiated power, and equating respective expressions for P_{rad} defines the radiation resistance. Voltage excitation, preferred for most printed patch antenna models, is also possible of course; the radiation power, for example, would $\frac{1}{2} G_{\text{rad}} |V_o|^2$, be where G_{rad} is the radiation conductance.

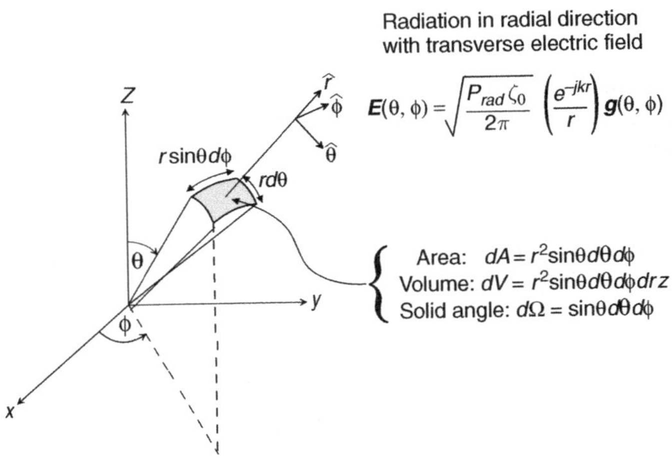

Figure 15.2 The far electric field, $E(\theta, \phi)$, decays as $1/r$ and has a phase of e^{-jkr}; its relation to the normalized antenna pattern is $g(\theta, \phi)$.

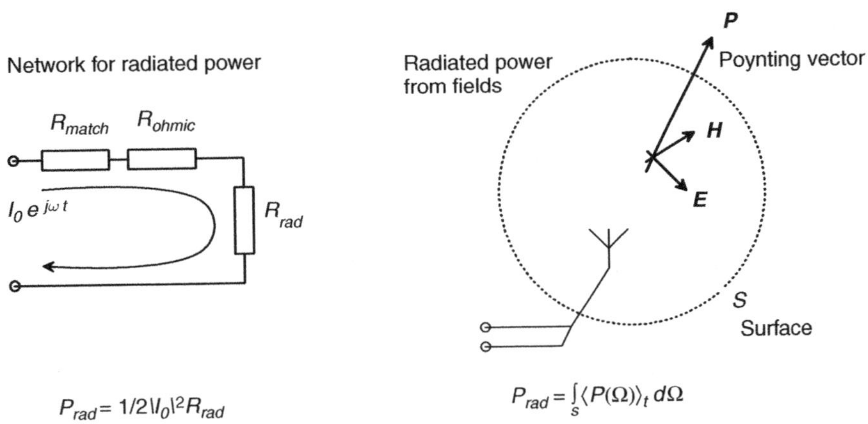

Figure 15.3 The radiated power links a circuit model and the physical situation. The match resistance includes losses in any balun present. The antenna efficiency includes only the ohmic losses resistor since the matching circuit is not strictly part of the antenna.

The *antenna efficiency*, sometimes called the *antenna radiation efficiency*, is now defined from the network circuit:

$$\eta_{\text{ant}} = \frac{R_{\text{rad}}}{R_{\text{rad}} + R_{\text{ohmic}}} \tag{15.10}$$

When there is a matching circuit required for the antenna, the impact on efficiency is included through the *radiation efficiency*:

$$\eta_{\text{rad}} = \frac{R_{\text{rad}}}{R_{\text{rad}} + R_{\text{ohmic}} + R_{\text{match}}} \tag{15.11}$$

The radiation efficiency is not strictly an antenna parameter, since it includes the matching network. However, in a link budget, which is of course necessary for antenna evaluation, it is necessary to account for matching losses, and the antenna matching is normally a part of antenna design. Any balun required for an antenna falls within the matching network here, so any balun losses are accounted for by the radiation efficiency.

For compact antennas, for example an electrically small printed small loop, the radiation resistance can be just fractions of an ohm and can have a balanced port. Matching this to a typical interface, say 50 ohms unbalanced, requires a network that can have a series resistance of several ohms, so that most of the power is dissipated in the matching network rather than the antenna radiation. Often the balun and matching are integrated with printed antennas, and for comparing different designs, care must be taken to include all the feed losses for a fair comparison between antennas. In terms of size, a compact antenna may need a matching and perhaps a balun network, each of which can be physically much bigger than the antenna itself. Unless the matching circuit and its losses are included, it is again an unfair comparison to pit the unmatched antenna size against a larger antenna that is already matched. Comparison of compactness between antennas is discussed further below.

In practice, the polarizations may not be aligned, and the *polarization efficiency*

$$\eta_{\text{pol}} = \left(\frac{\boldsymbol{p}_{\text{tx}} \cdot \boldsymbol{p}_{\text{rx}}^*}{|\boldsymbol{p}_{\text{rx}}|\,|\boldsymbol{p}_{\text{tx}}|} \right)^2 \tag{15.12}$$

is used in a link budget, where $\boldsymbol{p}$ denotes the vectors describing the antenna polarizations in the LOS directions. The tx and rx subscripts here can refer to the transmitter and receiver. Recall that the pattern is the transmit pattern, which is why the conjugate appears, or of course one of the polarization vectors could refer directly to the wave rather than an antenna.

For linear polarizations that are offset from each other by an angle γ, for example $\boldsymbol{p}_{\text{tx}} = \hat{\phi}, \boldsymbol{p}_{rx} = \cos\gamma\hat{\phi} + \sin\gamma\hat{\phi}$, then $\eta_{\text{pol}} = \cos^2\gamma$. The alignment accuracy is not critical in terms of power loss in the copolarization – a 0.5 dB polarization mismatch loss means that the polarization alignment should be within about 20°. However, for suppressing a signal in the unwanted polarization, the alignment accuracy is critical, the dependence being $\sin^2\gamma$. For example, when there is no cross-polar component, keeping the cross-polar measurement below −40 dB requires an alignment of better than about half a degree. For circular polarization, there is no such polarization alignment required, except in terms of getting the hands correct; if the transmit antenna has an RHCP transmit pattern then the receiver should have an LHCP pattern, i.e. an RHCP transmit pattern.

Finally, the *matching efficiency*, also called the impedance match gain or the mismatch factor, is $M = (1 - |\Gamma|^2)$, with Γ the voltage reflection coefficient at the antenna port. This is taken to be the same whether the antenna is transmitting or receiving.

15.2.3 Effective Aperture and Equivalent Circuits

The transmit antenna produces a power density at the receive antenna of $\tilde{S} = P_{\text{tx}} G_{tx}/(4\pi d^2)$ watts/m^2. It is insightful to view the receiver as a cross-section that intercepts this flow of power density. The power intersected by this area is the power delivered to the load at the receiver, which is why it is called an *effective* aperture, denoted A_{eff}. The received power is $\tilde{S}A_{\text{eff}}$. This can be equated with the received power calculated from a basic Friis link equation for free space (see Equations (15.25) and (15.26) below), which results in

$$A_{\text{eff}}(\theta, \phi; \eta_{\text{pol}}; \Gamma_{\text{rx}}; R_{\text{ohmic}}) = \frac{\lambda^2}{4\pi} M_{\text{rx}}(\Gamma_{\text{rx}}) \eta_{\text{pol}} \eta_{\text{rad}} D_{\text{rx}}(\theta, \phi) \tag{15.13}$$

where the dependencies on the receiving antenna are made explicit: direction, polarization match, load match and ohmic losses in both the antenna and any matching circuit. The latter inclusion means that the radiation efficiency term, η_{rad}, can be used in this equation.

By equating the received power in terms of the electric field, $\tilde{S}A_{\text{eff}} = \frac{1}{2}|E|^2/(\zeta_0 A_{\text{eff}})$, with the power in the load, $\frac{1}{2}|I|^2 R_{\text{L}}$, of a series equivalent circuit model comprising an open-circuit voltage source, an antenna impedance of $Z_{\text{A}} = (R_{\text{rad}} + R_{\text{ohmic}}) + \text{j}X_{\text{A}}$, and a load impedance Z_{L}, the relation between the effective aperture and the pattern is obtained as

$$\begin{aligned} A_{\text{eff}}(\theta, \phi; \eta_{\text{pol}}; \Gamma_{\text{rx}}; R_{\text{ohmic}}) &= \frac{R_{\text{L}}\zeta_0}{(R_{\text{rad}} + R_{\text{ohmic}} + R_{\text{L}})^2 + (X_{\text{A}} + X_{\text{L}})^2} \eta_{\text{pol}} |\boldsymbol{h}(\theta, \phi)|^2 \quad \text{(general)} \\ A_{\text{eff}}(\theta, \phi; \eta_{\text{pol}} = 1; \Gamma_{\text{rx}} = 0; R_{\text{ohmic}} = 0) &= \frac{1}{4} \frac{\zeta_0}{R_{\text{rad}}} |\boldsymbol{h}(\theta, \phi)|^2 \quad \text{(losless, matched)} \end{aligned} \tag{15.14}$$

Therefore, for matched and lossless conditions the pattern relates to the directivity (now the same as the gain) as

$$|\boldsymbol{h}(\theta, \phi)|^2 = 4 \frac{R_{\text{rad}}}{\zeta_0} \frac{\lambda^2}{4\pi} D(\theta, \phi) \quad \text{(lossless, matched)} \tag{15.15}$$

Because the effective aperture represents only the power absorbed by the load of the receiving antenna, it does not represent the power scattered by the antenna or any ohmic losses in the antenna and any matching circuit. The *capture area* is the name given to the aperture that represents the total power gathered by the antenna. For a minimum scattering antenna (the MSA is an idealized, lossless element, which is well approximated by single-mode antennas, such as a printed patch), the scattered power and the power in the load are the same, and the scattered power has the same pattern as the radiating pattern [3,4]. The equivalent circuit is the open-circuit voltage source in series with the self-resistance, R_{rad}, and the load, $R_{\text{L}} = R_{\text{rad}}$ (cf. Figure 15.11, below, with $M = 1$). The power dissipated in R_{rad} is the scattered power, or reradiation of the incident wave with the antenna pattern. Even if the antenna is lossless and matched in polarization and impedance, only half of the power captured reaches the load for signal detection, and the other half is scattered – a bad start for efficiently gathering power! For this reason, a more accurate name for MSAs would be equal scattering antennas [5]. Aperture antennas can do better, as noted in the following.

For larger, directive antennas, such as printed arrays, the efficiency of the *physical aperture* of the antenna is of interest. With such aperture antennas, the scattered power in any direction is not simply related to the aperture efficiency (the ratio of the effective aperture to the physical aperture) of that direction; the scattered power has a different pattern to the receive pattern. A limiting case is when there is no power scattered back in the direction of incident power, i.e. a zero radar cross-section. Aperture antennas have measured aperture efficiencies of up to 85 %. If the capture area is close to the physical aperture size, which is often the case for large arrays and reflectors, then this means that only 15 % of the power passing through the capture area is reradiated or dissipated in ohmic losses. As noted above, the corresponding value is never more than 50 % for an MSA. For non-MSA antennas such as aperture antennas, a more complicated equivalent circuit model is required, with multiple sources, than the single-source circuit for MSAs [5]. In summary, for a small, single-mode antenna such as a dipole or resonant cavity patch, the basic series, single-source, equivalent circuit holds. For aperture antennas, for example an antenna comprising a combination of elements, the situation is more complicated. MSA principles can be used for appropriate aspects of element evaluation, but not for combined arrays.

15.2.4 Impedance Bandwidth and the Q Factor

The bandwidth of an antenna is best discussed in relation to the quality factor, Q, a fundamental parameter of a resonant circuit [3, 6]. It is defined as the ratio of the energy stored to the energy dissipated in a circuit. In terms of calculating Q from an input impedance, there is no differentiation between the antenna losses and the antenna radiation. Only if the antenna is lossless, so that $R_A = R_{rad}$, does the impedance-measured Q refer to the radiation Q, a point discussed further below. For a current excitation, Q can be expressed in terms of the slope of the reactance as

$$Q = \frac{\omega}{2R_{rad}} \frac{\partial X_A}{\partial \omega} \tag{15.16}$$

The dual expression for a voltage excitation has G_{rad}, the radiation conductance, in place of R_{rad} and the antenna susceptance, B_A, in place of X_A. These exact relationships offer a way to estimate $Q(\omega)$ from an impedance measurement around a nominal operating frequency. Foster's reactance theorem is that the slope of the reactance is always positive for a lossless circuit. In a measurement, the slope can be negative owing to the finite loss, and the absolute value of the slope may be taken.

Most antennas operate at or near resonance. This allows a simple series RLC circuit to model the impedance behaviour. With the resonance frequency $\omega_0 = 1/\sqrt{LC}$, the impedance is

$$Z = R + \mathrm{j}\left(\omega L - \frac{1}{\omega C}\right) = R\left[1 + \mathrm{j}Q\left(\frac{\omega}{\omega_0} - \frac{\omega_0}{\omega}\right)\right] \tag{15.17}$$

Over a narrow bandwidth, the resistance is taken as fixed while the reactance changes, and the result is the classical relationship between Q and the half-power, or 3 dB, double-sided bandwidth

$$\frac{\Delta\omega_{3dB}}{\omega_0} = \frac{\Delta f_{3dB}}{f_0} = \frac{1}{Q} \tag{15.18}$$

In a measurement, there is not an open, or unloaded, circuit, of course. For a perfectly matched antenna, the resistance in the equivalent circuit doubles (because of the conjugate match) and so the loaded bandwidth is twice this value from the equation. The loaded Q, which is measured in a perfectly matched circuit, is half of the unloaded Q, viz. $Q = 2Q_L$.

It is convenient to use the VSWR or the voltage reflection coefficient, Γ, which is normally taken as the same as the measured single-port scattering parameter, S_{11}, to estimate

Q. The relation is VSWR $=(1+|\Gamma|)/(1-|\Gamma|)$. If the antenna is not perfectly matched, then the overcoupled solution can be written [7]

$$Q=\frac{2Q_{\mathrm{L}}}{1-|\Gamma(\omega_0)|} \tag{15.19}$$

Different bandwidth definitions can be used. A couple of useful cases are for $\Gamma=-10$ dB(VSWR $=1.925$) and VSWR $=2(\Gamma=-9.54$ dB), which are almost the same. The expressions are

$$Q_{(\Gamma=-10\ \mathrm{dB})}=\frac{2}{3}\left(\frac{\Delta\omega_{\Gamma=-10\mathrm{dB}}}{\omega_0}\right)^{-1}; \qquad Q_{(\mathrm{VSWR}=2)}=\frac{1}{\sqrt{2}}\left(\frac{\Delta\omega_{\mathrm{VSWR}=2}}{\omega_0}\right)^{-1} \tag{15.20}$$

For finding the relative bandwidth for other values of the VSWR for a single resonance, the following formula is used:

$$\left(\frac{\Delta\omega_{\mathrm{VSWR}}}{\omega_0}\right)=\left(\frac{\Delta\omega_{\mathrm{VSWR}}=2}{\omega_0}\right)\frac{\sqrt{2}(\mathrm{VSWR}-1)}{\sqrt{\mathrm{VSWR}}} \tag{15.21}$$

and a general relation is

$$Q=\frac{\mathrm{VSWR}-1}{Q\sqrt{\mathrm{VSWR}}}\left(\frac{\Delta\omega_{\mathrm{VWSR}}}{\omega}\right)^{-1} \tag{15.22}$$

As noted above, using impedance measurements for estimating the antenna bandwidth has a pitfall. The Q of an antenna comprises the radiation losses and the ohmic losses. If the antenna is particularly lossy, or has lossy material in its near-field, such as with a cell phone situation, the measurements can indicate a large bandwidth, but it must be remembered that this is the impedance bandwidth and does not reflect the radiation bandwidth. The relation between the impedance-measured Q and the Q contribution from the radiation and that from the ohmic losses is

$$\frac{1}{Q_{\mathrm{meas}}}=\frac{1}{Q_{\mathrm{rad}}}+\frac{1}{Q_{\mathrm{ohmic}}} \tag{15.23}$$

which is used below for compactness evaluation. In summary, a lossy situation can create the impression that an antenna is well matched and has a large bandwidth. Placing hands, or other lossy matter, around a microwave antenna and observing the impedance behaviour illustrates this – the impedance can become perfectly matched but virtually no radiation occurs!

In printed antennas such as patches, dipoles, and loops, the metallic and dielectric losses are readily estimated from calculations of modelled current distributions or fields of cavity modes, etc. For losses in the near-field where deterministic methods are difficult, numerical estimates of the loss must be made using FDTD-type solutions.

15.2.5 Compactness Evaluation

The Q factor is useful for benchmarking the bandwidth performance for compact antennas [8,9]. The size is taken from the electrical circumference of an inscribing sphere around the antenna, ka, where a is the radius. This may not be an ideal measure for printed antennas, where the design goal is probably a low-profile panel, rather than a small spherical shape. However, it serves as a benchmark since different antennas can be compared for compactness in this way. For printed patch-type antennas with a ground-plane, the ground-plane must be included as part of the antenna. For benchmarking, a way around accounting for an arbitrarily large ground-plane is to use the shape of the antenna plus its ground-plane image.

The Chu [8] and McLean [10] limits relate the minimum size of the antenna's inscribing sphere for a given Q. The Chu limit applies for a single-dipole mode, either TM or TE. The McLean limit corresponds to operation with the energy shared equally between both modes, allowing a lower limit. These limits are [10]

$$Q^{(\mathrm{TM})} = \frac{1}{(ka)^3} + \frac{1}{ka} \quad \text{(Chu)}, \qquad Q^{(\mathrm{TM+TE})} = \frac{1}{2(ka)^3} + \frac{1}{ka} \quad \text{(McLean)} \tag{15.24}$$

and for small ka, the cubic term shows the halving of the Q limit from using both modes. The limits are plotted in Fig. 15.4, along with some antenna examples [11]. Any antenna with its measured impedance giving an operating point below the appropriate limit must be lossy. Equation (15.23) can be used with the limits to indicate the amount of loss, as shown by the curves for antenna efficiencies of 50, 15, and 5 %. As an example, a so-called chip antenna which fits into a sphere with $ka = 0.2$, with an impedance-measured Q of 6, must

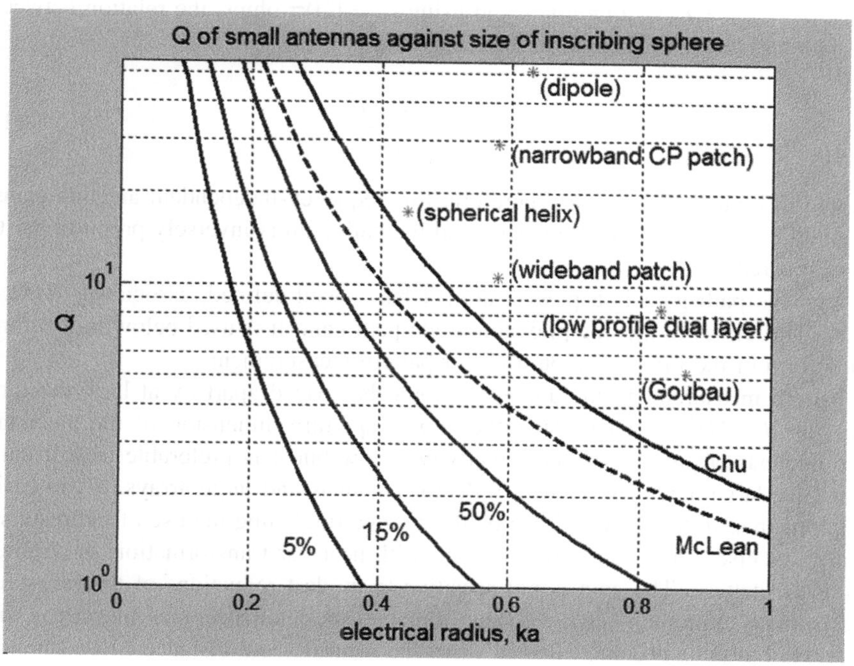

Figure 15.4 The bandwidth, expressed as a Q factor, against the size of the antennas expressed as the electrical circumference of the inscribing sphere, for compact antennas ($ka < 1$). The Chu limit applies for a single-mode antenna (TE or TM dipole mode), while the McLean limit is for the dual-mode or circularly polarized (equal TE and TM), antennas. Some antenna examples are included, with the wideband patch from [12]. The limits are for lossless antennas, and the other solid curves are labelled with the single-mode antenna efficiency.

be, at best, less than 5 % efficient (and likely much less), which means that it contributes a link loss of more than 13 dB.

15.2.6 Free-Space Links for Gain Measurement

A Friis link equation lays out the role of the antenna gain and the antenna impedance match. Here the ratio of received power to transmitted power is

$$P = P_{\text{rec}}/P_{\text{tx}} = \eta_{\text{pol}} G_{\text{tx}} M_{\text{tx}} G_{\text{rx}} M_{\text{rx}} G_{\text{path}} \tag{15.25}$$

where the subscripts denote transmit and receive. In general, the Friis equation defines the path gain. For free-space propagation, the path gain is the inverse square relation resulting from the spherical waves:

$$G_{\text{path}}^{(\text{FS})}(d) = \left(\frac{1}{4\pi d/\lambda} \right)^2 \tag{15.26}$$

Here d is the nominal distance between the antenna phase centres and λ is the wavelength at the centre frequency. By equating the power density (see Equation (15.13)) from an impedance- and polarization-matched transmit antenna with the copolar field strength at the receiver, $\frac{1}{2}|E_{rx}|^2/\zeta_0 = P_{\text{tx}} G_{\text{tx}} (4\pi d^2)^{-1}$, and using $\zeta_0 = 120\pi$ ohms, the relation between the antenna gain and the field strength at the receiver is

$$|E_{\text{rx}}| = \frac{\sqrt{30 P_{\text{tx}} G_{\text{tx}}}}{d} \tag{15.27}$$

as indicated in Figure 15.2 above. Therefore, for frequency-independent antenna gains, the field strength is independent of frequency and the path gain is inversely proportional to the frequency squared.

Although the path gain is a power ratio, it has a well-defined, associated phase term, $e^{-j2\pi d/\lambda}$, which relates to the complex amplitude path gain. It is used below in the complex amplitude form of the Friis equation used for pattern measurement.

In a pattern measurement, the antennas should be spaced apart by at least the far-field distance, say $d \geq 2D^2/\lambda$, where D is the largest aperture dimension of the antenna. For compact antennas ($D/\lambda \lesssim 1/\pi$), the directivity is low but it is preferable to still maintain $d > \lambda$, and for large antennas, such as reflector antennas and large arrays, d can easily be tens of kilometres. In this case, a large outdoor range (including the use of radio stars for a transmitter), or else a near-field measurement with near–far transformation, is required for evaluating the pattern. The gain is not sensitive to modest reductions of the measurement distance from its minimum far-field value, but pattern detail becomes inaccurate and, in particular, pattern nulls become filled in when the measurement distance is too small.

15.2.7 Gain evaluation

The deterministic nature of $G_{\text{path}}^{(\text{FS})}$ is used for evaluating antenna gain for reciprocal antennas. Free-space conditions are approximated by using a radio–anechoic chamber where the reflections of the walls of the chamber are minimized by using absorbing material. The measurement distance d is fixed at a far-field value.

In the *gain substitution method*, an antenna with a known gain is used with the unknown gain antenna. The measurement comprises pointing the two antennas at each other in their known maximum directions with their polarizations aligned, so that $\eta_{pol} = 1$. A calibrated transmitter and receiver is connected to the antennas so that the transmit power is known and the received power can be measured. From the Friis equation, the unknown gain can be estimated from the measurement using

$$G_{unknown}(\theta_0, \phi_0) = \frac{P}{G_{path}^{(FS)}(d) G_{known}(\theta_0, \phi_0)} \tag{15.28}$$

When no known-gain antenna is available, the *two-antenna method* uses two identical antennas with the same unknown gain. These are polarization-aligned, the transmit and receive power is measured, and then the Friis equation gives the gain as

$$G(\theta_0, \phi_0) = \sqrt{\frac{P}{G_{path}^{(FS)}(d)}} \tag{15.29}$$

The *single-antenna method* uses a single antenna pointing at a large flat reflecting surface so that the antenna's image acts as the second antenna. The distance to the screen is $d/2$, and the screen should be larger than the first Fresnel zone for a reasonable basic estimate of the gain. Because the single antenna is both transmitting and receiving, a circulator is required for separating the transmit and receive paths. The circulator is not part of the antenna, and it must be included in the calibration path of the test set (see below). For CP antennas, the transmit polarization is (say) RHCP and the reflected wave undergoes a change of hand, making it LHCP, which is the polarization of the receiving pattern.

Finally, the *three-antenna method* estimates three unknown antenna gains. Applying the combinations of the three antennas to the two antenna measurement arrangement, three measurements are obtained of the various products of gains. If these measured gain products are denoted X, Y and Z, these relate to the antenna gains as

$$\begin{gathered} G_1 G_2 = X; \qquad G_2 G_3 = Y; \qquad G_1 G_3 = Z \\ \text{or} \\ G_1 = \sqrt{XZ/Y}; \qquad G_2 = \sqrt{XY/Z}; \qquad G_2 = \sqrt{YZ/X} \end{gathered} \tag{15.30}$$

15.2.8 Nonreciprocal Antennas

Nonreciprocal antennas include those that use magnetized ferrites or plasmas, or 'active antennas', or where, for some reason, perhaps the integrated construction, the active components are electrically inseparable from the antenna port. Here there are problems with gain measurement. Only the gain substitution method will work, and in the case that they may be different, both the transmit gain and the receive gain of the known antenna are required. The reason for this is as follows. If the two-antenna method could be used, for example, with each antenna having unknown transmit and receive gains, a measurement of

$$G_{1tx} G_{2rx} = X; \qquad G_{1rx} G_{2tx} = Y \tag{15.31}$$

would be obtained, and even if the two antennas are identical, with $G_{1tx} = G_{2tx}$; $G_{1rx} = G_{2rx}$, and so $X = Y$, the transmit and receive gains cannot be separated from the measurements of their products. If the antennas are not identical, then there are four unknowns and two equations.

15.2.9 Complex Pattern Cut Measurement

For undertaking the complex pattern measurements, a traditional setup is to use a two-port microwave test set with an automated rotating pedestal. The test set has a vector network analyser, preferably with a scattering parameter test set. Normally, an automated measurement of all of the scattering parameters, as a vector of frequencies, is a straightforward capability of a commercial test set. The test set has a matched receiver load and generator, and the calibration plane is at the antenna port (see Figure 15.5).

Returning to reciprocal antennas, the Friis equation can be written as a complex amplitude quantity. For example, for the ith measurement of an azimuthal pattern cut, the complex amplitude form of the Friss equation (15.25) is

$$\left[\frac{S_{21}}{(1-|S_{11}|^2)^{1/2}(1-|S_{22}|^2)^{1/2}}\right] = \sqrt{\eta_{pol}}\, h_1(\theta_0, \phi_0)\, h_2^*(\theta_0, \phi_i)\sqrt{G_{path}^{(FS)}(d)}\mathrm{e}^{-j2\pi d/\lambda} \quad (15.32)$$

where all the terms are understood to be a function of frequency. Here, antenna 1 is the testing antenna and is transmitting. Antenna 2 is the antenna-under-test and is receiving. Normally the testing antenna polarization is set up to be the reference polarization (for

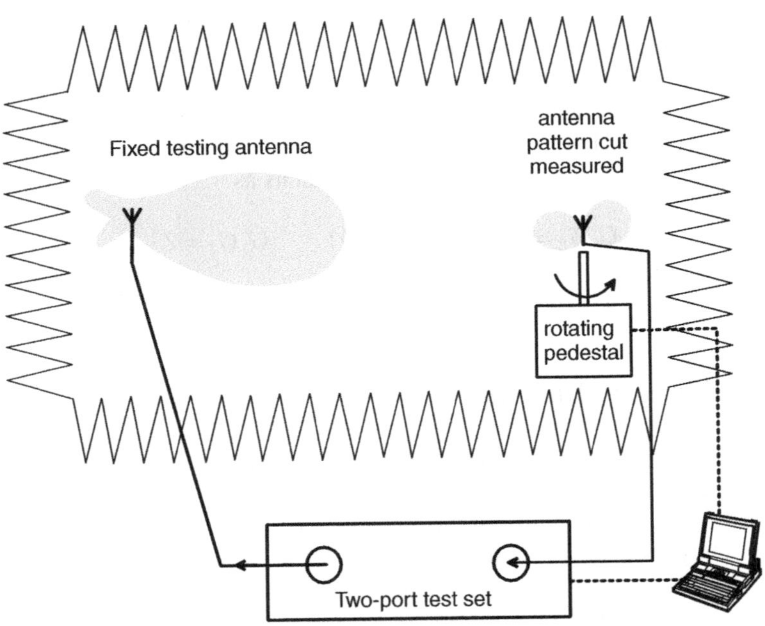

Figure 15.5 A pattern cut measurement is a coupling measurement between the testing antenna and the antenna under test. An anechoic chamber reduces the reflections to reduce inaccuracy from multipath.

both co- and cross-polarization pattern cuts), so that $\sqrt{\eta_{\text{pol}}}\underline{\underline{\Delta}}1$ in the pattern measurement. Simplifying further, a complex pattern cut is proportional to the measurement of S_{21}, i.e. $S_{21}(\phi_i) = \kappa\, h_2^*(\theta_0, \phi_i)$, where κ is complex.

For complete pattern cut information, a complex measurement is required for each polarization and the testing antenna needs to be able to be set up for transmitting co- and cross-polarizations. Each of the S parameters are sampled at, say, $k = 1$ to K_ω frequencies, denoted ω_k, and the $i = 1$ to I_ϕ azimuthal angles, ϕ_i, of the antenna-under-test, so that the dimension of each S parameter is complex (K_ω, I_ϕ) for each polarization. Ideally, the reflections, S_{11} and S_{22}, will not change with the changing azimuth angle of the antenna under test. However, this can happen in practice as the relative positions of the rotating antenna and the close-proximity feed cable change as the antenna rotates.

To improve the accuracy, reciprocity is used, and a complex average of S_{12} and S_{21} is taken, which replaces the S_{21}. The quantity

$$\epsilon = \frac{|S_{12} - S_{21}|}{\frac{1}{2}|S_{12} + S_{21}|} \tag{15.33}$$

represents the difference between the reciprocal measurements, which is typically kept below $-20\,\text{dB}$ to indicate good measurement accuracy. In the single-antenna method, this trick cannot be used since both transmit and receive are already needed for a single measurement.

For evaluating the distributed gain (see below), a measurement of the complete three-dimensional pattern is required. In a three-dimensional far-field measurement, several two-dimentsional pattern cuts need to be taken. This can be very time consuming, even with an automated system. One technique developed commercially for fast pattern acquisition is the Satimo system. This uses a ring bearing many testing antennas, which rotates about the antenna under test, with each testing antenna taking azimuthal cuts at different zenith angles. This spherical sampling can be either in the far-field or the near-field.

Near-field sampling is another technique for pattern measurement [13, 14]. This is particularly useful for large antennas, where the far-field distance is large. In a near-field measurement, the tangential $\boldsymbol{E}$ and $\boldsymbol{H}$ fields on a well-defined surface are sampled and stored. These are transformed to the far-field patterns using a knowledge of the surface shape. A limitation is that some part of the surface enclosing the antenna-under-test cannot be probed because of the mechanical support system and cabling. The measurement is usually undertaken in an anechoic chamber, and an automated probing system is required. Spherical surfaces can be used although this means expensive probing mechanics, but the Satimo system offers a solution to this. Cylindrical surfaces are the most popular for near-field measurement, where the antenna-under-test is rotated as the probe is moved up and down the cylindrical surface. Normally the cylinder ends are not probed, with at least one of the ends occupied with the feed cabling and mechanical support. Planar surfaces can also be used but such a surface is confined to directive, broadside patterns in a hemisphere only, as with a planar printed array, for example.

Finally, the frequency dependence of the gain requires comment. Normally in a measurement, the S parameters are taken automatically over a range of frequencies, and each measurement can be multiply-sampled and averaged in time for an improvement in the SNR. Both of these dimensions require more time for a given measurement of course. The SNR, after averaging, improves by 3 dB with each doubling of the number of samples. The issue of gain definition for an antenna operating over a wide bandwidth now arises. The directional gain of an antenna is frequency dependent and even operating over a modest relative bandwidth, say 10 %, the variation may be several dB. In the spirit of the single frequency gain definition, which is taken as the maximum over the set of directions, it seems reasonable to cite the maximum gain over the bandwidth. However, it is more useful to use the average gain over frequency for a given direction. In the absence of standards guidance, wideband gain specifications should include information on the gain definition used.

This completes the review of the antenna parameters through the concepts of impedance measurement and free-space point-to-point link measurement. For a line-of-sight link, the pointing direction of the maximum antenna gain is used for each antenna of course. However, if there are strong interference sources in certain directions, then the optimum pattern maximizes the SINR and the pattern shape is typically arranged to minimize the interference. The resulting maximum gain direction may not coincide with the direction of the link.

15.2.10 Line-of-Sight Links

In practice, most wireless links will have multiple-path propagation, so several directions, instead of one, become important. For a mobile antenna, the same propagation path will be seen by different pattern directions as the mobile terminal rotates and moves within fixed geographic directions-of-arrival (see Figure 15.6).

Fixed LOS radio links are analysed using only the LOS antenna gain. The primary antenna parameter in this situation is the copolar directive gain. (Recall that the antenna gain is defined by its maximum directive gain and that this includes both polarizations.)

Free-space conditions seldom apply in practice. Even LOS links using highly directive antennas experience some multipath propagation, for example from a ground bounce. At the receiving antenna, the arriving energy is not confined to one direction in the pattern, but instead occupies two or more different angles. In the ground bounce case, the two angles may not be resolvable, so there are two waves arriving within the same angular bin (see Figure 15.6(b)). The path gain factor, as d changes, experiences variable fading caused by the constructively and destructively interfering signal components combined in the receiving antenna. Now, instead of the antenna gain being a variable (Figure 6(a)), it is the path gain

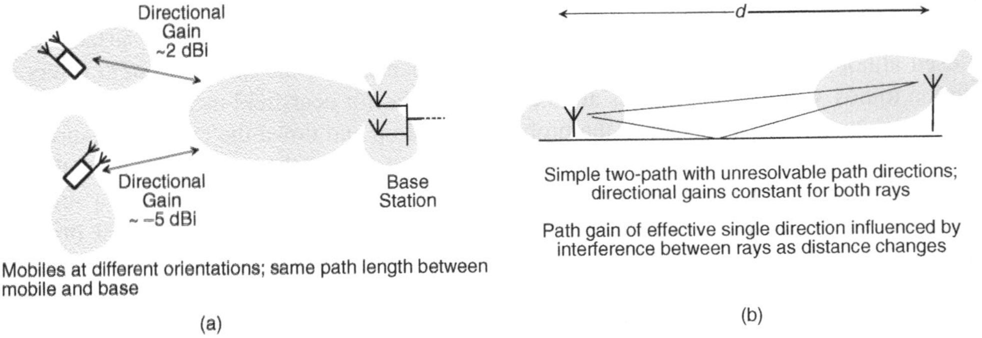

Figure 15.6 (a) A point-to-point mobile link, here depicted with the mobile antenna in two rotation positions, experiences changing received power because of the changing directive gain at the mobile. The path gain remains the same in this situation. (b) A point-to-point link with fixed directive gains but a path gain that changes with a changing separation distance. Here the received power is variable because the path gain, comprising two closely spaced interfering rays, is variable.

term that is a random variable. So the main impact of the multipath is that the path gain term is no longer a deterministic quantity except in the simplest of multipath scenarios, and that the antenna gain term is also variable. For simplification, new definitions for the path gain and antenna gains are required. Even a mobile LOS link with fixed antenna patterns illustrates the need for a stochastic approach to link analysis.

15.2.11 Mobile NLOS Link

Most wireless links are non-line-of-sight (NLOS) and so they rely only on non-line-of-sight propagation which is the multipath. Here, the path gain term is assigned to be the only variable – in the spirit of the classical Friis link equation – and the antenna gains and matching are fixed to their mean values, where the averaging is over all angles.

The transmit and receive antennas, and their impedance matching, are all considered to be mutually independent. Mathematically, the mean normalized received power can then be written

$$\langle P \rangle = \langle \eta_{\mathrm{pol}} G_{\mathrm{tx}} M_{\mathrm{tx}} G_{\mathrm{rx}} M_{\mathrm{rx}} G_{\mathrm{path}} \rangle = \langle G_{\mathrm{tx}} \rangle \, \langle M_{\mathrm{tx}} \rangle \, \langle G_{\mathrm{rx}} \rangle \, \langle M_{\mathrm{rx}} \rangle \, \langle G_{\mathrm{path}} \rangle \tag{15.34}$$

The angled bracket indicates averaging over the mobile channel, which in practice means time or spatial averaging, or both if required. The received power depends on directional gain, which changes with the mobile movement. The distributed gain, discussed below, replaces the directional gain in order to get round this issue, and at the same time it absorbs the polarization efficiency term. The reason for including the impedance matching gain here as a random variable is that, with personal terminals, in particular cellphones, the antenna is in close proximity to a changing environment – for example the head, hand, and body of a cellphone user or the changing desk-top environment next to a laptop antenna. The matching gain of a cellphone antenna can change by several dB as the user changes the handset position. At a cellphone base station, which is well isolated from close-proximity changes, the matching is of course unchanging. However, the mounting structure and environment may result in a different match to that measured in controlled conditions such as an anechoic chamber.

In stochastic situations, the averaging detail is a critical aspect for evaluation. As a wireless link sees a changing physical environment, the observed changes in the received power can be attributed only to the product of terms at the right-hand side of the Friis equation. The question arises as to how to allocate the changes to the different terms. For the changing matching at a hand-held terminal, the averaging is over the usage habits (the way the cellphone is held, etc.). The average matching gain depends somewhat on the particular user. A manufacturer designs the matching to suit an ensemble of users, rather than for a single-user habit.

The first step in analysis is to express the open-circuit voltage from a continuum of incoming waves cf. (Equation(15.4)):

$$V_{\mathrm{o}}(\omega) = \int_{0}^{2\pi} \int_{0}^{\pi} \boldsymbol{E}_{\mathrm{inc}}(\theta, \phi; \omega) \cdot \boldsymbol{h}^{*}(\theta, \phi; \omega) \sin\theta \, \mathrm{d}\theta \, \mathrm{d}\phi \tag{15.35}$$

which also shows the inseparable nature of the antenna pattern and the incident wave distribution in the received signal. To clarify the averaging, the average overangle in Equation (15.20) is denoted by

$$V_o(\omega) = \langle V_o(\omega; \theta, \phi) \rangle_{\theta,\phi} \tag{15.36}$$

and the dependence of the incoming waves, on position or time, for a mobile terminal is understood. The receiving antenna pattern is taken as fixed and its angular coordinates are referred to the receiving terminal orientation.

The incident waves are generally assumed to be uncorrelated in angle. This means that the fading multipath signal from one direction is uncorrelated, when averaged over space or time, with that from any other direction, relative to the receiving antenna coordinates. Similarly, the polarizations are assumed to be mutually uncorrelated for all directions. Dropping the frequency dependence for brevity, the average received power is proportional to the autocorrelation, $\langle V_o V_o^* \rangle$, where $V_o(\omega) = V_o$ is understood to be averaged over angle, and the time or space averaging (denoted with no subscript on the angle bracket) is

$$\begin{aligned} \left\langle |V_o|^2 \right\rangle &= \left\langle \left(\iint E_{\text{inc}}(\theta, \phi) \cdot h^*(\theta, \phi) \sin\theta \, d\theta \, d\phi \right) \left(\iint E_{\text{inc}}(\theta, \phi) \cdot h^*(\theta, \phi) \sin\theta \, d\theta \, d\phi \right)^* \right\rangle \\ &= \iint \left\langle |E_{\text{inc}\theta}(\theta, \phi)|^2 \right\rangle |h_\theta(\theta, \phi)|^2 \sin\theta \, d\theta \, d\phi + \iint \left\langle |E_{\text{inc}\phi}(\theta, \phi)|^2 \right\rangle |h_\phi(\theta, \phi)|^2 \\ &\quad \times \sin\theta \, d\theta d\phi \end{aligned} \tag{15.37}$$

Because the open-circuit voltage is used here, with no current flowing in the equivalent circuit model, no loss factor explicitly appears. The incident angular power is now represented as a distribution, denoted from now on as, for example, $\langle |E_{\text{inc}\theta}(\theta, \phi)|^2 \rangle) = S_\theta(\theta, \phi)$ (the proportionality constant is understood), and as the vector $\mathbf{S}(\theta, \phi) = S_\theta(\theta, \phi)\hat{\theta} + S_\phi(\theta, \phi)\hat{\phi}$ in watts/m^2 steradian, and as a scalar $S(\theta, \phi) = S_\theta(\theta, \phi) + S_\phi(\theta, \phi)$.

The received power of Equation (15.37) is here using the normalized pattern times the antenna efficiency and normalizing by the incident power

$$\frac{P_{\text{rec}}^{(n)}}{P_{\text{inc}}} = \frac{\eta_{\text{ant}} \iint \left(S_\theta(\theta, \phi) |g_\theta(\theta, \phi)|^2 + S_\phi(\theta, \phi) |g_\phi(\theta, \phi)|^2 \right) \sin\theta \, d\theta \, d\phi}{\iint \left(S_\theta(\theta, \phi) + S_\phi(\theta, \phi) \right) \sin\theta \, d\theta \, d\phi} \tag{15.38}$$

and the incident power, P_{inc}, can be viewed as the power received by a pair of colocated, omnidirectional, unity gain, lossless antennas, each of which receive an orthogonal polarization, viz. $g_\theta(\theta, \phi) = g_\phi(\theta, \phi) = 1$. (Such patterns are a concept only – an antenna cannot easily be built with these patterns.) This ratio of powers can be viewed as an antenna gain, which is formalized in the following.

This angular incident power is scaled in order to introduce probability density functions (PDF) for each polarization, denoted and $p_\theta(\theta, \phi)$ and $p_\phi(\theta, \phi)$, where

$$\iint p_\theta(\theta, \phi) \sin\theta \, d\theta \, d\phi = 1, \qquad \iint p_\phi(\theta, \phi) \sin\theta \, d\theta \, d\phi = 1 \tag{15.39}$$

(Note that this is a different normalization than that of the pattern, Equation (15.8), and it is also different, by a factor of sin θ from that of a standard mathematical two-variable joint PDF, which is $\int\int p_{\theta,\phi}(\theta, \phi) \, d\theta \, d\phi = 1$.) The scale factors are $\kappa_\theta, \kappa_\phi$, respectively, so that $p_\theta(\theta, \phi) = \kappa_\theta^{-1} S_\theta(\theta, \phi)$, for example, and the incident power normalization is expressed by

$$\iint S_\theta(\theta, \phi) \sin\theta \, d\theta \, d\phi = \kappa_\theta, \qquad \iint S_\phi(\theta, \phi) \sin\theta \, d\theta \, d\phi = \kappa_\phi \tag{15.40}$$

The total, incident, angularly distributed power at the terminal location is $P_{\mathrm{inc}} = \kappa_\theta + \kappa_\phi$. As noted above, to receive the total power available with an antenna would require lossless, omnidirectional reception in both polarizations.

The ratio of the averaged power in each of the polarizations, called the cross-polar ratio (XPR), can be expressed as a function of angle, averaged over the mobile channel (e.g. averaged over time) or, respectively, averaged over both time and the angle,

$$\mathrm{XPR}(\theta,\phi) = \frac{S_\theta(\theta,\phi)}{S_\phi(\theta,\phi)}; \qquad \mathrm{XPR} = \frac{\kappa_\theta}{\kappa_\phi} \tag{15.41}$$

If the receiving antennas were isotropic in each polarization, the XPR would also represent the time- and incident angle-averaged cross-polar ratio as seen from the power received by the polarized antenna ports. In general, however, the XPR is an incident power ratio and is independent of the antenna.

The received power by an antenna pattern in a distributed source environment, for example in multipath, normalized by the total incident power, is equivalent to the *distributed directivity* [15–17], and has also been called the *mean effective gain* [18], although the *mean effective directivity* would be a more correct name since this latter formulation contains no losses. The *distributed gain* is expressed as the power received by the antenna over the incident power, (15.38), here using the normalized antenna power pattern, $|g(\theta,\phi)|^2$, and the corresponding gain pattern, $G(\theta,\phi) = \eta_{\mathrm{ant}}|g(\theta,\phi)|^2$, as

$$\begin{aligned} G^{(\mathrm{D})} &= \langle G(\theta,\phi)\rangle_{\theta,\phi} = \eta_{\mathrm{ant}} D^{(\mathrm{D})} = \eta_{\mathrm{ant}} \langle D(\theta,\phi)\rangle_{\theta,\phi} \\ &= \eta_{\mathrm{ant}} \frac{\iint \left(\kappa_\theta p_\theta(\theta,\phi)\,|g_\theta(\theta,\phi)|^2 + \kappa_\phi p_\phi(\theta,\phi)\,|g_\phi(\theta,\phi)|^2\right)\sin\theta\,\mathrm{d}\theta\,\mathrm{d}\phi}{\kappa_\theta + \kappa_\phi} \\ &= \tfrac{\mathrm{XPD}}{1+\mathrm{XPD}} \iint p_\theta(\theta,\phi)\,G_\theta(\theta,\phi)\sin\theta\,\mathrm{d}\theta\,\mathrm{d}\phi + \tfrac{1}{1+\mathrm{XPD}} \iint p_\phi(\theta,\phi)\,G_\phi(\theta,\phi)\sin\theta\,\mathrm{d}\theta\,\mathrm{d}\phi \end{aligned} \tag{15.42}$$

and the antenna efficiency is here taken as the same in each polarization.

The distributed gain can interpreted as the fixed gain when the instantaneous incident power is fixed and given by $S(\theta,\phi)$. The antenna pattern is a spatial, or rather an angular, filter, and in this sense, for high distributed gain, the fixed pattern should be match as well as possible to the source distribution. Another interpretation is that if the source distribution is changing relative to the antenna coordinates and has a mean value of $S(\theta,\phi)$, then $G^{(\mathrm{D})}$ is the statistical mean of the gain.

To help fix ideas, consider the simplest distributed source distribution as omnidirectional, with the same power in each polarization; i.e. it is completely random, so the PDFs are $p_\theta = p_\phi = 1/(4\pi)$. This is often used as an averaged indoor scenario for mobile terminals. In this case, note that the distributed gain is half of the antenna efficiency, $G^{(\mathrm{D})} = \eta_{\mathrm{ant}}/2$, independent of the antenna pattern. Therefore, the maximum value of the distributed gain is $\frac{1}{2}$, which is the value of the averaged polarization efficiency. The explanation is that from an equal allocation of incident power to each polarization, a single-port antenna can only receive one polarization – that of the receiving pattern, which corresponds in this case to half of the incident power. This is how the polarization efficiency, which is explicit in the point-to-point case, is absorbed into the distributed gains of the multipath case.

Note that, in general, the distributed gain is dependent on the incident sources, a strange concept in some ways, but one that is necessary in order to have a fixed gain in the changing multipath environment of mobile communications. Including the averaged parameters of the multipath is helpful in Equation (15.42) because now the distributed gain is in terms of parameters of the averaged source scenario referred to the receiving antenna coordinates. The Friis equation (15.34) can now be written using the distributed gain terms, for example $\langle G_{rx}\rangle = G_{rx}^{(D)}$, and if the matching gains are fixed, the variation in the changing channel is attributed solely to the path loss:

$$\langle P\rangle = G_{tx}^{(D)} M_{tx} G_{rx}^{(D)} M_{rx} \langle G_{path}\rangle \tag{15.43}$$

15.2.12 Scenario Models for Distributed Gain Estimation

For antenna design and modelling, the average source scenario requires modelling. Recall that the averaging is over time, which can also be interpreted, in the case of mobile antennas, as averaging over space (the locus travelled by the mobile), including angular orientation changes. Outage statistics apply to each mobile terminal operating in the same coverage area. Another interpretation of the averaged scenario is to average over many fixed users in different fixed environments:

$$S_{av}(\theta, \phi) = \frac{1}{N_u} \sum_{i=1}^{N_u} S_i(\theta, \phi) \tag{15.44}$$

This is applicable for many fixed terminals in an area associated with a central base station system, for example. Here the average distributed gain of the terminal becomes a distributed gain averaged over many different users, and now the outage applies as a statistic over the users, rather than over time or space, for each user. In the mobile case, there may be an outage of, say, 90 %, meaning that each terminal has no connection for 10 % of the time, whereas, in the fixed terminal case, 10 % of the terminals are not connected.

It is clear that the averaged source distribution should play an important role in antenna evaluation. For modelling the signals of the mobile channels, the notion of an *effective source distribution* refers to the product of the source distribution and the antenna pattern. For fixed patterns, the receiver cannot separate the pattern effect from the source distribution effect. The true source distributions are seldom known and the source modelling approximation is a major limitation in analytic approaches to antenna evaluation.

15.2.13 Distributed Source Models

As noted above, for indoor mobile terminals, a convenient model is simply to have a uniform distribution over the whole sphere, so that $p_\theta = p_\phi = 1/(4\pi)$. For a wall- or ceiling-mounted antenna – typically printed antennas on a ground-plane structure – a hemispheric distribution can be used, for example $p_\theta = p_\phi = 1/(2\pi)$. If the signals from behind the antenna are important, for example for interference calculations when the interference is incident from behind the ground-plane of the antenna (coming through the wall or ceiling), then clearly the full sphere needs to be considered for the antenna pattern.

Other source scenario models that are sufficiently simple but are widely used include the Clarke model, which is the ring of sources in azimuth, here shown for one polarization only $(\kappa_\phi^{-1} = 0)$:

$$S(\theta, \phi) = S(\theta)\,\hat{\theta} \propto \delta(\theta - \pi/2)\,\hat{\theta} \tag{15.45}$$

The Clark model is useful for vehicular antennas, where the source directions are often close to the horizon and on average are essentially vertically polarized owing to the essentially conducting ground. A uniform distribution in azimuth and from 0° to 30° in elevation has been used to evaluate the distributed directivity, $D^{(D)}$, of roof-mounted patch antennas [19]. Variations include adding elevation dependence but maintaining a form that lends itself to integrations. For example [20], an elevation angular extent of $2\beta_\theta$, which is centred on the horizon so that the waves arrive from below the horizontal, can be expressed using the elevation angle $\beta = \pi/2 - \theta$:

$$\begin{aligned} S_\theta(\theta, \phi) = S_\theta(\theta) &\propto \frac{\cos[\beta(\theta)]}{\sin \beta_\theta}, \qquad |\beta| \leq |\beta_\theta| \leq \pi/2 \\ = 0 &\quad \text{elsewhere} \end{aligned} \tag{15.46}$$

A variation is to remove the discontinuity at the support edges, $\beta = \beta_a$, using a scenario proportional to $\cos^2[(\pi/2)(\beta/\beta_a)]$, for example.

For more directive situations such as for elevated base stations in an urban environment, where the transmitting mobile is seen as a cluster of scatterers, Gaussian forms are convenient. For a mean angle, (θ_m, ϕ_m), and angular spreads represented by $(\sigma_\theta, \sigma_\phi)$, the vertical polarization, for example, can be written

$$S_\theta(\theta, \phi) \propto \exp\left(-\frac{(\theta - \theta_m)^2}{2\sigma_\theta^2} - \frac{(\phi - \phi_m)^2}{2\sigma_\phi^2}\right), \quad 0 \leq \phi \leq 2\pi, \; 0 \leq \theta \leq \pi \tag{15.47}$$

The Laplacian shape (double-sided exponential) gives a better experimental fit in azimuth than the Gaussian in urban environments [21]. It is written, again here for just the vertical polarization, that

$$S_\theta(\theta, \phi) \propto e^{-(\sqrt{2}/\sigma_L)|\phi|}, \quad -\pi \leq \phi = \pi \tag{15.48}$$

and σ_L represents an azimuthal angular spread of the source distribution. These directive distributions suit the evaluation of arrays of printed antennas, etc., used at base stations.

There are many more models and these are often chosen as a trade-off between a likely realistic representation and a simple, for example an easily integrated, form. Often, the models need to be developed to suit a specific application.

15.3 NOISE

The noise power is as important as the signal power since it is the SNR that defines the system performance. It is of particular interest with satellite antennas [22]. The SNR is also required for the signal combination of diversity antennas. Here, the noise contribution from the front-end amplifier, the antenna noise, or noise from radiation incident on the antenna, and the feed cable noise contribution, are reviewed. All use the concept of temperature to describe their thermal contribution to the noise at the receiver. For detail on noise measurement in multielement antenna systems, see Reference [23].

The branch SNR can be estimated by first measuring the noise power and then the signal plus noise power, and then calculating their ratio which gives (SNR+1) (see Equations (15.52) below). The thermal noise of the front-end amplifier can be viewed as the branch power with the antenna terminals shorted. However, for measuring this, some amplifier specifications require a correct termination, say R_L, at their input. This resistor's available power to the amplifier load is $P_{av} = V_R^2/(4R_L) = k_B T_R \Delta f$ watts, where $k_B = 1.38 \times 10^{-23}$ joules/degree is Boltzman's constant, T_R is the temperature of the resistor in degrees kelvin, and Δf is the receiver bandwidth in Hz. This power is multiplied by the gain of the amplifier, G_{amp}, to give the noise power from the resistor at the output of the amplifier. The contribution from the terminating resistor can be calculated and subtracted from the measured noise at the amplifier output. If the total measured power in a matched output load is P_{meas} watts, then the estimate of the amplifier's thermal noise contribution, in watts, is

$$P_{amp} = P_{meas} - k_B T_R \Delta f G_{amp} \tag{15.49}$$

The antenna noise stems from blackbody radiation incident on the receiving antenna. The fields relate to the angular brightness distribution [15], $B(\theta, \phi)$, as $B(\theta, \phi)\Delta f = \frac{1}{2}\langle |E(\theta, \phi)_{inc}|^2\rangle_t / \zeta_0$. It is randomly polarized, i.e. it is the same for any polarization, and so it is not expressed as a vector. The brightness of a blackbody is proportional to the temperature as $B(\theta, \phi)\Delta f = 2k_B c^2 / \left[f^2 T_B(\theta, \phi)\right]$ watts/m^2 per steradian per cycle (Rayleigh–Jeans law). This leads to a useful parameter, the *antenna temperature*. It is the mean of the *effective temperature distribution* over the (normalized) distribution of the antenna directivity:

$$T_A = \langle T(\theta, \phi)\rangle_{|g(\theta,\phi)|^2} = 1/(4\pi) \int\int T_{EB}(\theta, \phi)|g(\theta, \phi)|^2 \sin\theta \, d\theta \, d\phi \tag{15.50}$$

T_A is the temperature of a source resistor that models the antenna noise power contribution. The term *effective* follows because the distribution includes the effect of any impedance mismatch and the radiation absorption coefficient of the antenna (which would include any antenna ohmic losses), etc.

For a matched antenna that absorbs all of the incident radiation ($\eta_{ant} = 1$) and a constant temperature distribution $T_{EB}(\theta, \phi) = T_0$, then $T_A = T_0$. This would apply to an indoor environment where all of the visible region is at a room temperature, usually taken as $T_0 = 290$ K.

As with the distributed signal scenario, models are used for the effective temperature distribution, which are not elaborated here. The terrestrial blackbody temperature is taken as 290 K and the sky is often taken as 3 K (not blackbody), but the effective sky temperature is significantly increased at higher frequencies (above 10 GHz) by atmospheric absorption. At low frequencies (less than a few hundred MHz) the galactic antenna noise normally dominates the receiver thermal noise, so at these frequencies the antenna noise must be accounted for in evaluating the performance of an antenna.

Normally, a front-end amplifier is located as close as possible to the receive antenna port, and with printed antennas it can be integrated within the antenna. However, if there has to be a long feedline between the receive antenna and the amplifier, then this will both attenuate the power from the antenna and also introduce its own thermal noise contribution. If the line is L metres long, is at temperature T_L kelvin, and has attenuation α nepers/m, then the temperature of the antenna, or a terminating resistor at the input of the amplifier, is $T_{AL} = T_A e^{-2\alpha L} + T_L\left(1 - e^{-2\alpha L}\right)$. Then T_{AL} is used for the temperature of the input terminating resistor for the amplifier and the available power to the receiver is $k_B T_{AL} \Delta f$.

15.4 GAIN FACTORS FOR NON-LINE-OF-SIGHT

Antenna diversity is a well-established technique for countering the fading power gain of the multipath channel. The basic idea is to have two or more antenna ports that receive uncorrelated fading signals and combining these signals so that the combined signal fades less. The reduction in the fading signal is illustrated in the eight-branch combination example of Figure 15.7 [11]. Signalling techniques also allow the diversity action to be undertaken at the transmit end. MIMO systems can be implemented using a form of diversity at each end of the link.

In diversity, the combination of two antenna branch signals provides a resultant signal that has less fading. Mathematically, this is explained as follows. Denote an equivalent baseband signal $s = re^{j\phi} = x + jy$. If, for example, two uncorrelated signals are added, then we have a resultant power signal $r_{tot}^2 = r_1^2 + r_2^2 = x_1^2 + y_1^2 + x_2^2 + y_2^2$ and this is distributed as a chi-square with 4 degrees of freedom (DOF). The normalized (by the mean squared) variance of the chi square reduces by the number of Gaussian components, so relative to one antenna signal (two real Gaussians), a two-branch combined signal (four real Gaussians), for example, has its normalized variance halved.

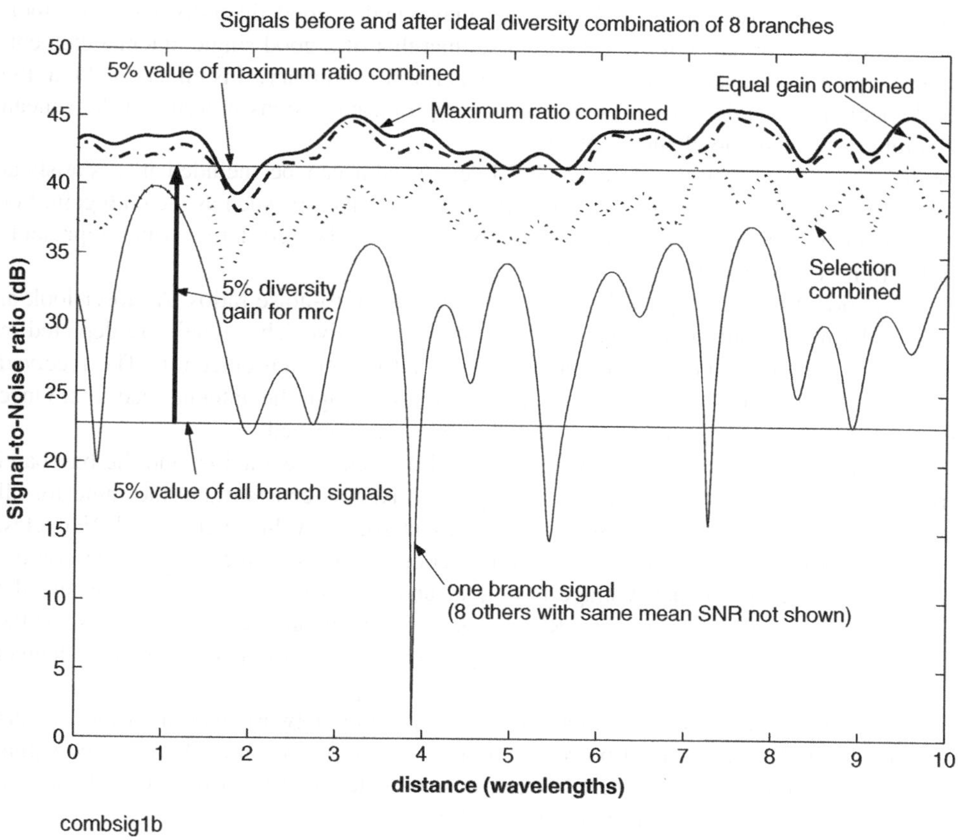

Figure 15.7 A single-antenna fading signal and various combinations of eight such antenna branches with uncorrelated fading and the same SNR. The dB difference between the 0.05 % outage levels of the two signals is the diversity gain. In this example, the diversity gain is about 18 dB.

The fading power level caused by the changing multipath prompts the link *outage* as a performance parameter. The outage is derived from the cumulative distribution function (CDF) of the link gain. The digital outage is the relative time (or relative number of terminals at a given time in a multiuser system, as discussed above) for which the BER or capacity falls below some acceptable level, whereas the analogue outage refers to the SNR or SNIR. For a given outage probability, a *diversity gain* can be defined, which is the increase in the SNR of the combined antenna signal relative to the best SNR of the single antennas. There is no standard for the probability value, but it is often taken as 0.5 % The diversity gain can be mathematically defined here in dB, as

$$G^{(\mathrm{DIV})}(\mathrm{dB}) = \left[\frac{\gamma_{\mathrm{C}}}{\Gamma_{\mathrm{C}}}(\mathrm{dB}) - \frac{\gamma_1}{\Gamma_1}(\mathrm{dB})\right]_{\text{Given probability}} \tag{15.51}$$

where γ_{C} is the SNR of the combined signal and $\Gamma_{\mathrm{C}} = \langle\gamma_{\mathrm{C}}\rangle$ is its mean SNR value, with the averaging over the mobile channel of interest; and similarly for γ_1, which refers to the best SNR of the antenna signals being combined. It is assumed here that γ_1 represents a sensible single element gain. Taking a poor single-branch performance for γ_1 may artificially inflate the diversity gain. This can happen with antennas near lossy matter, such as a cellphone, where, for example, two poor SNR branches may yield a good diversity gain, but their combined SNR performance may still be worse than that of a good single-antenna element, such as a dipole well separated from the lossy material. In summary, it is understood that the multielement antennas do not represent overly compromised designs in terms of their mean SNRs, relative to a single-element antenna.

With an outage probability defined, the diversity gain can be included in a statistical version of the Friis link equation along with the other gain factors, such as the path gain. For small percentage outage probabilities, the diversity gain can be very large, which represents the attraction of antenna diversity.

It is traditional to use a signal description for the signal-to-noise ratios. A closer look at the signal definitions during a signal combination is instructive. The signals are depicted in Figure 15.8, including their connection to a receiver for optimum detection. The received signal at the antenna has its SNR degraded as it passes through the antenna feed line. Once it has passed through a front-end amplifier, the SNR is established.

The baseband equivalent signal voltage at the ith branch is denoted α_i and the baseband equivalent noise voltage is n_i, with the noise power taken as constant and the same for all branches, $N_0 = N_i = \overline{|n_i|^2}$. Here the overbar indicates an average value over several RF cycles, and it is assumed that this average, for practical considerations, is the same as the ergodic mean, denoted N_0. The branch voltage is $\alpha_i + n_i$ and its SNR is $\gamma_i = \overline{|\alpha_i|^2}/N_0$, where the averaging, if necessary, is over many RF cycles to establish sampled power levels for the baseband signals. The SNR is a local, or instantaneous, value in terms of a fading channel (see below).

Now with the antenna connected, including any matching network, and the wanted signal only being received, the received power is $\overline{(\alpha_i + n_i)^2} = \overline{|\alpha_i|^2} + N_0$ since the received signal is uncorrelated with the noise. The SNR can now be calculated from the ratio of the two measured powers, averaging over many RF cycles:

$$\frac{\left(\overline{|\alpha_i|^2} + N_0\right)}{N_0} = \frac{\overline{|\alpha_i|^2}}{N_0} + 1 \Rightarrow \mathrm{SNR} = \frac{\left(\overline{|\alpha_i|^2} + N_0\right)}{N_0} - 1 \tag{15.52}$$

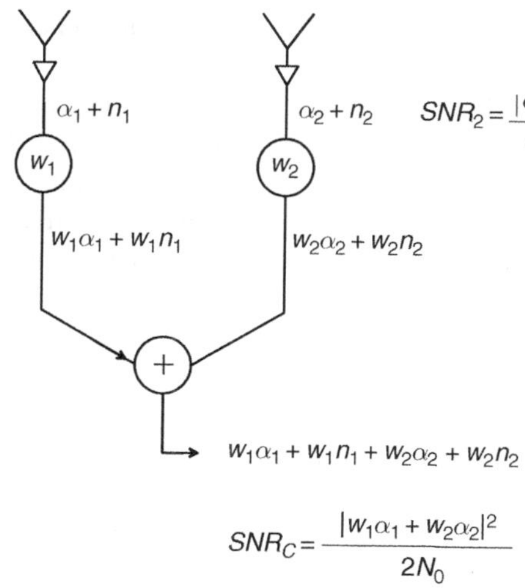

Figure 15.8 The combined signals and instantaneous SNR representations for signal combination. The branch SNR is assumed to be established after the front-end amplifier and it is not affected by the weight, although the weights determine the combined signal SNR.

From now on, the RF averaging for α_i is understood and the overbar notation is dropped.

In such a measurement of the SNR, interference should be avoided since it will spoil the accuracy of the measurement of the power in the wanted signal. This may require a shielded environment measurement with the wanted signal transmitted in controlled circumstances. (Normally, interference is present, but its power is also a random variable and the interference at the test site is likely to be different to that at an operational site.)

The signal-to-noise ratio of the combined signal, with the RF averaging understood, is

$$\mathrm{SNR_C} = \frac{|w_1\alpha_1 + w_2\alpha_2|^2}{|w_1 n_1 + w_2 n_2|^2} = \frac{|w_1\alpha_1|^2 + |w_2\alpha_2|^2 + 2\mathrm{Re}\{w_1 w_2^* \alpha_1 \alpha_2^*\}}{(w_1 + w_2)N_0} \tag{15.53}$$

where the branch noises are taken as mutually uncorrelated over several RF cycles.

The discussion of SNR above refers to its local, or instantaneous, value. If the fading of the channel is now averaged, correlation of the fading between the branch signals is introduced. Again the averaging over the mobile channel, for practical considerations, is taken to be essentially the same as the ergodic value, written $\langle \alpha_1 \alpha_2^* \rangle \approx R_{\alpha_1 \alpha_2} = E\{\alpha_1 \alpha_2^*\}$. The covariance is $C_{\alpha_1,\alpha_2} = E\{(\alpha_1 - \langle \alpha_1 \rangle)(\alpha_1^* - \langle \alpha_1^* \rangle)\}$. The correlation coefficient, or normalized covariance, is normally used:

$$\rho = \rho_{\alpha_1 \alpha_2} = \frac{C_{\alpha_1 \alpha_2}}{\sigma_{\alpha_1} \sigma_{\alpha_2}} \approx \frac{\langle \alpha_1 \alpha_2^* \rangle - \langle \alpha_1 \rangle \langle \alpha_2^* \rangle}{\sqrt{\langle |\alpha_1 - \langle \alpha_1 \rangle|^2 \rangle} \sqrt{\langle |\alpha_2 - \langle \alpha_2 \rangle|^2 \rangle}} \tag{15.54}$$

and the multipath is modelled with zero mean, complex, circular, normalized Gaussian signals (see Equations (15.56) to (15.60) below); $\alpha = x + \mathrm{j}y$, where $E\{\alpha\} = 0$, $E\{x^2\} = E\{y^2\} = \sigma^2$, $E\{xy\} = E\{yx\} = 0$, and the Gaussian correlation coefficients are $E\{x_1 x_2\} = -E\{y_1 y_2\} = \rho_{x_1 x_2}$, $E\{x_1 y_2\} = -Ex_2 y_1 = \rho_{x_1 y_2}$, so that $C_{\alpha_1 \alpha_2} = R_{\alpha_1 \alpha_2}$ and $\rho_{\alpha_1 \alpha_2} = 2\left(\rho_{x_1 x_2} - \mathrm{j}\rho_{x_1 y_2}\right)$.

15.4.1 Signal Description of Combining: No Interference

Maximum ratio combining (MRC) is a benchmark combining technique, offering the best combined SNR which is also the sum of the branch SNRs. The two-branch case is described in the following. The weighting is the ratio of the conjugate of the signal voltage-to-branch noise power, for example $w_1 = \alpha_1^*/N_0$. After the MRC weighting, the branch wanted signal voltage over the noise voltage and the corresponding signal-to-noise ratio are, for example,

$$\frac{\text{Signal voltage}}{\text{Noise voltage}} = \frac{w_1\alpha_1}{w_1 n_1} = \frac{|\alpha_1|^2/N_0}{n_1\alpha_1^*/N_0} = \frac{\alpha_1}{n_1}; \qquad \text{SNR}_1 = \frac{|\alpha_1|^2}{N_0} \tag{15.55}$$

The combined signal, here for uncorrelated signals, $\rho_{\alpha_1\alpha_2} = 0$, has an SNR of

$$\text{SNR}_\text{C} = \frac{\left(\dfrac{|\alpha_1|^2}{N_0} + \dfrac{|\alpha_2|^2}{N_0}\right)^2}{\left(\dfrac{n_1|\alpha_1|^2}{N_0^2}\right) + \left(\dfrac{n_2|\alpha_2|^2}{N_0^2}\right)} = \frac{|\alpha_1|^2}{N_0} + \frac{|\alpha_2|^2}{N_0} \tag{15.56}$$

which has been implicitly averaged over the fading of the mobile channel, i.e. $\langle \text{SNR}_\text{C}\rangle = \langle \text{SNR}_1\rangle + \langle \text{SNR}_2\rangle$. Therefore, for equal mean branch powers, a two uncorrelated branch MRC results in a signal with twice the mean SNR of each individual branch.

Using a signal notation for the complex amplitude of the SNR, $r_1 = |\alpha_1|/\sqrt{N_0}$, and its power signal, r_1^2, the algorithm for the M-branch MRC can be written as $r_\text{MRC} = \sqrt{\sum r_\text{m}^2}$. Other common combination algorithms include *equal gain combining* (EGC) (all weights the same), $r_\text{EG} = \sum r_\text{m}/\sqrt{M}$, and *selection combining* (SC) (one only non-zero weight at a time), $r_\text{S} = \max(r_1, r_2, \ldots, r_M)$. A variation of selection combining is *switched combining*, where a fading channel is tracked until it reaches an outage state, measured from, for example, the BER or from the instantaneous SNR or just the received power. For notation, the SNR as a power ratio is denoted $\gamma = r^2$ and its mean is $\Gamma_\gamma = \langle\gamma\rangle$.

With selection and switched combining, only one branch, or one fixed combination of the branches, is used at once. With MRC and EGC, etc., multiple channels are combined simultaneously, and even if the branch SNRs are too low for signal detection, the combined signal can have sufficient SNR to provide an acceptable outage. It is evident that any mutual coupling in the antennas will have an impact for MRC and EGC, but may not for selection and switched combining.

The diversity gains from these combining algorithms are approximately related for SNRs well below the mean, so if the diversity gain is known for MRC (the easiest to calculate), the others can be estimated, which can be difficult to calculate [24]. For M uncorrelated and equal mean branch SNRs, the ratio of the MRC diversity gain to that of SC, for $\gamma \ll \Gamma_\gamma$, is $\sqrt[M]{M!}$, and the diversity gain of EG falls between these. For example, for two ideal branches, the dB diversity gain difference is about 1.5 dB, and for eight such branches it is almost 6 dB.

15.4.2 Many-Branch Complex Gaussian Statistics

In general, there are M correlated branches with unequal mean branch SNRs. The statistical description is fundamental to the antenna performance so it is reviewed as follows. A signal description is used here with the complex amplitude SNR denoted by $\mathbf{z}^{\mathrm{T}}=[z_1, z_2, \ldots, z_M]$, and each of these distributed are given as complex Gaussians, written as $\mathbf{z} \sim CN(\mu_z, C_z)$, with $\boldsymbol{\mu}$ the vector of means. The covariance matrix is

$$\mathbf{C}_z=\mathbf{C}_x+\mathrm{j}\,\mathbf{C}_y+\mathrm{j}\left(\mathbf{C}_{xy}+\mathrm{j}\mathbf{C}_{yx}\right)=2\left(\mathbf{C}_x-\mathrm{j}\mathbf{C}_{xy}\right) \tag{15.57}$$

where, for example,

$$\mathbf{C}_x=E\left\{(\boldsymbol{x}-\boldsymbol{\mu}_x)(\boldsymbol{x}-\boldsymbol{\mu}_x)^{\mathrm{T}}\right\}=\begin{bmatrix} \sigma_1^2 & \sigma_1\sigma_2\rho_{12} & \cdots & \sigma_1\sigma_N\rho_{1N} \\ \sigma_1\sigma_2\rho_{12} & \sigma_2^2 & \cdots & \cdots \\ \cdots & \cdots & \cdots & \cdots \\ \sigma_1\sigma_N\rho_{1N} & \cdots & \cdots & \sigma_N^2 \end{bmatrix} \tag{15.58}$$

and the associated joint PDF for the real components is

$$p_x(x_1, x_2, \ldots, x_N)=\frac{1}{\left(\sqrt{2\pi}\right)^N\sqrt{|\mathbf{C}_x|}}\,\mathrm{e}^{-\frac{1}{2}(\boldsymbol{x}-\boldsymbol{\mu}_x)^{\mathrm{T}}\mathbf{C}_x^{-1}(\boldsymbol{x}-\boldsymbol{\mu}_x)} \tag{15.59}$$

with the complex form

$$p_z(\mathrm{z})=\frac{1}{\pi^N|\mathbf{C}_z|}\,\mathrm{e}^{-\mathrm{z}^{\mathrm{H}}\mathbf{C}_z^{-1}\mathrm{z}} \tag{15.60}$$

15.4.3 Sample Sizes for Estimating the Mean Power and the Gaussian Correlation

For estimating the diversity gain, the most accurate technique is to sample the branch SNRs, computer-combine these, and then calculate the diversity gain from these results. The question now arises as to how large should the sample be in order to get a statistically respectable result. The moments of interest are the mean and correlation, so the estimates of these, i.e. for the mean SNR and the correlation coefficient, are of direct interest. The mobile channel variation of the branch signal1 (Rayleigh-distributed amplitude fading) means that the power is modelled as exponentially distributed, a result directly from Gaussian PDF transformation. Therfore samples of the power (envelope squared) are distributed as

$$p_\gamma(\gamma)=\frac{1}{\Gamma_\gamma}\,\mathrm{e}^{-\gamma/\Gamma_\gamma} \tag{15.61}$$

and the usual mean estimate from N samples is $\bar{\gamma}=(1/N)\sum_{i=1}^{N}\gamma_i$, with variance $\sigma_{\bar{\gamma}}^2=\sigma_{\gamma}^2/N$ if all the SNRs have the same variance. It is now assumed that the mean estimate is Gaussian distributed and in the usual way the distribution of the estimate, $\bar{\gamma} \sim \mathrm{N}(\Gamma_\gamma, \sigma_\zeta=\sigma_\gamma/\sqrt{N})$, is normalized to $\zeta \sim \mathbb{N}(0, 1)$, with the transformation $\zeta = (\bar{\gamma} - \Gamma_\gamma)/(\sigma_\gamma/\sqrt{N})$. This allows a convenient confidence interval for the SNR mean estimate:

$$\text{Prob}\left[1 - \left(\frac{X\sigma_\gamma}{\sqrt{N}\Gamma_\gamma}\right) < \frac{\bar{\gamma}}{\Gamma_\gamma} < 1 + \left(\frac{X\sigma_\gamma}{\sqrt{N}\Gamma_\gamma}\right)\right] = \text{erf}\left(\frac{X}{\sqrt{2}\sigma_\zeta}\right) \qquad (15.62)$$

and putting $X = 2\sigma_\zeta$ gives a 95 % confidence level, i.e. the limits in this equation will hold for 95 % of the measurements. Figure 15.9 gives the limits in dB for the continuous wave (cw) or zero-bandwidth case, which is labelled $Bs = 0$, as well as some cases for a finite bandwidth. The finite bandwidth is not elaborated here except to note that a measurement over a wide bandwidth can increase the measurement accuracy [25]. Recall that these curves are approximations only, especially for the smaller numbers of samples. As an example, for 20 independent samples, the mean estimate for 95 % of the estimates will be from about −2.5 to 1.6 dB about the true mean, a spread of about 4.1 dB. After 30 samples, the curves are converging very slowly. The 95 % 3 dB spread is reached at 40 independent samples.

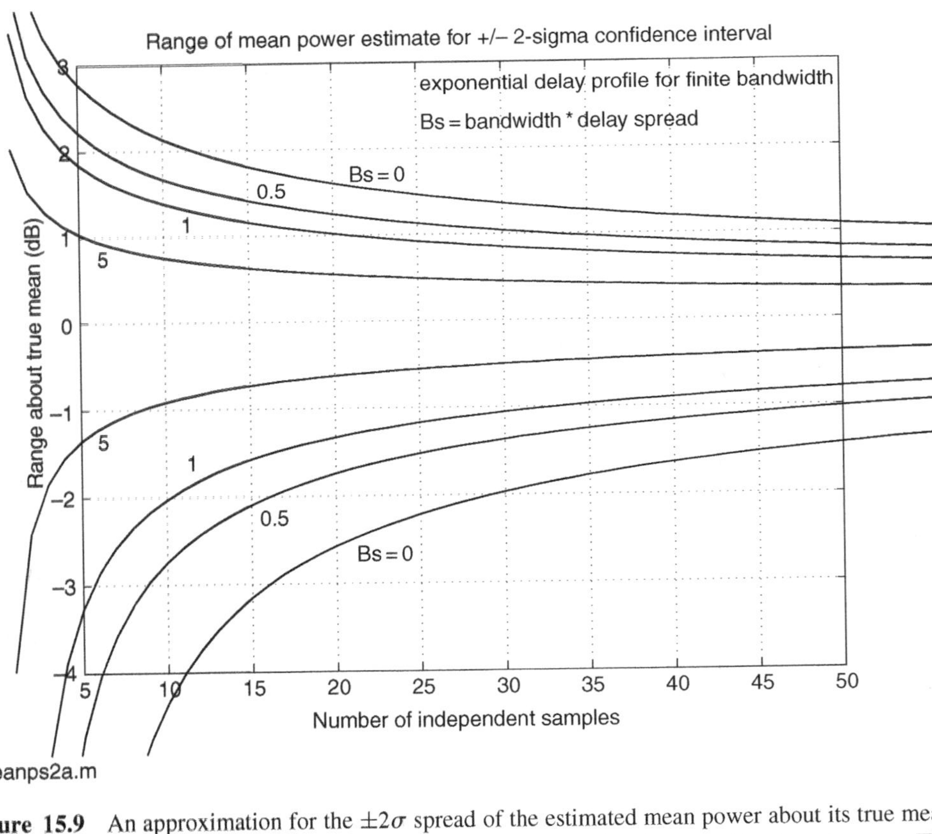

meanps2a.m

Figure 15.9 An approximation for the ±2σ spread of the estimated mean power about its true mean calculated from independent samples of the power. The narrowband case is the $Bs = 0$ curve. The effect of the nonzero bandwidth as a parameter is included here from Reference [25], but this case is not elaborated here except to note that using an increasing bandwidth allows an increasing accuracy in the estimate.

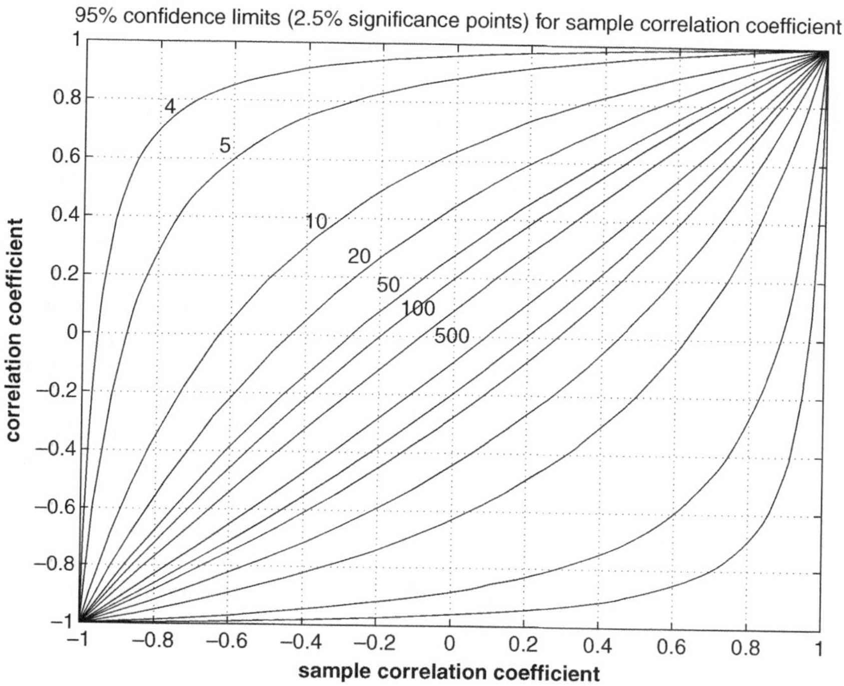

Figure 15.10 The 95 % confidence intervals for the sample correlation coefficient with the number of independent samples as a parameter.

The correlation coefficient estimate uses the same technique and was solved by R. A. Fisher in 1921 for the Gaussian case. The 95 % (i.e. $\pm 2\sigma$) upper and lower limits for N independent samples are

$$[\rho_U, \rho_L] = \left[\tanh^{-1}\left(\frac{-1.96}{\sqrt{N-3}}\right) + \tanh^{-1}\rho, \tanh^{-1}\left(\frac{-1.96}{\sqrt{N-3}}\right) + \tanh^{-1}\rho\right] \qquad (15.63)$$

which are given in Fig. 15.10. For correlations close to zero, more samples are needed than for high correlations, for a given accuracy of estimate. For high correlations, $\rho \to \pm 1$, the number of samples required becomes very small. As an example, for a sample correlation coefficient of 0.4 calculated from 10 independent samples, the 95 % interval of the estimate is that the true value is between 0.3 and 0.8. An estimate of 0 from 500 samples still gives a 95 % range of $\pm$ 0.1.

Using these results, the sample sizes required for estimating the mean SNR and the correlation coefficients between branches can be nominated. Measuring a low correlation coefficient is clearly the most challenging, since for reasonable accuracy it requires a larger sample than that required for large correlation coefficients or for the mean power estimate. It is possible to calculate the correlation coefficient from the envelope samples as well, using the relations $|\rho_{\alpha_1\alpha_2}|^2 = \rho_{|\alpha_1|^2|\alpha_2|^2} \approx \rho_{|\alpha_1||\alpha_2|}$. If complex samples are available, then the complex correlation calculation should be performed, which allows better accuracy than using the corresponding real envelope values for the calculation.

15.4.4 Mutual Coupling for Diversity Evaluation

The ideal diversity antenna system has equal mean powers (SNRs) and zero correlation coefficient. Mean powers can be achieved by designing for the same mean gain, using the techniques above. Designing for low correlation is more difficult because normally a compact antenna is sought, and, normally, the physically closer the elements, the more coupled they become. If the elements are mutually coupled then the signals become correlated and the diversity gain reduces. However, the mutual coupling is not the only mechanism causing correlated branch signals. The distribution of incident waves plays an integral role. For example, if the incident angular power distribution is a single incident wave, mathematically written as $p_{\theta,\phi}(\theta,\phi)=1/[\sin\theta_0\delta(\theta-\theta_0)\delta(\phi-\phi_0)]$, then the antenna signals will be correlated. This is not a multipath situation and antenna diversity is of no use, except in the sense of achieving array gain. However, close to this limiting example, there can be multipath with small angular spread, and the same situation can apply. When the angular spread of the multipath is comparable to or exceeds the extent of the antenna patterns, then another possibility arises, developed in the following.

In a generalized form of Equation (15.36), the correlation between the open-circuit voltages of two antennas can be expressed in terms of their patterns integrated over real space [17, 26]. Here abbreviating using $(\Omega)=(\theta,\phi)$, $\mathrm{d}\Omega=\sin\theta\mathrm{d}\theta\mathrm{d}\phi$, and normalized patterns, the correlation coefficient is

$$\rho_{V\mathrm{o}1V\mathrm{o}2}=\frac{C_{V\mathrm{o}1V\mathrm{o}2}}{\sigma_{V\mathrm{o}1}\sigma_{V\mathrm{o}2}}=\frac{\langle V_{\mathrm{o}1}V_{\mathrm{o}2}^{*}\rangle}{\sqrt{\langle|V_{\mathrm{o}1}|^2\rangle\langle|V_{\mathrm{o}1}|^2\rangle}}$$

$$=\frac{\int(p_{\theta}(\Omega)\,h_{1\theta}(\Omega)\,h_{2\theta}^{*}(\Omega)+p_{\theta}(\Omega)\,h_{1\phi}(\Omega)\,h_{2\phi}^{*}(\Omega))\mathrm{d}\Omega}{\sqrt{\int p_{\theta}(\Omega)\,|h_1(\Omega)|^2\,\mathrm{d}\Omega\int p_{\phi}(\Omega)\,|h_2(\Omega)|^2\,\mathrm{d}\Omega}}\quad\text{(general)}$$

$$=\frac{\int p_{\Omega}(\Omega)\,(h_{1\theta}(\Omega)\,h_{2\theta}^{*}(\Omega)+h_{1\phi}(\Omega)\,h_{2\phi}^{*}(\Omega))\mathrm{d}\Omega}{\sqrt{\int p_{\Omega}(\Omega)\,|h_1(\Omega)|^2\,\mathrm{d}\Omega\int p_{\Omega}(\Omega)\,|h_2(\Omega)|^2\,\mathrm{d}\Omega}}\quad\text{(polarizations with same PDF)}$$

$$=\int h_1(\Omega)\cdot h_2^{*}(\Omega)\,\mathrm{d}\Omega\quad\text{(also full homogenous PDF)}\qquad(15.64)$$

The last line follows for an environment in which the incoming power distribution is the same for each polarization, homogeneous, and covers the extent of the patterns, for example all of real space. In this case, the correlation is stated directly as the inner product of the normalized patterns. In multipath, one of the ideal conditions for diversity action, which is zero correlation, can be stated mathematically where the patterns should be orthogonal over the distribution of incoming waves. At the same time, for minimum scattering antennas, the normalized mutual resistance has the same form [27], allowing the following to be written [17, 26]:

$$\frac{R_{12}}{\sqrt{|R_{11}|\,|R_{22}|}}=r_{12}=\rho_{V\mathrm{o}1V\mathrm{o}2}\quad\text{(full homogeneous)}\qquad(15.65)$$

Under certain conditions – the antennas should be good approximations to MSAs and the imcoming power distribution should have the right properties – the correlation coefficient is the same as the normalized mutual resistance. The significance of this equality is that the correlation coefficient is a difficult-to-measure statistical parameter, whereas the mutual resistance is a well-defined antenna parameter, is relatively easy to measure, and of course is easily repeatable.

From a set of scattering parameter measurements, denoted with matrix **S**, the mutual resistance is found from the impedance matrix, **Z**, with the usual transformation:

$$\mathbf{Z} = Z_0 (\mathbf{I} - \mathbf{S})^{-1} (\mathbf{I} + \mathbf{S}) \tag{15.66}$$

The major simplification is not easy to obtain in Equation (15.65); the problems of the sample averaging for the correlation estimate are shifted to the estimate, or rather the assumptions, of the incoming power density distribution. Nevertheless, the mutual resistance will at least give a convenient estimation for the correlation coefficient for conditions that are commonplace for NLOS communications. In summary, $r_{ij} \approx \rho_{ij}$ can be written for the ith and jth elements of a multielement antenna.

15.4.5 Line Length Calibration

The calibration plane of an S-parameter test set is normally taken to be the open end of the test cables. The measurement plane for a printed antenna is normally taken right at the antenna itself, so there is at least a connector length for a coax-fed antenna (and more for stripline-fed antennas) to be considered between the calibration plane and the plane of interest at the antenna. Denote the electrical length of such a line with $k_g L$, where L is in metres and k_g is the guided wavenumber. For example, solid teflon (relative permittivity of about 2.1)-filled cables have $k_g \approx \sqrt{2.1}\, 2\pi/\lambda_0$. Then the two-way delay, plus any loss (k_g would be complex if there are losses in the line) through the line length $k_g L$, is expressed as $e^{-j2k_g L}$, and if all the delays are the same, then the S-parameters matrix used in the above equation are replaced by $\mathbf{S}e^{-j2k_g L}$.

15.4.6 Mutual Coupling: Network Model

To include the effect of mutual coupling, the network model for an M-branch multielement antenna reception is given in Figure 15.11. The load voltages, $\mathbf{V}_L = [V_{L1}, V_{L2}, \ldots, V_{LM}]^T$, is the relevant signal set for signal reception. These relate to the open-circuit voltages by $\mathbf{V}_0 = (\mathbf{Z}_A + \mathbf{Z}_L)\,\mathbf{I}_A$, where $\mathbf{Z}_A$ is the $M \times M$ matrix of multielement antenna impedances, $\mathbf{Z}_L$ is the $M \times M$ matrix of load impedances, which in general can include off-diagonal terms for descattering, and $\mathbf{I}_A = \mathbf{Z}_L^{-1}\mathbf{V}_L$ is the vector of antenna branch currents. For maximum power transfer, $\mathbf{Z}_L = \mathbf{Z}_A^*$, but for practical reasons, the load impedance is often simply set to a diagonal of 50 Ω loads.

15.4.7 Loaded Circuit and Open-Circuit Correlations

The correlation matrix of loaded circuit voltages is denoted $\mathbf{L}_L = E\{\mathbf{V}_L \mathbf{V}_L^H\}$ (the subscript H is a conjugate transpose) and similarly for the open-circuit voltages, $\mathbf{L}_o = E\{\mathbf{V}_o \mathbf{V}_o^H\}$. The two correlation matrices are related by a function of the antenna and load impedances:

$$\mathbf{L}_L = \mathbf{F}\mathbf{L}_o\mathbf{F}^H \tag{15.67}$$

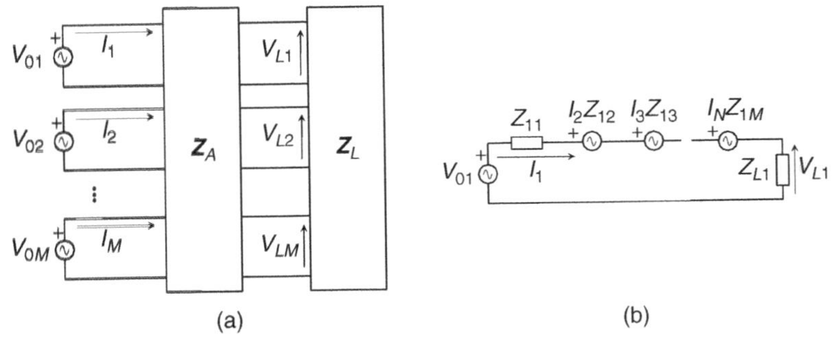

Figure 15.11 (a) The circuit for a terminated multielement antenna. (b) The equivalent circuit for a single MSA-like element (denoted 1) with a diagonal load impedance matrix.

where

$$\mathbf{F} = \mathbf{Z}_{\mathrm{L}} (\mathbf{Z}_{\mathrm{A}} + \mathbf{Z}_{\mathrm{L}})^{-1} \tag{15.68}$$

This relationship allows spatial correlations, calculated from the modelled distributions of the incident fields, to be transformed to the correlations seen at the antenna branches when the mutual coupling may be significant, i.e. in compact multielement antennas. The effect of different loads on an array can be found conveniently as well. Note that mutual reactances between elements can be tuned out.

15.4.8 Received Power for Resistive Loads and Mutual Coupling Power Loss

The received power, averaged over the mobile channel, is proportional to

$$P_{\mathrm{rec}} = \langle \mathbf{V}_{\mathrm{L}}^{\mathrm{H}} \mathbf{I}_{\mathrm{A}} \rangle = \langle \mathbf{V}_{\mathrm{L}}^{\mathrm{H}} \mathbf{Z}_{\mathrm{L}}^{-1} \mathbf{V}_{\mathrm{L}} \rangle = \langle \mathbf{V}_{\mathrm{o}}^{\mathrm{H}} \mathbf{Y} \mathbf{V}_{\mathrm{o}} \rangle \tag{15.69}$$

where

$$\mathbf{Y} = (\mathbf{Z}_{\mathrm{A}} + \mathbf{Z}_{\mathrm{L}})^{-1\mathrm{H}} \mathbf{Z}_{\mathrm{L}}^{\mathrm{H}} (\mathbf{Z}_{\mathrm{A}} + \mathbf{Z}_{\mathrm{L}})^{-1} = \mathbf{F}^{\mathrm{H}} \mathbf{Z}_{\mathrm{L}}^{-1} \mathbf{F} \tag{15.70}$$

If all the voltages are normalized, then this received power is, for M antenna elements,

$$P_{\mathrm{rec}}^{(\text{with mc})} = \sum_{j,k}^{M} (\mathbf{Y})_{jk}\, \rho_{0jk} \approx \sum_{j,k}^{M} (\mathbf{Y})_{jk}\, r_{jk} \tag{15.71}$$

For no mutual coupling (mc) ($\rho_{ij} \approx r_{ij} = 0$) and normalized average powers from the voltages, the received power reduces to

$$P_{\mathrm{rec}}^{(\text{no mc})} = \sum_{k}^{M} (\mathbf{Y})_{kk} \tag{15.72}$$

The reduction in the received power owing to the mutual coupling is expressed with the ratio $P_{\mathrm{rec}}^{(\text{with mc})} / P_{\mathrm{rec}}^{(\text{no mc})}$. This gain acts on the diversity gain and has an impact as the antennas become highly coupled. For real load impedances and fully correlated antennas, viz. $(I_1 = I_2 = \ldots = I_M)$, $r_{ij} \approx \rho_{0ij} = 1$, the matched load resistance is $R_{L1} = R_{11} + R_{12} + \cdots + R_1 M$, and the average received power is

$$P_{\text{rec}} = \left\langle \mathbf{V}_L^H \mathbf{Z}_L^{-1} \mathbf{V}_L \right\rangle = \frac{M}{R_L} \left\langle |V_L|^2 \right\rangle \tag{15.73}$$

With the load matched, $V_L = V_o/2$ and the gain from the mutual coupling is denoted

$$G^{(\text{mc})} = \frac{P_{\text{rec}}^{(\text{with mc})}}{P_{\text{rec}}^{(\text{no mc})}} = \frac{\frac{M}{R_{L\,(\text{with mc})}} \left\langle |V_o|^2 \right\rangle}{\frac{M}{R_{L\,(\text{with no mc})}} \left\langle |V_o|^2 \right\rangle} = \frac{1}{1 + r_{12} + r_{13} + \cdots + r_{1M}} \tag{15.74}$$

Note that in some circumstances the mutual resistances can be negative. There is another potential benefit of mutual coupling – it can be used to make more compactly spaced elements [28].

15.4.9 *Diversity Gain Estimate*

Recall that the diversity gain is from the difference between the cumulative distributions of the single branch SNR and the combined SNR. The cumulative distribution function (CDF) for a single branch (Rayleigh fading) is from the exponential PDF (15.60), and is

$$\text{Prob}(\text{SNR} \le \gamma) = \text{Prob}(\gamma) = 1 - e^{-\gamma/\Gamma} \tag{15.75}$$

For MRC with M uncorrelated branches with same mean SNRs, the CDF derives from the gamma function (chi squared with $2M$ degrees of freedom), and results in

$$\text{Prob}(\gamma_C) = 1 - e^{-(\gamma_C/\Gamma)} \sum_{m=1}^{M} \frac{(\gamma_C/\Gamma)^{(m-1)}}{(m-1)!} \tag{15.76}$$

which is used for reference in the plots below.

When the branches are correlated and have unequal mean powers, equivalent number of equal mean uncorrelated branches can be used to give the same diversity gain. This is called the *effective diversity order*.

Taking the measured normalized, mutual resistance matrix, and setting this to be the correlation matrix, $\mathbf{r} \approx \mathbf{p}$, $\mathbf{r}$ is decomposed in a singular value style, giving the diagonal matrix, $\mathbf{\Lambda}$, of the eigenvalues, λ_m, and a unitary matrix $\mathbf{V}$ containing columns of the eigenvectors, with $\mathbf{rD} = \mathbf{rV}$, i.e.

$$\mathbf{\Lambda} = \mathbf{V}^T \mathbf{r} \mathbf{V} \tag{15.77}$$

The eigenvectors are not required from now on. The number of nonzero eigenvalues is the equivalent number of uncorrelated branches, and the eigenvalues represent their powers. The

MR combining performance can be written as the CDF for uncorrelated, unequal branch powers, with the effect of the power loss from the mutual coupling:

$$\mathrm{Prob}(\gamma_C) = \sum_{m=1}^{M} \frac{1}{\prod\limits_{\substack{n=1 \\ n\neq m}}^{M} \left(1 - \frac{\lambda_n}{\lambda_m}\right)} \left(1 - \mathrm{e}^{-\{(\gamma_m/\Gamma_m)[1/(G^{(\mathrm{mc})}\lambda_m)]\}}\right) \tag{15.78}$$

From the ratio of the CDFs at a given probability, the diversity gain estimate can now be ascertained. This completes the theoretical basis for diversity gain evaluation, but examples will help fix ideas.

15.4.10 Examples

The results from some multielement antennas are as follows. The elements are air – substrate PIFAs which have a VSWR = 2 impedance bandwidth of 14 %. The PIFA dimensions are approximately a half-wavelength by a quarter-wavelength. A four-element antenna whose elements are arranged in two opposing pairs is shown in Figure 15.12. Two forms of the antenna are: (a) with one opposing pair of elements adjacent (a common cavity wall) and (b) with both pairs spaced by different amounts, as shown in the figure. These configurations offer compact arrays. A higher permittivity version will be more compact of course, but the bandwidth will be lower and the mutual coupling will be higher. A third configuration (c) has the same spacing for both opposing element pairs (circular array), but this is a less compact antenna with even lower mutual coupling and correlations than the first two configurations. In this latter configuration, the end slots of the both opposing PIFA element pairs are spaced apart by over a wavelength so in a conventional array application this antenna will have grating lobes for both element pairs.

At the centre frequency (1.8 GHz), the normalized, measured mutual resistance matrices for each configuration are

$$\mathbf{r}^{(\mathrm{a})} = \begin{bmatrix} 1.0000 & 0.7234 & 0.1098 & 0.1002 \\ 0.7234 & 1.0000 & 0.0913 & 0.1268 \\ 0.1098 & 0.0913 & 1.0000 & 0.1308 \\ 0.1002 & 0.1268 & 0.1308 & 1.0000 \end{bmatrix},$$

$$\mathbf{r}^{(\mathrm{b})} = \begin{bmatrix} 1.0000 & 0.0707 & 0.0366 & 0.0285 \\ 0.0707 & 1.0000 & 0.0226 & 0.0317 \\ 0.0366 & 0.0226 & 1.0000 & -0.0964 \\ 0.0285 & 0.0317 & -0.0964 & 1.0000 \end{bmatrix} \tag{15.79}$$

The adjacent elements have quite high coupling ($r_{12}^{(\mathrm{a})} \approx 0.7$), but the other couplings remain low. The resulting diversity gain performances are presented in Figure 15.13. Here the ideal MRC CDFs are drawn for different numbers of antennas. The dotted line is the CDF for the

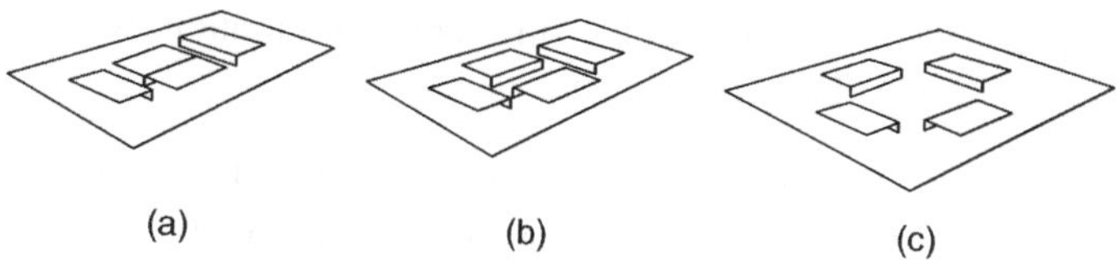

Figure 15.12 An antenna for MIMO comprising four PIFA elements. The distance between the opposing elements is different in general, in (a) one set of opposing elements share a wall and in (b) they are spaced by a fraction of a wavelength, but when made the same (c), the antenna becomes a circular array.

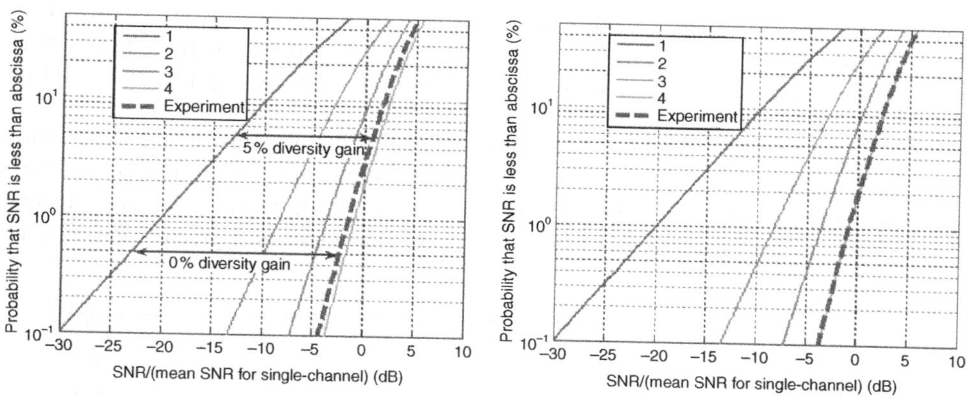

Figure 15.13 The CDF for the antenna configuration (a) at the left and (b) to the right. The MRC diversity gain (dotted line) provides a figure of merit for the antennas if they operated in a full, uniform uncorrelated multipath scenario. The solid lines are for ideal (1–4) branch MRC combining, which provide the order of diversity. The diversity gain is indicated for (a), where the effective order of diversity is 3.8.

MRC combination. For (a), the more compact antenna, the effective diversity order is about 3.8, so the mutual resistance, and the assumed-the-same correlations, reduces the number of diversity branches by just 0.2, representing a reduction in the diversity gain of about 1 dB. About half of this is from the term $G^{(mc)} = 0.42$ dB, which has the effect of moving the CDF curve to the left. The mutual coupling acts to change the slope of the curve from that of its original number of (correlated) branches. In this example, the ratio of the effective diversity order to the number of implemented branches is {3.8/4}.

A calculation using $\mathbf{r} = \mathbf{r}_Z = \Re\{|\mathbf{Z}_{12}|/\sqrt{|\mathbf{Z}_1||\mathbf{Z}_2|}\}$, which has all positive entries, makes a difference of about 1 more dB reduction in diversity gain, with the effective diversity order reducing to about 3.5. Most of this reduction in the diversity gain is from finite correlation, rather than from the mutual coupling term, $G^{(mc)}$. For the configuration (b), there is no drop in diversity gain from this calculation.

The larger antenna configuration (b) has a negligible loss of effective diversity order. This is because the mutual couplings (and the assumed correlation coefficients) are too small to have a significant impact. Similarly, configuration (c) also has a negligible loss of effective diversity order.

A compact 12-port MIMO antenna, with a different configuration, was evaluated in the same way (see Figure 15.14). The matrix **r** is:

1.00	0.11	−0.14	0.07	−0.04	0.04	0.01	0.00	0.03	−0.08	0.12	0.28
0.11	1.00	0.16	−0.12	0.06	−0.04	0.00	0.02	−0.06	0.14	0.36	0.08
−0.14	0.16	1.00	0.19	−0.14	0.07	0.01	−0.07	0.14	0.41	0.03	−0.06
0.07	−0.12	0.19	1.00	0.17	−0.12	−0.06	0.15	0.41	0.08	−0.05	0.02
−0.04	0.06	−0.14	0.17	1.00	0.16	0.13	0.38	0.03	−0.04	0.01	0.00
0.04	−0.04	0.07	−0.12	0.16	1.00	0.27	0.07	−0.05	0.02	0.01	−0.00
0.01	0.00	0.01	−0.06	0.13	0.27	1.00	0.10	−0.07	0.02	−0.04	0.03
0.00	0.02	−0.07	0.15	0.38	0.07	0.10	1.00	0.13	−0.06	0.06	−0.05
0.03	−0.06	0.14	0.41	0.03	−0.05	−0.07	0.13	1.00	0.17	−0.14	0.08
−0.08	0.14	0.41	0.08	−0.04	0.02	0.02	−0.06	0.17	1.00	0.16	−0.13
0.12	0.36	0.03	−0.05	0.01	0.01	−0.04	0.06	−0.14	0.16	1.00	0.14
0.28	0.08	−0.06	0.02	0.00	−0.00	0.03	−0.05	0.08	−0.13	0.14	1.00

This again features quite low correlations. The spacing of the elements is not well-related to the matrix numbering here; for example, the most highly coupled branches appear at the furthest entry from the diagonal. The larger dimension, and the decreasing returns of antenna diversity with the larger number of branches, means that there is a greater performance

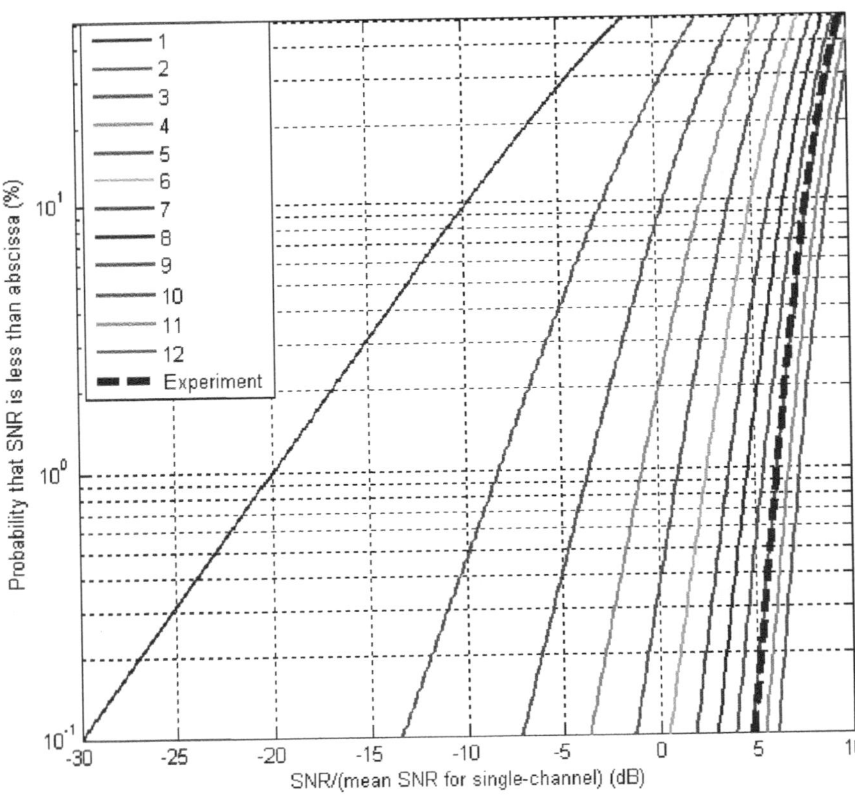

Figure 15.14 The CDF for the 12-branch antenna, showing that the effective order of diversity is about 10.3.

reduction in terms of the effective diversity order. From the figure, the effective order of diversity is 10.3, a reduction of 1.7 branches, i.e. a ratio of {10.3/12}. The reduction in diversity gain is about 1.3 dB, and about 0.9 dB of this is from the $G^{(mc)}$ component. A calculation using $\mathbf{r}_z$ results in the effective diversity order being about 10.1.

The assumptions for the above diversity gain evaluation are now summarized. Firstly, the incoming wave distribution is taken as homogeneous and occupies the full radiation space support (a hemisphere, both polarizations) of the patch antenna element patterns. It is taken that the radiation behind the ground-plane has a negligible impact, despite the finite size of the ground-plane. Secondly, the normalized mutual resistances are taken to correspond to the correlation coefficients. This relates to the antenna elements being MSA-like. The latter assumption is reasonable for the elements used in the examples, but the former depends on the propagation environment, and needs the careful interpretation required of any diversity gain, as discussed above. Under these conditions, the above antennas can have their diversity gain evaluated using the convenient multiport impedance measurement technique.

15.5 SUMMARY

Antenna performance evaluation is not only part of the communications system analysis and design but it is also part of the iterative process of compact antenna design. Consequently, convenient evaluation techniques are important. In this chapter, both theoretical and practical aspects of evaluating antennas for high capacity efficiency have been described. This calls for an emphasis on the efficiency and gain factors. A multipath communications link analysis leads to the antenna polarization efficiency becoming incorporated into the distributed gain of the antenna element. The impedance bandwidth and its trade-off with antenna compactness offer a helpful comparison for different types of antenna elements. For evaluating multielement antennas for multipath environments, the diversity gain offers a robust performance measure. However, the diversity gain is a statistical quantity and to produce a direct estimate of it requires a considerable measurement campaign with considerable signal processing. Moreover, the diversity gain does not provide the specific information on how to improve the configuration of the antenna elements. The required specific information is the set of mean branch powers and in particular the set of correlation coefficients between the loaded antenna signals. Direct estimates of these statistics would be made from sampling the antenna signals when the antenna is operating in different multipath scenarios. However, gathering and using such samples brings difficulties in the interpretation and repeatability, and the sample size required for accurate estimates is large, which means a large measurement effort. These factors highlight a need for a more convenient evaluation technique, especially for the antenna design process. Measuring the antenna impedances, together with modest off-line signal processing, offers such an alternative, although this approach requires certain conditions regarding both the propagation environment and the antenna elements. If these conditions hold, then the diversity gain can be estimated from the impedance measurement. For designing multielement antenna configurations, the diversity gain can be couched as an equivalent number of ideal branches, i.e. an effective diversity order. The effective diversity order is a performance metric that is suitable for the designer to trade off against compactness of the multielement antenna configuration. This has been illustrated using air–substrate design examples in which the antenna is made sufficiently compact to force the effective order of diversity to be reduced significantly from the physical number of branches.

ACKNOWLEDGEMENT

The author warmly thanks Mr Ezra Ip for assistance with multiport measurements and fun with π.

REFERENCES

1. R. G. Gallager, *Information Theory and Reliable Communications*, John Wiley & Sons, Ltd, Chichester, 1968.
2. W. L. Stutzman and G. A. Thiele, *Antenna Theory and Design*, John Wiley & Sons, Ltd, Chichester 1981.
3. C. G. Montgomery, R. H. Dicke and E. M. Purcell, *Principles of Microwave Circuits*, McGraw-Hill, New York, 1948.
4. W. K. Kahn and H. Kurss, 'Minimum scattering antennas', *IEEE Transactions*, **AP-13**, 671–675, September 1965.
5. J. B. Andersen and R. G. Vaughan, 'Transmitting, receiving, and scattering properties of antennas', *IEEE Antennas and Propagation Magazine*, **45**(4), 93–98, August 2003.
6. G. L. Matthaei, L. Young and E. M. T. Jones, *Microwave Filters, Impedance-Matching Networks, and Coupling Structures*, McGraw-Hill, New York, 1964.
7. E. Nyfors and P. Vainikainen, *Industrial Microwave Sensors,* Artech House, Norwood, Massachusetts, 1989.
8. L. J. Chu, 'Physical limitations of small antennas', *Journal of Applied Physics*, **19**, pp. 1163–1175, December 1948.
9. K. Fujimoto, A. Henderson, K. Hirasawa and J. R. James, *Small Antennas*, Research Studies Press/John Wiley & Sons, Inc., 1987.
10. J. S. McLean, 'A re-examination of the fundamental limits on the radiation limits on the radiation Q of electrically small antennas', *IEEE Transactions*, **AP-44**(5), 672–676, May 1996.
11. R. G. Vaughan and J. B. Andersen, *Channels, Propagation and Antennas in Mobile Communications*, IEE, London, 2003.
12. R. B. Waterhouse, 'Design and performance of small printed antennas', *IEEE Transactions*, **AP-46**(11), 1629–1633, November 1998.
13. D. Slater, *Near-Field Antenna Measurements*, Artech House, Norwood, Massachusetts, 1991.
14. H. Trzaska, *Electromagnetic field measurements in the near field*, Noble, March 2000.
15. R. E. Collin and F. J. Zucker, *Antenna Theory*, McGraw-Hill, New York, 1969.
16. W. C. Jakes (ed.), *Microwave Mobile Communications*, AT&T Press, 1974; also IEEE Press, 1993.
17. R. G. Vaughan and J. B. Andersen, 'Antenna diversity for mobile communications', in *Proceedings of the Nordic Seminar on Digital Land Mobile Radiocommunications*, Espoo, pp. 87–96, February 1985.
18. T. Taga, 'Analysis for mean effective gain of mobile antennas in land mobile radio environments', *IEEE Transactions*, **VT-39**, 117–131, May 1990.
19. R. G. Vaughan, 'Two port higher mode circular microstrip antennas', *IEEE Transactions*, **AP-36**(3), 1365–1374, March 1988.
20. T. Aulin, 'A modified model for the fading signal at a mobile radio channel', *IEEE Transactions*, **VT-21**(1), 27–38, 1972.
21. K. I. Pedersen, P. E. Morgensen and B. H. Fleury, 'Power azimuth spectrum in outdoor environments', *Electronic Letters*, **33**(18), 1583–1584, 28 August 1997.
22. D. Bell, P. Dessouky, P. Estabrook and M. K. Sue, 'MSAT-X antennas: noise temperature and mobile receiver G/T', *MSAT-X Quarterly*, **16**, 13–15, 1988.
23. U. R. Kraft, 'Gain and G/T of multielement receive antennas with active beam forming networks', *IEEE Transactions*, **AP-45**(12), 1818–1029, December 2000.

24. S. Stein, Part III of M. Schwartz, W. R. Bennett and S. Stein, *Communications Systems and Techniques*, McGraw-Hill, New York, 1966.
25. R. G. Vaughan, 'Finite sample estimates for mobile channels', in *Proceedings of the IEEE Vehicular Technology Conference (VTC 2000)*, Boston, Vol.2, pp.797–804, Massachusetts, September 2000.
26. R. G. Vaughan and J. B. Andersen, 'Antenna diversity in mobile communications', *IEEE Transactions*, **VT-36**(4), 149–172, November 1987.
27. W. Wasylkiwskj and W. K. Kahn, 'Theory of mutual coupling among minimum scattering antennas', *IEEE Transactions*, **AP-18**(2), 204–216, March 1970.
28. R. G. Vaughan and N. L. Scott, 'Closely spaced terminated monopoles', *Radio Science*, **28**(6), 1259–1266, November/December 1993.

Index

Printed Antennas for Wireless Communications Edited by R. Waterhouse

Printed and bound by CPI Group (UK) Ltd, Croydon, CR0 4YY
16/09/2026
14963387-0006